Current Problems in Electrophotography

Current Problems in Electrophotography

edited by

W. F. Berg and K. Hauffe

With 397 figures and 31 tables

W DE G

1972

Walter de Gruyter · Berlin · New York

smooth running of the meeting, including an enjoyable boat ride on the lake. We appreciated very much the cooperation with the Walter de Gruyter Press.

Finally, we wish to express our gratitude to Prof. Dr. K. V. Chibissov, Corresponding Member of the Academy of Science of the USSR, for accepting the position as Honorary President of the Colloquium, and to the Sessions' Chairmen Dr. P. M. Cassiers, Prof. Dr. G. Haase, Prof. Dr. K. Hauffe, Prof. Dr. E. Inoue, Dr. J. E. Lu Valle, Dr. S. R. Morrison, Dr. E. M. Pell, Prof. Dr. G. M. Schwab for running the Meeting.

August 1972 *W. F. Berg* *G. Haase* *K. Hauffe*

Current Problems in Electrophotography

edited by

W. F. Berg and K. Hauffe

With 397 figures and 31 tables

W DE G

1972

Walter de Gruyter · Berlin · New York

This 3rd European Colloquium at Zürich, August 1971 was organized by W. F. Berg, Prof. Dr., Department of Photography, Swiss Federal Institute of Technology (ETH-Z), Zurich.

G. Haase, Prof. Dr., Chairman, Institute for Scientific Photography, Technical University, München.

K. Hauffe, Prof. Dr., Institute of Physical Chemistry, University, Göttingen.

ISBN 3 11 003699 1

Library of Congress Catalog Card Number 72-83642

Printed in Germany by Mercedes-Druck, Berlin 61

Cover designed by U. Hanisch, Berlin.

Preface

The surveys and original papers in this volume were presented at a Colloquium held at the Department of Photography, Swiss Federal Institute of Technology (ETH-Z) August 1 to 6, 1971. This was the third meeting of its kind; the first was held at Hannoversch-Münden in 1967, the second at Munich in 1969.

The Colloquium had a special note in that it was timed to coincide with the celebrations in honour of

John Eggert

on the occasion of the 80th return of his birthday. Dr. J. H. Dessauer gave the Jubilee Lecture which is reproduced in full below, to an audience of some 140 people who had gathered to celebrate the occasion.

Although designated as Eurepean, the Colloquium had a strong intercontinental flavour, both as far as contributors and participants were concerned. If one compares the contents list with that of the First International Conference on Electrophotography, held at Rochester in 1969 (Appl. Optics, Suppl. 3, Electrophotography 1969), at which a similar number of papers was offered, the strong accent on the fundamental aspects is evident: we had at Zürich no papers concerned with equipment.

The Colloquium was held in 8 half-day sessions devoted to the following subjects

I Charging and Discharging of Electrophotographic Layers and of Single Crystals.
II Mechanism of Dye Sensitization.
III Organic Photoconductors as Electrophotographic Materials.
IV Mechanism of Liquid Development.
V New Approaches.

The present volume has been produced by fast production methods, in order to be able to publish it as soon as possible after the conference. This has meant fewer proof stages, with the attendant liability of the occasional misprint. All manuscripts have been edited in order to achieve some uniformity in style. All contributors, however, have their own responsibility to the content of their papers.

It has been possible to invite a number of speakers and participants from abroad, thanks to the generous assistance provided by a large number of individuals and organizations who are either interested in the subject or wished to manifest their desire to do honour to Professor Eggert. Thanks are due also to the President and Administration of the ETH-Z for placing at our disposal a new air-conditioned lecture hall – very welcome during an exceptionally warm spell. We warmly wish to thank also K. Pfister, Dr. F. Tomamichel and the whole staff of the Department who ensured the smooth running of the meeting, including an enjoyable boat ride on the lake. We appreciated very much the cooperation with the Walter de Gruyter Press.

Finally, we wish to express our gratitude to Prof. Dr. K. V. Chibissov, Corresponding Member of the Academy of Science of the USSR, for accepting the position as Honorary President of the Colloquium, and to the Sessions' Chairmen Dr. P. M. Cassiers, Prof. Dr. G. Haase, Prof. Dr. K. Hauffe, Prof. Dr. E. Inoue, Dr. J. E. Lu Valle, Dr. S. R. Morrison, Dr. E. M. Pell, Prof. Dr. G. M. Schwab for running the Meeting.

August 1972 *W. F. Berg* *G. Haase* *K. Hauffe*

Eggert Celebration

This was held on Aug. 1 at the Carlton Elite Hotel at Zürich, beginning with the talk by Dr. John H. Dessauer, reproduced below, a period for presentation of messages, cocktails and a luncheon, at which 139 guests took part. Messages were presented as follows: –

Prof. Dr. K. V. Chibissov, Corresponding Member of the Academy of Sciences of the USSR, on behalf of the Photographic Science Committee of the Academy.

Prof. Dr. O. Helwich presented the Certificate of Honorary Membership of the Photographic Society of Vienna and afterwards the Golden Voigtlander Medal of the same Society, the prerequisite of which is Honorary Membership. He greeted Prof. Eggert as long-standing advisor to his journal, the Photographische Korrespondenz.

Mr. B. Obrecht, President of the Photographers' Association of the Kanton of Zürich, transmitted greetings from the Federation of Swiss Photographers.

Prof. Dr. H. Frieser, until recently Head of the Institute for Scientific Photography, Technical University of Munich, brought greetings from the German Photographic Society.

Prof. Dr. E. Inoue, Head of the Imaging Science Dept. of the Tokyo Institute of Technology, congratulated on behalf of the Society of Scientific Photography of Japan.

Messages were read from: –

the RPS (Royal Photographic Society) of London, the oldest Photographic Society in the World;
the SPSE (Society of Photographic Scientists and Engineers);
Schweizerische Naturforschende Gesellschaft;
SGOEM (Schweiz. Gesellschaft für Optik und Elektronen-Mikroskopie).

Prof. Dr. E. Klein, Director at Agfa-Gevaert, of Leverkusen, spoke on behalf of his Company which Prof. Eggert had joined 50 years ago in order to establish their research laboratory.

Prof. Dr. G. M. Schwab, previously Head of the Institute of Physical Chemistry at the University of Munich and collaborator on the later editions of Eggert's Textbook on Physical Chemistry, brought congratulations from the Bavarian Academy of Science. Amongst many other signs of appreciation received by Prof. Eggert was the provisional establishment of an

Eggert Prize Fund.

A considerable sum of money has already accumulated. The intention is to offer prizes to honour good research papers by young scientists working in the field of the science of photography. Details will be worked out by a committee yet to be appointed.

Finally, Prof. Eggert expressed his appreciation in moving words. No more speeches were offered during the luncheon, but it should be placed on record that Prof. Chibissov toasted the birthday child with a jolly drinking song, which brought the celebration to a happy finish.

List of Authors

Albrecht Dr. W., Forschungsinstitut für Geochemie, Concordiastr. 28, D-86 Bamberg
Arneth Dr. R., Kalle AG, Postfach 9165, D-6202 Wiesbaden-Biebrich

Berg Prof. Dr. W. F., Photographisches Institut E T H, CH 8006 Zürich
Byrne Dr. J. F., Xerox Corp., Xerox Square, Rochester, NY 14603, USA

Camenisch H., Photographisches Institut E T H, CH 8006 Zürich
Comizzoli Dr. R. B., RCA Laboratories, Princeton, N.J. 08540, USA

Dannert Dr. H., Philips Forschungslaboratorium Aachen GmbH, Postfach 1980, D-51 Aachen
Dessauer Dr. J. H., P.O.Box 373, Pittsford, NY 14534, USA

Eckenbach Dr. W., Philips Forschungslaboratorium Aachen GmbH, Postfach 1980, D-51 Aachen
Epping Dr. R. H., Weidenweg 5, D-8051 Neufahrn

Fridkin Prof. Dr. V. M., Institute of Crystallography, Leninsky Prospekt 59, Moscow, B-333, USSR

Gaidelis Dr. V., Institute of Electrography, Vilnius, USSR

Hauffe Prof. Dr. K., Institut für Physikalische Chemie, Universität Göttingen, Bürgerstr. 50, D-34 Göttingen
Heiland Prof. Dr. G., Physikalisches Institut TH Aachen, Templergraben 55, D-51 Aachen
Hercock Dr. R. J., Ilford Research Laboratories, Ilford, Essex, GB
Hirsch Dr. H. J., Philips Forschungslaboratorium Aachen GmbH, Postfach 1980, D-51 Aachen
van Hove Dr. H., Fysico-Chemisch Laboratorium, Universiteit Leuven, Heverlee, Belgium, now at the "Afd. Fysica en Elektronica van de Halfgeleiders" K.U.L. Leuven

Inoue Prof. Dr. E., Tokyo Institute of Technology, O-Okayama, Meguro-Ku, Tokyo, Japan

Ionescu Dr. I., Institut für Physikalische Chemie, Academia Republicii Socialiste Romania Bukarest, Rumänien

Jung Dr. M., Philips Forschungslaboratorium Aachen GmbH., Postfach 1980, D-51 Aachen
Jurevieius D., Vilniaus Valstybinis V. Kapsuko Universitetas Vilnius, USSR

Kassel Dr. K. H., AEG-Telefunken, Schützenstr. 30, D-4785 Belecke
Kiess Dr. H., Laboratories RCA Ltd., Badenerstr. 569, CH-8048 Zürich
Kosche Dr. H. H., Renker GmbH., Postfach 445, D-516 Düren

Lanker W., Turlabor AG., Geissacherstr. 8, CH-8126 Zumikon
Lewis Dr. R. B., Xerox Corp., Xerox Square, Rochester, NY 14603, USA
Leysen Dr. R., Fysico-Chemisch Laboratorium, Universiteit Leuven, Heverlee, Belgium
Löhr Dr. B., Kalle AG., Postfach 9165, D-6206 Wiesbaden-Biebrich
LuValle Dr. J. E., SCM Research Laboratory, 3210 Porter Drive, Palo Alto, Calif. 94304, USA

Mammino Dr. J., Xerox Corp., Xerox Square, Rochester, NY 14603, USA
Matulionis Dr. A., Vilniaus Valstybinis V. Kapsuko Universitetas, Vilnius, USSR
Meier Dr. H., Forschungsinstitut für Geochemie, Concordiastr. 28, D-86 Bamberg
Mrs. Meskuotiene E., Institute of Electrophotography, Vilnius, USSR
Meyer-Laack Dipl.-Phys. A., Institut für Physikalische Chemie, Universität Göttingen, Bürgerstr. 50, D-34 Göttingen
Montrimas Dr. E., Vilniaus Valstybinis V. Kapsuko Universitetas, Vilnius USSR
Moraw Dr. R., Kalle AG., Postfach 9165, D-6202 Wiesbaden-Biebrich
Morrison Dr. S. R., Stanford Research Institute, Menlo Park, Calif. 94025 USA

Pazera Dr. A., Vilniaus Valstybinis V. Kapsuko Universitetas, Vilnius, USSR

Petrikat Dr. K., Institut für Physikalische Chemie, Universität Göttingen, Bürgerstr. 50, D-34 Göttingen

Pocius Dr. Z., Vilniaus Valstybinis V. Kapsuko Universitetas, Vilnius, USSR

Pusch Dr. H., Institut für Physikalische Chemie, Universität Göttingen, Bürgerstr. 50, D-34 Göttingen

Radler Dr. R. W., Xerox Corp., Xerox Square, Rochester, NY 14603, USA

Range Dr. J., Institut für physikalische Chemie, Universität Göttingen

Rein Dipl.-Phys. H., Institut für Physikalische Chemie, Universität Göttingen, Bürgerstr. 50, D-34 Göttingen

Rochlitz Dr. J., Kalle AG., Postfach 9165, D-6202 Wiesbaden-Biebrich

Schaffert Dr. R. M., 12771 Woodmont Drive, Saratoga, Calif. 95070, USA

Stark Dr. H. M., Xerox Corp., Xerox Square, Rochester, NY 14603, USA

Stockman Dr. D. L., Xerox Corp., Xerox Square, Rochester, NY 14603, USA

Stotz Dr. S., Philips Forschungslaboratorium Aachen GmbH., Postfach 1980, D-51 Aachen

Suys Dr. A., Gevaert - Agfa NV, B-2510 Mortsel

Tauraitiene Dr. S., Institute of Electrography, Vilnius, USSR

Tauraitis Dr. A., Institute of Electrography, Vilnius, USSR

Tschirwitz Dr. U., Forschungsinstitut für Geochemie, Concordiastr. 28, D-86 Bamberg

Verhille Dr. K., Gevaert-Agfa NV, B-2510 Mortsel

Viščakas Prof. Dr. J., Vilniaus Valstybinis V. Kapsuko Universitetas, Vilnius, USSR

Weigl Dr. J. W., Xerox Corp., Xerox Square, Rochester, NY 14603, USA

Whittaker Dr. G. L., Xerox Corp., Xerox Square, Rochester, NY 14603, USA

Winkelmann Dr. H., Kalle AG., Postfach 9165, D-6202 Wiesbaden-Biebrich

List of Participants

Arneth Dr. R., Kalle AG, D-6202 Wiesbaden-Biebrich

von Babo Dr. H., Lenggstr. 16, CH-8008 Zürich
Baumeister M., Dow Chemical Europe SA, CH-8810 Horgen
Baumeler R., TCL-ETH, CH-8006 Zürich
Beggiato Dr. G., Consiglio Nazionale delle Ricerche, I-40126 Bologna
Berg Prof. Dr. W. F., Photographisches Institut ETH, CH-8006 Zürich
Bethe Dr. G., Universität Karlsruhe, D-75 Karlsruhe
Beyer Dr. E., Günther-Wagner Pelikan-Werke, D-3 Hannover
Boroky J. S., Tetenal Photowerk, D-2 Hamburg-Norderstedt 1
Bosenick Dr. R., Renker GmbH, D-516 Düren
Boyer Mlle S., Kodak-Pathé SA, F-94 Vincennes

Camenisch H., Photographisches Institut ETH, CH-8008 Zürich
Cassiers Dr. P. M., Gevaert-Agfa NV, B-2510 Mortsel
Cremer Prof. Dr. E., Institut für Physikalische Chemie, A-6020 Innsbruck

Danzmann Dipl.-Chem. H.-J., Institut für physikalische Chemie, D-34 Göttingen
Dessauer Dr. J. H., Xerox Corp., Rochester, N.Y. 14603

Eckenbach Dr. W., Philips Forschungslaboratorium, D-51 Aachen
Epping Dr. R. H., Weidenweg 5, D-8051 Neufahrn b. München
Errede Dr. L. A., 3 M Research Ltd., Harlow, Essex
Forgo Dr. G., Turlabor AG, CH-8126 Zumikon
Frey M. P., Multitec AG, Wankdorffeldstr. 66, CH-3000 Bern
Frieser Prof. Dr. H., Hauberisserstr. 8, D-8 München

Gilbert J., Ozalid Ltd., Loughton, Essex
Goodmann A. M., RCA Laboratories, Princeton, J. J. 08540
Grossa Dr. M., Du Pont Fotowerke Adox GmbH, D-6078 Neu-Isenburg
Grossmann Dr. W., Gretag AG, CH-8105 Regensdorf
Grünzweig Dr. T., Zürcher Papierfabrik an der Sihl, 8045 Zürich
Günther Dr. E., Mimosa, Agfa-Gevaert, D-2300 Kiel 23
Günzburger P., Hasler AG, CH-3000 Bern 14

Haase Prof. Dr. G., Institut für wissenschaftliche Photographie TU, D-8 München 2
Hamada Dr., Ricoh SA, Schiphol Centrum, Amsterdam
Hansen Dr. N., Philips Forschungslaboratorium, D-51 Aachen
Hasek J., Rob. Viktor Neher AG, CH-8280 Kreuzlingen
Hauffe Prof. Dr. K., Institut für Physikalische Chemie der Universität Göttingen, D-34 Göttingen
Heiland Prof. Dr. G., Physikalisches Institut TH Aachen, D-51 Aachen
Heinzer Dr. P., Wifor AG, CH-8045 Zürich
Hercock R. J., Ilford Ltd., Ilford, Essex
Hirsch Dr. H. J., Philips Forschungslaboratorium, D-51 Aachen
Hirth Dr. H., Farbwerke Höchst, D-623 Frankfurt/M 80
Hoegl Dr. H., Institut Battelle, CH-1227 Carouge
Honjo Dr. S., Fuji Photo Film Co. Ltd., Asakashi, Japan
Hupfeld Dipl.-Phys. J., Institut für physikalische Chemie, D-34 Göttingen
Inoue Prof. Dr. E., Tokyo Institute of Technology, Meguro-ku, Tokyo
Ionescu Dr. N., z. Zt. Institut für Physikalische Chemie, Universität Göttingen, D-34 Göttingen, Bukarest Rumänien.
Ivanis Mrs. B., Lumoprint Zindler KG, D-2000 Hamburg

Jagrovic Dr. P., Multitec AG., CH-3000 Bern
Jshikawa Dr. Y., z. Zt. Institut für Physikalische Chemie, D-34 Göttingen, Tokio Japan

Kassel Dr. K., AEG-Telefunken, D-4785 Belecke
Kegelmann Dr. G., Hoffmann & Engelmann AG, D-673 Neustadt/Weinstraße
Kemme Dr. G., Felix Schoeller Jr., D-4501 Lüstringen, Krs. Osnabrück
Kern Dr. R., Turlabor AG, CH-8126 Zumikon
Kiess Dr. H., Laboratories RCA Ltd., CH-8048 Zürich
Kikuchi Dr. K., Photographisches Institut ETH, CH-8006 Zürich
Klein Dr. D., Du Pont Fotowerke Adox GmbH, D-6078 Neu-Isenburg

Knechtel Hr., ECE, D-6300 Gießen-Rödgen
Koch H.-C., Sinar AG, CH-8245 Feuerthalen
Kohlmannsperger Dr. J., Agfa-Gevaert AG, D-8 München 90
Kosche Dr. H., Renker GmbH., D-516 Düren
Kultscher W., Tonec SA, CH-1196 Gland

Lange P., ECE, D-6300 Gießen-Rödgen
Lanker Dipl.-Phys. W., Turlabor AG, CH-8126 Zumikon
Lanzavecchia Prof. Dr. G., Montecatini Edison S. p. A., I-20121 Milano
La Roche Dr. U., Gretag, CH-8105 Regensdorf
Lewis Dr. R. B., Xerox Corp., Rochester, N. Y. 14603
Lu Valle Dr. J. E., SCM Research Laboratories, Palo Alto, Calif. 94304

Maeder Dr. E., Kodak AG, CH-8041 Zürich
Mäder P. E., Neptunstr. 67, CH-8032 Zürich
Matulionis Dr. A., Vilnius University, Vilnius, USSR
Mehendru Dr. P. C., National Physical Laboratory, New Delhi 12
Meier Dr. H., Forschungsinstitut für Geochemie, D-86 Bamberg
Meyer-Laack Dipl.-Phys. A., Institut für physikalische Chemie, D-34 Göttingen
Meyer Dr. R., Renker GmbH, D-516 Düren
Mohn Dr. E., Turlabor AG, CH-8126 Zumikon
Moraw Dr. R., Kalle AG, D-6202 Wiesbaden-Biebrich
Moroson Mr., Saxon Development Corp., Miami Beach, Florida 33139
Morrison Dr. S. R., Stanford Research Institute, Menlo Park, Calif. 94025
Munder Dr. J., Kalle AG, D-6202 Wiesbaden-Biebrich
Myers Dr. W. T., St. Regis Paper Co., West Nyack, N. Y. 10994

Nishiyama S., Fuji Photo Film Co., Minamiashigara, Japan
Nissen H. F., Anemonevej 3, DK-2820 Gentofte

Ohta T., Ricoh Co. Ltd., Ohta-ku, Tokyo

Pasini Dr. G., National Research Council, I-40126 Bologna
Patschorke Dr. J., Celfa AG, CH-6423 Seewen-Schwyz
Pavlik V., Turlabor AG, CH-8126 Zumikon
Pell Dr. E. M., Xerox Corp., Rochester, N. Y. 14603
Pfister K., Photographisches Institut ETH, CH-8006 Zürich
Poganski Dr. S., AEG-Telefunken, D-4785 Belecke
Pusch Dr. H., Institut für physikalische Chemie, D-34 Göttingen
Patzke Dr., ABG, D-8 München

Range Dr. J., Institut für Physikalische Chemie, Universität Göttingen, D-34 Göttingen
Reichel Dr. D., Celfa AG, CH-6423 Seewen-Schwyz
Rein Dipl.-Phys. D., Institut für physikalische Chemie, D-34 Göttingen
Reiss Dr. K. H., Siemens AG, D-8520 Erlangen
Rochlitz Dr. J., Kalle AG, D-6202 Wiesbaden-Biebrich
Russel G. J., Dow Chemical Europe SA, CH-8810 Horgen

Schäfer D., ECE, D-6300 Gießen-Rödgen
Schaffert, Dr. R. M., 12771 Woodmont Drive, Saratoga, Calif. 95070
Schanz D., Haindl Papier GmbH., D-8961 Hegge/Allgäu
Schledermann M., Lumoprint Zindler KG., D-2 Hamburg 50
Schlesinger Dr. M., Institut für Wissenschaftliche Photographie TU München, D-8 München 2
Schmidt Dr. R., Philips Forschungslaboratorium, D-51 Aachen
Schwab Prof. Dr. G. M., Universität München, D-8 München 2
Schwerdtel, Dr. E., Hasler AG, CH-3000 Bern 14
Simm Dr. W., Farbenfabriken Bayer AG, D-509 Leverkusen
Sprong A., GAF (Nederland) N. V., Delft
Stark Dr. H. M., Xerox Corp., Rochester, N. Y. 14603
Stockman Dr. D., Xerox Corp., Rochester, N. Y. 14603
Stotz Dr. S., Philips Forschungslaboratorium, D-51 Aachen
Struck Dr. B., Dr.-Ing. Rudolf Hell, D-2300 Kiel 14
Svestka Dr. L., Turlabor AG, CH-8126 Zumikon
Szilard I., Turlabor AG, CH-8126 Zumikon

Tschibissov Prof. Dr. K. W., Academy of Sciences USSR, Lenin-Prospekt 61/1/43, Moskau W-333

Ulbrich Dr. K.-H., Zinkweiß-Forschungs-GmbH, D-42 Oberhausen

Van Auken J. A., Sacon Development Corp., Miami Beach, Florida 33139
Van Hove Dr. H., Leuven University, B-3030 Heverlee
Verhille K., Agfa-Gevaert N. V., B-2510 Mortsel
Viscakas Prof. Dr. J., Vilnius University, Vilnius USSR
Vokral Dr. P., Rob. Viktor Neher AG, CH-8280 Kreuzlingen
Volz Dipl.-Chem. H., Institut für physikalische Chemie, D-34 Göttingen

Weigl Dr. J. W., Xerox Corp., Rochester, N. Y. 14603

Zographos G., Gretag AG., CH-8105 Regensdorf
Zschokke-Gränacher Dr. D., Institut für Angewandte Physik, Universität Basel, CH-4056 Basel
Zwicky Dr. H., Typon AG., CH-3400 Burgdorf

List of Donors

AEG-Telefunken
Agfa-Gevaert AG
Battelle Institute, Geneva
Balzers AG
Bron Elektronik AG
Brown, Boveri & Cie.
Celfa AG
Ciba-Geigy Photochemie AG
Condor-Film AG
Du Pont de Nemours (Deutschland) GmbH
Folex AG
Gebrüder Fretz AG
Gelatinefabrik Winterthur
Heberlein & Co. AG
F. Hoffmann - La Roche & Co. AG
Imago AG
Kalle AG
Kodak SA
Kollmorgen AG
Mettler Instrumente AG
Multitec AG
Orell Füssli AG
Ozalid AG
Philips Forschungslaboratorium Aachen GmbH
Polaroid AG
Rank Xerox AG
Laboratories RCA Ltd.
Renker GmbH
Ringier & Co. AG
Dr. E. Rüst
Sandoz AG
Schmid + Co. AG
Sinar AG
Gebrüder Sulzer AG
Turlabor AG
Typon AG
I. Weinberger
Wild Heerbrugg AG
Wissenschaftliches Forschungsinstitut AG
Zürcher Papierfabrik an der Sihl

Contents

Image Technology for the Humanity

John Dessauer

Professor Eggert, Mrs. Eggert, Dr. Berg, honored guests, ladies and gentlemen.

It is with particular joy that we are gathered here to honor Professor Eggert on his eightieth birthday.

We honor him as a true friend of many years.

We honor him as our teacher and educator.

We honor him as a leading scientist.

We honor him as an author and pioneer in the field of physical chemistry of photography, yes, much broader, in the field of visual communication technology.

We have many of Dr. Eggert's students here in the audience who can attest to his extraordinary qualities as teacher and educator.

Drs. Frieser, Klein and Matejec published in the ninety seventh issue of the "Photographische Korrespondenz" in monograph form a survey of the many important scientific and technical achievements and contributions to the field of visual communications that Dr. Eggert made in a life-time of intensive scientific work. Dr. Frieser et al described it so well that I will not try to repeat what they have said. May I sum it up by saying that Dr. Eggert's contributions to the fundamental understanding of the photographic process as applied to visual communications as well as to x-rays have become corner stones in the evolution of this technology and so many of his basic ideas are in practical daily use, from his x-ray dosimeter to his studies of the photographic elementary processes that I would like to place him among the great technical innovators and movers of our times, with Thomas Edison, Alexander Graham Bell, Marconi, Roentgen and others.

My own long and happy association with Dr. Eggert goes back to a time when Haloid Company, the forerunner of Xerox Corporation, needed advice on photographic and photoimaging technology. With his knowledge, sound judgment, and friendship, he was able to be of great help to Xerox Corporation and to me personally. Just let me say, Thank you, to you, Dr. Eggert, with a clear voice for all the advice you have given.

On an occasion like this, I hope you will agree with me that it might be proper to take a *look back,* take a *look* at the *present,* and then take a *brief glance into* the *future.* I would like to do this by focusing on the tremendous evolution that has taken place in the field of photography, yes broadened, in the field of visual human communications and consider and discuss the impact it has had on all of us in our present time and then speculate for a few minutes on what might be ahead of us.

When Johann Zahns invented and developed his "Camera Obscura" in 1665, little did he dream what this simple optical device would look like or do in the twentieth century. When Arago announced to the French Academy the achievements of Daguerre and Nièpce, the public seemed to have become aware of the importance of their contribution. Fox Talbot's invention of the negative and positive process added to this immensely. It wasn't until the end of the nineteenth century and the beginnings of the twentieth century that George Eastman in America and Agfa, Lumière, and others in Europe, started to make this invention commercially available on a practical and large scale. But even these efforts were based more on empirical procedures than scientific data and it took men like Professor Eggert in Europe and Dr. Mees in America to organize scientific investigations to provide the fundamental understanding and definitions of the imaging processes which were so necessary as foundation for the impressive progress we all have witnessed within the past 25 years. Just think of instant color pictures at incredibly short exposures, the diagnosis our doctor can make based on x-ray pictures, think of newspapers and magazines giving us the news of the day in pictorial form, think of the cinema or television. Let us never forget the pioneers like Professor Eggert who were prime movers in the development of the technology of visual communication, who made all this possible and to whom we wish to pay our tribute to day.

1. Past

We live in an age of great technological advances in photography, atomic energy, antibiotics, supersonic transportation, and many others. Yet, as we recall our own school days or watch our children learn, we must very humbly realize that our own capacity to creatively discover, reason, or even retain, is very limited indeed. Perhaps this is why many of us experienced fears about school exams.

However, through visual communication, man has long been able to draw on the discoveries and the experiences of others–of our time and of times before us–in fact civilization, culture, and technology are based on knowledge communicated to us by others who lived before us.

Communication, moving a concept of one mind to another, is the cement of human civilization. Yet not all communication has the same quantitative effect. The epic poet, Homer, communicated most eloquently. Through his communication he was a civilizing agent helping to bind together the ancient world in his time.

Yet the size of what has been called the civilized world indicates the still prevailing limits of communication through the spoken word. Today Homer's poems in printed visual form can become part of the lives of millions of people cutting across space, culture and time itself. The beginnings for visual communications were modest indeed. One person could address only one other person or at best a few other persons at any one time. Chiseled in stone, Egyptian hieroglyphics provided even more limited audience communication than word to mouth communication. Greek and Roman papyrus scrolls reached a wider audience because it was seldom duplicated, the single message on the scroll could be communicated to only one or a few readers at one time.

To widen its effectiveness visual communication needed the ability to be duplicated. This advance finally began in the middle ages with the painstaking copy work being done by monks. Then during the fifteenth century Johann Gutenberg invented the letter press and mankind truly entered a new era. It was an era in which a single thought, a single piece of information, a single concept could replicated with relative ease, and it could be distributed widely enough to enable a multitude to absorb this thought at the same time.

It was so revolutionary a development, it made mass literacy possible and individual literacy a necessity.

Until the twentieth century nearly all duplication of visual communication was an application of the Gutenberg principle. Copper etching or gravure were followed by offset methods to duplicate or reproduce in book form, the writings and thoughts of others. Visual communications embrace now an almost infinite variety of media, a letter, a newspaper, a book, a painting, a projected slide, a photograph, a movie, a television program, a picture magazine, an X-ray. Today the process seems almost ubiquitous.

2. Present

It seems natural, therefore, that a number of industries have been developed to serve men and society through visual communications by a variety of techniques. The progress has been awesome. In the field of business communication for example, the United States produced two hundred and seventy five billion paper copies in 1967 and this figure merely covers copiers and duplicators and does not take into account the miles of copying spun from computers, nor microfilm copies, nor spirit and stencil machines.

With your permission I would like to dwell for a minute on the impact which visual communications have on our daily lives. As we get up in the morning we enjoy our newspaper. It contains information and pictures transmitted by wire photo from all over the world. Some of us, especially in America will turn on the television even before we have shaved to watch the "Today Show". Skilled clever advertising may even give us a list of things to buy downtown for our families. As we proceed to go to our office or laboratory, advertisements are visible on street signs and posters everywhere. Arriving in our office we face a stack of mail much of which consists of copies of correspondence we share with others. In the case of a physician, a stack of X-rays of his patients may also be there to be viewed. In reviewing correspondence we have to recall previous data, sometimes from several years back, and a microfilm reader is brought into our office which will enable us to quickly review correspondence of years past. A subsequently called conference with a number of associates is unthinkable in our present time without visual displays of charts, of pictures and of texts outlined for better audio-visual communication. There are few classroom lectures or scientific meetings taking place now without at least a few photo slides. Our lunch hour brings us again newspapers and picture magazines and quite often also a stop by the brokerage house to see a display of the quotations on the stock market. Before returning home for the evening, the thoughts come to mind about the gratitude we owe to our security departments in the city for protecting us from crime, for making our environment safer by aerial photography picking out polluted waters and other damage, to our military for having found ultra resolution photographic means of monitoring perhaps via satellite from incredible altitudes the activities of a potential troublemaker or more peacefully to our men in laboratories and technical institutions for their ability to accurately photograph and record their day's progress as well as their thinking placed on paper in the form of sketches and drawings. Finally arriving at home, we may be resorting again to the evening newspaper or to a good book or review family motion pictures or slides or colored photographs and even spend a little time watching colored television.

Yes, visual communication has such a profound impact on our lives that it is unthinkable that we could ever be without it and only a blind man could imagine what human society had before these advances were made. I myself am particularly fascinated by the aerial survey photography which will permit the discovery of minerals by satellite photography and which also quite successfully allows weather predictions on a more accurate basis; by underwater photography which is going to be the basis of a whole new technology, oceanography, and above all, what has been captured in pictures of what microscopes and electron microscopes show of the infinite world of microcosms. In fact there are so many fields where visual communication plays an important part, that it becomes impossible to do more than cite a few examples to emphasize the impact it has on our lives and society. Professor Eggert with his technical contributions has had a major part in forming these technologies.

3. Future

Further advances in communications technology are well underway and exciting new developments can be seen. I would like to share with you some speculations about our future, the potential new direction of visual communications. I would like to emphasize also that with all these new technological advances will come new obligations for our scientists, for engineers and for society to make sure these new things will accrue to our benefit and not to our detriment. The understanding of semiconductor behavior and the discovery of photoconductors in a wide range of sensitivity opened up vast new horizons in visual and graphic communication. It was Chester Carlson who put the understanding to practical use on October 22, 1938, thirty three years ago when he inked the date 10–22–38 Astoria on a glass microscope slide and rubbed the photoconductive sulfur coated metal plate in the dark with a handkerchief to charge it. The plate was immediately exposed to light passing through the microscope slide and was then sprinkled with dyed lycopodium powder, which clings readily to charged surfaces. The loose powder was then blown off and the remaining powder was transferred to a sheet of waxed paper. The paper was then heated to melt the wax to hold the powder to it. On the waxed paper, almost miraculously, was a near-perfect duplicate of the original. Electrostatic copying was born that day and xerography became a reality. The same technology and coincidently the same material that replaced sulfur as the plate coating, selenium, became an important element in the development of the television industry and the vidicon tube. I believe these two apparently distinctive technologies, electronic imaging and xerography, will merge ultimately and lead to a whole new family of dual visual and graphic means of communication. If that prospect is parenthetical at the moment, we have already entered a greater communication revolution and we are only beginning to know how to handle it. There are times when the paper mountain of information available overwhelms us.

Yet to put it in its real perspective we know that the information explosion has introduced us to new challenges for abstracting, storing, recording and distributing the output of artists, scientists, teachers, and writers. We really need to learn how to cope with these challenges. We have begun to tackle the problem through the computer which one could redefine as a new instrument for converting graphics and alphanumeric data into electrical signals and working these signals according to programmed instruction. Essentially the computer has introduced a new technology for handling data. We also have begun to cope with a challenge through optical reduction of stored data. Microfilm and microfiche are becoming widely applied methods of filing and indexing the phenomenal output of a civilization which has become virtually overarticulate.

However, simply storing, indexing and manipulating information does not solve the challenge. We need new dimensions which will make it easier to absorb knowledge more quickly. Future generations will need to retain a greater store of knowledge than was previously necessary.

Indication of the mounting pressures on people to be informed is in the staggering fact that man's recorded knowledge doubled between 1948 and 1960 and estimates are that it will have doubled again by now. It will be perhaps more than a little daydream of the future when I speak of some more possibilities.

A new way of phototypesetting will permit an author to feed his creative or editorial thinking into an electronic computer, the computer will work out, rectify, and then print out on a cathode ray tube a text to be photographed and put on a master printing plate. Thus we will be able to produce what graphic arts people call camera-ready copy in a matter of minutes.

Yes, we may even find that electronic methods will permit the storage of magazines and book data which can be retrieved on demand and typed or printed, a way of "demand printing".

The introduction of the really automatic picture taking methods such as Kodak's Instamatic Camera have added by an order of magnitude to the simplicity and ease with which beautiful images can be obtained in black and white and color. Just as color television has become an accepted household appliance, color photography and color printing are well underway to replace most black-and-white images.

Electrically produced facsimile is already commercially available. I can foresee that this process will become simplier, faster, and cheaper, and first class or preferential mail may well be made available through facsimile or digital printers to recipients subscribing to the service from the Post Office.

Education too is well underway to be affected by the communication revolution. Demand printing or duplicating enables the professor to distribute his classroom comments to his students as up to date as his lecture text. This will eliminate the need for notetaking in classes and stimulate students to think about instead of record the subject. Someday children may use a household computer terminal with accessory screen and audio receiver to supplement or even replace some classroom instructions. The youngster will simply dial a central education computer, identify himself by code and continue a lesson in language, mathematics, physics, etc., starting from the point where he left off the previous day. The computer will have a typewriter keyboard on the console, and lessons will be illustrated when necessary through the viewing screen. The youngster will communicate with the computer by using the typewriter keyboard or by drawing on the screen with a light pen. Improbable as this process may seem now, much of it is already in experimental stages in Florida and California where schools are testing computer-assisted instruction systems.

A checkless society is also around the corner. New ways of banking, charging, and paying bills will be with us as soon as proper protections can be developed to keep men honest.

More recently I read about an approach in Boston for remote diagnosis of patients between the Logan Airport and Massachusetts General Hospital. This is for patients who are stricken ill on a journey or in the large international airport area.

All these exciting possibilities are based on technologies in which Professor Eggert had an important hand. Enough fundamental knowledge has been acquired for further important new breakthroughs in many areas.

In fact, the services available to mankind are already so overwhelming and so broadly affecting us that we should not be surprised when social scientists, the economists, the political scientists have to work hard to keep up with fast advances. Like all things, new technological tools will not necessarily accrue only to the benefit of mankind but can also be abused by men. Pornographic pictures, poorly conceived and inaccurate or sensual motion pictures, over-dramatized news reporting of subjective opinions, untruthful television advertising are some of the examples. More recently we all should become concerned about the potential which data storage and retrieval has for crime control but also for abuse for men in political power. It is no longer a challenge to record a man's high school mistakes and middle age successes and have it available for almost instant retrieval. This involves an invasion of privacy, an obstruction of one of the basic human psychological needs which is to forgive and forget. An opportunity for power to hold over his fellow men which is staggering, which should be recognized and with which society must cope.

St. Augustine, the great philosopher, author, and teacher of the fourth century, said that time is a three-fold present, the present as we experience it, the past as a present memory, and the future as a present expectation. By that criterion the 1970's and 1980's have already arrived. In the decisions we make now, the future becomes committed. On this festival occasion which we celebrate Professor Eggert's 80th birthday and pay tribute to him as scientist, researcher, and educator, I believe it is appropriate that I should end up on a new challenge for the commitments we are about to make or will make tomorrow.

In analysing the impact of technological advances like the ones I have just discussed, visual communications, or for that matter the computer, the transistor, and other modern creations, a valid question must be raised whether further advanced technology will make us richer or whether we have reached the point where we must be concerned that it will make us poorer. There are many who maintain that we were happier in the days when we were not so sophisticated. Life then, they say, was simpler and more wholesome.

Are they right? After all the debate about overflowing files, air pollution, noise, invasion of privacy, and other concomitants of contemporary progress, I believe the answer remains as it has always been, "though the machine has in many ways made men immensely richer, it can also do harm and impoverish him unless he uses it with prudence."

The use of a new scientific discovery, of new adaptations of technology, is a challenge that faces all of us here and those who follow us.

I would like to end upon on the hope that we jointly will meet it with a sense of high responsibility not only to ourselves but to all mankind. As new scientific discoveries and new technologies become available at an ever faster pace, finding ways of applying them for the welfare of society will be the greatest opportunity that awaits men in the future.

Current Theories Concerning Charge/Discharge Phenomena in Xerography

R. M. Schaffert

Abstract

This is a review of the current status of knowledge and theories concerning the basic mechanisms of the charging and discharging of xerographic layers.

The conventional concepts of photoconductivity, which have evolved from semiconductor science, do not provide adequate explanations for the photodischarge of photoconductive insulating layers under the high-field, transient conditions prevalent in xerography.

The range-limited model, proposed by Jaenicke and Lorenz, and the Li and Regensburger model, based on field-controlled recombination and bulk trapping, are now considered to be inadequate in view of later investigations by Tabak and Warter. They found that the efficiency of photogeneration of free carriers in amorphous selenium is strongly field controlled.

Field-controlled photogeneration has now been observed in various other photoconductive materials including organic materials. However, the author has not found any publications reporting field-controlled photogeneration in ZnO-resin layers.

Photoconduction and dark conduction in amorphous materials can be explained in terms of the Poole-Frenkel effect at high fields. In some cases Schottky emission may also be involved. At low fields, trapping, recombination, and space charge effects become apparent, and conduction may be dominated by a hopping mechanism.

No clear-cut model appears to have been proposed to account for discharge mechanisms involved in ZnO-resin layers.

1. Introduction

The most important processes in the formation of electrostatic images in xerography are the charging and discharging of photoconductive insulating layers. Refined instrumentation is now available for accurate measurements on these processes, and research investigations have led us to a better understanding of the fundamental phenomena involved. However, there is still a lot to be learned, and many gaps to be filled before we can piece together a

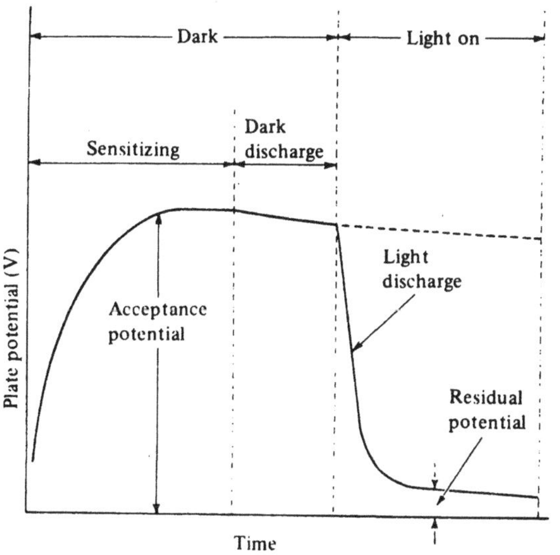

Fig. 1. Charging and discharging of a xerographic plate.

good phenomenological model that fully describes the behavior of photoconductive insulators during xerographic operations.

Fig. 1 illustrates the charging and discharging steps in terms of the plate potential as a function of time. There is a gradual rise of the plate potential determined primarily by the response time of the charging device. After charging stops, there is a slow discharge in the dark followed by a rapid discharge when the surface is exposed to light. In some cases the layer does not discharge completely, leaving a residual potential. The amount of surface charge the layer will sustain in the dark is usually referred to as the acceptance potential.

2. Corona Charging

The method of charging commonly used today is corona discharge. Other charging techniques have been used experimentally, such as conductive resilient rollers, radioactive ionization, and electrostatic induction [1]. Recently, Gallo and Leiga [2] have described a method of positive charging involving electron emission from the surface when the photoconductor is exposed to UV radiation.

The importance of corona charging characteristics in electrostatic imaging systems is sometimes overlooked. The performance of photoconductive layers in such systems depends, in part at least, upon the operational limitations of the corona charging device. Undercharging of the photoconductive layer will not permit utilization of maximum sensitivity and contrast, while overcharging can produce fatigue effects and surface defects.

In practice we encounter many different kinds of corona discharge devices with various arrangements of corona wires.

The most important measurement for a corona charging unit is the current flowing to the plate.

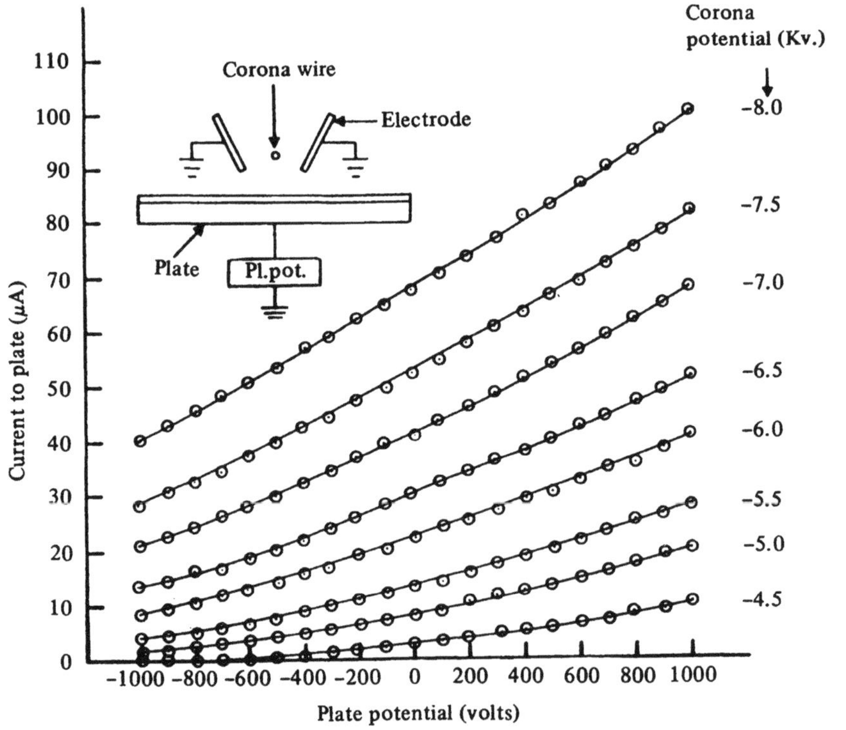

Fig. 2. Plate current vs. plate potential for a single-wire corona charging device with side plate electrodes.

this can be measured over a range of corona voltages and plate potentials. A typical set of curves showing corona current vs. plate potential is shown in Fig. 2. It can be seen that a plate potential of opposite polarity to that of the corona increases the plate current, whereas a plate potential of the same polarity decreases the plate current. As shown in Fig. 2, the curves slope downward as the bias potential becomes less positive and more negative. At a sufficiently high negative bias the plate current can be reduced to zero. The data here are for negative corona. Similar arguments hold for positive corona when the appropriate polarities are taken into account.

It is important to point out that the corona current is controlled by the surface potential. It is surface-potential limited, whereas the photoconductive insulating layer is field limited. It is necessary, therefore, in determining charge acceptance (acceptance potential) to make sure that the corona device is operating in a range capable of exceeding the acceptance potential of the layer.

We now know a great deal about corona charging. Shahin [3], using mass spectrometer techniques, has identified the ion species formed in negative and positive corona discharges in air, and Gallo, Germanos and Courtney [4] have determined the effect of humidity and temperature on the behavior of wire-to-plane coronas. There is still some uncertainty, however, concerning the nature of the surface charge formed when corona ions are deposited on the surface of a photoconductive layer. Shahin points out that the interaction of a corona ion with the surface is determined primarily by the recombination energy of the ion and the work function of the surface. The relative magnitude of these two factors will determine whether the ion is simply deposited on the surface, whether there is a charge exchange with surface atoms, or whether there is a charge exchange plus secondary electron emission from the surface.

In the negative corona charging of ZnO-resin layers there is an anomalous increase of corona current [4]. (Fig. 3.) Elder [5] proposed the

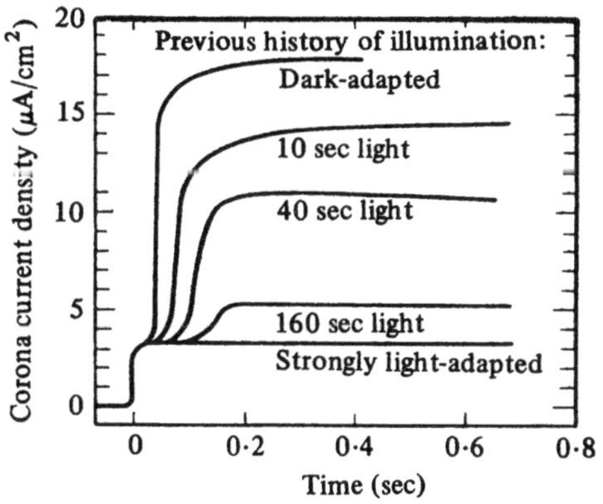

Fig. 3. Anomalous corona current during negative corona charging of a ZnO-resin layer at constant corona potential. Illumination just prior to charging was 2×10^{13} photons/cm² at the times indicated. (From Elder [5]).

hypothesis that negative corona produces small islands of excess charge on the ZnO-resin surface, and that the resulting highly localized potential gradients cause additional ionization which enhances the corona current. Smith and Behringer [6] have observed similar effects in CdS-resin layers and even in amorphous selenium. They attribute these anomalies to secondary corona discharge in the layer surface.

3. The Discharge Processes

Dark decay of surface charge will depend primarily upon the rate of injection of charge carriers at the surface or at the interface of the layer and the conductive substrate, while light decay (photodischarge) will depend upon the rate of generation of carriers and their ability to move through the layer.

The free surface of an insulating film normally acts as a blocking layer preventing injection of surface charges into the bulk of the film. The nature of the photoconductor-substrate interface is usually the most important factor in dark decay. It is essential that a blocking contact for at least one sign of carrier be effective at this interface. For positive charging the interface region must provide a barrier for electrons, and vice versa for negative charging. It is, of course, evident that we are dealing with layers under high electrical stress. Electric fields within these layers as high as 10^6 V/cm. are not uncommon. Therefore, the blocking layer must be effective at high fields in order to insure that a developable surface charge is retained in the image areas of the photosensitive layer.

Examples of dark decay curves are shown in Figs. 4 und 5, for amorphous selenium and for a polymeric photoconductor, polyvinyl-carbazole-trinitrofluorenone (PVCz-TNF), respectively. The effect of the substrate surface is quite apparent in Fig. 5.

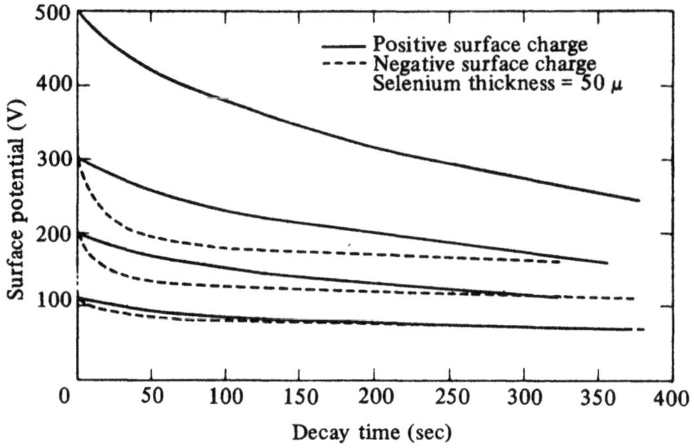

Fig. 4. Typical dark decay curves for an amorphous selenium xerographic plate, starting at different initial potentials.

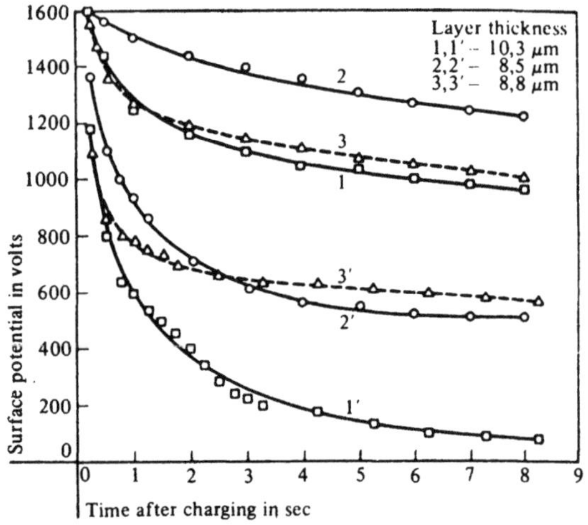

Fig. 5. Dark decay curves for 1 : 1 PVCz-TNF films on several different substrate metals. Negative (1) and positive charging (1') on aluminum; negative (2) and positive charging (2') on copper; and negative (3) and positive charging (3') on gold. (From Schaffert [18]).

Figs. 6, 7, and 8 are examples of light decay curves for three materials: amorphous selenium, ZnO-resin, and PVCz-TNF. The curves for amorphous selenium and PVCz-TNF show a rather sharp initial drop, whereas those for the ZnO-resin layer exhibit a slow initial decay, apparently due to the oxygen ion desorption process [7]. All curves show a slower rate of decay as the surface potential decreases. Thus, as the illumination proceeds, the photosensitive layer undergoes a change from high-field to low-field condition, and the mechanism of discharge changes accordingly.

Care must be exercised in measuring and interpreting light decay curves. Curves obtained by intermittent exposure with intermittent electrometer recording sometimes have shapes which differ from curves taken with a transparent probe during continuous exposure. It is important also to measure more than just surface potentials vs. time. It is much more meaningful to know the surface charge density, and such things as the dielectric constant and thickness. Also, the intensity and wavelength of the incident light should be measured accurately. The two quantities most frequently neglected in published papers on the light discharge of xerographic layers are light intensity and layer thickness.

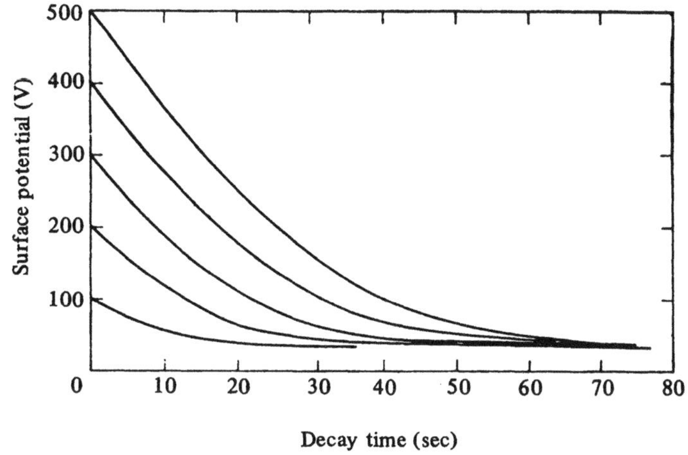

Fig. 6. Light decay curves for an amorphous selenium xerographic plate at different initial potentials. Constant intensity illumination at 450 nm. Positive surface charge.

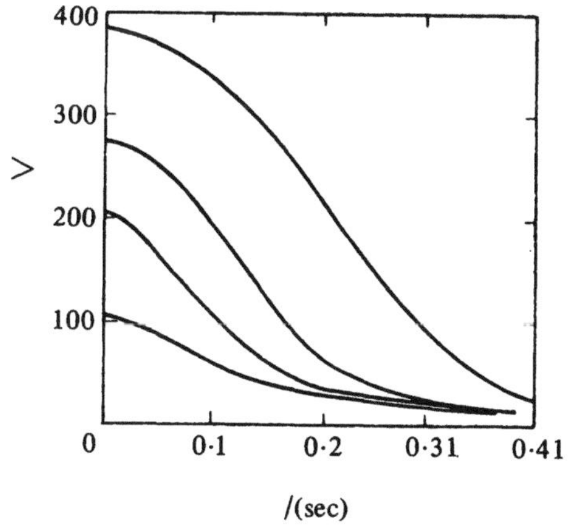

Fig. 7. Experimental curves for discharge of electrophotographic layers of zinc oxide on illumination with 600 lux. Resin binder: polyvinyl butyral. (From Levinson and Plavina [48]).

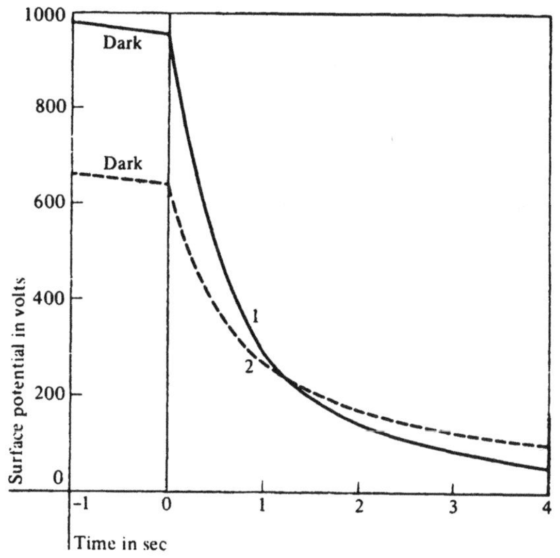

Fig. 8. Typical light-decay curves for a 1 : 1 PVCz : TNF sample. (1) Negative charge exposed to 3.02 $\mu W/cm^2$; (2) positive charge exposed to 9.79 $\mu W/cm^2$. $\lambda = 546.1$ nm. Film thickness, 15 μm. (From Schaffert [18]).

4. Discharge Mechanisms

Xerography started when the science of solid state physics was more or less in its formative years. Thus, in the beginning the sources of scientific information applicable to photoconductive insulators were very few. In fact, insulators were thought to be non-photoconductive. Today, the situation has changed considerably. However, there are still some riddles to be solved.

The photoelectric mechanisms that produce photodischarge in xerographic layers have long been subjects of discussion and speculation. Early attempts to explain these phenomena were either based upon conventional concepts of photoconduction applicable to semiconductor devices [8] or upon range limited models of charge-carrier transport [9], or a combination of both.

While it has long been recognized that the electrical charging of thin photoconductive insulating layers to high potentials produces high electrical stress within such layers, the effect of such high fields on photoelectric behavior was not fully appreciated until recently.

The range-limited model proposed by Li and Regensburger [10] about nine years ago included the concept of field-controlled recombination. However, the approximations imposed in the development of their model negated the effect of bulk trapping, but the range-limiting factor was nevertheless retained in their final equation. This equation, although reduced to a more or less empirical expression by the approximations imposed upon it, had several good qualities: (1) It could be fitted to experimental results very nicely, and (2) it started some arguments.

The pulse techniques used by Many [11] and by Tabak [12] for use with sandwich cell structures have provided independent measurements on insulating films which have added greatly to a better understanding of the photoconductive mechanisms involved.

The recent investigations of Tabak and Warter [13] on amorphous selenium have opened up new avenues for experimental and theoretical research on photoconductive insulators. They found that the efficiency of photogeneration of free carriers in amorphous selenium is strongly field controlled, and that the absence of a thickness effect indicated no appreaciable bulk trapping in this materials. (It should be pointed out, however, that the presence or absence of traps in amorphous selenium is highly dependent upon the purity and conditions of preparation of sample layers [14]. The samples used by Tabak and Warter may not have been comparable to samples used by other investigators.) More recent results by Tabak [15] indicate that recombination of electrons and holes in the absorption region of amorphous selenium is also absent, and that experimental evidence points to a monomolecular recombination within the bulk layer. However, Viščakas et al. [16] stress the importance of recombination in the absorption region of this material and explain the energy-dependent quantum efficiency in the short wavelength region on the basis of a recombination process.

Field-controlled photogeneration has been observed in anthracene by Batt, Braun and Hornig [17] and in a polymeric photoconductor by Schaffert [18]. This effect has also been found in amorphous As_2S_3 by Ing, Neyhart and Schmidlin [19] and in As_2Se_3 by Scharfe [20]. In the semiconductor area Stiles, Chang, Esaki and Tsu [21] found that the photoconductivity edge in amorphous films of GeTe and GeSe shifted toward lower photon energies with the application of high electric fields.

Numerous other indications of high-field effects on photoconductivity and dark conductivity in thin films have been reported in the literature. The impetus for research in this field has come primarily from recent interest in thin-film devices involving metal-insulator structures and amorphous semiconductors. The electric field strengths employed are in some cases about of the same magnitude as those encountered in electrically charged xerographic layers. This opens up a whole new area of experimental data and theory potentially applicable to photoreceptor materials used in xerography. The out-pouring of pertinent literature from these sources is already upon us. We have only to read some of the recent papers and reviews by Jonscher [22] and his associates at the University of London to recognize the importance now attached to this subject.

5. The Poole-Frenkel Effect and Schottky Barriers

Field-enhanced conductivity is usually explained by utilizing the so-called Poole-Frenkel effect [23] if the carriers originate within the bulk of the layer, and by the Schottky effect [24] if they are emitted from electrodes into the layer. The two effects are somewhat analogous. The Poole-Frenkel concept involves the lowering of the potential barrier surrounding a localized charge under the influence of an applied electric field, whereas the Schottky effect involves emission over a field-lowered barrier, as illustrated in Fig. 9.

The Poole-Frenkel situation is applicable to defect centers, traps, and thermally activated band gaps. (Fig. 10.)

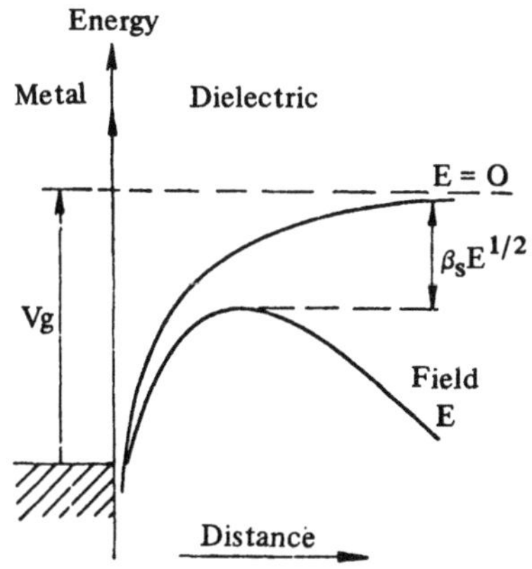

Fig. 9. The Schottky effect with a planar cathode. (After Jonscher [27]).

Mathematical equations for the two effects are usually expressed as follows:

$$I = I_{OS} \exp\left(-\frac{\phi_S}{kT}\right) \exp\left(\frac{\beta_S V^{1/2}}{kT}\right) \quad \text{(Schottky)}$$

$$I = I_{OF} \exp\left(-\frac{\phi_{PF}}{kT}\right) \exp\left(\frac{\beta_{PF} V^{1/2}}{kT}\right) \quad \text{(Poole-Frenkel)}$$

where

$\beta_S = (e^3/4\pi\epsilon\epsilon_0 d)^{1/2}$,

$\beta_{SF} = 2\beta_S$,

ϕ_S is the barrier height at the metal-insulator interface, ϕ_{PF} the potential barrier surrounding a localized charge, e the electronic charge, ϵ the dielectric constant, ϵ_0 the free space permittivity, and d the layer thickness.

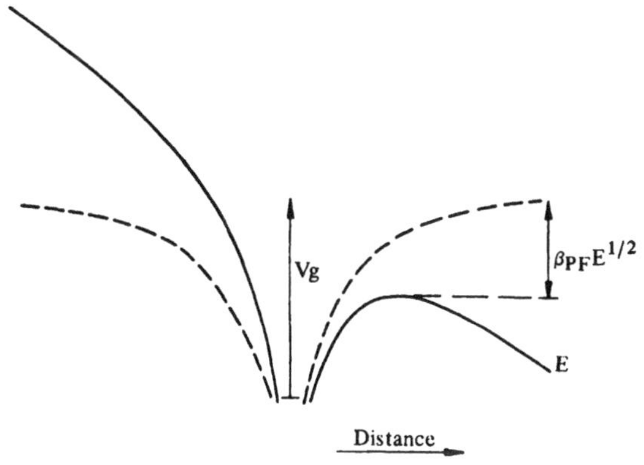

Fig. 10. Poole-Frenkel effect for electron emission from a point defect. (After Jonscher [21]).

Actually, the original equation derived by Frenkel [23] was given in terms of conductivity rather than current. However, many authors have plotted log I vs. $\sqrt{V}$ or $\sqrt{E}$ (field), as in the above equations, to test agreement of their data with either the Poole-Frenkel or the Schottky model. There seems to be no agreement among authors as to which one to use. If we stick to the original Frenkel formula, we should plot log I/V vs. V rather than log I vs. $\sqrt{V}$. It is conceivable that the effect of this is lost in fitting the experimental data to a straight line.

When the current is generation-limited, it is perhaps meaningless to talk about conductivity since mobility no longer plays a significant role in the conduction process except when its value is very low.

Figs. 11 through 19 show some samples taken from the literature where the logarithms of photocurrents and dark currents, as well as several other related factors are plotted as a function of the square root of the applied electric field.

Fig. 11 was taken from a paper by Pai and Ing [25]. It shows the pulse height vs. $\sqrt{E}$ for an amorphous selenium sandwich cell. Pulse height in this case is proportional to the number of carriers drifting across the sample as a result of a light pulse through a transparent electrode. It can be seen that the experimental points depart from the straight-line relationship at the lower field strengths, beginning in this case at about 10^4 V/cm. At shorter wavelengths this departure begins at somewhat higher field strengths. The authors suggest that this low field departure from linearity may be due to recombination.

Figs. 12 and 13 are for amorphous As_2S_3 as reported by Ing, Neyhart and Schmidlin [19]. It will be noted in Fig. 12 that there are two ranges of linearity for the plots of the log of initial xerographic discharge rate vs. $\sqrt{E}$ for the charged layers. At the lower fields ($\sim 10^4 - 10^5$ V/cm) all experimental points fall approximately on the same straight line independent of thickness. At the very high fields ($\sim 10^5 - 10^6$ V/cm) the slope is considerably less than for the lower field range, and a thickness dependence becomes apparent. The authors attribute this to a trapping effect or schubweg. At a critical field, Ec, the carrier range becomes equal to the thickness of the sample, and at fields greater than this the discharge rate is essentially generation limited. The authors propose that at the lower fields the discharge rate is controlled by two pheno-

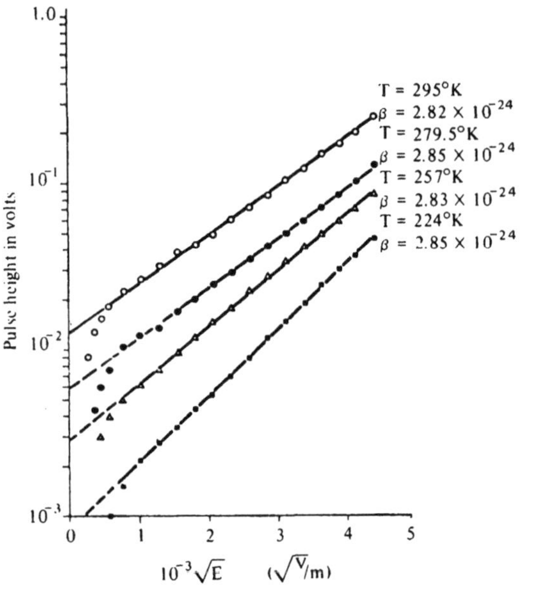

Fig. 11. Voltage pulse height versus square root of the applied electric field for amorphous selenium. λ = 560 nm. (After Pai and Ing [25]).

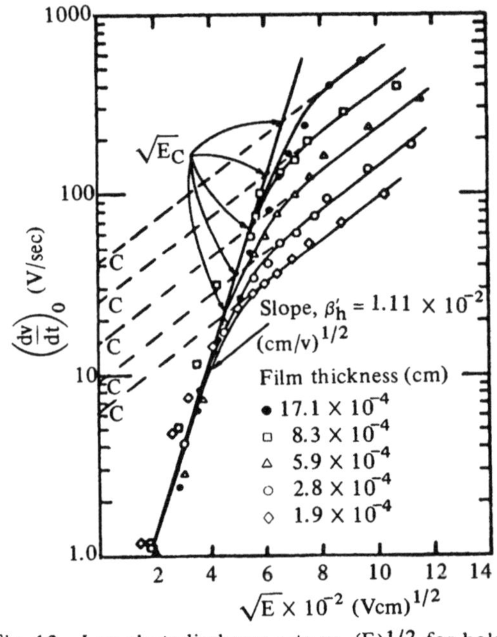

Fig. 12. Log photodischarge rate vs. $(E)^{1/2}$ for holes. (After Ing, Neyhart and Schmidlin [19]).

mena: (1) a field dependent photogeneration, and (2) a field dependent range limitation. The latter is shown in Fig. 13, where the log C (proportional to thickness) is plotted as a function of $\sqrt{E}$. The linear relationship suggests a Poole-Frenkel type of effect on bulk trapping.

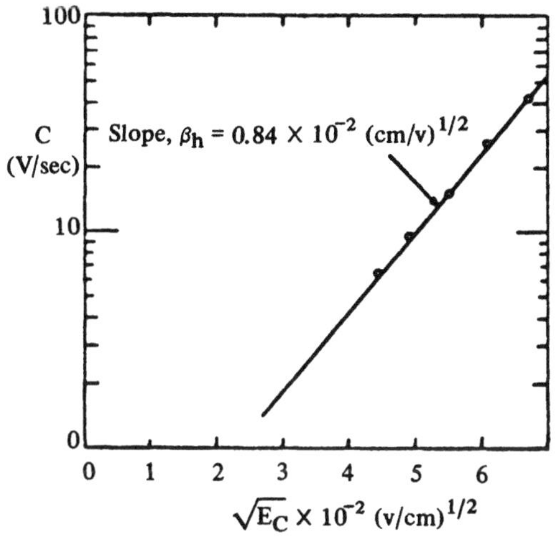

Fig. 13. Log C vs. $(E)c^{1/2}$. (After Ing, Neyhart and Schmidlin [19]).

Perhaps the first investigator to suggest the Poole-Frenkel effect to explain the mechanism of conductivity in amorphous selenium was Shidlovskii [26]. He employed pulse techniques to measure dark currents through sandwich type cells and determined resistivity through a range of field strengths and temperatures. His results are shown in Fig. 14, where log ρ is plotted vs. $\sqrt{E}$. Linearity is apparent over a wide range of electric field strengths (~10^3 to 10^5 V/cm) and temperatures. Shidlovskii also determined an activation energy from the resistivity-temperature relationship over a range of field strengths. The results are shown in Fig. 15. A linear relationship when activation energy is plotted against $\sqrt{E}$ is again apparent at high fields, particularly in the case where corrections are made for "high voltage polarization" (space charge). Since metal electrodes were used, it is not possible to determine from the data presented

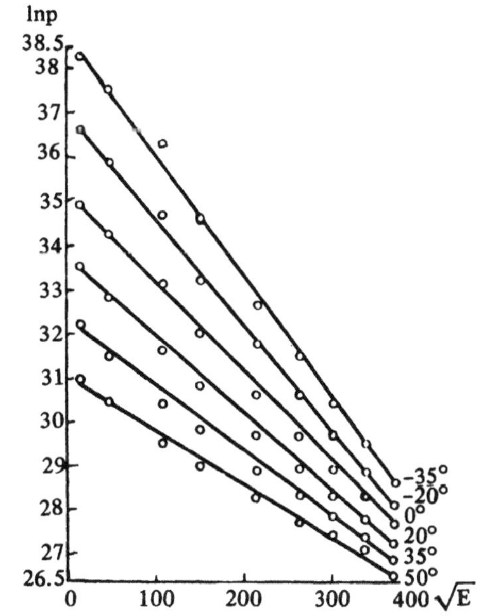

Fig. 14. Log of dark resistivity vs. $\sqrt{E}$ for amorphous selenium (Layer thicknesses 0.2 to 1 mm. Temperature in °C.) (After Shidlovskii [26]).

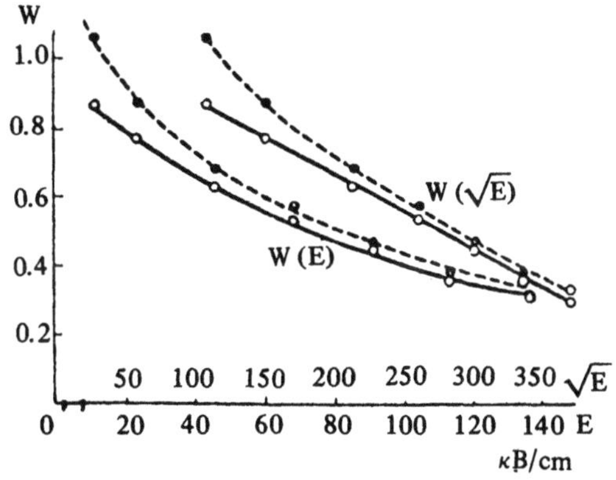

Fig. 15. Activation energy W vs. E and $\sqrt{E}$ for amorphous selenium. Broken lines are for values of W prior to correction for high-voltage polarization. (After Shidlovskii [26]).

whether the conductivity is a result of emission from electrodes or bulk effects or both.

Fig. 16 is an example of Schottky emission as distinct from Poole-Frenkel. The data were taken from Jonscher's paper [27], and were attributed to Mead [28]. The data are for a very thin (57 A) film of ZnO. With films of this thickness it is not expected that the current would be limited by bulk phenomena.

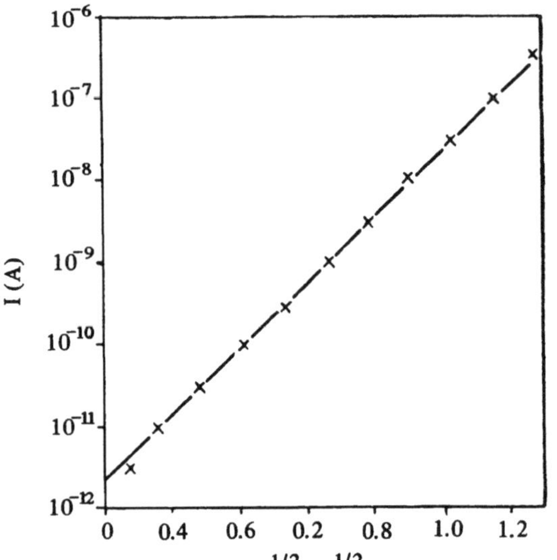

Fig. 16. Schottky emission at 300 K in a thin ZnO sample, Thickness 57 Å. (After Jonscher [27], attributed to Mead [28]).

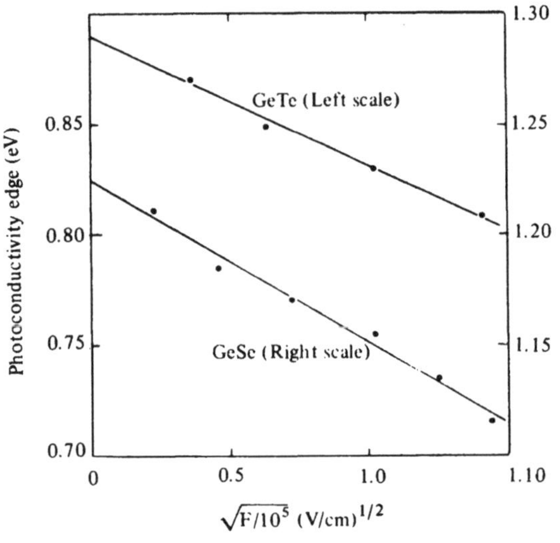

Fig. 17. A plot of the photoconductivity edge versus $F^{1/2}$ for both GeTe and GeSe. (After Stiles, Chang, Esaki and Tsu [21]).

Examples showing evidence of the Poole-Frenkel effect in amorphous semi-conductors are shown in Fig. 17. The data given are for thin films (400 to 2000 A) of amorphous GeTe and GeSe. The photoconductivity edge shifts toward lower photon energies with the application of high electric fields. The authors [21] indicate that the shift is a bulk effect and is not due to junction effects at the Al-semiconductor interface.

Fig. 18 shows a plot of log *dark current* vs. $\sqrt{V}$ using the data of Hartman et al. [29] on SiO. This figure also contains the calculated values (points) of Schug, Lilly and Lowitz [30] using the model proposed by Frank and Simmons [31]. This model is based on Schottky emis-

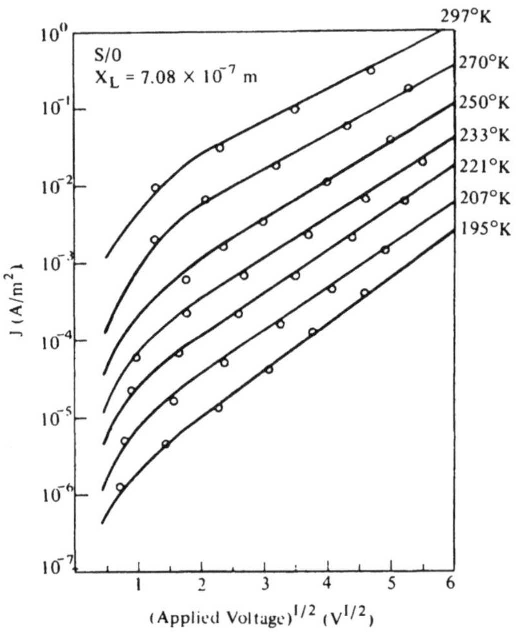

Fig. 18. I-V calculations from the theory (points) fitted to the data of Hartman et al. [29] on SiO (solid line) at different temperatures. (After Schug, Lilly and Lowitz [30]).

sion from the cathode coupled with Poole-Frenkel ionization of traps in the bulk of the SiO layer, the departure from linearity at low fields being accounted for by space-charge effects. In fitting these calculated values to the experimental data for SiO the above authors found it necessary, however, to employ a temperature-dependent work function.

The data shown in Fig. 19 indicate a probable Poole-Frenkel or Schottky barrier effect in thin polymer films. This figure was taken from a paper by Colburn and Christy [32]. The polymer is tetramethyltetraphenyltrisiloxane deposited by electron-beam bombardment. The data show a linearity of log *conductivity* vs. $\sqrt{E}$ in a quite high range of field strengths ($\sim 10^5$ to 10^6 V/cm). The authors attribute the dark current to a field-assisted hopping transport and apparently rule out Schottky emission on the basis that the value of the slope of the straight line portion of the plot is in closer agreement with the Frenkel formula.

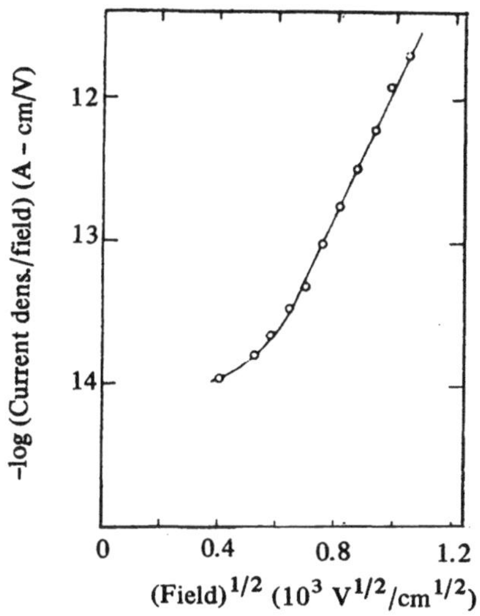

Fig. 19. Dark current as a function of applied voltage for 250-A thick polymer film. (After Colburn and Christy [32]).

Various authors [33–37] have investigated the theoretical significance of the ionization of defects in dielectrics and amorphous solids by applied electric fields. The subject is treated at length by Hill [33]. He finds that the Poole-Frenkel effect is a limiting case of a more general analysis that can be extended to cover the experimentally available range of field strengths and temperatures. He also shows that Poole's law [38] (log *conductivity* proportional to E) occurs when the density of defects is sufficiently great to cause overlapping of the coulomb potentials of the defects, and that the Poole-Frenkel effect applies when the ionized centers are non-interacting.

Pulfrey, Shousha, and Young [39] have developed a model for electronic conduction in amorphous insulating films in which electron injection occured according to the Schottky effect. Trapping was assumed to be independent of field and detrapping was controlled by the Poole-Frenkel effect. Their model indicates that at fields high enough to neutralize the negative space charge at a position close to the cathode, the electric field changes from the Schottky value at the metal-insulator junction to a uniform value in the bulk. When this occurs the conduction mechanism will be due entirely to the Poole-Frenkel effect.

Jonscher [27] points out that the probability of escape of a carrier in the Poole-Frenkel case is not uniform for all directions. He derives a modified formula to account for the change of the potential barrier with direction. A similar approach has been used by Hartke [37] in developing his three-dimensional model of the Poole-Frenkel effect. The equations derived by the two authors are not quite the same, however, but they reduce to approximately the same result at high fields.

Jonscher [27] states that the Poole-Frenkel formula still gives the best fit with experimental results at high fields, and that these results cannot be explained in terms of tunnelling and space charge-limited currents, and only most unlikely in terms of hot electrom phenomena.

Seki [40] has derived a formula combining field-dependent recombination with the Poole-Frenkel effect. His calculated results provide fairly good agreement with Pai and Ing's data on amorphous selenium throughout the range of fields used in the experiments.

6. Low Field Effects

The foregoing discussions have been primarily concerned with high field effects. The complete discharge of a charged xerographic layer involves a full range of field strengths from very high to very low values. Perhaps we need not be concerned with the very low values, but we must certainly be concerned about what happens at values in between high and low. When discharge has reduced the field within the layer to something less than about 10^4 V/cm the photogeneration rate decreases, recombination and trapping increase, detrapping is reduced, and space charges may build up. It is conceivable that photoinjection out of the absorption layer may cease entirely at low fields. This can happen if the rate of recombination within this layer becomes equal to the rate of photogeneration. If this occurs before the surface charge has been completely neutralized a residual potential will be observed, even without trapping.

Jonscher [27] indicates that hopping conduction may be the dominant mechanism at low fields. Such a mechanism would show a linear dependence of current on voltage. The rather extensive theoretical treatment by Hill [33], concerning ionization of local defects by an applied field, includes conduction at "low fields".

In recent years various investigators have prosposed theories to account for the photodischarge of xerographic layers particularly with application to amorphous selenium. Some of these [41, 42] show good agreement with experimental results in the high-field range of the discharge, i.e., the initial part of the discharge curve. However, we have yet to see a theory that follows a photodischarge curve all the way from the initial potential to near zero. This is true also of dark-decay curves. However, in dark decay the layer remains in the high field region for a greater period of time, and it may be sufficient for practical purposes to treat dark decay as essentially a high-field phenomenon.

Goodman and Rose [43] have derived an approximate theory to account for the double extraction of uniformly generated electron-hole pairs from an insulting photoconductor layer with non-injecting contacts. The theory predicts four regions of current vs. applied voltage behavior: (1) At low voltage, $I \propto V$; (2) at higher voltage, a transition region between $I \propto V$ and $I \propto \sqrt{V}$; (3) at still higher voltage, $I \propto \sqrt{V}$; and (4) at very high voltage (above a saturation value), I = constant. The results are in good agreement with experiments on lead oxide layers.

7. Zinc Oxide

Thus far we have been discussing mainly the amorphous and polymeric materials—the so-called molecular solids. Zinc oxide-resin binder layers constitute a different type of photoconductor. These layers charge differently and they apparently discharge differently. The mechanism of photogeneration is also somewhat different because of the oxygen adsorption-desorption process.

The author has not found any published material indicating field-controlled photogeneration in ZnO-resin layers. Williams [44] reports tentative evidence that electron multiplication at high fields can contribute to the discharge in ZnO single crystals. However, Kiess [45] did not observe any carrier multiplication in his investigation of ZnO single crystals even at fields as high as 1.6×10^6 V/cm. He attributes the increase in photocurrents to a photoinduced increase in the tunelling rate of charge carriers from surface states. In his measurements on dark currents, Kiess considered the possibility of Schottky emission but did not find a linear plot of log j vs. $\sqrt{V}$.

Although the experiments with single crystals shed some light on the basic photoelectric phenomena in ZnO, the results are not directly applicable to ZnO pigments dispersed in a resin binder. Kiess [46] compared the photodischarge characteristics of pigment-resin layers with those of single crystals and found them to be quite different. He concludes that the discharge current in pigment-resin layers is a primary saturated photocurrent independent of surface potential at high potentials.

He also cautions that the interpretation of discharge curves is only meaningful if the type of discharge current and the voltage dependence of the layer capacitance are known. Comizzoli [47] finds evidence of a bimolecular recombination process for photocurrent in dye-sensitized ZnO-resin layers. However, he found that UV exposure (ZnO band-gap) provides a discharge consistent with monomolecular recombination and a simple capacitor model.

There is now a large amount of published literature on ZnO-resin layers. The greater portion of this material has been directed toward the practical usage of such layers in electrophotography and much of the data is of little value in the development of theoretical models.

Summary

We have attempted here to review current theories concerning the charging and discharging of the photosensitive layers used in xerography. Pertinent literature relating to recent theories on high-field effects in amorphous dielectric and semiconductor films has been discussed. The Poole-Frenkel effect has been emphasized because of its apparent success in explaining high field conduction in amorphous films and dielectric layers. This effect appears to account for the photo and dark discharge in xerography at high fields, but needs to be modified to explain what happens at low fields.

ZnO-resin layers are treated separately. The behavior of such layers differs considerably from that of amorphous films. There appears to be no published information indicating field-controlled photogeneration in ZnO-resin electrophotographic layers. More good scientific data on the photoelectric behavior of this material are needed.

References

[1] R. M. Schaffert, Electrophotography, Focal Press (London) 1965, p. 26.

[2] C. F. Gallo, A. G. Leiga, Phot. Sci. Eng., **15**, 158 (1971).

[3] M. M. Shahin, Appl. Opt. Suppl. 3: Electrophotography, p. 106 (1969); J. Chem. Phys., **47**, 4392 (1967).

[4] K. Kinjyo, S. Nagashima, K. Shimazu, K. Yoshitake, Denshi Shashin Electrophotography-Japan **3**, No. 3, 33 (1961).

[5] T. Elder, J. Appl. Phys., **33**, 2804 (1962).

[6] M. Smith, A. J. Behringer, Phot. Sci. Eng., **12**, 165 (1968).

[7] K. Hauffe, J. Phot. Sci., **10**, 321 (1962).

[8] Ref. 1, p. 268.

[9] W. Jaenicke, B. Lorenz, Zeits. Elektrochem., **65**, 493 (1961).

[10] T. H. Li, P. J. Regensburger, J. Appl. Phys., **34**, 1730 (1963).

[11] A. Many, J. Phys. Chem. Solids, **26**, 575 (1965).

[12] M. D. Tabak, Trans. Met. Soc. AIME, **239**, 330 (1967).

[13] M. D. Tabak, P. J. Warter Jr., Phys. Rev., **173**, 899 (1968).

[14] Ref., 1 pp. 213, 223, 228.

[15] M. D. Tabak, Appl. Opt. Suppl. 3: Electrophotography, p. 4 (1969).

[16] J. Viščakas, V. Gaidelis, A. Matulionis, E. Montrimas, A. Satas, Appl. Opt. Suppl. 3: Electrophotography, p. 27 (1969).

[17] R. H. Bhatt, C. L. Braun, J. F. Hornig, Appl. Opt. Suppl. 3: Electrophotography, p. 20 (1969).

[18] R. M. Schaffert, IBM J. Res. Develop., **15**, 75 (1971).

[19] S. W. Ing, Jr., J. H. Neyhart, F. Schmidlin, J. Appl. Phys., **42**, 696 (1971).

[20] M. E. Scharfe, Phys. Rev. B, **2**, 5025 (1970).

[21] P. J. Stiles, L. L. Chang, L. Esaki, R. Tsu, Appl. Phys. Letters, **16**, 380 (1970).

[22] e. g., A. K. Jonscher, Thin Solid Films, **1**, 213 (1967); J. Electrochem. Soc., **116**, 217C (1969); also R. M. Hill, Phil. Mag., **23**, No. 181, 59 (1971).

[23] J. Frenkel, Phys. Rev., **54**, 647 (1938).

[24] W. Schottky, Phys. Z., **15**, 872 (1914); also D. A. Wright, Semi-Conductors, Methuen (London), Wiley (New York) 1950, p. 21.

[25] D. Pai, S. W. Ing, Jr., Phys. Rev., **173**, 729 (1968).

[26] M. K. Shidlovskii, Zh. Tekhn. Fiz., **24**, No. 5, 837 (1954).

[27] A. K. Jonscher, Thin Solid Films, **1**, 213 (1967).

[28] C. A. Mead, Basic Problems in Thin Film Physics, ed. R. Niedermayer and H. Mayer, Vandenholck and Ruprecht (Göttingen) 1966.

[29] T. E. Hartman, J. C. Blair, R. Bauer, J. Appl. Phys., **37,** 2468 (1966).
[30] J. C. Schug, A. C. Lilly, D. A. Lowitz, Phys. Rev. B, **1,** 4811 (1970).
[31] R. I. Frank, J. G. Simmons, J. Appl. Phys., **38,** 832 (1967).
[32] R. H. Colburn, R. W. Christy, J. Appl. Phys., **40,** 3958 (1969).
[33] R. M. Hill, Phil. Mag., **23,** No. 181, 59 (1971).
[34] G. A. Dussel, K. W. Boer, Phys. Stat. Sol., **39,** 375 (1970).
[35] F. W. Schmidlin, J. Appl. Phys., **37,** 2823 (1966).
[36] L. Onsager, Phys. Rev., **54,** 554 (1938).
[37] J. L. Hartke, J. Appl. Phys., **39,** 4871 (1968).
[38] H. H. Poole, Phil. Mag., **42,** 488 (1921).
[39] D. L. Pulfrey, A. H. M. Shousha, L. Young, J. Appl. Phys., **41,** 2838 (1970).
[40] H. Seki, Phys. Rev. B, **2,** 4877 (1970).
[41] I. P. Batra, K. K. Kanazawa, H. Seki, J. Appl. Phys., **41,** 3416 (1970).
[42] H. Seki, I. P. Batra, W. D. Gill, K. K. Kanazawa, B. H. Schechtman, IBM J. Res. Develop., **15,** 213 (1971).
[43] A. M. Goodman, A. Rose, J. Appl. Phys., **42,** 2823 (1971).
[44] R. Williams, Appl. Opt. Suppl. 3: Electrophotography, p. 15 (1969); also R. Williams, A. Willis, J. Appl. Phys., **39,** 3731 (1968).
[45] H. Kiess, J. Phys. Chem. Solids, **31,** 2387 and 2391 (1970).
[46] H. Kiess, J. Appl. Phys., **40,** 4054 (1969).
[47] R. B. Comizzoli, J. Appl. Phys., **41,** 4148 (1970).
[48] I. B. Levinson, I. Z. Plavina, Dokl. Akad. Nauk., **143,** 822 (1962); Soviet Phys. Dokl., **7,** 313 (1962).

The Function of Depth in Electrophotographic ZnO-Binder Layers

W. F. Berg, and H. Camenisch

Abstract

In order to discover the function of the lower layers in bringing about light sensitivity of ZnO-binder electrophotographic sheets, coatings were made up consisting of 4 "elementary" layers (EL), some of which were dye-sensitized (S), other non-sensitized (U). All exposures were to light absorbed by S, and, e.g. an EL-S could be arranged at various depths to act as a probe.

A coating consisting of several identical ELs behaved exactly like one coated to the same total thickness by a single pass, provided the ELs were air-dried at room temperature for a short time.

The coatings would charge up to much the same limiting potential with any combination of U and S ELs.

The light sensitivity of the coatings was characterized by $t_{0.5}$ and $t_{0.25}$, the times required to discharge them to 0.5 and 0.25 their original charge by light of constant intensity.

With an EL-S on top, the nature of the underlying ELs was unimportant. If an EL-U was first and an EL-S the second EL, the coating behaved similarly to one with EL-S on top, but the lower the EL-S the slower was the discharge by light. Part of this may be due to loss of light in the upper ELs.

1. Introduction

The long-term aim of the investigation, of which the present report is a preliminary part, is the determination of the optical characteristics of ZnO-binder layers and of the correlation of these characteristics with image quality, in particular from the point of view of understanding the function of depth of the layers. Similar work has been done on silver-halide photographic layers (1), whereby the layer is imagined to be subdivided into elementary layers (ELs), whose contribution to the optical characteristics add up to global ones, such as modulation transfer function and granularity.

For photographic layers after full development, the individual ELs react to exposure in identical fashion where-ever they may be situated in the depth of the layer. With ZnO layers this is likely to be quite different. The optical modulation transfer function for the ELs at various depths in the layer will have any meaning only if we understand how the various ELs respond to light in the electrophotographic discharge process. In other words, the first task would be to determine the electric response function for exposing light distribution of spatial frequency zero, for ELs at various depths.

The pursuit of this kind of question leads us straight to the problem or why ZnO layers have to be so uncomfortably thick in practice. Whilst the data to be described below give no answer, the present approach is hoped to lead to an understanding of that problem, which does not appear to have been discussed in the literature so far.

2. Experimental approach

The layers investigated here were made up by multiple coatings, the ELs being either unsensitized (U) or dye-sentized (S). As a typical experiment, four ELs may be coated on top of one another, only one of which may be sensitized and absorb actinic light and be placed at different depths, acting as a probe, as it were. Other combinations of U- and S-ELs also give useful information.

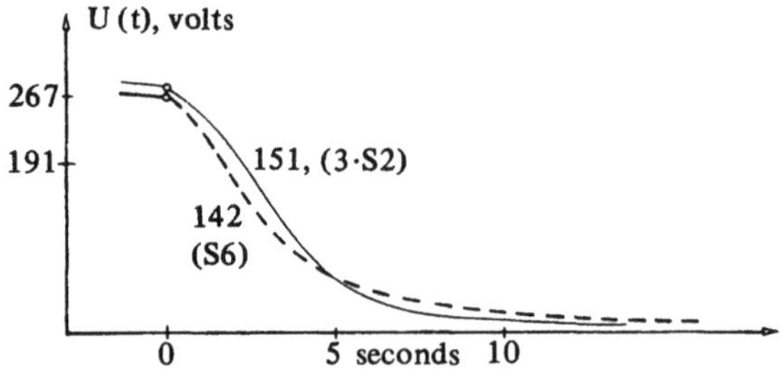

Fig. 1. Comparison of discharge characteristics of a multi-layer coating and a single coating of the same thickness of ca. 35 μm, both being dye sensitized. The surface potential is plotted against time for a constant illumination of 5 μW/cm² at λ = 576 nm.

3. Experimental procedure

The ZnO material was "Fotofax" (Imperial Smelting Co., Avonmouth, G.B.), the binder a copolymer of ca. 95% vinyl acetate plus 5% crotonic acid, polymerized approx. 90% in ethyl acetate solution*, and the dye was 1–2-hydroxyphenyl azo 2-naphtol** which had been determined by Brändli (2) to be firmly adsorbed to zinc-oxide. The recipe of preparing the slurry was as follows:

Dissolve 29 g of 50% resin solution in 100 g of a mixture of 80% toluene and 20% of isopropanol by weight:
Add 100 g ZnO;
Ballmill with 100 ml of 2-mm glass beads for 10 min;
Add 62 ml of dye solution (4 ml of 0.01 molar solution in methanol in 100 ml of toluene).

The amount of dye used corresponds to a molar ratio of 1/50'000 of dye to zinc oxide (63 ppm dye by weight).

The coatings were made on aluminum foil as a substrate by means of wire rods, the coating weight of the ELs being ca. 2.1 mg/cm² and the coatings being 11 μm thick.

The individual ELs were air dried at room temperature for 10 minutes before the next EL was applied. The completed multilayer was air dried at 70° C for 20 minutes. The resulting multilayer was ascertained to behave like a layer of similar thickness produced by single coating and dried in the same manner as the final multilayer, as shown in Figure 1.

For testing, the layers were charged to a limiting potential by moving them at a rate of 2.2 cm/sec under a single-wire corotron and then placing them under a transparent probe connected to a Cary Type Model 31 vibrating reed electrometer. There the sample was illuminated with the green-light mercury line, the illumination energy at the sample being ca. 5 μW/cm², at λ 576 nm.

Figure 2 shows the absorption curve of the dye in solution together with the mercury line utilized which was chosen to be at as long a wavelength as possible so as to minimize any intrinsic absorption by the zinc oxide.

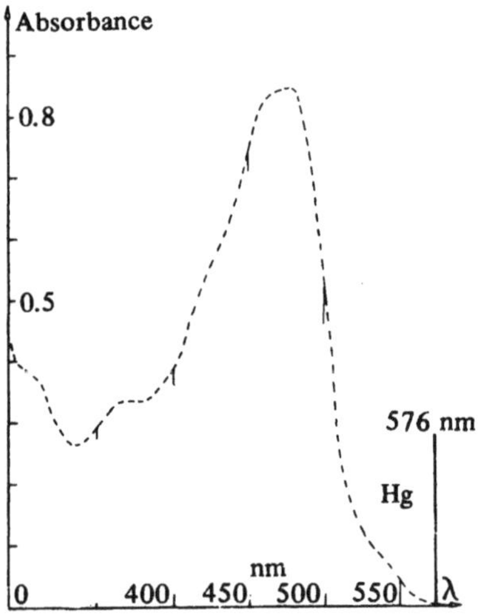

Fig. 2. Absorption curve of the sensitizing dye in solution. The position of the mercury line at λ = 576 nm is also shown.

* Kindly supplied by Turlabor (Zumikon, Switzerland).
** and the Dept. of Technical Chemistry, ETH-Z).

The handling of the samples is illustrated in Fig. 3 which also demonstrates the definitions to be used below. A period of 5 seconds elapses before exposure begins. This both gives the charged layer a chance to settle down to reach a reasonably stationary "starting potential" U_0 and is easy to manipulate reproducibly. There may be further "dark decay"; if so this

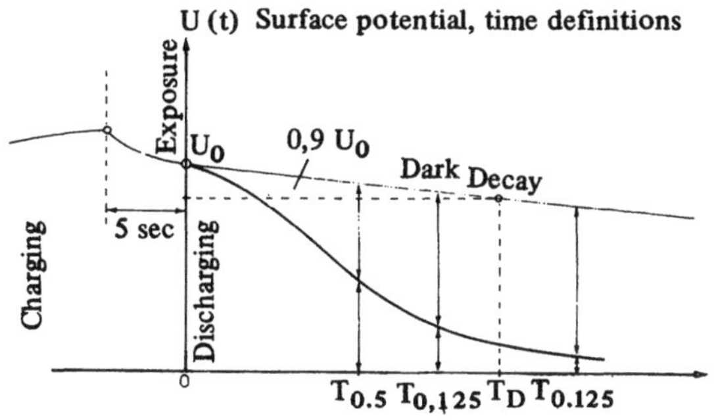

Fig. 3. Diagrammatic representation of testing procedure.

is characterized by the time t_D requires for the surface potential U(t) to drop to 90% of U_0 or, if t_D exceeds one minute, the value of U is noted after that time.

The dark decay is assumed to be essentially linear, being used to correct the sensitivity characteristics for which the times $t_{0.5}$, $t_{0.25}$, $t_{0.125}$ etc. are used at which the initial potential has dropped to 1/2, 1/4, 1/8 etc. of U_0, when the layer is exposed. For our layers, however, the dark decay was so small that no corrections were required.

4. Results

4.1. Reciprocity failure

Reciprocity failure tests were run originally in order to discover an exposure regime where results did not depend critically on the value of intensity. In such a region variations in intensity can be compensated for by changing the time of exposure provided the intensity is monitored.

Although in the present work no changes in intensity were made, the reciprocity curves (Fig. 4) are interesting on their own account.

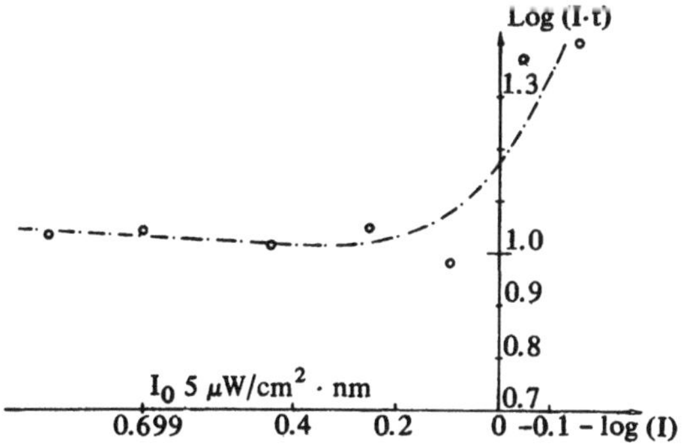

Fig. 4. Reciprocity failure of a dye-sensitized layer.

Over a certain range of intensities, there is no reciprocity failure; at increased intensities, there is pronounced short time failure. This would seem to indicate increased recombination and presumably is to be accounted for by trapping processes. This being an isolated result, however, we will refrain from speculating until further data are available.

4.2. Multilayers

The results on a series of multilayer coatings are shown in Figure 5, in which the values of $t_{0.5}$, $t_{0.25}$ and $t_{0.125}$ are shown against the make-up of the coatings. The values of U_0 also varied in an unsystematic fashion, as is shown by the dotted curve.

These data may be studied from the following points of view:

a) effect of an S-EL moving from the substrate to the top of the coating (123–126);
b) effect of an U-EL moving in the same manner (131–128);
c) effect of the thickness of a sensitized layer increasing from the top downwards (126, 148, 149, 146);
d) effect of the thickness of an unsensitized layer increasing in the same manner (131, 144, 123);

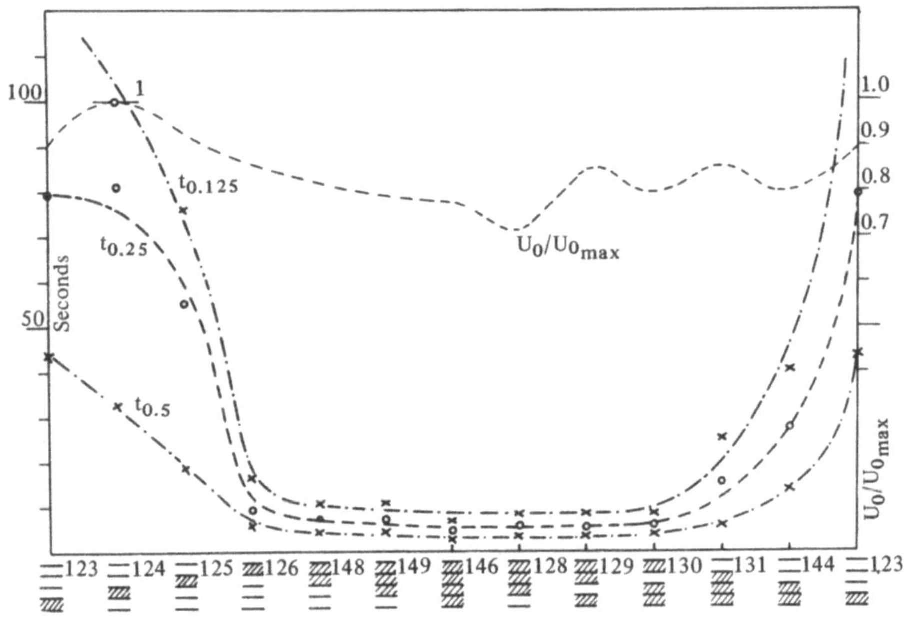

Fig. 5. Dependence of electric behaviour of ZnO multilayers of different compositions. Sensitivity values as characterized by those of $t_{0.5}$, $t_{0.25}$, and $t_{0.125}$ are plotted against the composition of the layers, the elementary layers being 11 μm thick. The maximum charging potential U_0 varies unsystematically from coating to coating, as shown by the dotted curve.

the total thickness of the coating always remaining constant.

The large changes of sensitivity seen presumably are the result of three interplaying factors: optical, complicated by the strong scattering of light; positional, in that light is effectively absorbed in the sensitized layers only; electronic, in that sensitization may alter the trapping of positive holes and electrons. At this stage, a complete analysis is impossible, in particular of the optical aspects.

Figure 5 shows (a) that position of a single S-EL is the most important factor. As long as one S-EL is on the surface, the layer has near-maximum sensitivity, regardless of whether the underlying layers are S or U. It is remarkable, however, that there is still some sensitivity left with a single EL right at the bottom (123), and if it is situated second (125), useful sensitivity is still obtained.

A question one would ask is that of the additivity of the effects of different ELs. Thus: can the behaviour of 149 be understood from those of 148 and 124? This question can be tackled by considering that discharge through exposure may be regarded as essentially that of a capacitor through different resistances which are characterized by the discharge times $t_{0.5}$, $t_{0.25}$ etc. The discharge curves cannot in fact be described by a simple R-C model, which would lead to expotential decay, but the simplified model yields a useful description of additivity, as is shown in the following Figures 6 and 7. Here the expected discharge times of the composite layers are calculated by adding up the reciprocals of the discharge times of the various component layers. The correction factor C is defined by

$t_{meas.} = C \cdot t_{calc.}$

On the whole the measured times are a little shorter than the calculated; in other words there is some interaction between the contributions of the various ELs. This may be optical or electronic in nature. The agreement is, however, surprisingly good in view of the crude nature of the model.

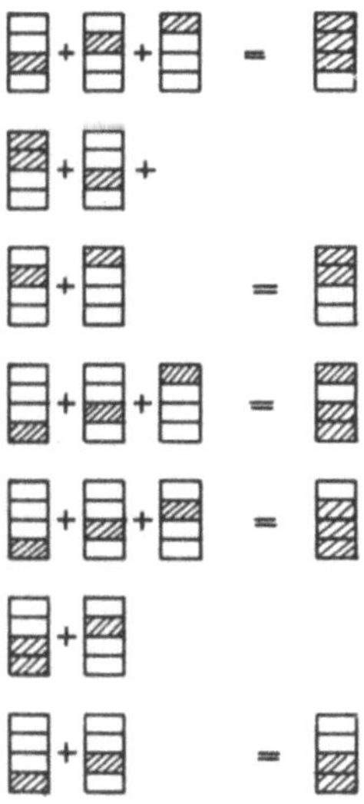

Equations	$(t_{0.5})$	$(t_{0.25})$	$C_{0.5}$	$C_{0.25}$
$1/t_{0,5} = 1/30.3 + 1/19\ 3 + 1/6.1$ $1/t_{0.25} = 1/81{,}3 + 1/55.5 + 1/9.3$	4.01	7.25	0.91	1.01
$1/t_{0.5} = 1/4.6 + 1/30.3$ $1/t_{0.25} = 1/7.3 + 1/81.3$	3.99	6.7	1.1	1.09
$1/t_{0.5} = 1/19 + 1/6.1$ $1/t_{0.25} = 1/55.5 + 1.93$	4.62	7.97	1	0.92
$1/t_{0.5} = 1/40.4 + 1/30.3 + 1/6.1$ $1/t_{0.25} = 1/79.9 + 1/81.3 + 1/9.3$	4.51	7.56	0.9	0.77
$1/t_{0.5} = 1/40.4 + 1/30.3 + 1/19$ $1/t_{0.25} = 1/79.9 + 1/81.3 + 1/55.5$	9.06	23.4	0.75	0.66
$1/t_{0.5} = 1/14 + 1/19$ $1/t_{0.25} = 1/27.3 + 1/55.5$	8.06	18.3	0.86	0.84
$1/t_{0.5} = 1/40.4 + 1/30.3$ $1/t_{0.25} = 1/79.9 + 1/81.3$	17.3	40.3	0.81	0.68
	Calculated, sec.		Correction	

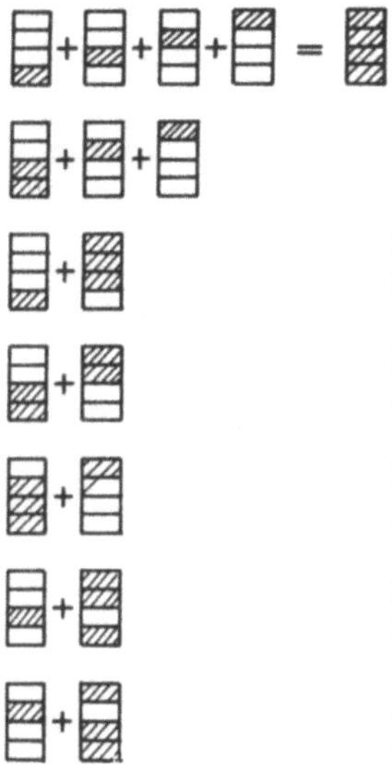

Equations	$(t_{0.5})$	$(t_{0.25})$	$C_{0.5}$	$C_{0.25}$
$1/t_{0.5} = 1/40.4 + 1/30.3 + 1/19 + 1/6.1$ $1/t_{0.25} = 1/80 + 1/81 + 1/56 + 1/9$	3.65	6.7	0.77	0.66
$1/t_{0.5} = 1/14 + 1/19 + 1/6.1$ $1/t_{0.25} = 1/27.3 + 1/55.5. + 1/9.3$	3.47	6.17	0.81	0.71
$1/t_{0.5} = 1/40.4 + 1/3.5$ $1/t_{0.25} = 1/79.9 + 1/5.5$	3.18	5.1	0.88	0.86
$1/t_{0.5} = 1/14 + 1/4.6$ $1/t_{0.25} = 1/27.3 + 1/7.3$	3.4	5.76	0.81	0.76
$1/t_{0.5} = 1/6.9 + 1/6.1$ $1/t_{0.25} = 1/15.4 + 1/9.3$	3.24	5.8	0.86	0.76
$1/t_{0.5} = 1/30.3 + 1/3.7$ $1/t_{0.25} = 1/81.3 + 1/5.5$	3.3	5.15	0.85	0.85
$1/t_{0.5} = 1/19 + 1/4.1$ $1/t_{0.25} = 1/55.5 + 1/5.8$	3.34	5.25	0.84	0.84
	Calculated, sec		Correction	

Figs. 6,7. Simulation of behaviour of multi-coatings from measurements on component layers, characterized by $1/t_{0.5}$ and $1/t_{0.25}$. The "calculated values" are based on a simplified model with exponential decay characteristics. The correction factor C by which the calculated data have to be multiplied to obtain the measured value is shown on the right.

5. Summary and conclusions

In electrophotographic ZnO-binder layers, the uppermost elementary layers are the most effective from the point of view of transfer of negative charge from the surface to the base support through exposure to light. Even the lowest layers still display some sensitivity. One has to assume, therefore, that positive charges are capable of traversing the whole thickness of an electrophotographic layer.

References

(1) Strübing, H., Die Verteilung der Belichtung im Innern einer photographischen Emulsion, Diss. 4008, ETH (Zürich 1968); Prof. Korr. **104,** 5, 26, 53, Riva, cap c, (1968). Vergleich zwischen verschiedenen Methoden zur Messung der Körnigkeit photographischer Schichten, Diss. 4218 ETH (Zürich 1968); Phot. Korr. **105,** 111, 128, 143, 159 (1968). Berg, W. F., J. Phot. Sci. **16,** 263 (1968); idem. App. Optics **8,** 2407 (1969).

(2) Brändli, R., Beitrag zur Sensibilisierung von Zinkoxyd mit komplexbildenden Azo-Farbstoffen, Diss. 4392, ETH (Zürich 1969).

Discussion

G. Heiland: The proposed explanation requires the motion of positive carriers through the layer. The creation of holes by photons with much less than band gap energy might be surprising. In this connection I would like to mention that we have observed photolysis at the surface of ZnO crystal with an adsorbed organic dye for the sensitized longer wavelength and there is some relation between photolysis of ZnO and the creation of holes.

R. Arneth: How did you make sure that there was no diffusion of the sensitizer from the lower sensitized layer into the upper layer during coating? Has the lower sensitized layer not been dissolved during coating?

W. F. Berg: Thickness of elementary layers was around 2.1 mg/cm^2, 11 μm thick. The dye was a strongly complexing azo dye; if it did diffuse we would not obtain the strong positional effects observed.

J. E. Lu Valle: I wish to confirm the observation of low intensity reciprocity failure. We have observed it in both polyvinylcarbozole and selenium layers. It appears to be rather commonplace.

Mehendru: What is the difference in the arrangement of elementary layers numbered as 123 and 148 in the bottom of your slide (last but one)? The elementary layer arrangements seem to be similar.

W. F. Berg: They represented different coatings with identical layer arrangements.

G. Heiland: Is there any observable discharge by light of the long wavelength if four layers without any sensitization are used?

W. F. Berg: No.

R. H. Epping: Regensburger reported about doping of PVCA with Selenium by evaporation on the surface. Photoconductivity occurs in the visible region in spite of insulating characteristics of PVCA in the spectral area.

It seems to be the same process, which allows the electron transport between doped and undoped layers.

H. Kiess: In the model proposed for the explanation is it required that light absorbed by the dye excites both free electrons and holes?

W. F. Berg: Our model was that e^- and e^+ are formed where light is absorbed, the e^+ travelling to the surface, the e^- to the base. The experiments indicate that e^- travel readily through the layer, the e^+ with difficulty; this would be the explanation of the positional effects described.

H. Kiess: The comment to the answer of Prof. Berg was, that firstly holes are strongly trapped in ZnO and secondly there are experiments which can only be interpreted on the basis that the light absorbed by the dye produces only free electrons and no mobile holes.

H. Moroson: The absorption spectrum of the dye employed appeared not to absorb any light at the wavelength 576 nm, illuminating the zinc oxide layer, thus violating the first law of photography.

W. F. Berg: The wavelength chosen lies on the very edge of the absorption band so as to be as far away as possible from the absorption of the ZnO proper.

Meier: Have you observed any change of the possibility of charge retention by the dye, that means the possibility for a positive charging of the surface ZnO by the dye?

W. F. Berg: Our layers would not charge positively.

Dark Decay in Electrophotographic Zinc Oxide Layers. Mechanism and Mathematical Model

K. Verhille and A. Suys

Abstract

The dark decay in zinc oxide electrophotographic layers is considered to be caused by charge transport following Ohm's law due to the presence of e.g. water molecules in the layer, and tunneling of charge carriers through potential barriers at the zinc oxide surface. Theory and experiment have been checked in 3 sets of data. The fit seems promising. Such a mechanism predicts that $V^{-1/2}$ vs. time should be a straight line, except at higher values of V, which is confirmed by experiment. Data on the behavior of a few zinc oxide materials at various binder/pigment ratios and at high relative humidity are presented and discussed.

1. Introduction

Fig. 1 shows the sequence formed by charging, dark decay and light decay of an electrophotographic layer.

The light decay has been treated by Cassiers et al [1]. This paper will deal with the dark decay.

Fig. 2 reproduces the actual registration of a dark decay curve. It differs in many respects from a light decay curve showing photocurrent as a function of time: such photocurrents always exhibit a maximum whereas the dark decay shows a continuous decline. There seems to be, as far as we know, no exception to this rule.

Light decay [1] was considered as the result of thermal lifting of charge carriers over potential barriers, located at the ZnO surface. Dark decay

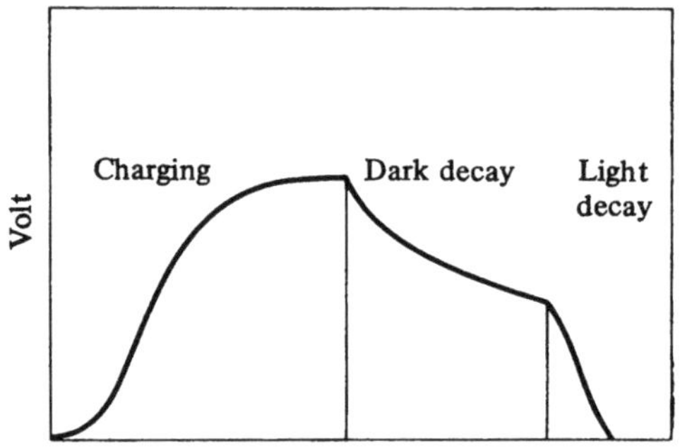

Fig. 1 - Sequence formed by charging, dark decay and light decay

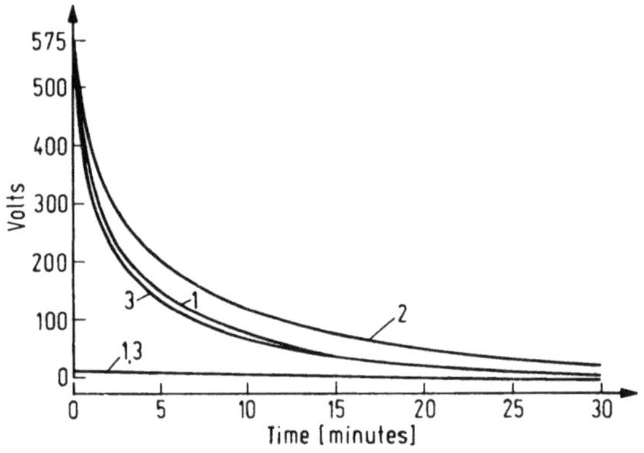

Fig. 2. - Actual registration of some dark decay curves

was assumed to be negligible which, according to the present study, excluded the presence of tunneling effects.

2. Theoretical

Dark decay, in general, can be seen as the result of the following concurrent phenomena:

2.1. Charge transport through proton transfer

Water, adsorbed at the ZnO surface, causes increased dark decay. This mechanism is probably dependent on the binder content of the layer and on the nature of the adsorption of the binder on the ZnO particles. One obtains for this part of the dark decay:

$$I_d^{(1)} = a_1 V \qquad (1)$$

whereby a_1 stands for a conductivity constant and V for the voltage applied to the layer. One can except influences of pigment volume concentration, particle distribution, relative humidity and a few other factors on the value of a_1.

2.2. Electrostatic pressure between adsorbed ions

Charge transport through this mechanism will be strongly dependent upon the presence of physical defects in the coated layer. The electrostatic pressure being proportional to $1/r^2$, and the mutual average distance r between ions being given by $1/V^{1/2}$, leads to

$$I_d^{(2)} = a_2 V \qquad (2)$$

a_2 is a proportionality constant.

2.3. Field emission

Although previous studies show that field emission can only be important at field strengths of about 10^5 Volt/cm or more, it must be held possible that in electrophotographic materials field emission effects can occur at relatively low charging voltages since this voltage is a sort of integrated average value for a larger area. Microscopically local field strengths may be well above the average value calculated for the whole layer.

The literature cites the following possible mechanisms:

2.3.1. Electrode emission

The emission of electrons from the cathode into the conduction band can be treated in a way similar to that given by Franz [2]

$$I = \frac{q^3 F^2}{8\pi h \cdot \Delta\epsilon} \cdot \Psi \cdot \exp\left[-\frac{4}{3}\,\frac{\sqrt{2m^*}}{hq} \cdot \frac{\Delta\epsilon^{3/2}}{F} \cdot \varphi\right] \qquad (3)$$

The field strength F is unknown for a corona-charged electrophotographic material. Only the voltage across the layer can easily be measured. If one is inclined to replace the value for F by the voltage, in itself proportional to the average field strength in the layer, one obtains:

$$I_d^{(3)} = b_1 V^2 e^{-const/V} \qquad (4)$$

2.3.2. Interband transitions

The literature [3] cites for the possibility of electrons tunneling directly from the valence band into the conduction band at extreme field strengths:

$$I_d^{(4)} = b_2 V^2 e^{-const/V} \qquad (5)$$

Hereby, again, the field strength F has been replaced by the voltage across the layer.

2.3.3. Field ionization

Tunneling of electrons from an impurity center into the conduction band is quoted [3] to follow the equation:

$$I_d^{(5)} = b_3 V e^{-const/V} \qquad (6)$$

whereby, once more, the same approximation of replacing F by V must be allowed.

The summation of the previous three processes leads to an equation of a rather empirical nature:

$$I_d = b_i V^n e^{-const/V} \qquad (7)$$

Experimentally n varies from 1 to 3, if the correct value for the field strength F is used.

Another theoretical possibility of describing the tunneling of charge carriers in electrophotographic material is by a treatment entirely similar to the one given by Zalm [4] in his study of luminescence in ZnS.

$$I_d = b_j e^{-K/V^{1/2}} \qquad (8)$$

It seems however more appropriate to replace the values for $\Delta\epsilon$ and F in equation (3) by ones more closely pertaining to the case at hand. It can be argued that the adsorbed negative charge is compensated by positive countercharges located within the crystal lattice of the semiconductor. The conductivity of the base paper is thought to be several orders of magnitudes higher than that of the photolayer.

This implies that the positive charges, in the case of a negatively charged ZnO material, are located in depletion layers, for which it can be shown, theoretically [5] and experimentally (see below) that the equation is fulfilled:

$$V = A \cdot Q^2 \quad (9)$$

The value for $\Delta\epsilon$ can here be associated with the depth of the energy level of the adsorbed negative ion beneath the conduction band. This level is, of course, slightly dependent upon applied field strength. According to Chynoweth [3] this variation is no more than a few tenths eV. It can, in the first approximation, be neglected, if one accepts the point of view [6] that adsorbed negative ions in the case of ZnO material are placed some 0.8 eV below the conduction band.

The value for F itself can be deduced along the following lines. For a semiconductor junction, with a sudden jump in impurity concentration, at the place of the junction, it has been shown by Moll [7] that the capacity can be expressed as

$$\frac{1}{C^2} = -\frac{2(V + V_b)}{qX^2}\left(\frac{1}{N_d} + \frac{1}{N_a}\right) \quad (10)$$

V_b being the contact potential.

For ZnO, N_d is much greater than N_a, thereby simplifying the expression, which in turn, together with the equation $C = dQ/dV$, leads to:

$$I_d = -KV^{-1/2} \cdot \frac{dV}{dt} \quad (11)$$

which can be deduced equally formally from the relation between voltage and charge in a depletion layer:

$$V = A \cdot Q^2$$

It will be shown further in this review that equation (11) is experimentally confirmed. Therefore straight-forward mathematics leads to

$$F = -\text{grad } V \quad (12)$$

$$V = F_{x=0} \frac{x^2}{2W} - x + \frac{W}{2} \quad (13)$$

W standing for the depth of the depletion layer, in turn proportional to the square root of the voltage V. The maximum field strength in the depletion layer is to be found for $x = 0$

$$F_{x=0} = \text{const.} \cdot V^{1/2} \quad (14)$$

It is this highest value for F which leads to tunneling. Substitution of this value and the one for $\Delta\epsilon$ in (3) finally leads us to:

$$I_d = b \cdot V \cdot e^{-K/V^{1/2}} \quad (15)$$

The parameters b and K contain the functions Ψ and φ, respectively. They are strictly speaking not constant but slowly diminishing in absolute value at increasing field strengths. In a first approximation, however, they are deemed to remain constant.

Hence, one has to verify the experimental evidence against the following three equations, of which the last one is believed to be more reliable:

$$I_d^{total} = a^{**}V + b^{**}\, V^n e^{-K^{**}/V} \quad (1) + (7) = (16)$$

$$I_d^{total} = a^*V + b^* e^{-K^*/V^{1/2}} \quad (1) + (8) = (17)$$

$$I_d^{total} = aV + bVe^{-K/V^{1/2}} \quad (1) + (15) = (18)$$

Integration of the latter, taking into account all the previously cited dependencies, one easily derives:

$$V^{-1/2} = \frac{1}{K} \ln \left[Me^{aA^{1/2}Kt} - \frac{b}{a}\right]$$

$$\text{with } M = \left(e^{KV_0^{1/2}} - \frac{b}{a}\right)^{\frac{1}{aA^{1/2}K}} \quad (19)$$

which, upon assumption of b/a being negligible with respect to the exponential term, simplifies into:

$$V^{1/2} = \frac{1}{K} (\ln M + aA^{1/2}Kt) \qquad (20)$$

This approximation is the more justified when the value for t is rather large; at lower values of time the deviation is such that the calculated values for $V^{-1/2}$ are too high.

3. Experimental

The experimental data on dark decay voltage and current as a function of time are tabulated below (Tables I to IV) for three different zinc oxide formulations. Graphically represented in Fig. 3, 4 and 5 they show very little deviation in comparison with the theoretical values derived from a best fit to equation (18) by trial and error.

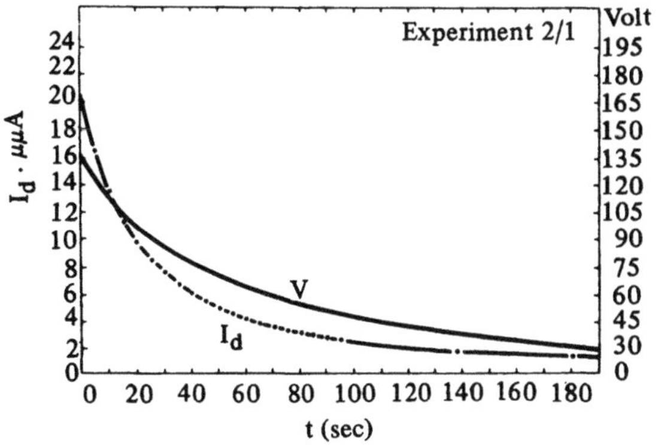

Fig. 3. - Dark decay of ZnO electrophotographic material
V = voltage. I_d = current

Table I. Dark decay of voltage and current

Experiment 2/1			Experiment 2/2			Experiment 2/8		
t(sec)	V	I_d	t(sec)	V	I_d	t(sec)	V	I_d
0	120	21	0	210	25	4.3	142.5	24.2
5	106	16.3	2	202.5	22.3	6.5	135	22.1
10	96	13.3	4.5	195	20	10	127.5	19.2
15	88	11.2	7	187.5	18.1	13	120	17
20	81	9.6	10	180	16.4	16.7	112.5	15
25	75	8.4	17	165	13.5	21	105	13
30	70	7.45	25	150	10.9	25.7	97.5	11.3
35	65	6.65	35.5	135	8.9	31.5	90	9.8
40	61	5.9	48	120	7.1	38	82.25	8.3
45	57	5.4	64.5	105	5.7	46	75	7
50	54	4.9	74	97.5	5.1	55	67.5	6
55	51	4.5	84.5	90	4.5	66.5	60	5
60	48	4.1	96.5	82.5	4	80	52.5	4.1
65	46	3.8	110.5	75	3.5	97	45	3.3
70	43.5	3.6	126	67.5	3.1	117.5	37.5	2.6
75	41	3.35	144	60	2.7	144	30	2
80	39	3.1	188	45	2.1	178.5	22.5	1.5
85	37.5	2.9						
90	35.5	2.7						
95	34	2.6						
100	32.5	2.5						
(Fig. 3)			(Fig. 4)			(Fig. 5)		

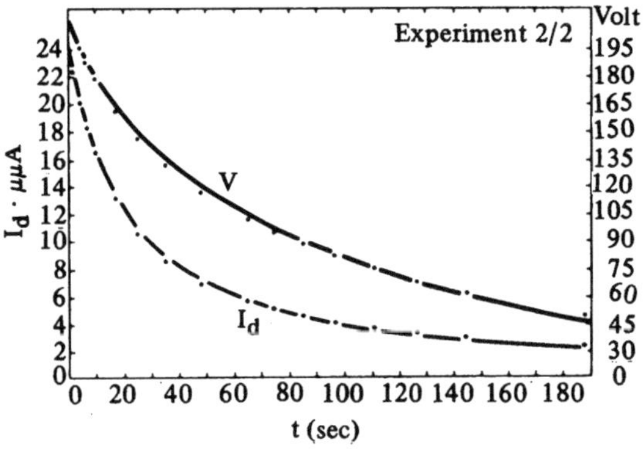

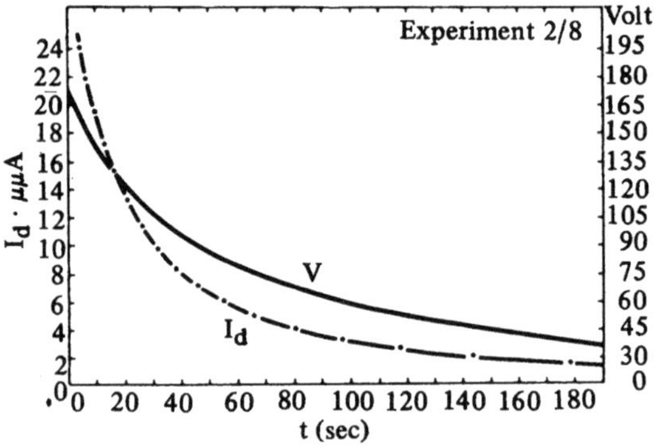

Figs. 4 and 5 (see caption under fig. 3)

Table II. Dark decay voltage and current: calculated and experimental

$I_d = 0{,}08\ V + 5{,}84\ V \cdot e^{-45/V^{1/2}}$ Experiment 2/1

V	$= I_{d,calc.}$	$I_{d,exp}$	t(sec)
120	= 21	21	0
106	= 16.34	16.3	5
96	= 13.28	13.3	10
88	= 11.32	11.2	15
81	= 9.65	9.6	20
75	= 8.63	8.4	25
70	= 7.48	7.45	30
65	= 6.62	6.65	35
61	= 6.01	5.9	40
57	= 5.42	5.4	45
54	= 5.01	4.9	50
48	= 4.26	4.1	60
43.5	= 3.73	3.6	70
39	= 3.28	3.1	80
35.5	= 2.94	2.7	90
22.5	= 1.8	1.7	138
15	= 1.2	1.25	180

Table III.

Dark decay voltage and current: calculated and experimental

$I_d = 0.0490\ V + 28 \cdot V \cdot e^{-87/V^{1/2}}$ Experiment 2/2

V	$I_{d,calc.}$	$I_{d,exp}$	t(sec)
210	24.7	25	0
202.5	22.44	22.3	2
195	20.25	20	4.5
187.5	18.32	18.1	7
180	16.52	16.4	10
165	13.39	13.5	17
150	10.80	10.9	25
135	8.73	8.9	35.5
120	7.07	7.1	48
105	5.75	5.7	64.5
97.5	5.17	5.1	74
90	4.67	4.5	84.5
82.5	4.2	4.0	96.5
75	3.77	3.5	110.5
67.5	3.36	3.1	126
60	2.95	2.7	144
45	2.21	2.1	188

Table IV.

Dark decay voltage and current: calculated and experimental

$I_d = 0.068\ V + 3.39 \cdot V \cdot e^{-42/V^{1/2}}$ Experiment 2/8

V	$I_{d,calc.}$	$I_{d,exp}$	t(sec)
142.5	24.2	24.2	4.3
135	21.56	22.1	6.4
127.5	19.05	19.2	10
120	16.89	17	13
112.5	14.89	15	16.7
105	13.01	13	21
97.5	11.32	11.3	25.7
90	9.78	9.8	31.5
82.5	8.35	8.3	38
75	7.08	7	46
67.5	5.96	6	55
60	4.97	5	66.5
52.5	4.11	4.1	80
45	3.35	3.3	97
37.5	2.68	2.6	117.5
30	2.09	2	144
22.5	1.54	1.5	178.5

4. Discussion

Difference between calculated and experimental values for dark current are small enough to warrant evaluation over a larger voltage and current range.

Similar calculations, based on equations (16) and (17) show that the deviation between theoretical and experimental data is of the same order of magnitude.

Here again, further experimental evidence is needed in order to be able to distinguish between the various formulae as to their validity.

4.1.

The values of Table I can also be used to ascertain the linear dependency of $V^{-1/2}$ versus t. This is clearly shown in Fig. 6.

4.2.

From the current data in Table I

$$I_d = 0.0490\ V + 28\ Ve^{-87/V^{1/2}} \quad \text{(experiment 2/2)}$$

is obtained. The constants used in this equation should also be valid for the voltage-time curve.

Transformation of these constants into the proper ones for equation (19):

$$V^{-1/2} = \frac{1}{87} \ln \left(942e^{0.0292t} - 572\right)$$

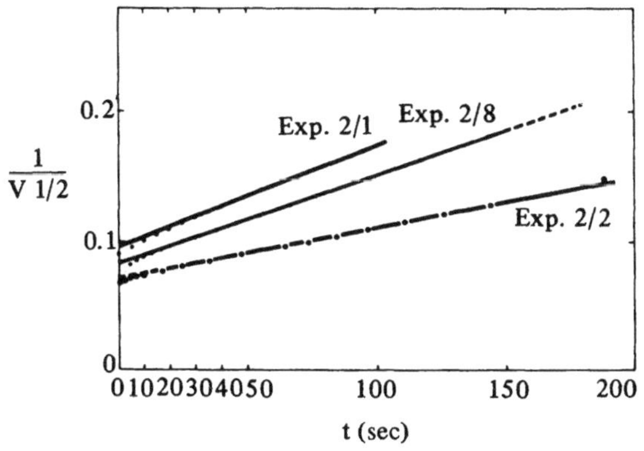

Fig. 6 - Linear dependency of $V^{-1/2}$ versus time

In the following Table V theoretical and experimental values are compared:

Table V.

Dark decay: experimental and calculated voltages V (equation 19)

Experiment 2/2

t	V_{exp}	V_{calc}
0	210	216
2	202.5	206
4.5	195	195
7	187.5	187
10	180	178
17	165	159.5
25	150	145
35.5	135	130
48	120	116
64.5	105	101.5
74	97.5	95
84.5	90	88.5
96.5	82.5	81.5
110.5	75	75
126	67.5	68.5
144	60	61
188	45	49.5

4.3.

Data are given in the next Table VI concerning the dark decay voltage for a number of zinc oxide materials with increasing binder content. *Electrox* 2030 was used as zinc oxide and poly(vinylacetate-acrylate) as a binder.

From these data it can be inferred that:

- at low binder/zinc oxide ratios, a second charging of the same sample leads to a slower dark decay (with respect to the first charging).
- this behaviour is reversed at higher binder/zinc oxide ratios. It can be shown that, at the same time, other photographic as well as physical characteristics of the coated layer will change. It is believed, though not strictly proved, that the change-over point is situated at the critical pigment concentration. The

Table VI. Dark decay for ZnO-binder layers with different composition ratios.

Number	1	2	3	4	5	6	7	8	9	10	11
Binder/ ZnO-ratio	40/1000	60/1000	80/1000	100/1000	120/1000	140/1000	160/1000	180/1000	200/1000	220/1000	240/1000
			Dark decay, 70% RH, in voltage, 1st and 2nd charging of the same specimen:								
t = 0 min	600–600	575–575	600–600	700–700	700–720		700–700			625–625	
0.5	407–467	390–430	393–480	470–525	505–580		470–585			465–385	
1	323–395	315–350	317–412	375–445	425–505		385–515			395–300	
1.5	275–350	265–305	270–367	320–390	370–450		330–455			345–240	
2	240–317	230–267	242–333	285–352	330–405		295–415			315–205	
2.5	215–290	205–243	216–305	255–323	300–375		270–380			290–180	
3	195–267	183–217	195–283	230–300	275–348		245–350			270–157	
3.5	175–250		178–267	212–275	255–325		227–325			252–140	
4	160–232	150–185	165–243	195–257	237–308		210–300			239–125	
5	135–205	125–157	140–213	170–226	205–275		185–265			212–100	
6	115–180										
7	100–162	92–120	107–167	133–183	160–235		140–210			175– 70	
10	67–117	63– 83	75–121	100–142	120–192		100–160			130– 42	
15	38– 75	40– 47	47– 75	67–100	85–132		63–115			90– 20	
20	24– 52	27– 30	31– 50	45– 70	60– 80		45– 85			67– 12	
25	16– 35	18– 18	22– 34	32– 50	42– 60		32– 60			50– 7	

inversed behavior of dark decay at higher binder/zinc oxide ratios is believed to be a direct consequence of a change in the number of donors per cm^3 and/or in the dielectric constant. Both are in fact inversely propor-

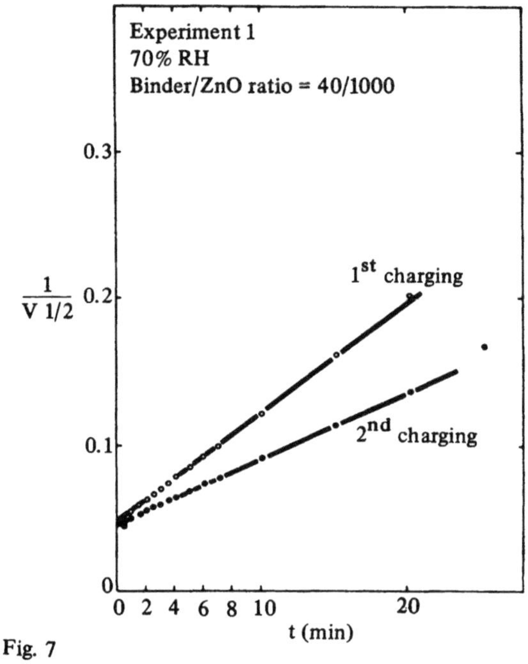

Fig. 7

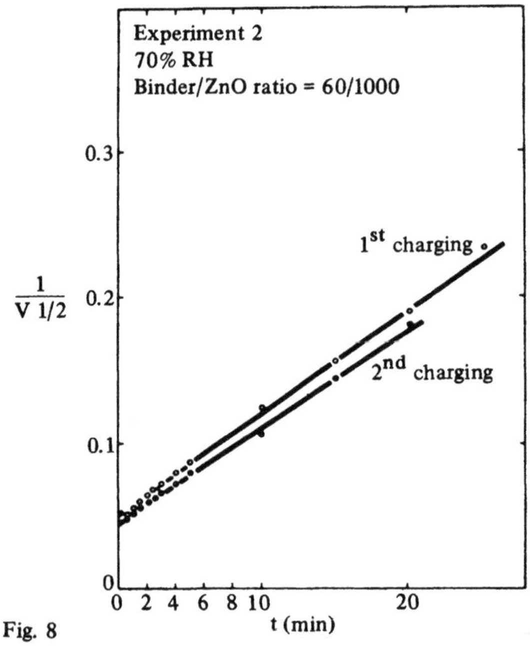

Fig. 8

tional to the value of A, connected with the slope of the straight line $V^{-1/2}$ versus t, as expressed in equation (20).

– linearity is observed for all experiments when plotting $V^{-1/2}$ versus t; deviations at lower values of time are in the expected direction.

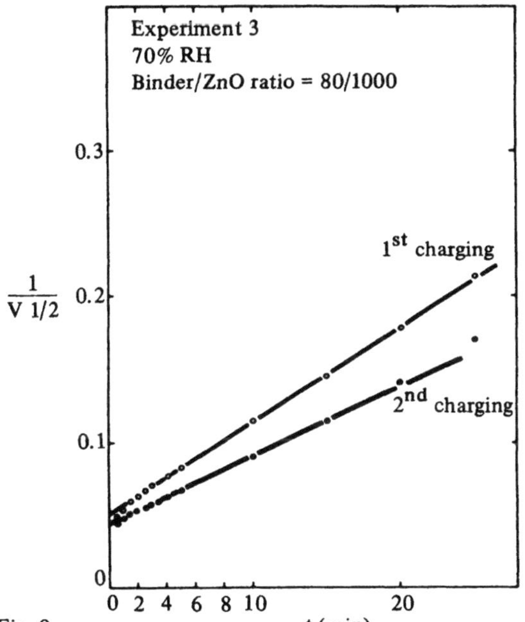

Fig. 9

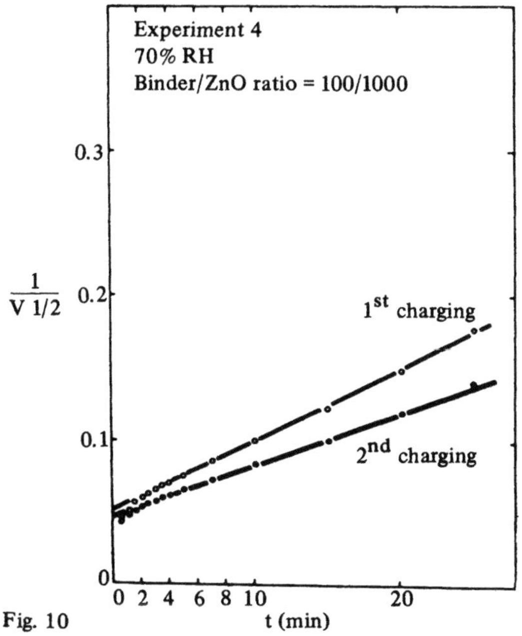

Fig. 10

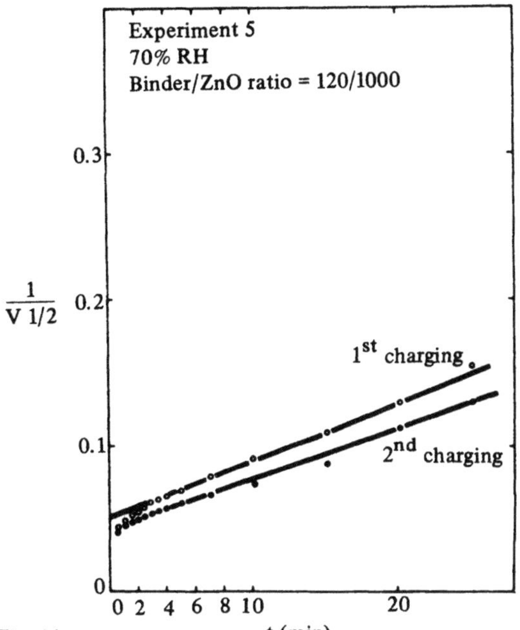

Fig. 11

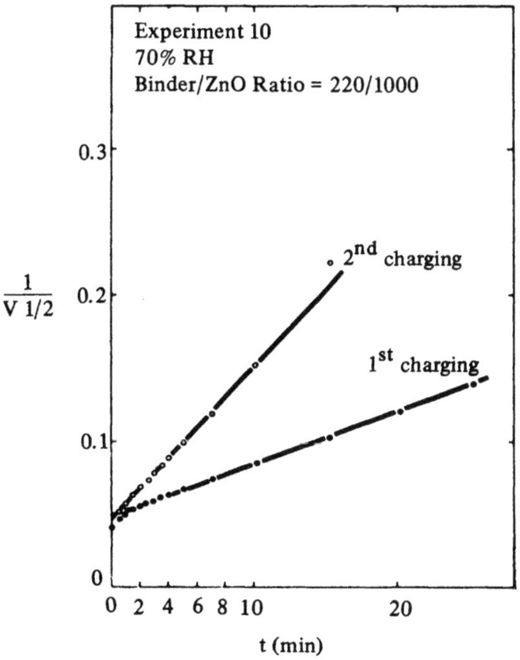

Figs. 7 to 12 (see caption of fig. 6)

5. Comparison with the literature

5.1.

Amick [8] gives experimental evidence with respect to the equation

$$\frac{dV}{d(Q^2)} = \text{const.} \qquad (21)$$

The mathematical treatment given above is based upon this same equation. By transforming (20) into (21) a more useful form is derived for numerical evaluation against our experimental data.

$$\frac{dV}{dt} = -2\left(\frac{V}{A}\right)^{1/2} I_d \qquad (22)$$

The curves in Figs. 13 and 14 constitute new evidence supporting this point of view. At higher voltages, above some 100 volts, a slight deviation from linearity is observed. We believe this to be caused by a variation in the value of (XN), inversely proportional to A.

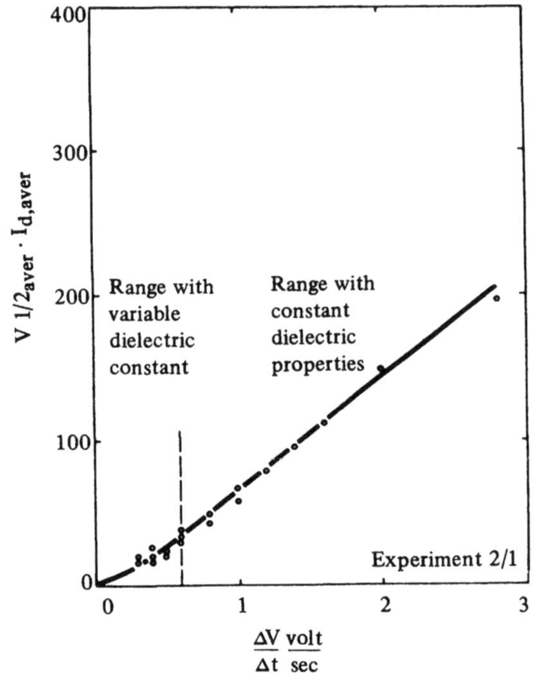

Fig. 13 - Experimental evidence supporting the formation of depletion layers in the ZnO material.

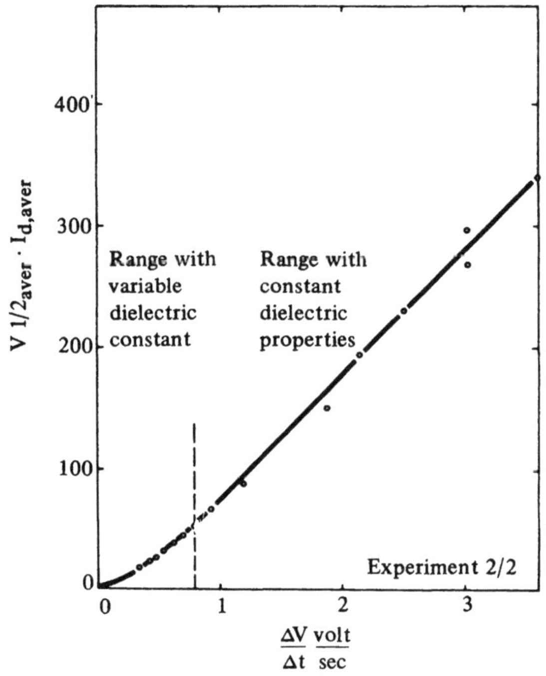

Fig. 14 (see caption fig. 13)

The literature mentions few cases of experimental data on a change in dielectric constant of a semiconductor at extreme field strengths. Viswanathan [9] gives proof of a 30% reduction in the value of the dielectric constant of GaAs at field strengths within the range of dielectric breakthrough.

5.2.

Hosterey [10] cites, without experimental or mathematical evidence, that

$$V_s = At + Bt^2 \qquad (23)$$

or

$$\frac{dV_s}{dt} = A + 2Bt \qquad (24)$$

A linear relation between t and dV/dt should therefore hold. In our experimental data such an interdependence could not be detected. (Fig. 15).

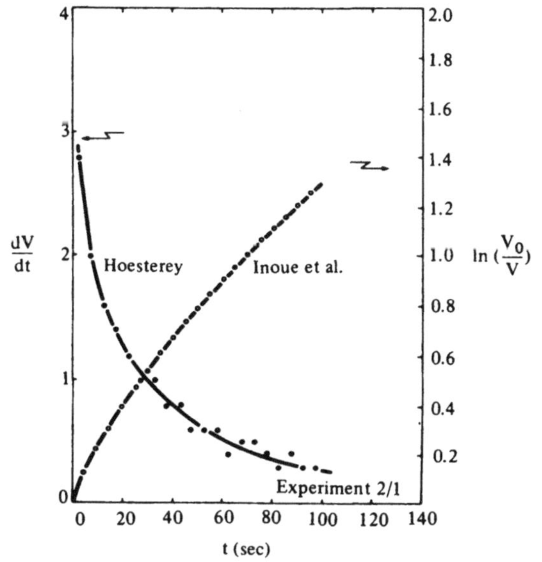

Fig. 15 - Failure to obtain a linear dependency between
a) dV/dt and .t (ref. 10) and
b) log $((V_0)/V)$ and t (ref. 11)

5.3.

Inoue [11] suggests that the dark decay follows the equation

$$V = V_0 e^{-at^{1/2}} \qquad (25)$$

This would result in linear dependence of $\ln(V_0/V)$ on time. Again, such behavior is not found in our experimental data (Fig. 16).

5.4.

Cassiers [12] Presents the expression

$$V = K + a\,e^{-bt} \qquad (26)$$

based upon a phenomenological treatment of dark decay. For the data at our disposal, (26) seems to be a rather good approximation.

5.5.

Hauffe and Stechemesser [13] cite:

$$V = V_1 - V_2 \ln(t + t_0) \qquad (27)$$

No mathematical derivation is given. Applied to the data at hand, a linear relationship between V and $\ln(t + t_0)$ could not be obtained (Fig. 17).

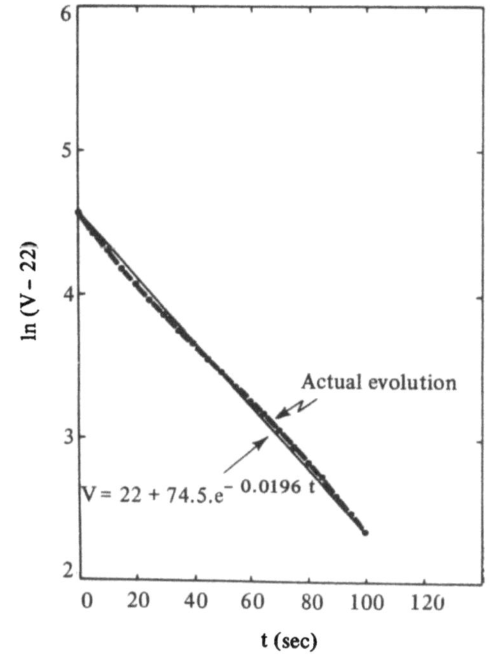

Fig. 16 - Comparison between experimental data and other results (ref. 12)

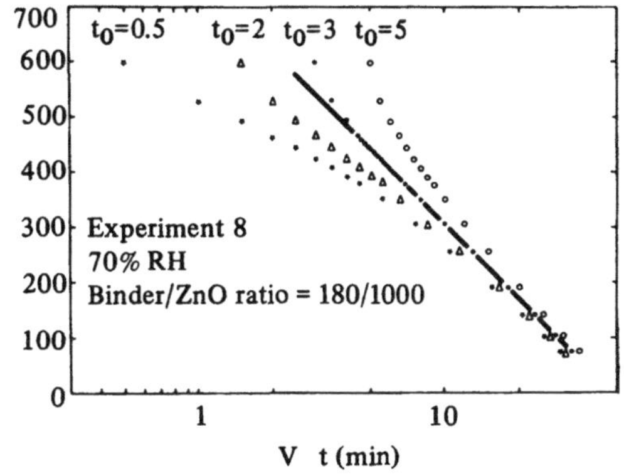

Fig. 17 - Comparison between experimental data and other results (ref. 13)

6. Conclusions

(1) Experimental evidence is given to support the relation

$$V = A \cdot Q^2$$

whereby V represents the voltage across an electrophotographic material. This signifies that the countercharge Q is contained within depletion layers inside the ZnO lattice.

(2) Theory and experiment sustain one another sufficiently in order to permit us to consider the dark decay of electrophotographic material as the combined result of Ohmic transport of charge carriers, resulting from water adsorbed on the ZnO surface, and of tunneling.

The resulting equations are:

dark current $I_d = aV + be^{-K/V^{1/2}}$

dark voltage $V^{-1/2}$

$$= \frac{1}{K} \ln \left[M e^{aA^{1/2}Kt} - \frac{b}{a} \right]$$

$$\text{with } M = \left(e^{KV_0^{-1/2}} - \frac{b}{a}\right)^{\frac{1}{aA^{1/2}K}}$$

(3) A zinc oxide layer can be characterized by three constants a, b and K, if variations in dielectric constant and concentration of donor levels are excluded.

References

[1] Cassiers, Verhille and Van Engeland, Symp. Rochester, Sept. 1968.

[2] Franz, W., „Handbuch der Physik", vol. 17, Springer Verlag, Berlin, 1956.

[3] Chynoweth, A. G., "Internal Field Emission" in "Progress in semiconductors", Vol. 4, Edited by Gibson, A. F. Heywood & Co, London 1960.

[4] Zalm, P., Philips Res. Rep., **11**, 353 (1956); ibid., **11**, 417 (1956).

[5] Bardeen, Phys. Rev., **71**, 717 (1947).

[6] Hauffe, K., Angew. Chemie, **72**, 730 (1960).

[7] Moll, L. J., "Physics of Semiconductors", Mc Graw Hill, 1964, p. 125.

[8] Amick, RCA Review, dec. 1959, 771.

[9] Visnawathan, K. S., Callaway, J. Phys. Rev., **143**, 564 (1966)

[10] Hoesterey, J. Appl. Phys., **33**, 992 (1962).

[11] Inoue, Nippon Insatsu Gakkai Ronbunshu, **2**, 43-8 (1959).

[12] Cassiers, P. M., Photogr. Science Symp. Zürich, Ed. W. F. Berg, **1963**, Focal Press London.

[13] Hauffe, K., Stechemesser, R., Helv. Phys. Acta, **41**, 872–878 (1968).

Discussion

K. Hauffe: Which percentage is the contribution of the various mechanisms on the dark-decay?

K. Verhille: ZnO bonder layers were coated at ±25 grams/m², the binder being of the poly (vinylacetate-acrylate) type.

Tunneling adds to dark decay to a larger extent than the ohmic part, when voltages are higher thân approximately 80 volts.

K. Hauffe: I may mention that the mathematical derivation of the logarithmic time law of discharge which you miss in our paper in Helvetica Physica Acta, is available in the Ph. D. thesis of R. Stechemesser (Göttingen 1967).

The Discharge Mechanism of Se Electroradiographic Layers Irradiated by X-rays

E. Montrimas and J. Rakauskas

Abstract

Amorphous selenium is generally used as the image carrier in electroradiography. The sensitivity of selenium layers to X-rays and the quality of electroradiograms depend essentially on layer structure, thickness, selenium purity, as well as on electroradiography conditions, involving initial potential, hardness of radiation etc. All these relationships are determined by the efficiency of carrier generation by X-rays and by recombination and capture in deep traps. The explanation of their role in the latent image formation process by X-rays or gamma radiation is the object of this investigation.

The theoretical description of layer discharge by X-radiation is given, taking into account the carrier capture by deep levels and their recombination. The relationship between layer sensitivity and its thickness is obtained for radiation of various degrees of hardness. Theoretical relationships are confirmed by experimental results. The spectral distribution of sensitivity calculated for X-rays (from 0.06 A to 1.0 A) fits fairly well the corresponding experimental data on selenium layers of different thickness, i.e. the sensitivity maximum shifts to shorter wavelengths while the layer thickness is increased. The utmost sensitivity in the spectral maximum is obtained for layers 350–450 μm in thickness. The dependence of X-ray sensitivity on initial potential of charging is presented for layers of various thickness.

The absorbed energy of X-rays required to neutralize an elementary charge on the surface is found to depend essentially on the electric field strength and layer thickness. For fields decreasing from 10^5 to 10^4 V/cm the energy required increases from 15 to 70 eV for thin layers (30 μm). Thicker layers require higher energies: e.g. for 400 μm thick layers the energy is 115 eV at 10^5 V/cm and 1300 eV at 10^4 V/cm.

The relationship between neutralization energy and sensitivity to X-rays on the thermal conditions of preparation of Se layers is also determined.

1. Introduction

Amorphous selenium layers are generally used as the photosensitive material in electroradiography. The main problem in developing the electroradiographical process for industrial defectoscopy is to improve the radiographic contrast [1–3], while for its use in medicine–to increase the photosensitivity of the electroradiographic layers [4]. The sensitivity of Se layers to X-rays and gamma radiation as well as the quality of the electroradiograms depend esentially on their structure, preparation technology, thickness, purity of Se as well as the electroradiographic conditions, i.e. the hardness of X-ray radiation and other factors [5–7]. On the other hand many of the processes going on in these layers during the formation of the latent image by X-rays and gamma irradiation have been investigated very little. There is no data on the maximum photosensitivity of Se ERLs. That is why investigation of the discharge process by X-ray radiation, i.e., the photogeneration of carriers, their capture by deep levels and recombination as well as the determination of the X-ray energy necessary to neutralize an elementary charge by discharge through the layer are dealt with in this paper.

2. Experimental Technique

Amorphous Se layers of various thickness were prepared by evaporating Se (99.992% purity) in 10^{-4} torr vacuum on oxidized aluminium substrata. The temperature of the substrata during evaporation was held constant and varied from 20° to 90°C.

Two mobile crucibles were used for evaporation at a coating speed of 3 to 4 micron/min. A trigonal Se sublayer was found by X-ray structure analysis in the layers when substrates were held at above 50°C temperature during evaporation. The amount of crystalline modification was insignificant in the volume. The effective lifetime of holes in these layers depends on the preparation temperature and with a rise in temperature from 20° to 80°C increases from 0.5 to between 10 and 15μsec, while the electron lifetime is between 2 and $5 \cdot 10^{-4}$ sec and any tendency to increase with temperature is less pronounced [11].

X-rays of various hardness were used in electroradiographic experiments. The continuous radiation spectrum was monochromatized with copper filters to h = 1.2 for investigations of neutralization energy and to h = 1.4 to 1.6 for other investigations. In some experiments the tungsten lines $\lambda_{K\alpha_1} = 0.2135$ A and $\lambda_{K\alpha_2} = 0.2086$ A separated by tantalum filters were also used with no difference in the experimental results on the neutralization energy [4].

The number of X-ray quanta absorbed was calculated according to:

$$N = N_0[1 - \exp(-\mu_0 l)] \frac{\gamma_{Se}}{\mu},$$

where N_0 is the number of incident quanta, μ-attenuation coefficient, γ_{Se}-absorption coefficient. The number of incident quanta N_0 was obtained from the dose rate measurements [8]:

$$N_0 = \frac{0.548 \cdot 10^{14}}{\gamma_a h\nu} P_0,$$

where γ_a is the absorption coefficient for air; energy of quanta is expressed in keV and the dose rate in R/sec. In the case of non-monochromatic radiation the effective value of $h\nu$ corresponds to the effective wavelength λ_{eff}.

3. The Energy for Neutralization of an Elementary Charge

A photogenerated carrier can be captured by deep levels during its drift across the layer. If it cannot be released, another free carrier must be generated nearby in order to continue the transport of an elementary charge across the layer from the surface to the substratum. Such cascade-type transport is completed by a number of generated carriers in a multiple process of drift, capture and recombination. Therefore the total amount of absorbed energy required for neutralization of one elementary charge on the surface and in the substratum may be much greater than the free-carrier photogeneration energy. This difference should be greater in thick layers.

To check these hypotheses experimentally, the neutralization energy was determined in thick layers and the data were compared to those in very thin layers where no carrier capture by deep levels was observed.

Layers were corona charged to various initial potentials and irradiated by X-rays. The relationship between the initial potential decay rate $dU/dt|_{t=0}$ and the initial potential was determined from photodischarge curves. Assuming that the initial capacity of a charged layer corresponds to the geometrical one, the charge neutralization rate was calculated from the equation

$$dQ/dt|_{t=0} = dU/dt|_{t=0} \cdot \frac{\epsilon\epsilon_0}{l}.$$

The results for layers of various thickness obtained for a constant absorbed energy ($6.7 \cdot 10^{11}$) eV/cm^2) are given in Figure 1. The surface charge neutralisation rate depends on the electric field and layer thickness.

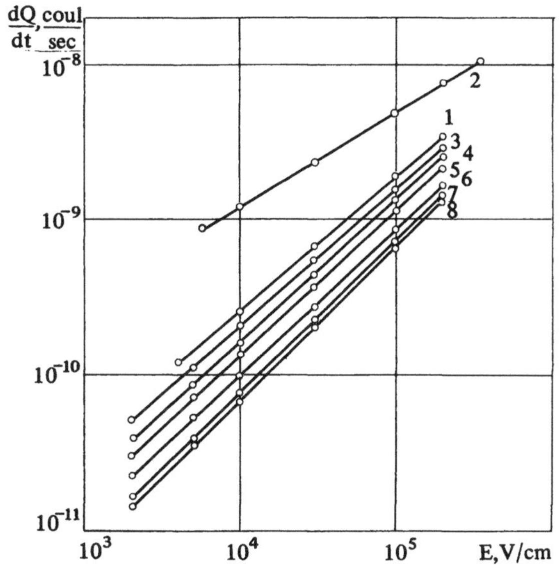

Fig. 1. Initial discharge rate $\frac{dQ}{dt}\Big|_{t=0}$ against the electric field for constant absorbed X-ray energy $6.7 \cdot 10^{11}$ eV/cm^2 for layers of various thickness (in microns): 1–68; 2–33; 3–138; 4–213; 5–275; 6–422; 7–670; 8–820.

Knowing the charge neutralized by a given absorbed X-ray energy, it is possible to determine the neutralization energy (Fig. 2). For Se layers of any thickness, as can be seen from Figure 2, the neutralization energy increases as the field decreases. It was also found that this energy in contradiction to [9] depends strongly on layer thickness.

In the case of a very thin layer the electrons and holes of the generated pairs reach the corresponding surfaces without being captured. It was shown earlier [10, 14], that in the case of simultaneous double injection of electrons and holes into a thin layer of amorphous Se, no recombination of carriers in the layer volume will occur. In such a case recombination of a pair can take place only in the photogeneration site before carrier separation. Taking into account what has been said above, it can be stated that the absorbed X-ray radiation energy necessary for the transport of an elementary charge across the layer and designated above as neutralization energy, is equal to the energy of free-carrier generation. This energy at a field of $3.5 \cdot 10^5$ V/cm (Fig. 2, curve 1) is found to be 7 eV, which is about three times the gap in amorphous Se, and possibly, coincides with the photogeneration energy by X-rays. The photogeneration energy of free carriers increases as the field decreases (Fig. 2, curve 1). The electric field is required for pairs dissociation in the same way as in the case of photogeneration by visible light [12–14].

With an increase of thickness, the neutralization energy substantially increases (Fig. 2, curve 2–5).

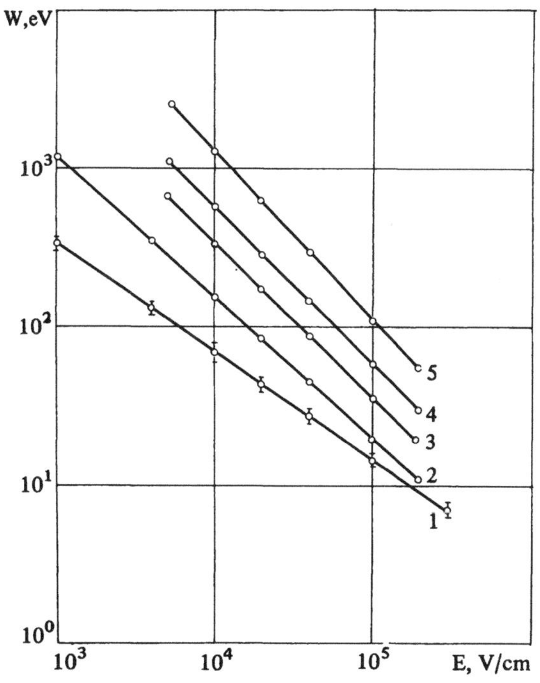

Fig. 2. Dependence of the elementary-charge neutralization energy on electric field for layers of various thickness (in micron): 1–30; 2–32; 3–68; 4–213; 5–413–422.

This is because the drifting carriers are captured by deep levels, which later on become recombination centers for newly generated carriers. Neutralization energy and field are more closely related for thick layers by the increase in the ratio of layer thickness to drift range as well as by non-uniformity of generation [15].

300 to 400 micron-thick layers and 2 to $3 \cdot 10^4$ V/cm fields are used for medical and industrial needs; therefore the neutralization energy is more than 10 times greater than the photogeneration energy for the same field. Consequently, any more favourable conditions of carrier drift may substantially improve the photosensitivity of Se electroradiographic layers.

4. Photodischarge of Se ERLs

The above results and other data [4, 15] suggest a physical model to describe the discharge and photosensitivity of Se layers by X-rays. It takes into account photogeneration, capture and recombination.

Let us assume that there are levels in the layer, which can capture holes and electrons, thus becoming recombination centers for carriers of opposite sign. The levels are energetically deep enough so that no thermal generation takes place. The energy transition scheme is given in Figure 3 where γ_{1n} and γ_{1p} are the coefficients of electron and hole capture, respectively, γ_{2n} and γ_{2p} are the recombination coefficients,

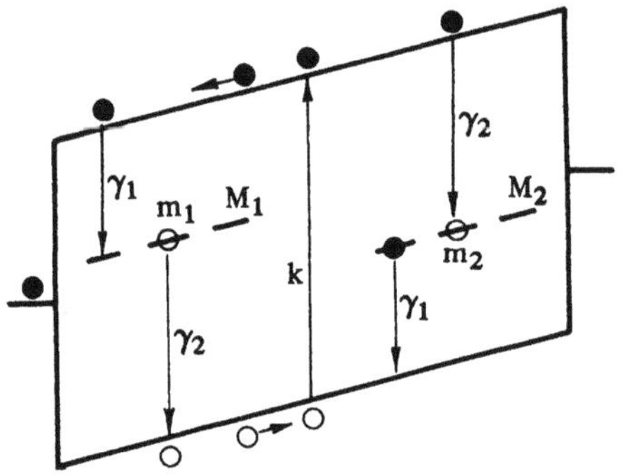

Fig. 3. A scheme of electronic transitions.

M_1 and M_2—the concentration of capture centers for electrons and holes, m_1 and m_2—the concentration of captured electrons and holes, and k—the generation rate of electrons and holes by X-rays. As the average attenuation length of X-rays is much longer than that of high-energy electrons it is possible to say that the generation distribution function approximately coincides with the X-ray absorption function. Therefore the equations have the following form:

$$\frac{\partial E}{\partial x} = -\frac{e}{\epsilon\epsilon_0}(m_1 - m_2 + n - p)\,, \tag{1}$$

$$j_n = e\mu_n nE, \quad j_p = e\mu_p pE\,, \tag{2}$$

$$\left.\begin{aligned} \frac{\partial m_1}{\partial t} &= \gamma_{1n} n(M_1 - m_1) - \gamma_{2p} m_1 p, \\ \frac{\partial m_2}{\partial t} &= \gamma_{1p}(M_2 - m_2) - \gamma_{2n} m_2 n, \end{aligned}\right\} \tag{3}$$

$$\begin{aligned} \frac{\partial n}{\partial t} &= k_0 \exp(-\mu x) + \frac{1}{e}\frac{\partial j_n}{\partial x} - \\ &\quad - \gamma_{1n} n(M_1 - m_1) - \gamma_{2n} m_2 n\,, \end{aligned} \tag{4}$$

$$\begin{aligned} \frac{\partial p}{\partial t} &= k_0 \exp(-\mu x) - \frac{1}{e}\frac{\partial j_p}{\partial x} - \\ &\quad - \gamma_{1p} p(M_2 - m_2) - \gamma_{2p} m_1 p\,, \end{aligned} \tag{5}$$

$$\epsilon\epsilon_0 \frac{\partial E}{\partial t} = -j_n - j_p\,, \tag{6}$$

where $k_0 = \frac{\gamma_{Se}}{W \cdot \gamma_a} \cdot 0.548 \cdot 10^{14} P_0$; W—is the field dependent electron and hole photogeneration energy, $W(E) = 4 \cdot E^{-2/3}$ (Fig. 2, curve 1), P_0—exposure dose rate, E—field, j_n and j_p—electron and hole current density.

Equations (1)–(5) are complicated and contain many unknown parameters. Let us assume that $M_1 = M_2 = M_0 = \text{const}$. Without lack of accuracy we may omit n and p in Equation (1), $\frac{\partial n}{\partial t}$ and $\frac{\partial p}{\partial t}$ in Equation (4) and Equation (5).

For further operations the following dimensionless variables are chosen:

$$m_i \to \frac{m_i}{M_0}\,;\; x \to \frac{x}{l},\; t \to \frac{t}{t_0}\,;\; j \to \frac{j_p}{j_0}\,;\; j_n \to \frac{j_n}{j_0}\,;$$
$$E \to \frac{E}{E_0}, \tag{7}$$

where

$E_0 = E(x, t = 0) = \sigma_0/\epsilon\epsilon_0$

and σ_0 is the initial density of the surface charge, $j_0 = \sigma_0/t_0$, l–layer thickness, t_0–time unit (for example, t_0 is the time during which the surface charge decreases by half).

Consequently:

$$\frac{\partial E}{\partial x} = -\,a(m_1 - m_2); \tag{8}$$

$$\frac{\partial m_1}{\partial t} = \Gamma_{1n}\frac{1-m_1}{E}\frac{\partial E}{\partial t} - -\frac{j}{E}\left[\Gamma_{2p}m_2 + \Gamma_{2n}(1-m_2)\right], \tag{9}$$

$$\frac{\partial m_2}{\partial t} = \Gamma_{2n}\frac{m_2}{E}\frac{\partial E}{\partial t} + +\frac{j}{E}\left[\Gamma_{2n}m_2 + \Gamma_{2p}(1-m_2)\right], \tag{10}$$

$$\frac{\partial j}{\partial x} = aQ\exp(-\eta x) - -\,a\,\frac{j}{E}\left[\Gamma_{2p}m_1 + \Gamma_{1p}(1-m_2)\right], \tag{11}$$

$$\frac{\partial E}{\partial t} = -j - j_n. \tag{12}$$

In (8) to (11) the dimensionless parameters have the following meanings:

$\eta = \mu l;\ Q = k_0 t_0/M_0,\ a = elM_0/\epsilon\epsilon_0 E_0$

$$\Gamma_{1n} = \frac{\gamma_{1n}\epsilon\epsilon_0}{e\mu_n}\,;\; \Gamma_{2n} = \frac{\gamma_{2n}\epsilon\epsilon_0}{e\mu_n}\,;\quad \Gamma_{1p} = \frac{\gamma_{1p}\epsilon\epsilon_0}{e\mu_p}\,;\; \Gamma_{2p} = \frac{\gamma_{2p}\epsilon\epsilon_0}{e\mu_p}. \tag{13}$$

The initial and boundary conditions have the following form:

$$E(x,0) = 1,\; m_i(x,0) = 0;\; j(0,t) = 0;\quad j_n(1,t) = 0,\; i = 1, 2. \tag{14}$$

The variables x and t in Equations (8 to 12) are independent and the solutions $E(x, t)$; $m_i(x, t)$; $j(x, t)$; $j_n(x, t)$ can be found in the form of a series in t power. Mathematical details can be found in [15]. If $a\Gamma_{1n} = g_n$; $a\Gamma_{1p} = g_p$ then the first terms for field are:

$$\epsilon_1(x) = \epsilon(0)\exp(g_n x) + aQ\frac{g_n+g_p}{g_p-\eta}\left[\frac{\exp(-g_n x)-\exp(-\eta x)}{g_n+\eta} - \frac{\exp(-g_n x)-\exp(-g_p x)}{g_n+g_p}\right]$$

$$j_0(x) = \frac{aQ}{g_p-\eta}\left[\exp(-\eta x)-\exp(-g_p x)\right];\quad \epsilon_1(1) = -j_0(1) \tag{15}$$

with the further result for the potential

$$U(t) = E_0 l\sum_{i=0}^{\infty} t^i U_i;\; U_i = \int_0^1 \epsilon_i(x)\,dx,$$

$$U_1 = \frac{aQ}{(g_n+\eta)(g_p-\eta)}\left\{\frac{g_p-\eta}{g_n}\left[1-\exp(-g_n)\right]\exp(-\eta) + +\frac{g_n+\eta}{g_p}\left[1-\exp(-g_p)\right] - -\frac{g_n+g_p}{\eta}\left[1-\exp(-\eta)\right]\right\}. \tag{16}$$

To determine the photosensitivity S of a layer from the inital decay rate Equation (16) is used:

$$S = -\frac{U_1}{t_0} = \frac{ek_0}{\epsilon\epsilon_0 U_0}\frac{l^2}{(g_p-\eta)(g_n+\eta)}\left\{\frac{g_p-\eta}{g_n}\left[1-\exp(-g_n)\right]\exp(-\eta) + +\frac{g_n+\eta}{g_p}\left[1-\exp(-g_p)\right] - -\frac{g_n+g_p}{\eta}\left[1-\exp(-\eta)\right]\right\};$$

$$g_n = \frac{\gamma_{1n}M_0 l^2}{\mu_n U_0}\,;\quad g_p = \frac{\gamma_{1p}M_0 l^2}{\mu_p U_0}\,. \tag{17}$$

The dependence of photosensitivity on layer thickness, calculated according to Equation (17) at various hardnesses of X-rays is shown in

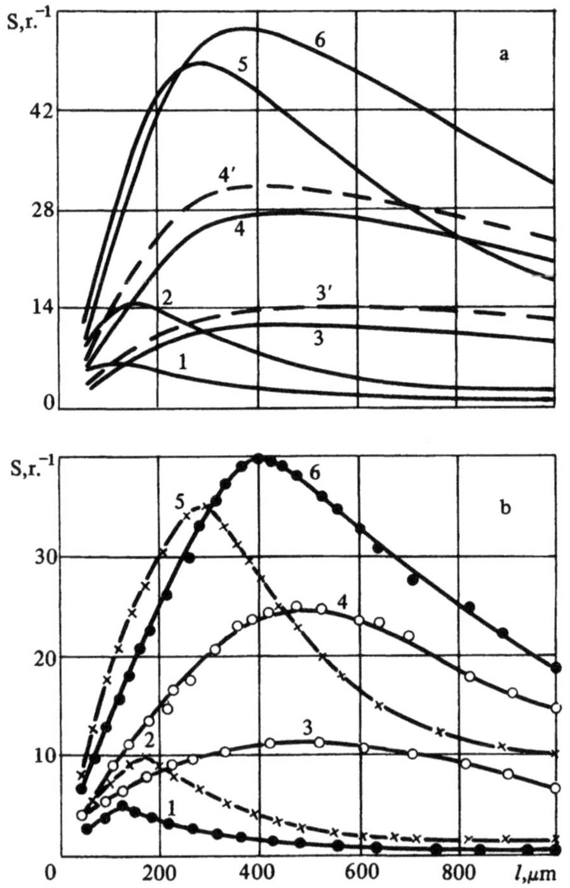

Fig. 4. Dependence of photosensitivity on layer thickness calculated (a) and experimental (b). Wavelengths of monochromatic radiation (a) and effective wavelengths (b) (in A): 1–0.8; 2–0.6; 3 and 3′–0.1; 4 and 4′–0.14; 5–0.32; 6–0.22. Substrate temperature –60° C. Potential U_0 = 800 V. (3′ and 4′ take into account the energy of scattered quanta).

Figure 4a. Calculations are made for parameters $g_p = 3 \cdot 10^6\ l^2\ U_0^{-1}\ V/cm^2$, $g_n = 9 \cdot 10^5\ l^2\ U_0^{-1}\ V/cm^2$ at a constant initial potential of 800 V. The experimental results obtained for the same hardnesses are given in Figure 4b. As can be seen from Figure 4 the theoretical relationship between photosensitivity and layer thickness at various X-ray hardness agrees well with the experimental data. Photosensitivity increases and goes through a maximum with an increase in layer thickness. The position of the maximum and the photosensitivity there depend on the X-ray hardness in a specific way. With an increase in hardness an increase in photosensitivity at the maximum is observed, while the maximum itself shifts towards thicker layers.

The parameters g_p and g_n which characterize the capture of holes and electrons were chosen to fit some of the experimental curves of the family, while the other theoretical curves were calculated for the same values of g_p and g_n. Calculations have shown that on changing the values of g_p and g_n theoretical and experimental data for the absolute value of photosensitivity, for the maximum positions and for the relative photosensitivity at various hardnesses will disagree badly. For example, if $g_p > 5 \cdot 10^6\ l^2\ U_0^{-1}\ V/cm^2$, then at any g_n the theoretical photosensitivity maximum at all hardnesses strongly shifts towards thin layers, while the absolute photosensitivity value is reduced. If $g_p, g_n < 1 \cdot 10^6\ l^2\ U_0^{-1}\ V/cm^2$ the photosensitivity is very high and its maximum for various hardnesses is strongly shifted towards thicker layers.

It follows that through a change in the parameters g_p and g_n it is possible to shift the maxima. But a fair fit of experimental and calculated data gives a strict limit for the parameter variation, namely $10^6\ l^2\ U_0^{-1}\ V/cm^2 < g_p \leqslant 3 \cdot 10^6\ l^2\ U_0^{-1}\ V/cm^2$; and

$$0.5 \cdot 10^6\ l^2\ U_0^{-1}\ V/cm^2 < g_n < 4 \cdot 10^6\ l^2\ U_0^{-1}\ V/cm^2.$$

The complex relationship of photosensitivity to layer thickness at a given hardness is determined by a number of competing factors. In thicker layers the amount of absorbed energy increases, but because of lower electric fields and longer distances of transit across a layer, photogeneration decreases, and carrier capture plays a more important role. The increase in photosensitivity with an increase in layer thickness is explained by the dominant influence of the first factor, while the decrease in photosensitivity in the thicker layer region is due to the field-controlled recombination of pairs and the capture of free carriers in weak fields.

In Figure 4b the parameters g_n and g_p were the same for the whole family of calculated curves, although there may be some variation of the

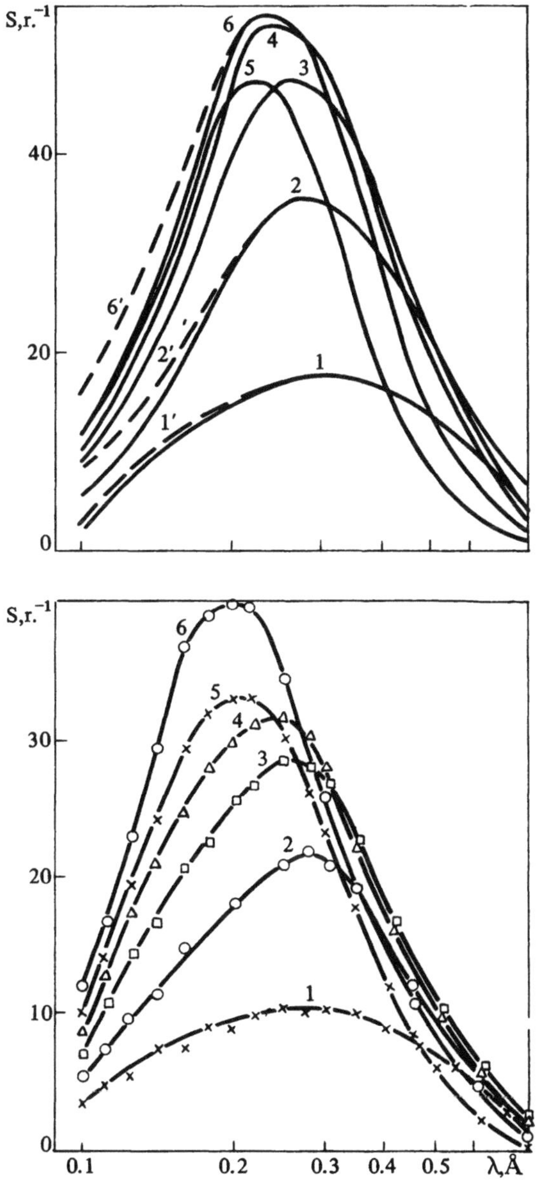

Fig. 5. Calculated (a) and experimental (b) spectral distribution of photosensitivity for Se ERLs to X-radiation for different thicknesses of layers: 1 and 1′ –70 μm; 2 and 2′ –140 μm; 3 –213 μm; 4 –275 μm; 5 –600 μm; 6 and 6′ –422 μm. (1′, 2′and 6′ – taking into account the energy of scattered quanta). Potential U_0 = 800 V.

parameters in thinner and thicker layers in the experimental studies (Fig. 4a). This may explain why the agreement of experimental and theoretical results is not as good as desirable. In addition to this, the field-controlled photogeneration was determined in the very thin layers, but it may be somehow different in thicker ones.

The spectral distribution of photosensitivity was calculated according to (17) for the same values of g_n and g_p as in the previous case, with a reasonable fit to experimental data (Fig. 5). Photosensitivity for soft X-rays is low because of non-uniformity of generation, while for relatively hard X-rays sensitivity decreases because the amount of absorbed energy gets less.

It was found experimentally that an increase in substrate temperature during evaporation results in an increase in photosensitivity (Fig. 6). Simultaneously, the maxima shift towards thicker layers. Similar behaviour is obtained if selenium of various purity is used for evaporation. Such correlated shift of the maxima is explained theoretically as a decrease of g_n and g_p. As g_p

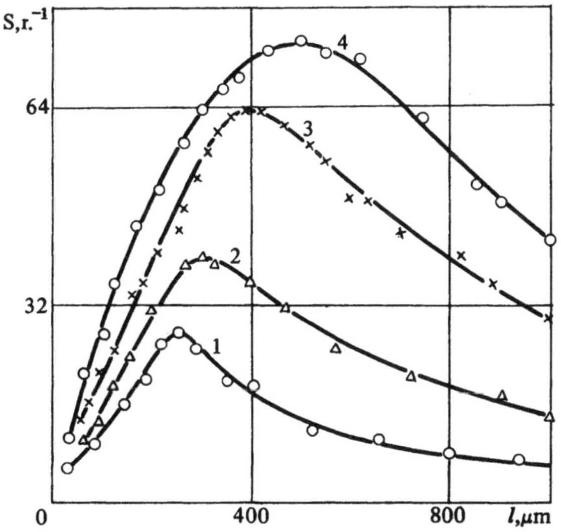

Fig. 6. Dependence of photosensitivity on thickness of layers prepared at various temperatures (°C): 1–40; 2–50; 3–60; 4–70, U_0 = 800 V. Effective wavelength –0.22 A / (Here photosensitivity was evaluated from the potential half decay time).

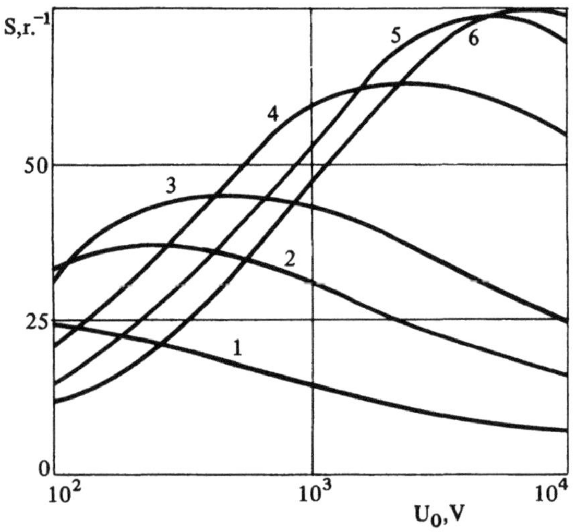

Fig. 7. Dependence of photosensitivity to 0.22 A-X-rays on initial potential. Layer thickness (in microns): 1–70; 2–140; 3–213; 4–422; 6–800; 7–275.

and g_n are proportional to the product of capture coefficient and center density, an improvement by technological means can be foreseen.

A very important factor which determines the photosensitivity is the initial potential. In Figure 7 calculated photosensitivity is plotted against initial potential. It is clear that for a given layer thickness an optimum initial potential exists. The increase in photosensitivity with increasing initial potential value is caused by a rise in photogeneration quantum yield and by an increase in range of the carriers. But with an increase in initial potential, there is an increase in surface charge density to be neutralized which in turn requires a larger number of generated carriers. The dominating role of the latter in high fields brings about a decrease in photosensitivity. These relationships are confirmed experimentally [15].

A number of theoretical and experimental results which govern the main factors determining the photosensitivity of selenium layers to X-rays makes the physical processes clearer and will help in the development and applications of selenium layers for electroradiography.

5. Summary

Optimum photosensitivity of Se ERLs at given radiation hardness can be reached by chosing the appropriate layer thickness and initial potential. Since for practical purposes layers of 300–400 μm thickness and fields of $2.3 \cdot 10^4$ V/cm are used for medical and industrial purposes, the energy to discharge the layer by neutralization of one elementary charge is more than one power of ten greater than the photogeneration energy for the same fields (Fig. 2); therefore the photosensitivity can be substantially increased by improving the carrier drift conditions.

References

[1] R. E. Cofield, Nondestructive Testing, **12** (2), 39 (1954).

[2] R. A. Kavaliauskas, Defektoskopia, **6**, 60 (1967).

[3] V. I. Gorbatchev, Defektoskopia, **2**, 124 (1969).

[4] E. Montrimas, b. Rakauskiene, J. Rakauskas, Lietuvos fiz. rinkinys, **10** (6), 94 (1970).

[5] A. Kaminskas, R. Kavaliauskas, J. Rakauskas, Zh. Nautchn. Prikl. Fotogr. Kinematogr. **9** (3), 189 (1964).

[6] A. Kaminskas, R. Kavaliauskas, J. Rakauskas, Voprosy radioelektroniki, **12** (25), 47 (1964).

[7] J. Neyhart, J. Schotmiller, Frontiers of Photography (preprints), Soc. Photogr. Scient. and Engrs, Washington, D.C. 1965, p. 63.

[8] H. E. Johns, The Physics of Radiology, Thomas, Springfield, USA, 1961.

[9] J. L. Donovan, J. Appl. Phys., **41** (5), 2109 (1970).

[10] V. Gaidelis, J. Kalade, E. Montrimas, A. Satas, Lietuvos fiz. rinkinys **9** (2), 325 (1969); **9** (2), 312 (1969).

[11] J. Kalade, E. Montrimas, A. Pazera, Lietuvos fiz. rinkinys, **11** (4), 579 (1971).

[12] D. M. Pai, S. W. Ing, Phys. Rev., **173** (3), 729 (1968).

[13] M. D. Tabak, Appl. Optics, Suppl. **3**, 4 (1969).

[14] J. Viščakas, V. Gaidelis, A. Matulionis, E. Montrimas, A. Satas, Appl. Optics, Suppl. **3**, 27 (1969).

[15] J. Kalade, E. Montrimas, J. Rakauskas, Lietuvos fiz. rinkinys, **12** (4), (1972), (to be publ.).

Discussion

S. R. Morrison: Have you found decreased photosensitivity with time which can be attributed to radiation damage by the X-rays?

A. Matulionis: Two types of fatigue are observed in the conventional electroradiographic plates: (1) Capture of holes and electrons causes a recovering temporary fatigue that results in faint development of the latent images that have been formed much earlier. (2) Gradual diminishing in the photosensitivity is also present, possibly because of structural changes under action of X-rays (damage).

In addition to this comment on conventional electroradiographic plates of Se, I want to tell you that the field-non-controlled photogeneration observed by us in a high-polymeric amorphous selenium disappears after irradiation to X-rays and to any other radiation with the energy of quanta above 2.5 ev.

W. F. Berg: What is the mechanism of multiple carrier formation?

A. Matulionis: An absorbed or scattered X-ray quantum produces a high-energy electron, which, in its turn, produces other electrons and so on and so forth. The cascade-type process of transportation of an elementary charge across the layer does not mean the above multiple photogeneration.

R. M. Schaffert: 1. What was polarity of surface charge? (Probably positive). 2. Did the authors consider the emission of carriers that results from X-rays absorption in substrata?

A. Matulionis: 1. Positive probably. 2. The present model does not include the mentioned phenomenon.

H. Moroson: Can you compare the X-ray exposure time required to make equal quality pictures between selenium electrophotography and silver halide photography?

A. Matulionis: The exposures required for a decrease of the potential by a half of its initial value are 1/70 R at best. The quality of electrophotographic pictures is usually much better because of better contrast.

H. Meier: 1. Do you see any possibility for a practical use in medicine in the near future?

A. Matulionis: The experimental electroradiographic aparatus are in use in many hospitals in Lithuania and elsewhere in the Soviet Union.

X-Ray Induced Discharge of Amorphous Selenium Layers

H. Dannert, H.-J. Hirsch and M. Jung

Abstract

The discharge properties of amorphous selenium layers from 10 to 1000 μm thick irradiated with different X-ray spectra have been studied under applied fields of 10^4 to $2 \cdot 10^5$ V/cm.

The light decay measured by an electrometer probe is nearly exponential. The time constant increases with increasing applied field and is inversely proportional to the layer thickness.

A model is proposed to interpret these experimental results. The absorption of a photon leads to the creation of a certain number of charge carriers. After thermalization these are located in a distinct part of the bulk. The charge carriers either recombine or participate in the discharge carrier flux. The relative distribution between these two competitive processes is determined by the applied field. During carrier drift no recombination occurs, the "schubweg" $(\mu\tau E)'$ being greater than the layer thickness.

Using this model and dosimetric measurements the pair production efficiency is calculated.

1. Introduction

The X-ray discharge of electrophotographic amorphous selenium layers has been treated in detail only in a few papers published in the literature [1, 2].

On the other hand a number of very extensive publications exists about the photoconductivity under irradiation with visible light. Several authors [3–6] showed that the photoconductivity cannot be explained by a range limitation model. For example, bulk trapping of electrons at applied fields of 10^5 V/cm occurs only for a layer thickness greater than 300 μm, while the analogous value for holes is a factor of 20 higher. To interpret the field dependence of the quantum efficiency these authors used models based on the Poole-Frenkel effect, i.e. field assisted thermal ionisation of bound electron-hole-pairs. However the generation of bound electron-hole-pairs under irradiation with X-rays is not yet clearly understood.

In this paper it is assumed that immediately after thermalization of a fast photoelectron, free electrons and holes are created. A certain number, depending on the electric field, recombines; while the rest, drifting across the layer, contribute to the discharge. This model is proposed to interpret the experimental results which clearly show that the amount of energy, which must be absorbed for neutralisation of one surface charge, decreases as the applied field increases.

2. Experimental Procedure and Results

A normal experimental arrangement was used (Fig. 1): On a rotating disk the selenium layer and a reference plate was fixed. The layers were charged by means of a corona source, and the decay of the surface potential, under irradiation with X-rays, was measured by an electro-

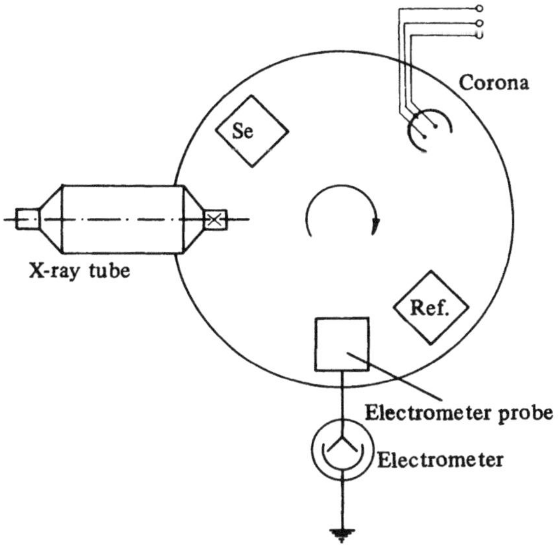

Fig. 1. Experimental arrangement used for the X-ray discharge of amorphous selenium layers.

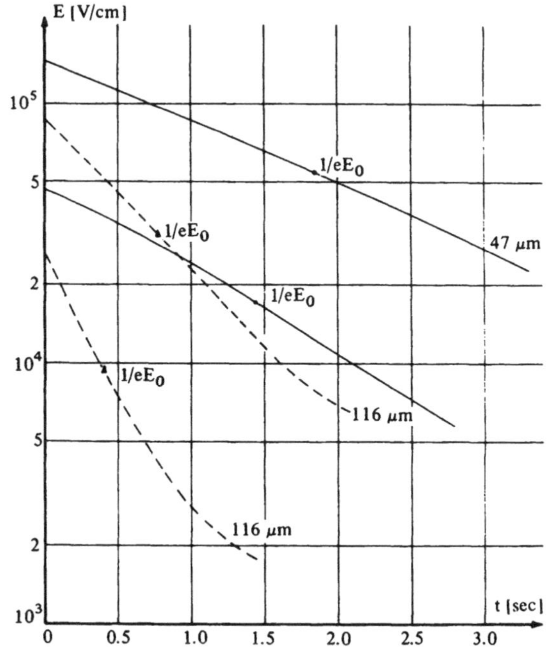

Fig. 2. Decay of the electric field under X-ray irradiation. 89 kV X-ray spectrum. Filter: 258 μm Yb plus 1 mm Al plus 3 mm Be.X-ray intensity 5 R/min.

meter probe. Several X-ray spectra were used although the main experimental features did not vary. An R-chamber was used to determine the X-ray intensity. The selenium layers were prepared by vacuum evaporation of 5 N selenium, obtained from Koch-Light, using aluminium plates as the substrate.

Figure 2 shows the time dependence of the electric field during X-ray-induced discharge. Obviously the curves are nearly exponential and the decay time to 1/e, called time constant in the following, depends on the layer thickness and on the initial electric field. These two parameters have been varied, both over one order of magnitude.

Fig. 3 shows the dependence of the time constant on the reciprocal layer thickness. The different samples were charged up to the maximum available surface potential. In spite of the scatter it is apparent that the time constant is inversely proportional to the layer thickness. These results were obtained with 75 kV bremsstrahlung spectrum. The analogous experiments with nearly monochromatic X-rays show the same typical behaviour.

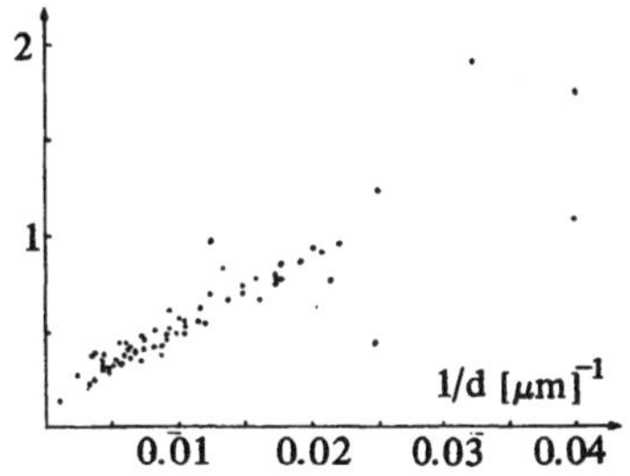

Fig. 3. Time constant vs. reciprocal layer thickness. 75 kV X-ray spectrum. Filter: 3 mm Be.

In Figure 4 the time constant is plotted against the initial electric field. The other parameter is the layer thickness. Depending on the layer thickness, the time constant varies up to a factor of two on increasing the initial electric field one order of magnitude.

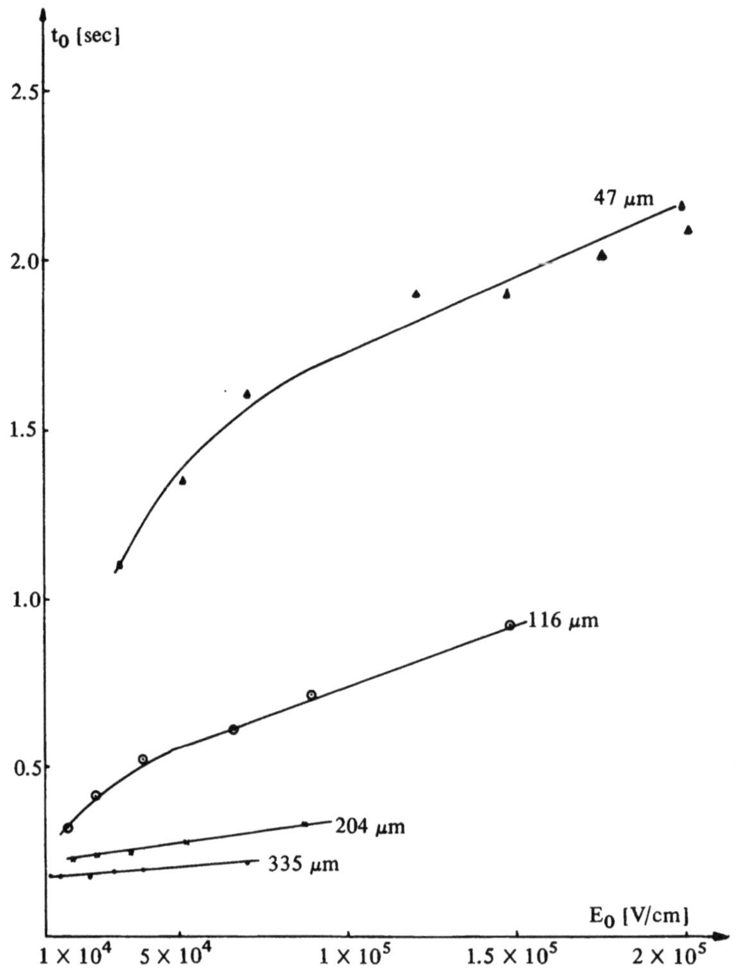

Fig. 4. Time constant vs. initial electric field for X-ray irradiated selenium. 89 kV X-ray spectrum. Filter: 258 μm Yb plus 1 mm Al plus 3 mm Be. X-ray intensity 5R/min.

3. Theoretical Model

The following model is used to interpret these experimental results:

The discharging process can be broadly divided into four steps:

1) Absorption of photons and generation of fast photoelectrons.
2) Thermalization of these photoelectrons and creation of electron-hole-pairs.
3) Separation of carriers of opposite sign by the electric field, taking recombination into account.
4) Drift to the surface and compensation of the surface charge.

The third step is the most important one. The carriers of opposite sign will either recombine or participate in the discharging carrier flux.

The mathematical treatment is simplified because the processes following the absorption of one single photon are finished before the next photon is absorbed. This means there is no interaction between the processes relating to different photons. This situation is true for the low X-ray intensities used in our experiments.

For one single photon it is relatively easy to calculate the number of carriers drifting to the surface. The following assumptions are made: After thermalization of the fast photoelectron electron-hole-pairs are generated in a distinct part of the bulk alongside the track of the fast photoelectron. Inside this volume a homogeneous distribution of free electrons and holes is assumed. The electric field shall be considered constant during the time in which recombination occurs, considering only one absorbed photon.

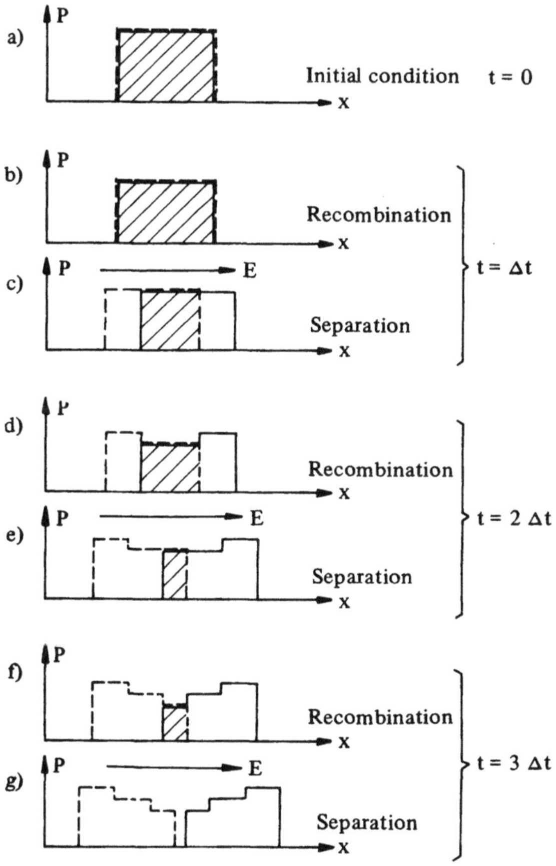

Fig. 5. Action of recombination and separation of charge carriers in the electric field.

Figure 5 illustrates the action of recombination and separation of charge carriers in the electric field. The carrier concentrations of electrons and holes are plotted versus the distance to the surface for several time intervals. Within each time interval, recombination and separation by the field are considered to be independent of each other. Part (a) shows the initial condition.

Within the first time interval (Part b) the carrier concentration decreases by recombination. In part (c) the carriers of opposite sign are shown separated somewhat by the field. The region in which recombination is possible gets smaller. This procedure is repeated for the following time intervals until the charge carriers are separated completely (Part g) and drift to the surface and the substrate respectively.

For the recombination process in the regions indicated a second-order recombination law is assumed

$$\frac{dp}{dt} = -\gamma p^2 \qquad (1)$$

where p is the concentration and γ the recombination coefficient.

By integrating this equation twice we have for the number of carriers, drifting to the surface, the following expression

$$Z = Z_0 \frac{E}{E^*} \ln \left(1 + \frac{E^*}{E}\right) \qquad (2)$$

Here Z is the number of carriers drifting to the surface and E the electric field. E* is a constant with the units of electric field, mainly determined by γ, the recombination coefficient and μ, the mobility of the holes. Z_0 is the total number of electron-hole pairs generated by absorption of one photon. Z_0 can be expressed by the photon energy divided by the conversion energy, that is the amount of energy which must be absorbed for the generation of one electron-hole-pair.

Summing over the number of absorbed photons per cm^2 per sec, the differential equation for the decay of the electric field can be derived. Its solution is given by:

$$t = -\frac{\epsilon\,\epsilon_0}{q}\,\frac{1}{\Phi_0\,(1 - e^{-\alpha d})} \times \chi_c\,E^* \int_{E_0}^{E} \frac{dE}{E\,\ln(1 + E^*/E)} \qquad (3)$$

Here t is the time
$\epsilon\,\epsilon_0$ is the dielectric constant
and q the electronic charge.

$\Phi_0\,[1 - \exp(-\alpha d)]$ is the absorbed energy per cm^2 per sec with Φ_0 the incident X-ray intensity, α the absorption coefficient and
d the layer thickness
χ_c is the conversion energy.

Replacing E by $\frac{E_0}{e}$, t becomes t_0, the time constant.

$$t_0 = \frac{\epsilon\,\epsilon_0}{q}\,\frac{1}{\Phi_0\,(1 - e^{-\alpha d})} \times \chi_c\,E^* \int_{E_0/e}^{E_0} \frac{dE}{E\,\ln(1 + E^*/E)} \qquad (4)$$

This solution applies to the case when trapping in deep traps can be neglected and is valid if the schubweg ($\mu E r$) is greater than the layer thickness. In accordance with the experimental results the time constant is almost inversely proportional to the layer thickness considering the following approximation

$$\Phi_0\,(1 - e^{-\alpha d}) \approx \Phi_0\,\alpha d$$

The influence of the initial electric field on the time constant is given by the integral in equation (4).

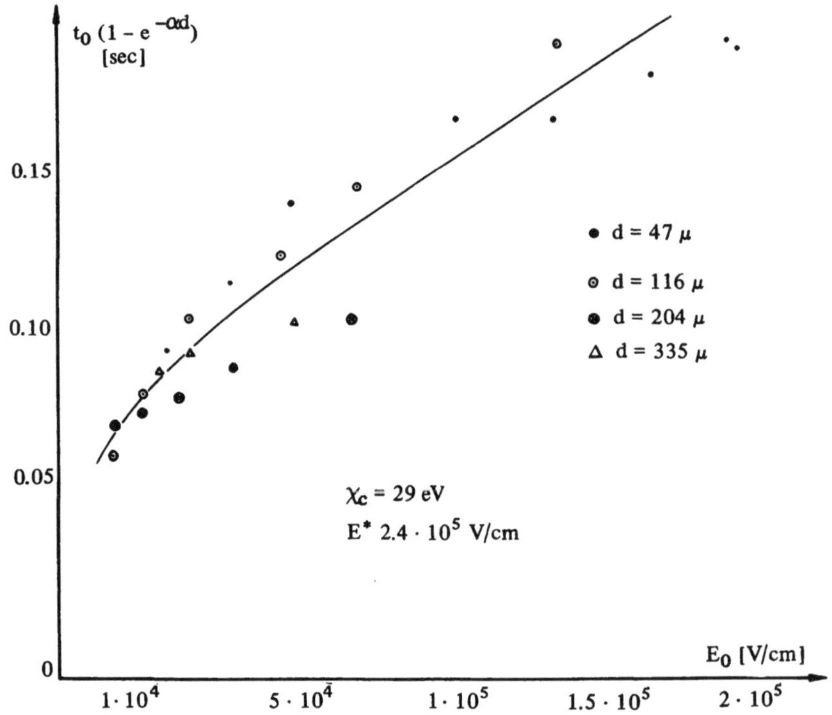

Fig. 6. Plot of $t_0\,(1 - e^{-\alpha d})$ against the initial electric field: Experimental points from Figure 4 and theoretical curve.

For the calculation of χ_c and E* the value of the X-ray intensity Φ_0 is very important. We used a calibrated dosimeter and obtained a value of 5 R/min.

Values for E* and χ_e were obtained by a fit as shown in Fig. 6. The function $t_0 \cdot (1 - e^{-ad})$ of the time constant and the layer thickness is plotted against the initial electric field E_0. The experimental data are the same as shown in Figure 4. The best fit obtained is with

χ_c = 29 eV
E* = $2.4 \cdot 10^5$ V/cm

The value of E* agrees numerically with the maximum available initial electric field and probably with the dielectric breakdown point. The value of 29 eV for the conversion energy is high compared to the normally reported value of 7 eV (3 × E_g). However, other loss processes are not considered in this model, such as secondary X-ray and fast secondary electron emission.

Taking the values for E* and χ_c, the theoretical decay of the field under irradiation is compared with experimental data (Fig. 7). The experimental decay (dashed lines) fits reasonably well the form of some representative theoretical curves (solid lines). This model shows, that it is possible to interpret the experimental results by a relatively simple recombination model of free electrons and holes.

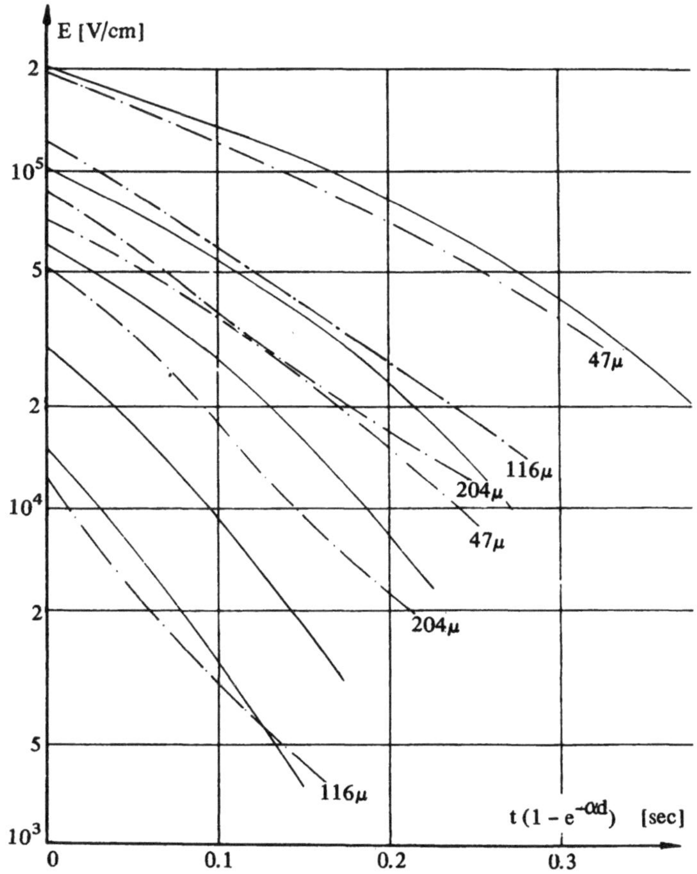

Fig. 7. Decay of the electric field for various initial electric fields. Comparison between experimental data (dashed lines) and a set of theoretical curves (solid lines).

Acknowledgement

The authors wish to thank Dr. N. Hansen and Dr. H. G. Junginger for numerous helpful discussions.

References

[1] J. L. Donovan, J. Appl. Phys. **41** 2109 (1970).
[2] S. Tsutsumi und H. Nagami, Denshi Shashin **4** 7 (1962).
[3] M. D. Tabak, Trans. AIME **239** 330 (1967).
[4] M. D. Tabak und P. J. Warter, Phys. Rev. **173** 899 (1968).
[5] D. M. Pai und S. W. Ing, Phys. Rev. **173** 729 (1968).
[6] H. Seki, Phys. Rev. **2** 4877 (1970).

Discussion

A. Matulionis: I want to make a short contribution in order to make it clear that the recombination of free electrons and holes is not efficient in amorphous selenium. In 1968 we reported our experimental results on single and double injection*. The main idea is as follows. A layer of selenium, supplied with a semitransparent substratum, was discharged by a certain exposure to a strong source of light from the surface; then it was additionally discharged by a similar exposure from the substratum. The next experiment was made with both exposures present at the same time, so that there was a moment when the moving "clouds" of the free carriers of the opposite signs overlapped. No difference in the results was obtained.

H. Kiess: Matulionis claims that recombination of electrons and holes is negligible in Se. He quotes an experiment with single and double injection into Se.

Question is: Are transit time and mean free time before recombination known, since the experiment is only meaningful if, in comparison to Hirsch and Jung's experiments, these data are known. If the transit time in Hirsch and Jung's experiment is larger than in the injection experiment than recombination may play a role.

H.-J. Hirsch: In the given reference recombination in the generation region is also assumed. Recombination during the drift-process has also been excluded in our work.

* J. Viščakas, V. Gaidelis, A. Matulionis, E. Montrimas and A. Satas. Proc. IX Int. Conference of physics of semiconductors, Moscow 1968, volume 2.

The Measurement of the Charging and Discharging Characteristics of Photoconductive Layers

R. J. Hercock

Abstract

An instrument is described for measuring a voltage proportional to the charge on insulating or photoconductive sheets.

The sample is mounted on a rotatable disc. The charge induces a voltage pulse on a probe as it passes underneath it. This pulse is sampled and stored on a condenser. To correct for drift, a zero reading is taken once per revolution and also stored on a condenser. The zero voltage is subtracted from the signal voltage and the resultant plotted on a recorder.

Provision is made for charging and discharging the layer.

Readings up to several thousand volts can be taken. From 0 to 1000 V the linearity is better than 1%. The noise is equivalent to ±0.1 V on the sample or ±0.2% of the sample voltage, whichever is the greater. The drift is equivalent to 3 V on the sample in the first hour and thereafter slowly fluctuates by ±0.1 V.

1. Introduction

In order to measure the charging characteristics of a photoconductive layer, the charger has to be separated from the measuring probe to prevent interference.

This can be achieved by mounting the sample on a rotatable disc, so that it is continually passed under the charger and the probe. Each time it passes under the charger, the sample receives an increment of charge which induces a voltage pulse on the probe as it passes underneath it.

The pulses can be displayed on a cathode ray oscilloscope when the envelope of the pulses will give the required charging characteristic.

However, it is more convenient to plot the response on a chart recorder. This can be done by sampling the pulse once per revolution by a switch synchronised with the rotation of the disc and storing the sampled voltage on a condenser between pulses.

Furthermore by slightly advancing the synchronization for each revolution of the disc, the sample can be scanned from one edge to the other. This can be used to plot the charge as a function of light intensity instead of time, by charging the sample, stopping the disc, exposing the sample through a step wedge and then starting the disc again.

If V_s is the voltage on the sample, then the voltage V_p induced on the probe is given by

$$V_p/V_s = C_{ps}/(C_{ps} + C_{pe})$$

where C_{ps} = probe-to-sample capacity

C_{pe} = probe-to-earth capacity

As C_{ps} is much less than C_{pe}, V_p is only a fraction of V_s and for this instrument the ratio is about 0.001. The actual ratio required is determined by the maximum likely voltage on the sample and the maximum input voltage that the head amplifier will accept. In the present case, the maximum input voltage for the amplifier is

±7 V, which gives nearly 9 kV as the maximum sample voltage. This is more than enough to cope with average samples.

With a single channel, handling only the signal, drift due to the head amplifier proved very troublesome. For this reason, a second channel was introduced, which takes a zero reading for every revolution of the disc. As the head amplifier is common to both channels, subtracting the zero reading from the signal reading effectively eliminates the drift.

2. Mechanical Details

The sample is mounted near the edge of an aluminium disc, 1/4″ thick, which can be rotated at 4 r.p.s. on ball races about a fixed vertical axis. The charger, light source and the probe are positioned at intervals of 90° round the circumference of the disc. Thus the probe and the charger are at opposite ends of a diameter and about 45 cm apart.

The probe consists of a brass plate 2 cm × 1 mm, fixed at the end of a brass rod, and arranged with the longer dimension radial to the disc. The rod is held inside a 1 in diam. brass tube with two PTFE washers. The lower end of the tube is closed with a cap in which is a slit of the same dimensions as the probe and parallel to it. The distance from the probe to the slit and from the slit to the sample is about 3 mm. At the upper end of the tube is a small box which contains the head amplifier.

The sampling of the voltage pulse induced on the probe is synchronised with the rotation of the disc by a magnet attached to the disc which closes a reed switch at the correct moment.

The reed switch is mounted on a disc which can be rotated by a synchronous motor about an axis co-axial with the disc. By this means the sample can be scanned from one edge of the sample to the other and a charge density profile plotted.

A second similar reed switch is also provided which closes when the sample is not under the probe and establishes a zero reference.

3. Electrical Details

A block diagram of the electrical circuits is given in Figure 1.

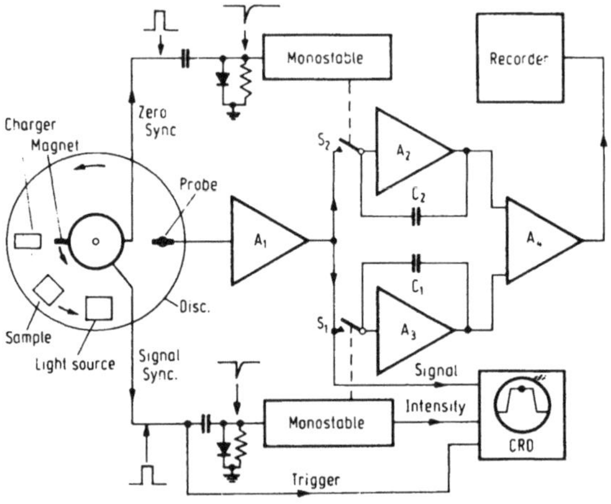

Fig. 1. Block diagram of circuits

A head amplifier, A_1, acts as an impedance convertor, coupling the high impedance probe to the sampling condensers. S_1 is closed when the sample is under the probe and S_2 when it is not. As a result C_1 is charged to a voltage proportional to the signal and C_2 to a voltage proportional to the zero-reference. The voltages on these condensers are subtracted in the differential amplifier, A_4, and the difference applied to the recorder.

The magnetically operated switches yield a square voltage pulse which is differentiated. The trailing edges of the pulses yield short negative-going pulses which trigger mono-stable multivibrators which in turn control the reed relays which close the switches S_1 and S_2.

The positive going voltage from the leading edge of the signal-synchronising pulse, triggers the time base of a cathode ray oscillograph. The output of the head amplifier is connected to the Y plates. A square pulse is also taken from the output of the signal multivibrator which brightens the trace whilst the signal switch S_1 is closed. The tube therefore displays the charge profile across the sample with a bright spot which indicates the point at which the reading is being taken.

4. Head Amplifier

The probe is directly connected to the head amplifier, which consists of a high-input-resistance operational-amplifier (*Ancom* 15 A-33) connected in the non-inverting mode as shown in Fig. 2. The gain is given by $(R_3 + R_2)/R_3$ which in this case is 2.

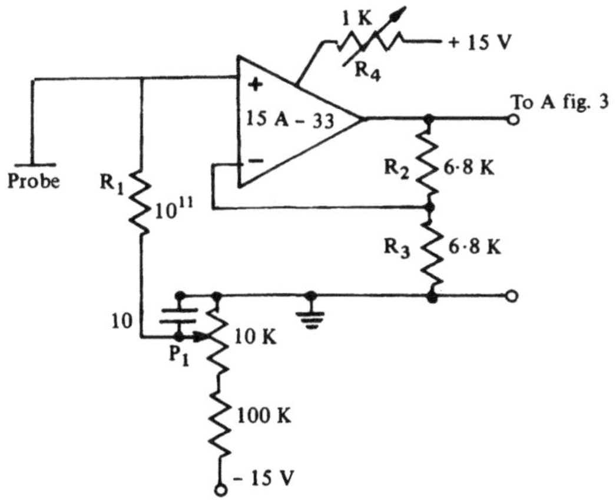

Fig. 2. Head amplifier

The high resistance R_1 connected to the probe is taken to the potentiometer P_1, so that the current through it can be adjusted to equal the input bias current. R_4 can be adjusted to give zero output volts when the input is shorted to earth.

5. Sampling Amplifiers

There are two of these, one for the signal channel and one for the zero-reference channel. Again high-input-resistance amplifiers are used but this time connected in the inverting mode. (Fig. 3).

* All resistor values are in ohms and capacitor values in μF.

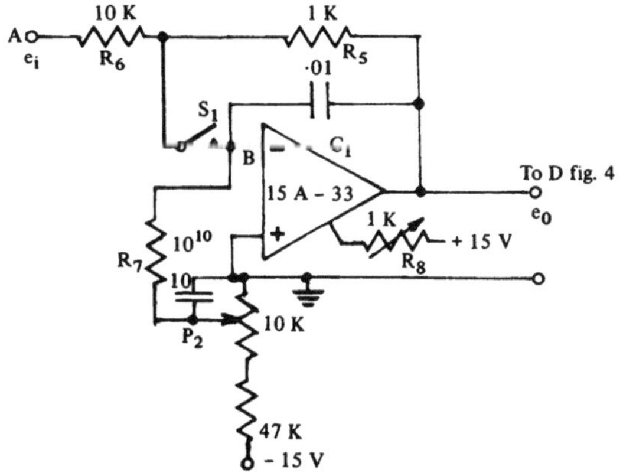

Fig. 3. Sampling amplifier

The gain is given by R_5/R_6 which equals 0.1. The sampling condenser is a polystyrene type and has a very low leakage current. It is connected between the output and the inverting-input B, which is therefore maintained at nearly earth potential.

The signal is applied to the input terminal, A, but it does not affect the amplifier until the switch S_1, controlled by the magnetic switch on the disc, is closed. The condenser C_1 is then charged to a voltage proportional to the input.

When S_1 is open, the amplifier is isolated from the input, but the output is held constant by the voltage on C_1, connected between the output and input.

The relationship between the output voltage, e_0, and the input voltage, e_i, is given by

$$e_0 = e_i \, [1 - \exp-(t/C_1 R_5)] \, R_5/R_6$$

If t (the time the switch, S_1, is closed) is much greater than $C_1 R_5$, this reduces to

$$e_0 = e_i R_5/R_6$$

As t is adjusted to be about 1 to 2 msec and the time constant $C_1 R_5$ is only 10 μsec this condition is satisfied.

In practice, this means that the output voltage is proportional to the input just before the switch opens and is independent of the time the switch is closed. The resolution of the system is, therefore, defined by the size of the probe.

In order to prevent drift of the output voltages when S_1 or S_2 is open, a high resistance, R_7, is connected from the input of the amplifier to P_2, which can be adjusted to cancel the input bias current.

The variable resistance, R_8, is used to give zero output voltage for zero input.

The other amplifier circuit is exactly similar except that it is controlled by S_2 instead of S_1. During construction the gains are made equal by the correct choice of R_5 and R_6.

6. Comparison Amplifier

The two voltages from the sampling amplifiers are substracted by a differential amplifier (Fig. 4).

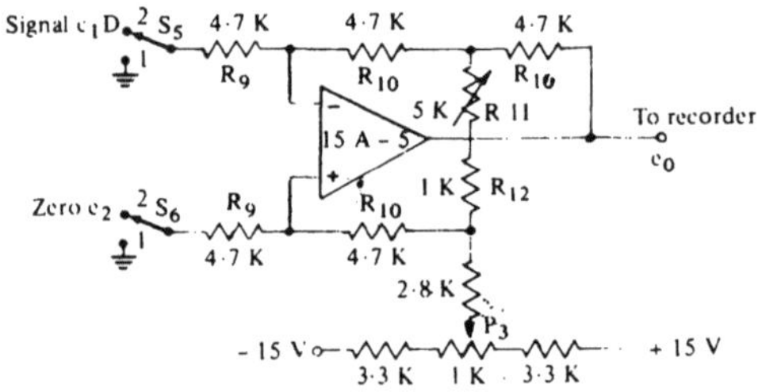

Fig. 4. Differential amplifier

The gain of this circuit is given by

$$e_0/(e_2 - e_1) = 2\ (1 + 1/k)\ R_{10}/R_9$$

where $k = (R_{11} + R_{12})/R_{10}$

As R_{11} is variable the gain can be set at any value between 3.5 and 10.

The zero of the recorder is set by adjusting P_3. The network attached to P_3 is designed to have the same value as R_{10}, irrespective of the setting of P_3. Hence the zero setting can be adjusted with negligible effect on the gain of the amplifier.

When adjusted so that 1 V on the sample gives 1 mV at the recorder the overall gain of the system is 1.25.

7. Switch Circuits

The switches S_1 and S_2 are controlled by two identical monostable multivibrators. The circuit of one of them is given in Figure 5.

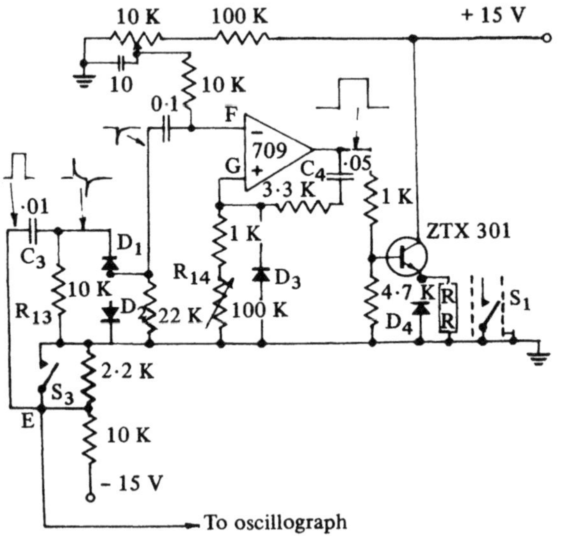

Fig. 5. Mulitvibrator

When the switch, S_3 (Fig. 1), is operated by the magnet on the disc, a square positive pulse appears at E. This is differentiated by C_3R_{13} and the positive pulse from the leading edge is suppressed by the diodes D_1 and D_2. The inverting input, F, of the 709 amplifier is normally held positive by the potentiometer P_4. As a result the output is negative. The non-inverting input, G, is at earth potential. When the negative pulse arrives at F, the output goes positive and applies a positive pulse through C_4, to the input, G, which drives the output more positive and speeds up the switching action. As C_4 charges, the curent through R_{14} decreases, hence the voltage on G falls until it is less than that on input F. When this happens the output falls rapidly from a positive voltage to a negative one and the circuit is again in equilibrium. The diode, D_3, suppresses the negative pulse from the output via C_2 and increases the rate of discharge of the condenser. The duration of the positive pulse at the output is determined by the value of R_{14} and C_4.

The square pulse from the amplifier is applied to the base of a *Ferranti* ZTX 301 transistor which energizes the reed relay, RR, in the emitter circuit and closes S_1. The diode D_4 suppresses the inductive kick which occurs when the transistor switches off. An earthed copper screen surrounds the glass envelope of the switch to screen it from the coil.

The circuit used for switching the signal channel also provides a positive pulse for synchronising the oscilloscope as shown.

The sampling amplifier and the comparison amplifier are enclosed in an earthed metal box and are connected to the head amplifier through a screened cable.

8. Adjustment

Before making measurements the circuit is adjusted as follows.

The probe is earthed and R_4 (Fig. 2) adjusted to give zero volts at the output. The probe is then unearthed and P_1 adjusted to give again zero volts at the output.

Next with everything running and both the inputs of the differential amplifier (Fig. 4) earthed (switches S_5 and S_6 in position "1"), P_3 is adjusted to give a convenient deflection on the recorder. Then with zero volts on the sample, switch S_5 in the signal channel is opened (position "2") and the signal sampling amplifier adjusted to give the same deflection as before with zero drift.

S_5 is then earthed and S_6 opened and the process repeated for the zero-sampling amplifier.

The recorder should now give the same deflection whether the switches S_5 and S_6 are earthed or not.

Finally, the instrument is calibrated by applying known voltages to a metal plate in the sample holder and recording the output. R_{11} (Fig. 4) is then adjusted until 1 V on the sample gives 1 mV at the output. Range switching is accomplished by altering the sensitivity of the recorder.

9. Performance

Noise. With zero volts on the sample, the noise referred to the input is ±0.1 V.

With a voltage on the sample, there is an additional noise equal to about ±0.2% of the sample voltage, probably due to the vibration of the disc. Above 50 V, the second source of noise is obviously the more important, but even so, it is not large enough to cause any worry.

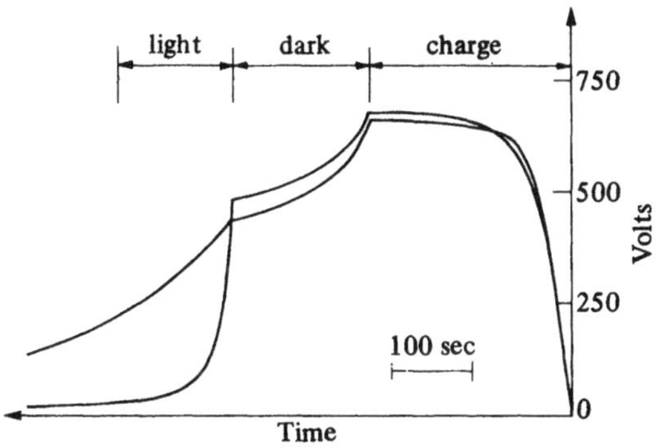

Fig. 6. Dark and light decay

The figures for the noise were taken from chart records.

Drift. The output voltage drifts by about 3 V (referred to the sample) during the first hour and slowly fluctuates thereafter by about ±0.1 V.

Linearity. Over the range 0 to 1000 V the deviation from linear response is less than 1%.

Discussions

R. M. Schaffert: Does this give an intermittent exposure? (Yes but one can expose in one pulse and then record).

But this does not provide the same effect as continuous exposure through a transparent probe.

R. J. Hercock: My apparatus measures either

1) Voltage as a function of time for constant intermittent illumination

2) Voltage as a function of light intensity with constant time of exposure

3) Continuous exposure through a transparent probe gives voltage as function of time for constant continuous illumination

For an image-wise exposure (2) is the measurement of importance.

Lange: ECE-Giessen has a measuring device which uses the Dyntest 90 apparatus; this device integrates impulses from a rotating probe to a DC voltage.

W. F. Berg: What is the difference between your instrument and that produced by ECE?

R. J. Hercock: I am not familiar with the ECE instruments.

Surface and Related Phenomena on Zinc Oxide

S. Roy Morrison

Abstract

Selected chemical and physical processes associated with zinc oxide are reviewed. Process leading to the generation, capture, and motion of charged particle are emphasized, as such phenomena are of particular importance in electrophotography. The origin of avalanche breakdown in ZnO which can control the charge acceptance, surface and bulk impurities which may act as electron sources in dark decay, and ions which may move and cause dark decay are all discussed. Photochemical process are also reviewed, emphasizing effects that can lead to electron trapping (decreasing the sensitivity of the elctrophotographic layer) or can lead to unstable electron donors (preexposure effects).

1. Introduction

In an electrophotographic layer, dark processes involving generation or trapping of mobile charged species within the layer are a serious problem which can drastically affect the quality of the photographic image. Thus when discussing the properties of ZnO with electrophotography in mind, it is appropriate to focus on those imperfections that can lead to such processes. Carrier generation processes are detrimental in electrophotography because they cause dark decay [1, 2, 3] of the electrostatic potential. If charged current carriers are generated spontaneously during or after the corona charging, they will move under the influence of the electric field, and the displacement of the charge will reduce the field. Mobile ions if available will have the same effect. The reason why carrier trapping is a problem in electrophotography is because it can reduce the optical sensitivity of the paper. For maximum sensitivity (maximum reduction of the surface potential per photoelectron), the photoproduced electrons must be able to traverse the entire layer (15×10^{-4} cm or so). If traps are present, the electrons are captured in a shorter distance and the sensitivity is reduced. We will discuss the properties of ZnO emphasizing imperfections contibuting to the generation, movement, and trapping of charged species. Thus for example we will discuss the generation of carriers by avalanche breakdown. We will discuss both the origin of bulk energy levels which can act as electron sources or sinks, and the origin of surface species which can act as electron sources or sinks. We will also discuss the availability of mobile ions in the ZnO that can lead to dark decay.

2. Carrier Generation Due to Breakdown

The first step of the photographic process is, of course, the corona charging of the layer. The rate of ion deposition at the surface during this step is rapid, and most carrier generation sources will not affect the attainable electrostatic potential.

Avalanche breakdown, however, provides a limitless source of carriers. When a high electric field is present in a semiconductor, as is produced by the corona charging, impact ionization occurs. An electron is accelerated in the electric field and accumulates sufficient energy to strike an immobile electron and free it. Then the second electron also moves in the field. When such multiplication occurs, an avalanche of charge carriers can be produced. When avalanche breakdown occurs at a certain spot, the field and hence the potential at that spot can increase no further, as all incoming ions are immediately neutralized by the breakdown currents. Actually a spot of local breakdown will lower the attainable potential over a substantial surrounding area, as pointed out and experimentally shown by Hoesterey [4]. Because of the low potential at a breakdown spot, ions from the corona which would normally strike a neighboring area are deflected to the breakdown spot.

Avalanche breakdown is most conveniently discussed in terms of the electric field attainable, but in electrophotography we are interested in the corresponding surface potential. The surface potential V corresponding to a given electric field E at the surface is found by solving Poisson's equation in the layer. We let the charge concentration be N_D, the density of donors in ZnO, and use the boundary conditions that $dV/dx = E$ at the surface, and $V = 0$ at the substrate. This yields

$$V = Ex_0 - eN_D x_0^2/2\,\epsilon\epsilon_0. \tag{1}$$

where ϵ is the dielectric constant, ϵ_0 the permittivity of free space, and x_0 the thickness of the photographic layer. Equation (1) applies if $N_D x_0/\epsilon\epsilon_0 < E$, that is, if the electrons are all exhausted from the layer. If the donor density is so high that the exhaustion region penetrates only part away, Eq. (1) reduces to

$$V = \epsilon\epsilon_0 E^2/2\,eN_D \tag{2}$$

If $E = E_B$, where E_B is the breakdown field, Eq. (1) or Eq. (2) gives the maximum electrostatic voltage attainable.

We will discuss first breakdown characteristics of the solid as a whole, and then discuss reasons why lower voltage breakdown can occur in certain localized imperfect regions.

Each solid has a well characterized avalanche breakdown field. This is the electric field necessary to accelerate charge carriers to the bandgap energy before an electron-phonon scattering event occurs, and in principle is the same value for single crystals as for powder crystallites.

Although theoretical calculations of such breakdown have been made [5], for our purposes an experimental measurement for ZnO is more satisfactory. Williams and Willis [6] measured the breakdown field for ZnO using electrochemical measurements. The method is as follows. With the ZnO as an anode, Pt as a cathode in an electrochemical cell, and KCl as the electrolyte, very little current will flow in the dark because electrons cannot make the transition from ions in the electrolyte to the conduction band of the crystal. Because the solution is a good conductor, most of the applied voltage V appears across the ZnO exhaustion region just as in the electrophotographic case. From Eq. (2), we can calculate the electric field present. In the electrolytic technique we have the advantage that we can increase the electric field simply by increasing the applied voltage V. Current will flow if the sample is illuminated. The current arises due to the photoproduced holes flowing to the surface, and the photoproduced electrons flowing to the bulk, Williams and Willis measured the photocurrent due to a constant intensity lamp. When the electric field became high the apparent photocurrent increased, which they assumed to be due to the beginnings of avalanche multiplication.

They report serious avalanche multiplication when the field is in the order of 1.5×10^6 V/cm.

As discussed above, this limiting field should be characteristic also of ZnO in an electrophotographic layer. Thus a layer 15×10^{-4} cm thick limited by this breakdown field should from Eq. (1) support at most an electrostatic voltage of about 2000 volts, decaying rapidly to perhaps 1000 V. Beyond 2000 V it appears that normal avalanche breakdown will generate unlimited carriers and completely prevent higher fields.

However, there are few sources of zinc oxide which when made into such a photographic layer will support even 1000 volts. There is apparently some other short time constant source of electrons which has the characteristics of breakdown and prevents the charging of the zinc oxide. We discuss two possible factors, dislocations and impurities.

In studies in our laboratory [7] we have obtained qualitative evidence that dislocations, arising from mechanical strains (milling) lead to very low apparent breakdown. The method of measuring breakdown was similar to that of William and Willis. Dislocations were introduced by lapping the (0001) surface of a ZnO single crystal with an alumina abrasive. The damaged region was gradually removed by etching with an 85% H_3PO_4 etch, and the breakdown field measured. The field at breakdown (defined as that field producing twice the normal photo current) increased from 10^5 V/cm with no etching, to 2×10^5 V/cm after a three minute etch, and to 7×10^5 with 33 min of etching. It was greater than 10^6 V/cm for a well etched sample, in agreement with the work of Williams and Willis. We conclude, therefore, that the presence of dislocations in high density can reduce the breakdown field up to at least a factor of 15. The poor quality electrophotographic layers observed when excessively milled ZnO is used [8, 9] may be associated in part with this effect.

A second factor which can lead to low voltage breakdown for certain ZnO particles is excessively high impurity densities, as is indicated in Eq. (1). The second term of Eq. (1) becomes 10% of the first term when $N_D = 10^{16}$ donors/cm^3, using $E_B = 1.5 \times 10^6$ V/cm, $x_0 = 15 \times 10^{-4}$ cm, $\epsilon = 8$. Presumably reasonably pure zinc oxide as commercially produced has fewer than 10^{16} donor impurities, and the second term of Eq. (1) may be neglected, but this may not be true for low quality ZnO and other semiconductor materials.

We have discussed the very rapid generation of charge associated with breakdown, and the resulting limitation of charge acceptance during the corona discharge. We do not, of course, suggest that breakdown is the only limitation on the corona charging. Other processes of carrier generation, which below we associate with dark decay, can, of course, also affect the corona charging if the reservoir of carriers is very large and the rate of carrier injection high.

3. Electron Sources in and on ZnO

In this section we discuss energy levels due to imperfections or impurities in and on ZnO which are normally occupied by electrons. When the conduction band electrons are stripped from the ZnO, as occurs in the electrophotographic process when the electric field is applied, these electrons can be thermally excited into the conduction band and contribute to the dark decay of the surface potential.

Because steps in the electrophotographic processes normally require time intervals in the order of several seconds, we are particularly interested in electron generation processes with time constants τ of this order of magnitude. Processes with τ in the order of milliseconds will be over before the corona charging is completed, so unless there is a large reservoir of carriers such processes will have little effect; processes with τ in the order of many minutes or more will not have time to occur. Thus we will first look to the theory for time constants.

From the theory of electron exchange with the conduction band, we have [10] for the net rate of electron capture U_n:

$$U_n = R - G = \sigma c\,(n_b^* p_t^* - n_1 n_t^*) \qquad (3)$$

where R is the rate of electron capture at a given set of levels, G the rate of electron generation. Here σ is the capture cross section of an electron at a trap, c is the mean electron velocity, n_b^* is the density of electrons in the conduction band, p_t^* is the density of empty levels; n_t^* is the density of occupied levels. The constant n_1 depends on the energy level of the electron source:

$$n_1 = N_c \exp\left\{-(E_c - E_t)/kT\right\} \tag{4}$$

where E_c is the conduction band energy, E_t the energy of the traps, and N_c is the equivalent density of states in the conduction band (in the order of 10^{19} cm^{-3}). At equilibrium, U_n must equal zero. This defines the value of n_t^*, the number of electrons stored in the traps. During the corona charging process, the free electrons are swept out of the layer (n_s^* becomes zero), and

$$\begin{aligned} U_n &= -\sigma c\, n_t^*\, n_1 \\ &= -\sigma c\, n_t^*\, N_c \exp\left\{-(E_c - E_t)/kT\right\} \end{aligned} \tag{5}$$

and electrons are injected at a rate dependent on the number of electrons available, n_t^*, the capture characteristics of the trap, σ, and the depth of the trap below the conduction band. Thus the trap will supply n_t^* electrons to neutralize the electrostatic voltage, with a time constant τ:

$$\tau^{-1} = \sigma c N_c \exp\left\{-(E_c - E_t)/kT\right\} \tag{6}$$

We can evaluate the time constant associated with a given trap depth, $(E_c - E_t)$. The value of c, the average electron velocity, is in the order of 10^7 cm/sec at room temperature, and σ, the capture cross section, can range from 10^{-22} to 10^{-12} cm^2, but is often near 10^{-15} cm^2, the geometrical area of a defect. If we use this value, and substitute $N_c = 10^{19}/cm^3$, and $\tau = 1$ second, we find that levels of the order of o.5 eV below the conduction band are expected to provide the most difficulty, that is to provide traps with τ in the order of seconds. However, with the wide possible variation in σ, in principle any trap can provide an electron source.

Bulk levels which can provide such a long time constant are rare. The donor levels in zinc oxide are normally considered to be associated with excess zinc. In sintered powder, Hahn [11] measured an energy level at 0.04 eV, a value close to the value expected for a hydrogenic donor. Other substitutional donors, hydrogen and indium [12, 13, 14, 15] show a similar very shallow level. Such levels should not be a problem with respect to dark decay, as τ should be much to small.

Undoped crystals show evidence of a much deeper energy level, variable in depth [14], which is of the order of a few tenths of an eV below the conduction band. The observations may be due to a trap associated with dislocations or some other unknown impurity in the zinc oxide.

Two acceptor impurities, lithium [16] and copper [12, 14, 17] provide deep levels in the zinc oxide. Copper is associated with a level about 0.3 to 0.5 eV. The level associated with lithium is not clearly defined [16], but it is extremely deep.

Lithium and copper are not normally expected to be present in photographic grade zinc oxide. Electron generation from dislocations is a possibility but it has not been explored. Thus in ZnO, bulk sources of electrons with τ of the order of a second have not been studied.

Many sources of electrons associated with surface species have been studied. We have made measurements [18, 19] of electron injection (electron sources) under conditions very analogous to the electrophotographic situation. A single crystal was used with various impurities of interest deposited on the surface. The surface was electrostatically charged by a corona wand, either in oxygen or in a mixture of Ar plus CCl_4, and electron injection was measured by the dark decay of the surface potential, V.

Figure 1 shows the decay due to K_3IrCl_6 molecules on the surface [19], as a curve typical of that observed with such additives. The ordinate used is $V^{1/2}$ because by the Schottky relation, $V^{1/2}$ is proportional to the surface charge. If the impurity were present as a uniform layer, the theory predicts that the electron injection should be linear. The non-linearity in Figure 1 was assumed to be primarily due to non-uniformity of the deposit. Regions with a heavy deposit lose their charge rapidly, those with a light deposit retain the charge for long periods of time.

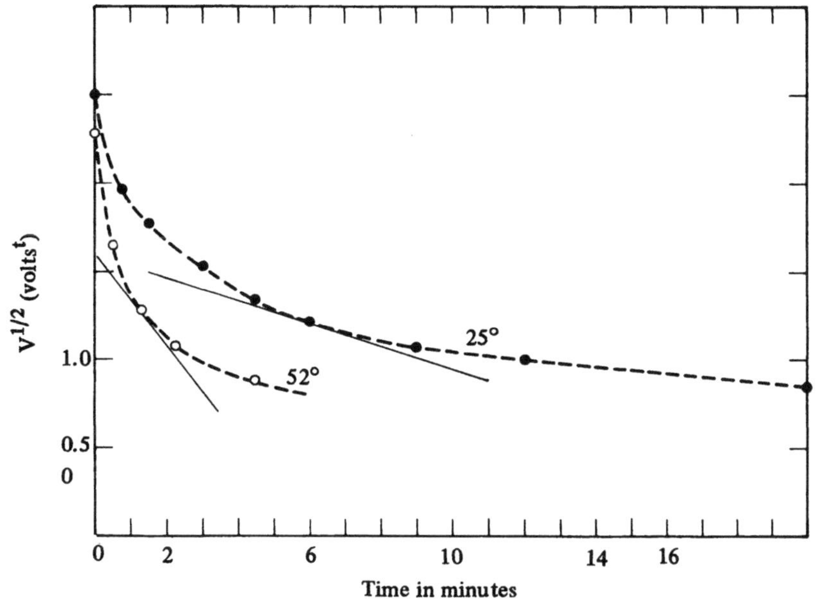

Fig. 1. Time decay of the square root of the surface potential V due to the presence of 2 x 10^{16} molecules/cm^2 of $IrCl_6^{-3}$ on the ZnO surface. Solid lines show decay rate at V = 1.4 as plotted in Figure 2.

In order to estimate the energy level of the electron source, the rate of decay is plotted against temperature, where the rate of decay is determined at a particular surface potential (such as the tangent lines in Figure 1). The justification for this procedure, based on the non-uniformity assumption, is discussed in the reference. Figure 2 shows such an Arrhenius plot for K_3IrCl_6, leading to a trap depth (from Eq. 3) of 0.65 eV below the conduction band. Similarly $K_4Fe(CN)_6$ was determined to have a trap depth of 0.1 eV, K_2MnO_4 a depth of 0.4 eV, and $Cr(NO_3)_3$ of 1.1 eV. The trap depth for oxygen could be determinated by permitting the reaction

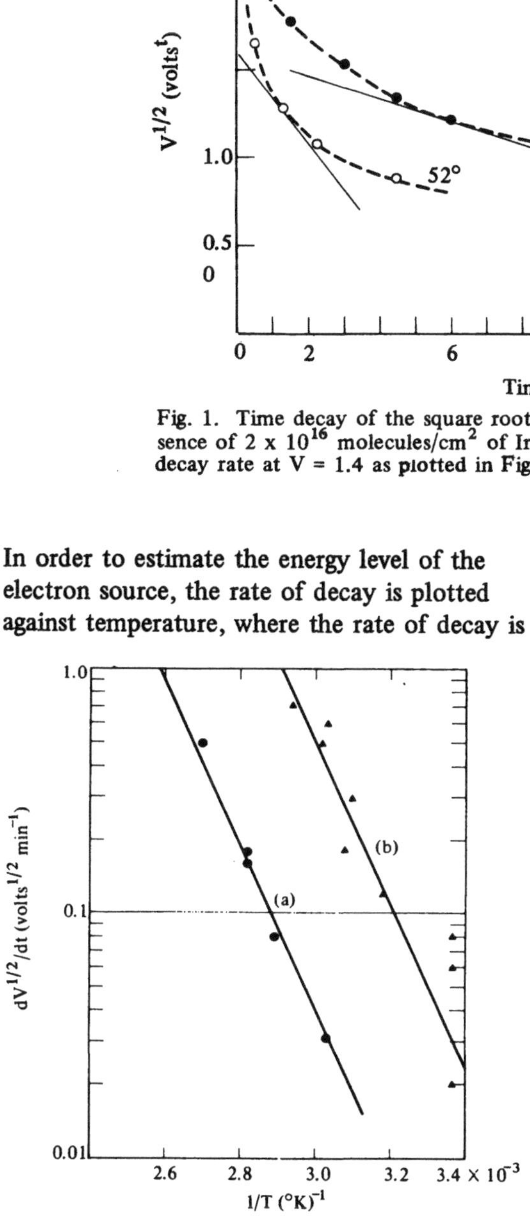

Fig. 2. Surface potential decay rate versus temperature with $IrCl_6{}^{-3}$ surface states. Curve (a): 2 x 10^{15} molecules/cm^2; curve (b) 2 x 10^{16} molecules/cm^2.

$$K_4Fe(CN)_6 + O_2 \rightarrow K_3Fe(CN)_6 + K^+ + O_2^-$$

to proceed. With oxygen present the trap depth was observed to be 0.9 eV below the conduction band.

Were these species present on photographic zinc oxide, particularly the ferrocyanide, they would certainly be detrimental. However, large quantities of these particular ions are not to be anticipated. Other surface states may act similarly, however.

Electron injection from methanol has also been reported [18]. Here the surface state energy is 1.3 eV below the conduction band. This energy level is relatively deep, but it is of interest because organic materials are present both as solvents and as the resin in an electrophotographic layer, and the results suggest that injection from adsorbed solvents or other organics, especially if present in high concentration as a liquid, could provide a source of electrons.

Another experimental technique used to measure electron injection into ZnO from surface is the electrochemical cell. As discussed earlier, it is found that in general electron injection does not appreciably occur into ZnO from aqueous solution, so that if the ZnO is made the anode of an electrochemical cell, negligible current is observed. However, if strong reducing agents are present in the solution, current flows and this is interpreted to mean that the energy level of the species is higher than the ZnO conduction band, so rapid injection can occur even at room temperature. Electron injection has been observed by Freund, Gomes, and Morrison [20] from V^{2+}, Cr^{2+}, Sn^{2+}, and ascorbic acid, and by Freund [21] from Eu^{2+}, and Ti^{3+}.

In summary, there are indications that adsorbates on the zinc oxide coming from the atmosphere or from the binder could act as a source of electrons for dark decay. Further studies are necessary to identify if there are particular solvents and polymeric materials which should be avoided.

4. Electron Sources due to Photocatalyzed Reactions

It is well known that the illumination of an electrophotographic layer prior to the corona charging, pre-exposure of the film, has a strong influence [22] on the charging characteristics, the dark decay, and the photosensitivity of the layer. In this section we explore some of the reactions that can occur under illumination and can provide sources of electrons or ions.

Neglecting dye excitation, the major direct influence of illumination is to produce holes in the zinc oxide. Because of the wide bandgap, the hole density without illumination in ZnO is negligible. These photoproduced holes are very strong oxidizing agents [23]. Many and varied reactions have been observed associated with oxidation of species by the photoproduced holes and the simultaneous reduction of species (usually oxygen) by the corresponding electrons.

If there is no reducing agent on a ZnO surface to be oxidized by the holes, the holes can oxidize the lattice oxygen and hence decompose the zinc oxide. This corresponds to reactions such as

$$h\nu \rightarrow p + e \qquad (7)$$

$$H^+ + ZnO + p \rightarrow 1/2\ H_2O_2 + Zn^{++} \qquad (8a)$$

or

$$ZnO + 2p \rightarrow (1/2)\ O_2 + Zn^{++} \qquad (8b)$$

Equation (8) has been observed in aqueous solution; H_2O_2 is formed and zinc ions are found in solutior [24]. A corresponding reaction occurs if the ZnO is in vacuum, where the reduced zinc dissolves in the solid [25, 26]. By these reactions, therefore, with no reducing agent available, the photolysis could provide a source of zinc ions for subsequent dark decay (see next section).

If reducing species, including many organic molecules, are present in modest concentrations, the photoproduced holes will react with them [23] rather than decompose the zinc oxide. This will often lead to active intermediate species in the vehicle which are electron sources.

It has been known for many years that if zinc oxide is illuminated in the presence of many organic species [21] and oxygen, the organic species becomes oxidized. For example the formate oxidation reaction is

$$HCOOH + O_2 \xrightarrow[h\nu]{ZnO} H_2O_2 + CO_2 \qquad (9)$$

Morrison and Freund [23] studied the formate oxidation using a single crystal of ZnO as an electrochemical cell. With no formate present in the solution the anodic current (the current due to holes) is proportional to the light intensity. This is in accordance with expectation for the rate of hole generation should be determined by the light intensity. When the organic formate was added, the current almost doubled at constant light intensity ("current doubling"). This result was very surprising because the addition of formate could not increase the rate of hole generation in the ZnO. However, the result could be explained by electron injection from a new chemical species, free radicals. The reaction with formate present became, instead of (8)

$$p + HCOOH \rightarrow HCOO^{\cdot} + H^{+} \quad (10)$$

$$HCOO^{\cdot} \rightarrow CO_2 + H^{+} + e \quad (11)$$

Thus the one-equivalent oxidation of the organic by the hole leads to a radical, in this case the formyloxy radical. This radical is very active chemically, and decomposes according to Eq. (11) by electron injection into the ZnO.

Such a process is of potential interest in electrophotography, for such radicals can diffuse into the resin and be stored, or can initiate chain reactions in the resin, leading to chemical changes and thus sources of electrons for subsequent dark decay. Thus we will examine the information available about the possible reactions.

On TiO_2 there is evidence [28] that the $OH^{\cdot}$ radical is formed by the oxidation of hydroxyl groups by holes. The $OH\cdot$ radicals were detected upon irradiation of coatings (TiO_2 + vehicle) using ESR and redox titration. This is of proven importance in coatings, as these radicals initiate chain reactions oxidizing the vehicle and causing chalking. Studies of the shorter range effects of importance in electrophotography have not been reported.

On ZnO there has been no evidence of $OH\cdot$ radical production. No similar measurements have been reported. Holes from ZnO, however, do oxidize many species with the production of radical intermediates as observed by current doubling. These include many alcohols, and formate. If we assume that the many photocatalytic reactions like (9) which have been reported in the literature [27, 29] always involve hole capture and thus radical formation, the list is extended to some phenols, amines, amides, and some aromatic compounds, such as benzene, toluene, and xylene. Recent measurements have been made in our laboratory designed to study [30] the oxidation of binder components by holes from ZnO. This work was supported by NASA with the object of exploring and eliminating such photoeffects on thermal control coatings. The results and techniques are directly transferrable to electrophotographic layers, however. The approach for studying hole reactions for the components of the paint vehicle was that suggested [23] and used [29] in the earlier photocatalytic work. The hole reactivity is compared to the hole reactivity of standards (such as methanol or ferrocyanide) using the current doubling effect. Thus for example the Owens Illinois 650 resin was not observably oxidized by holes in these preliminary experiments. Of its possible solvents dioxane was found to have a hole reactivity at least 30 times greater than acetonitrile and ten times greater than ethyl acetate. Such results would thus suggest that dioxane should be avoided as a solvent in the formulation of coatings based on ZnO.

In summary there are many possible photochemical reactions which can occur leading to new chemical species which can donate electrons to the ZnO conduction band. These may provide sources of electrons for the pre-exposure effects in zinc oxide electrophotography.

5. Dark Decay by Ion Movement

If there is a substantial density of positive ions dissolved in the ZnO when the electrostatic potential is applied, it is possible that these ions can move under the influence of the field and cause an apparent dark decay.

There are only three ions which are known to diffuse through ZnO rapidly enough to cause discharge at room temperature. These are hydrogen, zinc, and lithium, when introduced interstitially. Of the three, lithium is probably least interesting, as there are no obvious sources of lithium present.

The diffusion of hydrogen at low temperature has been studied by Thomas and Lander [13]. Between 200 and 650 °C they find the diffusion coefficient to be given by D = .03 exp (–0.91 e/kT) cm^2/sec with e the electronic charge. If we assume this expression is valid to room temperature, this leads to $D = 5 \times 10^{-18}$ cm^2/sec at 23 °C. The solubility of hydrogen is low at equilibrium, but there is a large reservoir of protons on the pigment surface, because the surface will normally be hydrolyzed at room temperature.

If we examine the numbers available for ZnO, we find dark decay due to bulk diffusion of protons is just possible. The velocity v of the particles under the influence of the applied field is

$$v = \mu E = DE/kT$$

where μ is the ion mobility, related to D as shown by the Einstein relation. With $E = 10^6$ volts/cm, the particle velocity at room temperature becomes 2×10^{-10} cm/sec.

If we assume a monolayer of adsorbed protons (2×10^{15} cm), then the proton density ρ is about 3×10^{19} $protons/cm^3$. For complete polarization (removal of the surface potential due to a surface charge N_s of 5×10^{12} $charges/cm^2$) the protons must move

$$x_0 = N_s/\rho = 10^{-7} \text{ cm}$$

and at the calculated velocity this requires a time of 10 minutes.

This simple calculation ignores any activation energy required to dissolve the protons. Thus the process may or may not be a problem in ZnO. However, with the high fields and small particle size desired for electrophotographic layers, proton diffusion is potentially a serious problem in any pigment.

In zinc oxide interstitial zinc is another potentially troublesone ion. Excess zinc can be produced at the surface of the particles by reduction of the oxide, either by photolysis (see last section) or by some reducing agent in the vehicle. Also a zinc compound forming a separate phase at the surface can be a source of interstitial zinc ions. From the work of Thomas [31], the diffusion coefficient of interstitial zinc is 2×10^{-4} exp (–0.55 e/kT) cm^2/sec, measured between 190 and 350° C. Assuming this law holds to room temperature, we find at 23 °C, $D = 4 \times 10^{-14}$ cm^2/sec, which we note is much higher than the value of D for protons. As in the preceding calculation we can calculate the mean velocity in the electrostatic field as 2×10^{-6} cm/sec. The effective concentration of available zinc is not clear, so it is difficult to anticipate an effective rate for the dark decay due to zinc ion movement.

We have made experimental measurements which bear on the problem of ion movement. These were measurements of the dark decay on a single

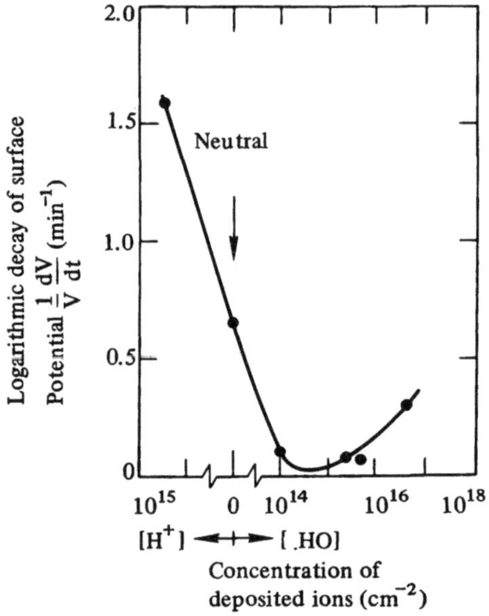

Fig. 3. Rate of dark decay on the ZnO (0001) face as a function of excess H^+ or OH^- ions deposited.

crystal (0001) face. In these measurements no active reducing agent was present and as suggested earlier [18] the most likely cause of decay is ion movement. Varying amounts of excess KOH or H_2SO_4 were deposited on the surface. With a large excess of acid or base, a second phase involving $KZn(OH)_3$ or $ZnSO_4$ + H_2SO_4 is expected, a potential source of zinc ions or protons. The sample was heated at 175° C for 5 minutes to diffuse the ions into ZnO, then the sample was cooled to 50 °C, charged by a corona, and the dark decay measured. The results are shown in Figure 3, where the initial rate of dark decay of the voltage, dV/dt, is plotted vs. the excess hydroxyl ions or protons deposited on the surface.

The minimum in the rate of dark decay for a surface covered with a monolayer of hydroxyl ions is of interest. The exposed face, the (0001), is the oxygen face. We suggest that hydrogen or zinc diffuses into the lattice at 175 °C, thus providing donor ions to diffuse out and cause dark decay of the surface potential at 50 °C. If hydrogen ions are the source, the minimum at one monolayer of OH^- ions may be associated with the displacement of the protons normally hydrolyzing the surface oxygen ions.

$$\rangle O_L^= -H^+ + KOH \rightarrow \rangle O_L^= -K^+ + H_2O \uparrow$$

where $O_L^=$ represents a surface oxygen ion.

On the other hand, zinc ions may be the source, and the effect may be associated with zinc made available by the presence of a separate phase as discussed above. The separate phase arises because the ZnO is dissolved by the deposited acid or base, then re-deposited when the sample is dried. A minimum dark decay would be predicted because the etching rate of ZnO is minimum under slightly basic conditions.

6. Trapping Centers

There is evidence [32, 3] that trapping of photoproduced electrons does occur in ZnO photographic layers, and does decrease the initial sensitivity of the layer until the traps are filled. Thus it is of interest to examine some of the possible sources of these traps.

The conditions under which trapping should be a problem deserve comment. As discussed earlier, the donor levels in zinc oxide powder are expected to be very close to the conduction band. Thus any bulk levels deep enough to trap an electron will tend to be already occupied at thermal equilibrium. Surface trapping effects are more complicated, but again any levels accessible for trapping will already be occupied in general by electrons from the donors at steady state. Trapping effects are expected, therefore, only when the levels are disturbed from thermal equilibrium (ionized) by some electrical, optical, or chemical process. For example, when the electrostatic field is applied, some potential trapping levels will tend to empty (dark decay) as discussed above, and these will be available as traps during the exposure cycle. Traps could also be ionized during a pre-exposure.

In materials with deeper donor levels than in ZnO (for example TiO_2) it is possible that at equilibrium there will be unoccupied levels deep enough to act as traps. Also if the zinc oxide is of very high resistivity (due to acceptor impurities, such as copper, dislocations, or perhaps adsorbed oxygen) then trapping can be anticipated.

With these reservations in mind, we will discuss a few cases of electron capture by bulk and surface species.

Imperfections which could act as bulk trapping levels in ZnO include substitutional lithium, and dislocations. Dislocations may be the most serious problem in electrophotography, as they can be produced by excessive milling. However, no thorough studies have been made. We have observed electron trapping on dislocations produced by lapping a single crystal. The damaged layer has an extremely high resistance, indicating the electrons are trapped. Electrochemical measurements of the charge in the surface region, by the standard capacity measurement [33] indicate a low positive charge compared to etched ZnO, again indicating electrons trapped in the surface region. Such trapping behavior of dislocations has been studied [34, 35, 36] in other materials, particularly germanium.

Capture of electrons from the ZnO conduction band by foreign ions at the surface has been studied by electrochemical methods, with a ZnO crystal as a cathode, and the ions dissolved in solution. Hauffe and Range [37] demonstrated electron capture by quinone. Freund and Morrison [33] and Morrison [38] studied one-equivalent ions, and showed the rate of electron capture is first order in the density of dissolved ions and first order in the density of electrons at the ZnO surface. Morrison, expanding on an earlier suggestion of Beck and Gerischer [39] suggested that for one-equivalent species there should be a relation between the standard redox potential of an ion and its energy level in solution. It was concluded [38] that for a one-equivalent ion to act as an electron trap on ZnO its standard redox potential should be below –0.1 eV.

Oxygen acts as an electron trap on zinc oxide, but there is little quantitative information available. To explain the many catalytic [40] photocatalytic [23] redox reactions on ZnO in solution, rapid reduction of oxygen by conduction band electrons (rapid electron capture by oxygen) has been postulated. Morrison and Freund [23] have examined semiquantitatively electron capture by oxygen using electrochemical measurement, but the results cannot be analyzed in terms of the several reaction steps, going from O_2 to O, and capture cross sections for the various oxidation states. A good study, electrochemical or other, of electron capture by oxygen on the ZnO surface, would be of significant value.

In the electrophotographic systems, trapping by dye molecules is particularly important, as these are always present. It seems clear that in many cases if the dye molecules are oxidized, an electron trap results. For example Langaton [41] has studied trapping by dyes, where the electron traps are prepared by pre-exposure of an electrophotographic layer at the dye wavelength (no holes). The pre-exposure removes the electrons from many dye molecules and trapping centers are formed. The photosensitivity of the layer is significantly decreased by the pre-exposure, the loss of sensitivity indicating that the photoproduced electrons are indeed captured by these traps.

Trapping by dyes has more recently been studied by Dudkowski et al. [42] and by Comizzoli [43]. In the latter studies no pre-exposure was used, but the trapping by dyes was observed by a photocurrent. It was found that the photocurrent in a film varies as the square root of the light intensity. Without trapping the photocurrent should vary linearly; the square root dependence suggests a bimolecular trapping process–the trap density equals the electron density and the trapping rate varies as the product. Further evidence of the trapping effect was found in measurement of the range of photoinduced electrons. When the electrons were excited from the valence band, and the dye molecules not ionized, the electrons moved about 10^{-2} cm/unit electric field, whereas when the electrons were excited from Rose Bengal, leaving the ionized Rose Bengal ions as traps, the carriers only moved 10^{-4} cm/unit electric field.

7. Summary

We have attempted in this review to summarize some of the possible sources and sinks for electrons or mobile ions which could deleteriously affect the characteristics of a ZnO electrophotographic layer. It is clear the possibilities are very numerous, and one cannot in any practical system avoid all the solid state defect problems and surface chemistry problems discussed. Experimental work is indicated to explore the types of reactions discussed with electrophotography in mind–to examine carrier generation, carrier trapping and mobile ion sources due to the chemical species normally found in electrophotographic layers.

References

[1] K. Hauffe and R. Stechemesser, Helv. Phys. Acta **41,** 872 (1968).
[2] J. A. Amick, RCA Review, 770 (1959).
[3] R. M. Schaffert, Electrophotography, Focal Press, London (1966).
[4] D. C. Hosterey, J. Appl. Phys. **33,** 992 (1962).
[5] P. A. Wolff, Phys. Rev. **95,** 1415 (1954).
[6] R. Williams and A. Willis, J. Appl. Phys. **39,** 3731 (1968).
[7] These unpublished measurements were made by the author in collaboration with Dr. T. Freund.
[8] K. Hauffe, J. Phot. Sci. **10,** 321 (1962).
[9] Y. Harazaki, H. Hasegawa, and K. Kirigana, J. Chem. Soc. Japan, Ind. Chem. Section **66,** 432 (1965).
[10] See, for example, A. Many, Y. Goldstein, and N. B. Grover, Semiconductor Surfaces, Wiley-Interscience Inc., New York (1965).
[11] E. E. Hahn, J. Appl. Phys. **22,** 855 (1951).
[12] G. Bogner and E. Mollwo, J. Phys. Chem. Solids **6,** 136 (1958).
[13] D. J. Thomas and J. J. Lander, J. Chem. Phys. **25,** 1136 (1956).
[14] G. Bogner, J. Phys. Chem. Solids **19,** 235 (1961).
[15] D. G. Thomas, J. Phys. Chem. Solids **9,** 31 (1958).
[16] J. J. Lander, J. Chem. Phys. Solids **15,** 324 (1960).
[17] M. Sumita, Jap. J. Appl. Phys. **6,** 418 (1967).
[18] S. R. Morrison, Surface Science **13,** 85 (1969).
[19] S. R. Morrison, Surface Science (inPress).
[20] T. Freund, W. Gomes, and S. R. Morrison, Proceedings of the IVth Int. Congress on Catalysis, Moscow (1968).
[21] T. Freund, J. Phys. Chem. **73,** 468 (1969).
[22] A. J. O. Axford and J. Gilbert, J. Phot. Sci. **12,** 20 (1964).
[23] S. R. Morrison and T. Freund, J. Chem. Phys. **47,** 1543 (1967).
[24] M. C. Markham and M. C. Upreti, J. Catalysis **4,** 229 (1965).
[25] R. J. Collins and D. G. Thomas, Phys. Rev. **112,** 388 (1958).
[26] S. R. Morrison, J. Vac. Sci. Tech. **7,** 84 (1970).
[27] For lists of references, see references 23 and 29.
[28] H. G. Volz, G. Kampf, and H. G. Fitzky, in X. Fatipec Kongress, Verlag Chemie GmbH (Weinheim, 1970).
[29] T, Freund and W. P. Gomes, Catalysis Reviews **3,** 1 (1969).
[30] K. M. Sancier and S. R. Morrison, Proceedings of AIAA (October, 1971).
[31] D. G. Thomas, J. Phys. Chem. Solids **3,** 229 (1957).
[32] P. M. Cassiers, K. Vechille, and J. Van Engeland, Applied Optics, Supplement on Electrophotography, p. 83 (1969).
[33] T. Freund and S. R. Morrison, Surface Science **9,** 119 (1968).
[34] M. Jastrzebska dn T. Figielski, Phys. Stat. Solidi **14,** 381 (1966).
[35] S. R. Morrison, Phys. Rev. **104,** 619 (1956).
[36] W. T. Read, Jr., Phil. Mag. **45,** 775, 1119 (1954).
[37] K. Hauffe and J. Range, Ber. Bunsengesell. Phys. Chem. **71,** 690 (1967).
[38] S. R. Morrison, Surface Science **15,** 363 (1969).
[39] F. Beck and H. Gerischer, Z. Electrochem. **63,** 943 (1959).
[40] K. Hauffe and H. Pusch, Ber. Bunsengesell. Phys. Chem. **72,** 669 (1968).
[41] D. J. Langston, Tappi, **49,** 16 (1966).
[42] S. J. Dudkowski, A. G. Kepka, and L. I. Grossweiner, J. Phys. Chem. Solids **28,** 485 (1967).
[43] R. B. Comizzoli, J. Appl. Phys. **41,** 4148 (1970).

Discussion

G. Heiland: Is there any direct experimental evidence for an appreciable movement of ions?

S. R. Morrison: In an earlier discussion (Surface Science, 1968) we have reported experiments which provide evidence of ion movement causing dark decay on single crystals.

One of the experiments reported, for example, was measurement of low temperature dark decay with undoped ZnO and Indium-doped ZnO. The former showed dark decay (presumed in that work to be due to interstitial zinc) whereas the Indium-doped crystal showed none. The reason for the latter was presumed to be low diffusion coefficient of the indium.

J. Range: 1) If you assume that the dark decay is caused by migration of interstitial ions, you should get a dependence of dark decay with the duration of corona charging. Did you observe it?
2) Did you find whether the equations for a Schottky-depletion layer are fulfilled or not?

S. R. Morrison: 1) Yes. After a high temperature cycle of the sample, (assumed to introduce a new supply of interstitial zinc), the longer the integrated time of corona, the slower the decay of the surface potential (see reference 18).

2) The decay of the surface potential due to ion movement would involve a very difficult solution of the field-aided diffusion equation, and I have not been able to solve the equation analytically.

Heiland: By heating to 500 °C for about 15 minutes in hydrogen a diffusion of donors is induced from the surface towards the interior of the crystal. The thickness of the diffused layer does not exceed the width of a depletion layer induced by adsorbed oxygen. Therefore all the excess conductance from the hydrogen treatment can be compensated by adsorbed oxygen. However, after heating in a vacuum to 300 °C for 10 min the oxygen is desorbed and a high surface conductance appears. The crystals keep this memory of the hydrogen treatment at least for months during storage in air, therefore not all the donors are extracted by the space charge field.

[G. Heiland, E. Mollwo and F. Stoeckmann in *Solid State Physics.* Eds. F. Seitz and D. Turnbull, Vol. 8 Academic Press, New York, 1959, p. 270, Fig. 64.]

K. Hauffe: The out-diffusion of protons is roughly 3 orders of magnitude smaller than the diffusion of zinc ions in interstitial position. Therefore, the "*memory* effect" of hydrogen-treated zinc oxide is understandable.

Surface Properties of Zinc Oxide Crystals

G. Heiland

Abstract

On clean cleaved *prism surfaces* of ZnO single crystals a p-type sign of the field effect has been found at low temperature. Spectral sensitization of photoconductivity for the visible spectral region can be achieved by evaporation of organic dyes onto clean prism surfaces. The quantum yield referred to absorbed photons exceeds that of the intrinsic absorption region by a factor up to five.

Cleavage normal to the polar c axis exposes two *polar surfaces*, the so called Zn face and O face. Annealing of the clean surfaces in ultrahigh vacuum causes no measurable change of the low surface conductivity on the O face, but creates a high conductivity on the Zn face. It has been shown that the surface donors can evaporate from the Zn face onto the O face. The density of surface states is studied with the field effect after different temperature treatments. Also irradiation with band gap light causes an irreversible increase of conductivity by six orders of magnitude on the Zn surface but not on the O face. LEED studies show that no superstructure appears on polar surfaces cleaved in ultrahigh vacuum and annealed at different temperatures up to 850° C. Extra spots are due to step arrays.

1. Introduction

The aim of this presentation is not a complete review [1]. Only a few comments are given on new results and old problems. Zinc oxide is a semiconductor with a wide band gap of about 3 eV. A band structure calculation has been published [2]. Due to an excess of zinc also crystals without intentional additions always shows n-type conductivity. Arguments have been given, that the donors are oxygen vacancies occupied by electrons [3].

In electrophotography zinc oxide is used in the form of a powder. To separate bulk and surface phenomena may be difficult with microcrystals. For a more basic understanding it seems helpful to study surface properties of large single crystals. The space charge layers penetrate into the material from the surface and the photoconductivity of zinc oxide is a surface effect in the intrinsic absorption region as well as in the range of longer wave lengths with sensitization by organic dyes.

Figures 1 and 2 show crystals as they are grown in our laboratory in Aachen from the vapor

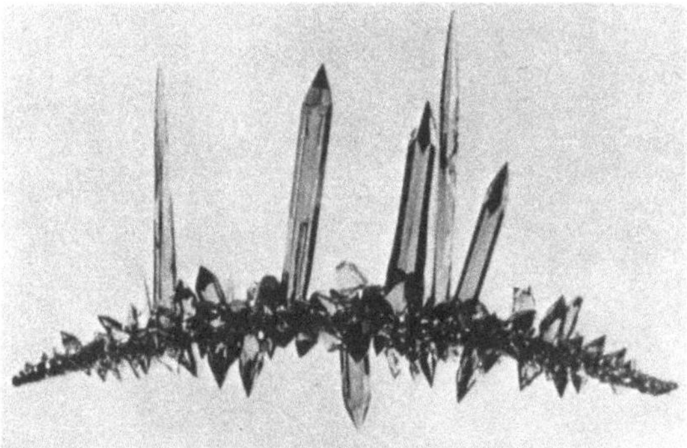

Fig. 1. ZnO crystals as grown on a quartz filament. They exhibit a blue colour by an addition of Indium.

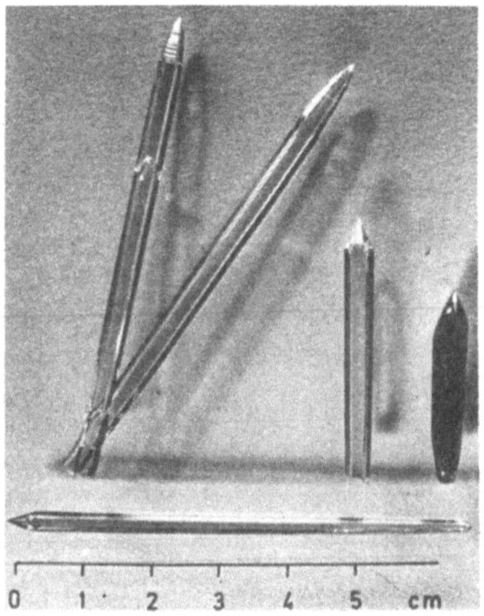

Fig. 2. ZnO crystals. The short crystal is doped with Indium.

phase. The axis of these hexagonal prisms is parallel to the c axis of the wurtzite lattice (Fig. 3). This structure has no center of symmetry and allows piezo- and pyroelectricity. From a crystallographic point of view three different types of surfaces are distinguished: The prism surfaces $\{10\bar{1}0\}$ are oriented parallel to the c axis, and in the ideal case they hold equal numbers of Zn and O atoms in the uppermost

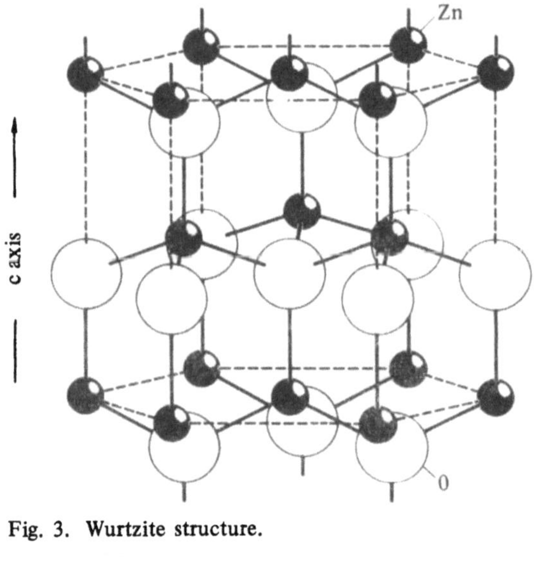

Fig. 3. Wurtzite structure.

layer. They can be prepared by growth and by cleavage. The other two types of surfaces (0001) and $(000\bar{1})$ are formed after cleavage normal to the $\underset{\sim}{c}$ axis. Because of the polar character of this axis one of the two surfaces should expose preferentially Zn atoms in the uppermost layer, the other O atoms [6]. It has been shown by LEED observations, that steps on these polar surfaces have a height of one c axis unit or a multiple [4]. In this way the character of the uppermost atomic layer is conserved (Fig. 12).

2. Properties of Prism Surfaces

These surfaces have been studied for a long time [1]. The surface conductivity can be changed through several orders of magnitude by adsorption of gases. Electronic surface states are created or removed during this process. Figure 4 gives two examples: surface donors from atomic hydrogen causing an accumulation layer and surface acceptors from oxygen creating

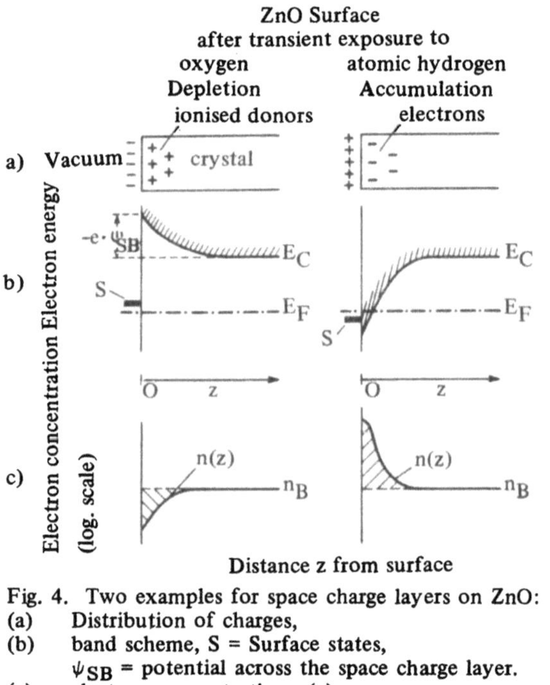

Fig. 4. Two examples for space charge layers on ZnO:
(a) Distribution of charges,
(b) band scheme, S = Surface states,
ψ_{SB} = potential across the space charge layer.
(c) electron concentration n(z),
n_B = electron concentration in the bulk.

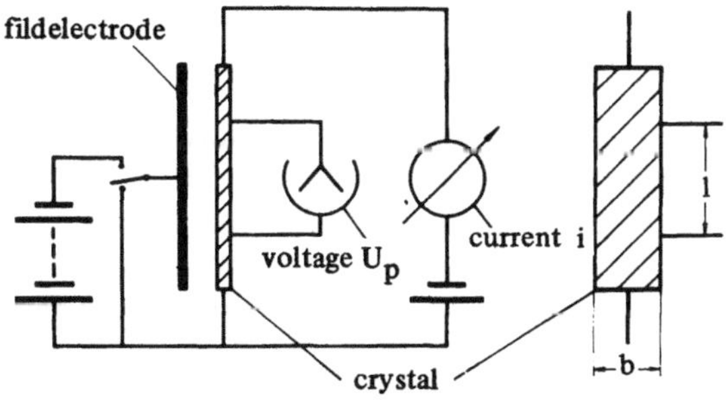

Fig. 5. Arrangement for field effect measurements. U_P voltage drop between potential probes with distance l, b width of the crystal surface.

a depletion layer. No surface acceptor is known so far which will drive the space charge layer to p-type conduction. Usually the adsorbed gases and the related surface states tend to lock the surface potential. The situation seems to be different with a clean cleaved prism surface in ultrahigh vacuum, which exhibits a p-type sign of the field effect at low temperature [5].

From Figure 5 the arrangement of a field effect measurement can be seen. The surface conductivity is changed by the charge which is induced by the field electrode on to the crystal surface. The sign of the change provides information on the sign of the charge carriers already present. The amount of the change allows under certain conditions an estimate to be made of the density of surface states [17].

The conductance of the crystal is defined as

$$g = i/U_P = g_B + g_S.$$

An induced charge varies only the surface conductance (see Fig. 5)

$$dg = dg_S; \quad d(\Delta\sigma) = \delta\sigma = dg \times l/b$$

Furthermore a sheet conductance is defined as [6]:

$$g_{\square} = e \int_{z=0}^{d} \mu n \, dz = \sigma_B d + \Delta\sigma.$$

d = thickness of the sheet contributing to the conductance g,
g_B = bulk conductance,
g_S = surface conductance,
$\Delta\sigma$ = surface conductivity,
μ = electron mobility,
n = electron concentration,
σ_B = bulk conductivity.

The integration is done in a direction normal to the surface or to the "sheet" (Fig. 4).

Photoconductivity of ZnO crystals has been observed in the intrinsic absorption region as a surface phenomenon which is closely connected with the surface conductivity in the dark. The explanation assuming electron traps is confirmed by field effect measurements [7]. At room temperature the photoconductivity is accompanied by photodesorption of oxygen and by photolysis [16], increasing the surface conductivity in the dark. Irradiation at low temperature results also in a rise of the dark surface conductivity, but transient warming to about 200 K restores the initial situation. This can be explained by trapping of holes at low temperatures without desorption of oxygen. Spectral sensitization of photoconductivity for the visible spectral region can be achieved by adsorption of organic dyes. The quantum yield exceeds that of the intrinsic absorption region by a factor of up to five. In these experiments the dye is evaporated in ultra-

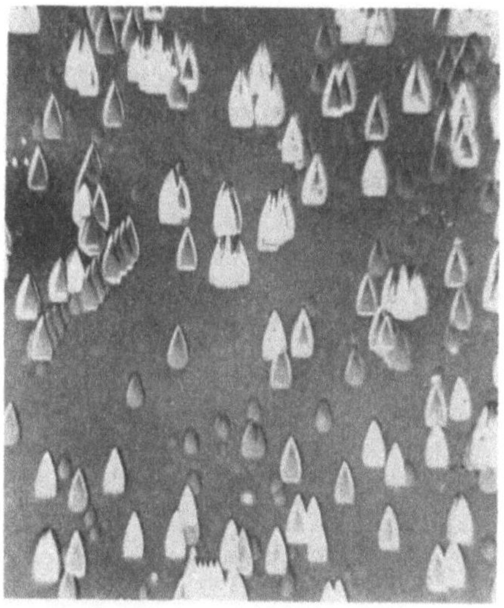

Fig. 6. Prism surface etched in HF, 40%. The etch pits point in the positive direction of the $\underline{c}$ axis. 50 x. After Klein [9].

high vacuum on to a clean crystal surface [8]. No solvent, no binder and no adsorbed oxygen contribute to the process of sensitization. More details will be given at the Conference at Sarlat [8].

The influence of the polar axis is not limited to the polar surfaces as can be seen from the etch pattern of a prism surface (Fig. 6).

3. Properties of Polar Surfaces

After cleavage normal to the c-axis the etch pattern of the two polar surfaces is very different. Figure 7 gives an example. The scanning electron microscope reveals the destruction of the oxygen surface by etching (Fig. 8). For electrical surface studies the crystals have been cleaved in ultrahigh vacuum (Fig. 9). Annealing in the same vacuum creates a high surface conductivity of the Zn side, whereas no noticeable change is observed on the O side (Fig. 10). Surface donors are created during the annealing procedure. Finally the density of conduction electrons in the accumulation layer amounts to the order of 10^{13} cm^{-2}. The decrease in surface conductivity above 250° C is attributed to the evaporation of surface donors. This was proved by the following experiment: After annealing at 250° C the oxygen side was moved close to the Zn side. During a subsequent heating to 400° C of this side alone a fraction of the donors was evaporated across a vacuum gap of 0.3 mm onto the O side and there increased the surface conductivity to a value which corresponds approximately to the loss on the Zn side (point e in Fig. 10). By heating of the O side these donors could be removed again (point h). It is well known that adsorbed Zn creates donors on prism surfaces at temperatures low enough to exclude diffusion processes [15], but the mechanism of donor generation on the clean Zn surface is still unknown. Furthermore gold contacts evaporated on to pairs of clean polar surfaces exhibit a surprising difference between the slopes of the capacity plot $1/C^2$ (U) as a function of the re-

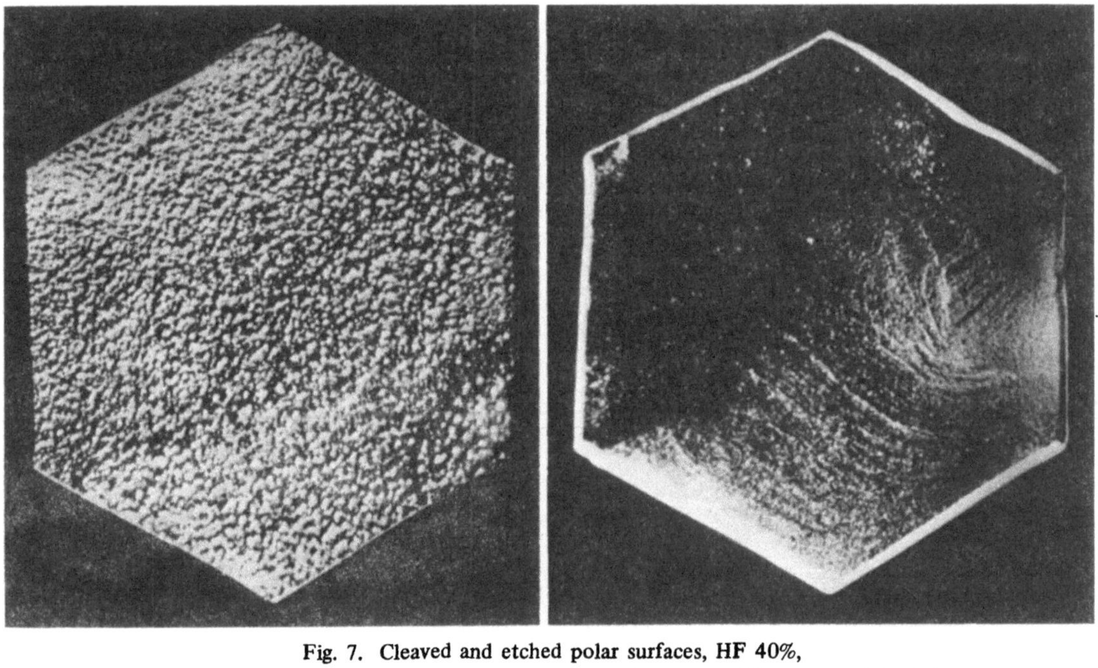

Fig. 7. Cleaved and etched polar surfaces, HF 40%,

(a) Zn surface. (b) O surface. 50 x, after Klein [9].

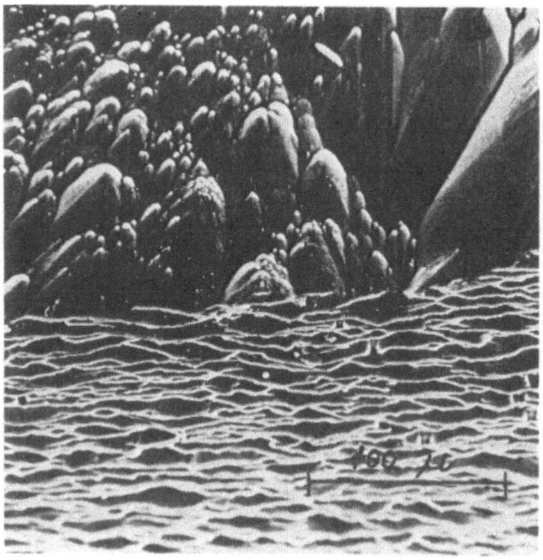

Fig. 8. ZnO crystal with Li doping after etching in HCl. Lower: part the prism surface, upper part: the O surface. Scanning electron microscope.

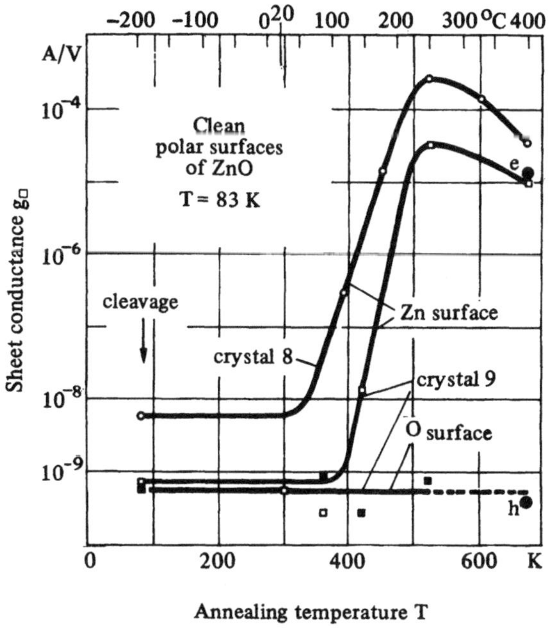

Fig. 10. Annealing curve for polar surfaces in ultrahigh vacuum. The bulk conductivity is decreased by a small addition of Cu. The two extra points e and h correspond to the conductance of the O surface after the reception of donors from the Zn side and after subsequent heating respectively. After Kohl [5].

verse bias U [12]. An explanation seems possible by the hypothesis of a quite unusual exchange of charge between surface states and the crystal across the depletion layer or by the assumption that during evaporation donors are created on the Zn surface which diffuse into the crystal at least to the depth of the space charge layer.

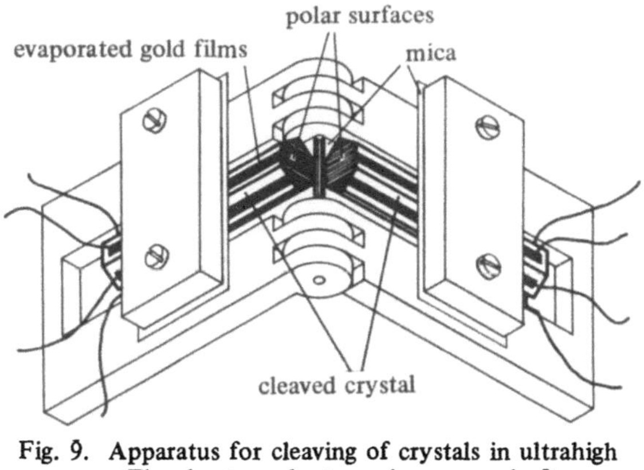

Fig. 9. Apparatus for cleaving of crystals in ultrahigh vacuum. The sheet conductance is measured after van der Pauw [11] with four evaporated gold electrodes. Before cleavage the whole crystal is covered by an SiO_2 film to stabilize the prism surfaces [10].

From field effect experiments (Fig. 5) at 86 K on clean cleaved polar surfaces a density of states near the *Fermi* level at the surface has been derived [5]. It changes along the annealing curve (Fig. 10) between very low values in the order of 10^{12} and high values of more than 10^{16} cm^{-2} $(eV)^{-1}$. No unique relation between surface conductivity and density of states has been found so far. The knowledge as to how to prepare a surface which is nearly free of surface states may be of some importance for the application of photoconductivity, since besides rise and decay also the efficiency of photoconductance may depend to a large extent on surface traps [7].

There is also a difference between the polar surfaces in the intrinsic photoconductance (Fig. 11). The O side behaves similarly to prism surfaces; also without photolysis a fraction of the photoconductance at low temperature does not vanish

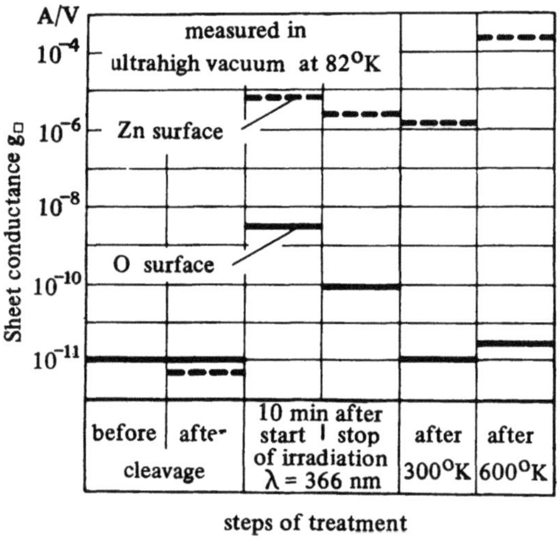

Fig. 11. Photoconductance of polar surfaces. Irradiation with light of the intrinsic absorption region. Cleavage, measurement and irradiation are performed in the same ultrahigh vacuum at 82 K. Light intensity 5×10^{16} quanta/cm^2 sec. The bulk conductivity is decreased by a small addition of Cu. After Kunstmann [10].

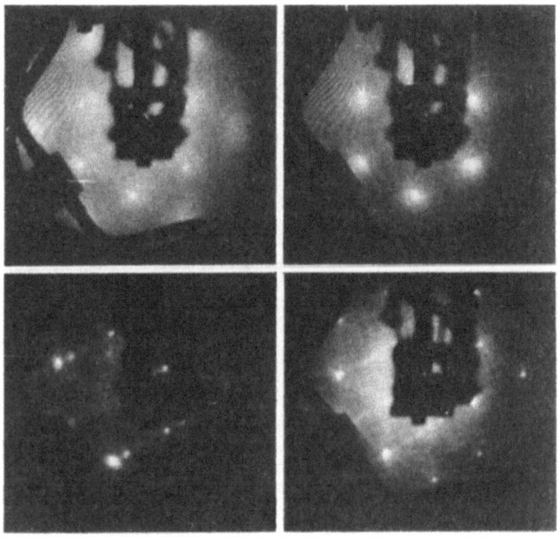

Fig. 12. LEED patterns of the polar faces after cleavage in ultrahigh vacuum.
(a) and (b) O surface immediately after cleavage. The sharpness of the spots varies with the electron energy of 95 eV (a) and 87 eV (b). This variation shows the presence of randomly oriented steps.
(c) O surface after annealing at 400 °C for 5 min. 92 eV. The extra spots not seen in (a) and (b) reveal regular step arrays with one orientation. No superstructure.
(d) Zn surface after annealing at 400 °C for 5 min. 150 eV. Regular step arrays in three directions [4].

with the irradiation. This has been explained with the trapping of holes. During subsequent warming the additional conductance vanishes below room temperature. However, the Zn surface suffers an irreversible increase of conductance by irradiation which remains also at room temperature in vacuum. This is not a small effect. The conductance increases by 6 orders of magnitude. During the final heat treatment the photoinduced conductance is concealed b the annealing process (Fig. 10). Perhaps the light induces a similar transition at the Zn surface as the annealing does.

On clean surfaces of other materials like germanium and silicon, superstructures have been observed with LEED (low energy electron diffraction). These superstructures are explained with the assumption of a surface atom rearrangement with a multiple periodicity distance of the bulk strructure. For the polar surfaces of ZnO such structures have been predicted [13] and also observed [14]. In our laboratory crystals have been cleaved in ultrahigh vacuum and no superstructure has been detected with LEED either before or after annealing treatments up to 850° C (Fig. 12). The observed patterns are explained by steps [18] which form regular arrays during annealing. No relation has been

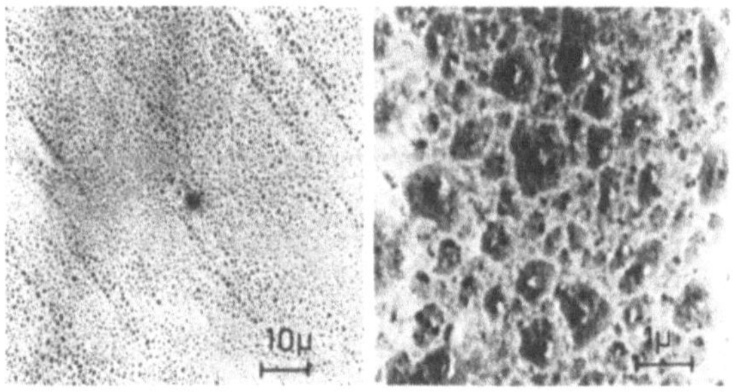

Fig. 13. Zn surface transferred to a scanning electron microscope after heating to 800 °C for 5 min in ultrahigh vacuum. (a) and (b) show two different magnifications.

found between the electric transition of the Zn surface at 200 °C and the diffraction pattern [4].

After an annealing at 800 °C in ultrahigh vacuum a Zn surface has been transferred into the scanning electron microscope. Figure 13 shows photographs of about the same region of the Zn surface. There might be a correlation between the formation of regular arrays of steps during the heat treatment (Fig. 12), the evaporation of Zn (Fig. 10) and the structures in Figure 13. At present however, we have only some preliminary results. A more extended study is in progress.

4. Conclusion

Some unsolved problems of the ZnO surface have been presented in this paper. There are no methods to determine the flat band potential, which is very important for many surface experiments. The minimum of surface conductivity cannot be used for this purpose like in other semiconductors whose surface conductance can be driven from n to p-type. Some information on the absolute value of band bending might be derived from surface photovoltage and from electroreflectance. A high irradiation intensity of band gap light, however, may cause photolysis and change the object which is to be measured. Thus the surface of ZnO still offers many possibilities for research. An improved understanding will not only increase our knowledge of II–VI compounds, of chemical bonding and of electronic surface states, it may also help in such important fields as heterogeneous catalysis or electrophotography.

Acknowledgements

The author wishes to thank Priv.-Doz. Dr. M. Henzler and Dipl.-Phys. D. Kohl for helpful discussions. Ing. F. Wetzels made the work possible by growing large and perfect crystals. The scanning electron micrographs were taken by Dipl.-Phys. L. Balk.

References

[1] G. Heiland, E. Mollwo, F. Stoeckmann, in Solid State Physics, Eds. F. Seitz and D. Turnbull, Vol. 8. Academic press, New York, 1959, p. 191.
[2] U. Rössler, Phys. Rev. **184**, 733 (1969).
[3] A. Hausmann, Z. Physik **237**, 86 (1970).
[4] G. Heiland, M. Henzler, D. Kohl, unpublished.
[5] D. Kohl, Aachen, unpublished.
[6] G. Heiland, P. Kunstmann, Surface Sci. **13**, 72 (1969).
[7] G. Heiland, J. Phys. Chem. Solids **22**, 227 (1962).
[8] W. Bauer, G. Heiland, J. Phys. Chem. Solids **32**, 2605 (1971).
[9] A. Klein, Z. Phys. **188**, 352 (1965).
[10] P. Kunstmann, Thesis, Aachen, 1967.
[11] L. J. van der Pauw, Philips Res. Rept. **13**, 1 (1958).
[12] H. Harreis, G. Heiland, Surface Sci. **24**, 643 (1971).
[13] R. W. Nosker, P. Mark, J. D. Levine, Surface Sci. **19**, 291 (1970).
[14] M. F. Chung, H. E. Farnsworth, Surface Sci. **22**, 93 (1970).
[15] D. G. Thomas, J. J. Lander, J. Phys. Chem. Solids **2**, 318 (1957).
[16] R. J. Collins, D. G. Thomas, Phys. Rev. **112**, 388 (1958).
[17] G. Heiland, J. Phys. Chem. Solids **6**, 155 (1958).
[18] M. Henzler, Surface Sci. **22**, 12 (1970).

Discussion

Weigl: Is the efficiency of dye sensitization greater for the O-surface or the Zn surface of the ZnO crystal?

Eckenbach: In which way did you determine the quantum efficiency of your ZnO crystals, and which values did you obtain?

Heiland: Sensitization of photoconductivity on polar surfaces has not yet been studied so far. The quantum efficiency was derived from the initial slope of photoconductivity. The absolute values depend on the trap density at the Fermi level. Therefore it is more helpful to compare the quantum yield for the spectrally sensitized and for the intrinsic photoconductivity. Both are measured after evaporation of the dye and referred to the absorbed photon flux. The comparison includes the assumption that the same density of electron traps is active for the sensitized and for the intrinsic photoconductivity. The ratio of the yields has been found as high as five. More details are given elsewhere; G. Heiland, W. Bauer and M. Neuhaus, International Conference on Photosensitization in Solids, Sarlat (France) 1971, and Ref. [8].

Electronic and Structural Properties of the ZnO polar Surfaces

R. Leysen and H. van Hove

Abstract

Surface conductivity variations were measured on the ZnO polar surfaces essentially for the oxygen-vacuum cycle of ambients in the dark. Freshly prepared (0001) surfaces showed an increase in surface conductivity under oxygen. Correlating those measurements with LEED results on similar samples led to the conclusion that oxygen partially takes over the surface compensation by intrinsic surface states and causes them to emit their charge into the conduction band. After long waiting times in oxygen at R.T. or after annealing in oxygen at 600° C, the surface compensation is definitively taken over by oxygen and the increase in conductance under oxygen changes to a decrease. The surface state density was estimated to be $6 \cdot 10^{11} cm^{-2}$ and they are situated at least 130 mV below the conduction band.

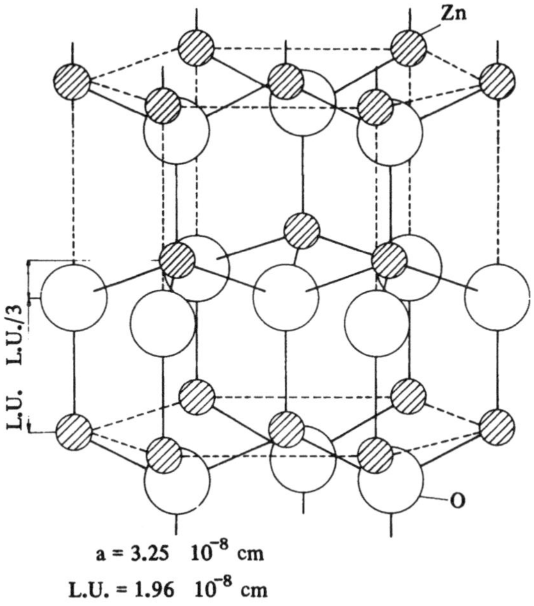

Fig. 1. Conventional unit cell of the wurtzite structure.

1. Introduction

The hexagonal wurtzite structure of ZnO is noncentrosymmetric and consequently exhibits a polar c-axis. In Figure 1 the conventional unit cell is shown: each atom is tetrahedrally surrounded by the others. The nearest anion-cation distance generally is denoted as a lattice unit (L.U.) and equals $1.96 \cdot 10^{-8}$ cm. The structure can be thought of as a series of stacked double layers. Measured along the c-axis, the distance between the O and the Zn atoms in the same double layer is 1/3 L.U. The distance between an O atom and a Zn atom from the next nearest double layer is 1 L.U. Finally, the distance between the atoms in a polar plane $a = 3.25 \cdot 10^{-8}$ cm.

The density of bonds is greater between the planes of the same double layer, and three fourths of the total binding energy is involved in this interaction. It is therefore expected that the preparation of polar planes would cause the crystal to separate between the double layers. Thus the (0001) plane should consist of Zn atoms in the outer layer and the $(000\bar{1})$ plane of O atoms.

The crystals as grown were in the form of long needles with hexagonal cross section, broken from a substrate at one end, and presenting steep pyramidal forms at the opposite end. X-ray absorption edge data established [1] that the steeply terminated end corresponds to the [0001] direction. Since a fast growing plane generally tends to disappear, leaving behind slower growing forms with lower surface energy, the (0001) plane as such should represent a rather unstable high-energy configuration.

Recently, Nosker et al. [2] calculated from electrostatics and a simple ionic model, that a polar surface termination is indeed highly unstable: a potential gradient exists, limited to the surface region, that calls for compensation by surface charge of one fourth of the bulk layer charge density, negative on the (0001) side and positive on the $(000\bar{1})$ side. Uncompensated surface configurations have no physical meaning because the Zn atoms on such a surface would be in a local positive potential and the O atoms in a negative potential. In the same paper different ways to achieve this surface compensation on clean surfaces are discussed, including surface reconstruction and faceting, and evaluated according to their ionic surface energy.

Reconstruction and faceting on the ZnO polar surfaces was observed by Chung and Farnsworth [3] after ion bombardement and annealing by LEED techniques. In a previous paper [4] some LEED evidence was given for the cleaved polar surfaces to be not reconstructed and consequently it was suggested that surface stabilization there is achieved by charge storage in intrinsic surface states. Theoretical calculations [5] have shown that intrinsic surface states indeed should exist on such clean polar surfaces and recently some indirect experimental evidence for their existence was found from field emission work [6].

Further it has been shown [4] that pre-annealing in oxygen of otherwise "real surfaces" causes them to diffract relatively easily and causes $(\sqrt{3} \times \sqrt{3})$ superstructures to appear on both sides. This $(\sqrt{3} \times \sqrt{3})$ pattern was very stable on the (0001) face as it could not be removed by heating or by weak argon-ion bombardement, while on the $(000\bar{1})$ face, it was easily removed by a short heat treatment. Also the intensity vs. energy modulation was different: the $\sqrt{3}$ beams on the (0001) face modulate nearly together with the substrate beams, while on the $(000\bar{1})$ face the modulation of the $\sqrt{3}$ beams and the (01) was completely different. It was concluded that oxygen on the $(000\bar{1})$ face was simply chemisorbed while on the (0001) face a stronger interaction occurs: oxygen takes part in the stabilisation of this surface by storage of the negative charge.

The polar surfaces thus show up to be relatively clean (diffracting easily) and stabilised differently after annealing in oxygen: therefore surface conductivity measurements were felt worthwile. The results of these measurements are reported in this paper.

2. Experimental

The material was in the form of long needles with hexagonal cross section. The polarity of the c-axis can be determined from the different etching characteristics of the polar surfaces [1]. Some of the needles showed up to be twinned along the c-axis: the cross section consists partially of (0001) and $(000\bar{1})$ areas. Figure 2 shows a scanning electron microscope (SEM) picture of such a cross section: those needles were disregarded in this work. Among the others, those with a homogeneous bulk resistivity of 5 Ohm cm were selected. After optical aligning, thin slabs were spark-cut to within one degree perpendicular to the c-axis. Those slabs were shaped in a final rectangular form of about 0.5 x 0.3 cm and mechanically polished in the usual way. The final thickness was a few 10^{-2} cm. The polishing damage was removed by chemical etching, HCl 30% for 15 seconds on the $(000\bar{1})$ and H_3PO_4 85% for 20 minutes on the (0001) face. At this stage, especially the $(000\bar{1})$ face is completely covered by facets. Chemical polishing, H_3PO_4 85% for 60 minutes on the $(000\bar{1})$ and HCl 5% for 5 minutes on the (0001) finally restores the polar surfaces

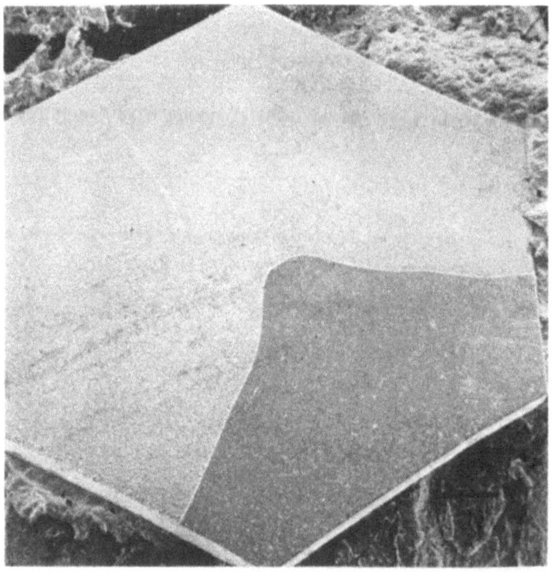

Fig. 2. S. E. M. photograph of a twinned cross section of a ZnO needle. The dark area is [0001] oriented. The HCl etch only removes material from the (000$\bar{1}$) part and makes it faceted. 20 x.

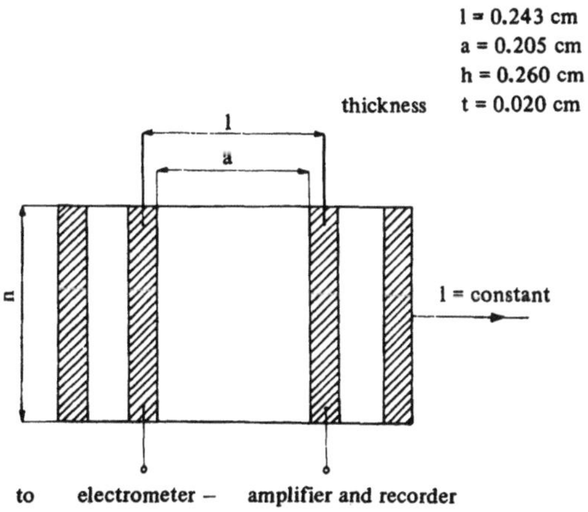

Fig. 3. The four-contact arrangement for $\Delta\sigma$ measurements.

and gave surfaces of mirror-like quality except for some macroscopic etchpits on the (0001) face: Figures 5 and 6. Prior to the measurement, four gold strips are evaporated on the surface, the back surface of the same sample being shielded from ambient changes by a layer of SiO, (Fig. 3).

A temperature-resistant sample holder was prepared consisting of a hollow copper block where any desired temperature between R.T. and 95° C could be obtained by thermostatted water circulation. Four adjustable tungsten wires (10^{-2} cm) pressed the sample on to the gold strips against the heat source. The thermal contact between the sample and the copper block was improved by a layer of vacuum grease when working in the 10^{-3} Torr range: temperatures were measured by a thermocouple on the copper block. A quartz window allowed for eventual ultraviolet irradiation.

For the actual measurement, the voltage developed on the two probe contacts by a constant current flowing through the two outer contacts, was picked up in an electrometer amplifier and the changes recorded: Figure 3. The measurements performed this way are entirely free from possible errors due to changes in contact resistance and reversing the current therefore had no effect on the results; photovoltages were found to be absent.

3. Experimental results

The surface conductivity variations were measured for the (oxygen – vac. or nitrogen – light) cycle of ambients. Very soon it became evident that the interesting pressure range for the actual measurements was 10^{-3} to 760 Torr and that vacuum of 10^{-3} to 10^{-8} Torr or nitrogen were equivalent for the reported results.

The usual behavior, measured intensively on the prism faces of ZnO, shows with intrinsically absorbed light (366 nm) an increase in surface conductivity due to hole trapping by chemisorbed O_2^- ions [7]. This behaviour will be referred to as a "positive" effect. Some of the surface preparations used in the course of this work cause the sample to show a so called "negative" effect. This behaviour is shown in Figure 4 for a (0001) sample at 80° C prepared by chemical etching and chemical polishing as described previously. When the surface is in oxygen, the

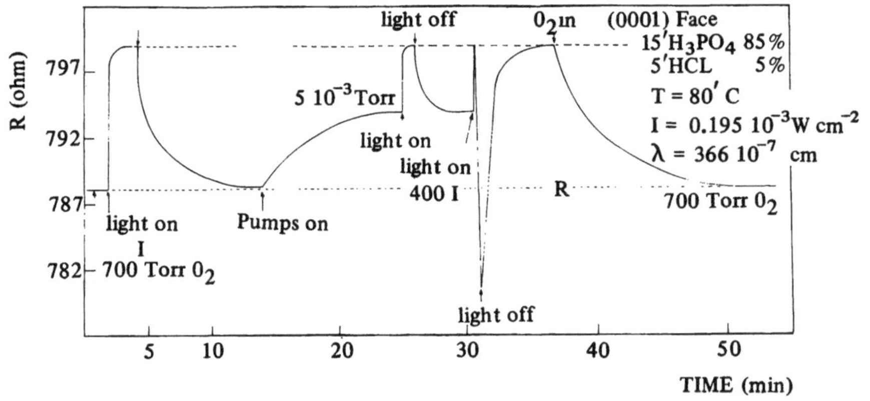

Fig. 4. (0001) face, 80° C. The negative photo surface conductivity and the corresponding decrease in resistance ΔR for the Vac-oxygen cycle of ambients.

photosurface conductivity is negative: under light the resistance increases over a certain amount ΔR and the subsequent relaxation is complete. Changing the ambient to a vacuum of 10^{-3}Torr produces a positive ΔR in the same direction and the surface photoconductivity change stays negative over the remaining part of ΔR. Increasing now the light intensity while still in vacuum, finally causes R to decrease, and after the subsequent relaxation the dark level is definitively changed by ΔR: the true equilibrium level in vacuum is R + ΔR and the corresponding negative effect has vanished.

When going back to the oxygen ambient, R + + ΔR → R and the dark resistance decays again. Therefore, light is not essential for the observed phenomena; the true equilibrium levels are R in oxygen and R + ΔR in vacuum: *the presence of oxygen in the dark makes the resistance decrease.*

From the experimental point of view, the correlation between the dark- and photo-surface conductivity changes is interesting: in the dark eventual photovoltages at the contacts are excluded while under light eventual bulk temperature changes of the sample are excluded, therefore both of these possible sources of error are negligible.

The experimental results as far as the sign of R changes for the cycle of ambients $N_2 - O_2$ in the dark is concerned, are summarised in Table I, and the SEM photographs of the surfaces after the different surface pretreatments are given in Figures 5 to 9.

After the combined etching with HCl and H_3PO_4, the surfaces show up to be flat: the (0001) face (Fig. 6) shows large flat areas separated by steps and large etchpits; the (000$\bar{1}$) surface (Fig. 5) shows a quite important amount of globular impurities on an otherwise flat background.

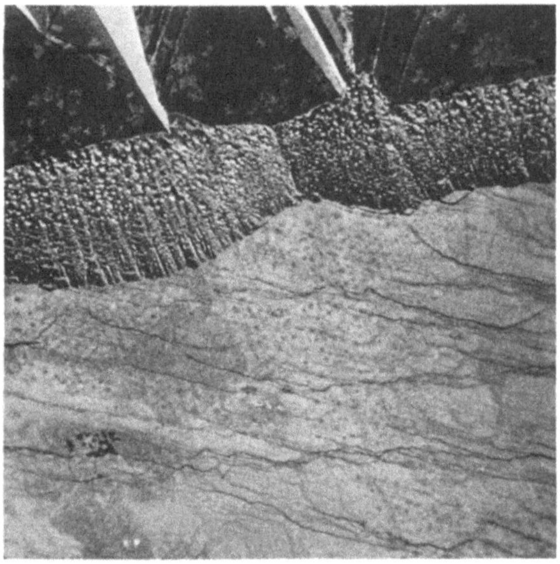

Fig. 5. (000$\bar{1}$) face. A large amount of globular impurities on a flat background. 1030 x.

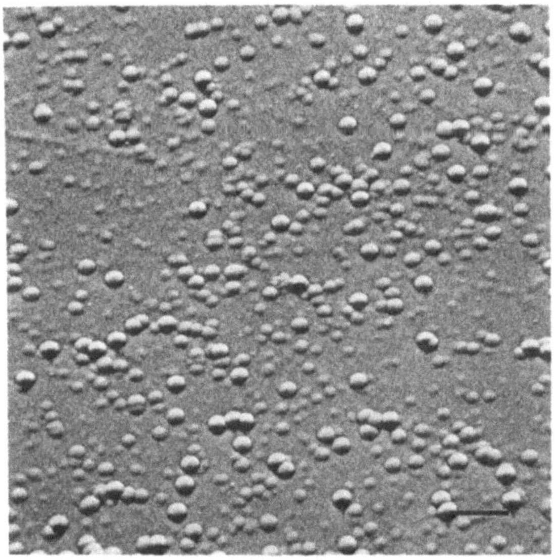

Fig. 6. A (0001) face near the border of a macroscopic etchpit. Large flat areas separated by steps. 180 x.

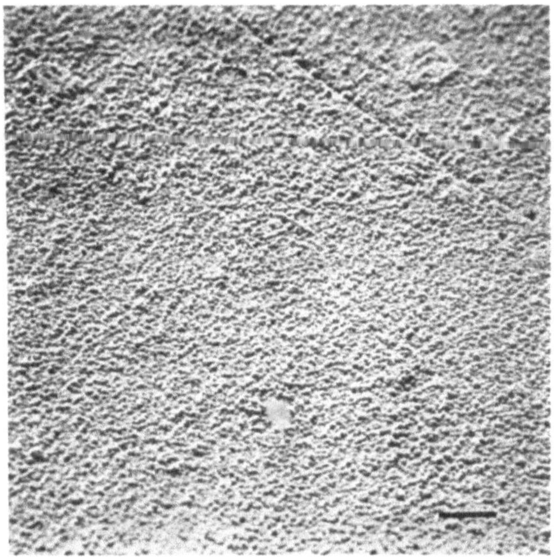

Fig. 8. Facets on $(000\bar{1})$ produced by an HCl etch. 240 x.

Guided by the LEED results [4] showing that after annealing in oxygen those surfaces were very easy to make diffracting and quite stable, the same samples were similarly annealed: the amount of globular impurities decreases drastically on the $(000\bar{1})$ face (Fig. 7); on the (0001) faces no change could be detected by SEM.

On those annealed surfaces, the two steps of the initial etching were repeated separately, again leaving the (0001) face practically un-

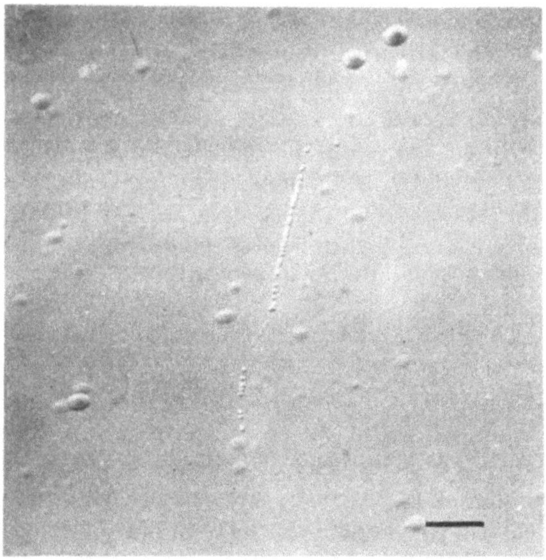

Fig. 7. The same as Fig. 5 but after oxygen annealing: most of the impurities disappeared. 220 x.

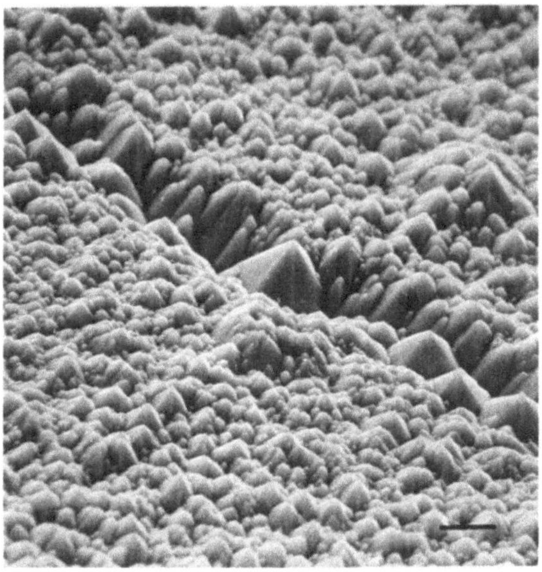

Fig. 9. The same as Fig. 8 but 1100 x.

changed, but the HCl etching on the $(000\bar{1})$ face produced very pronounced faceting as can be visualised on Figures 8 and 9 taken on the same surface with increasing magnifications.

The only situations where a negative ΔR is recorded are the faceted $(000\bar{1})$ face and the (0001) faces after H_3PO_4 etching. On these however, the ΔR changes slowly to positive in about 24 hours storage time in oxygen.

Table 1. Surface Resistance Changes for Various Treatments

$(000\bar{1})$ oxygen face		(0001) Zn face	
Pretreatment	ΔR	Pretreatment	ΔR
HCl, 30%, 15″ H_3PO_4, 85%, 60′ Figure 5	+	H_3PO_4, 85%. 20′ HCl, 5%, 5′ Figure 6	(*) $- \to +$
idem + annealing, O_2, 760 Torr, 600° C, 6 hrs Figure 7	+	idem + annealing, O_2, 760 Torr, 600° C, 6 hrs	+
idem + annealing + HCl, 30%, 15″ Figures 8 and 9	–	idem + annealing + HCl, 5%, 5′	+
idem + annealing + H_3PO_4, 85%, 60′	+	idem + annealing + H_3PO_4, 85%, 20′	(*) $- \to +$

(*) The behaviour changes to positive with increasing time in oxygen. Different surface treatments and the corresponding sign of the resistance changes for the N_2-O_2 cycle of ambients in the dark.

4. Discussion

It was shown [4] by LEED work that on the (0001) faces, the lattice compensation for the stability criterion to be fulfilled, can probably be taken over by oxygen. On cleaved surfaces, the compensating negative surface charge is trapped in intrinsic surface states, whereas annealing in oxygen at the same temperature and for the same time as used here, was shown to produce a very stable $(\sqrt{3} \times \sqrt{3})$ oxygen pattern that resisted heat treatment and weak argon-ion bombardement. If indeed the oxygen-stabilised (0001) surfaces are more stable than the surface state compensated surfaces, one should expect a gradual transfer of the compensating charge from the surface states to the adsorbed oxygen, when a freshly prepared surface is put in contact with an oxygen ambient, and a complete transfer when it is annealed.

Annealed (0001) surfaces, which are thus completely oxygen stabilised, always show a positive effect (Table 1). Therefore a negative ΔR corresponds to a surface situation where the compensation is still at least partially achieved by surface states. Such negative ΔR values are encountered on the (0001) surfaces only when previously etched in H_3PO_4 and not after etching in HCl. This is what one should expect, because only H_3PO_4 is known to remove some material from these surfaces thus producing "new surfaces" while HCl only polishes them. Those surfaces being relatively flat after the usual pretreatment (Fig. 6), HCl actually does nothing.

Therefore the gradual change from negative to positive ΔR with increasing time in oxygen is indicative for the stabilization of these surfaces to change from surface state compensated to oxygen compensated. The negative ΔR itself is to be attributed to the presence of negative charge in surface states. There are thus two different oxygen interactions: a reversible chemisorption that causes the actually existing sign of ΔR to appear, and a more stable one that causes the sign of ΔR to change from negative to positive, gradually when held at room temperature and more rapidly when annealed.

The change in surface charge is given by:

$$\Delta N = \frac{l^2}{e\,A\,\mu}\,\frac{\Delta R}{R^2} = 8.0 \cdot 10^{11}\,\mathrm{cm}^{-2}$$

where l is the distance between the potential probes, A is the surface area actually measured and μ is the mobility of the electrons taken as 140 $cm^2 V^{-1} sec^{-1}$ [8]. ΔR is measured (Fig. 4) and found to be 10.05 Ohm while R was 790.0 Ohm.

The maximum negative surface charge that could exist on ZnO of this bulk resistivity ($n_\infty^- = 6.6 \; 10^{15} cm^{-3}$) is $1.7. \; 10^{11} cm^{-2}$ (7). Therefore a surface charge change of 8.0. $10^{11} cm^{-2}$ brings the surface into enrichment conditions.

At this stage, contact potential difference (CPD) measurements were undertaken for the same cycle of ambients (N_2-O_2) on a freshly H_3PO_4 etched (0001) sample in a classical vibrating reed system and found to be 100 ± 10 mV. It should be noted that on annealed and HCl etched samples the CPD was indeed of the opposite sign for the same ambient cycle thereby confirming the $\Delta\sigma$ results of Table 1.

Knowing that the CPD is about 100 mV, that the surface is in enrichment conditions and that the corresponding ΔN is 8.0 $10^{11} cm^{-2}$, allows the surface potential in oxygen to be calculated from [9]:

$$\Delta N = \left(\frac{2\,kT\,\epsilon\epsilon_0}{e^2}\,n_\infty^-\right)^{1/2}\left(\exp\frac{e}{kT}(\varphi_0 - \varphi_\infty)\right)^{1/2}$$

where ϵ is the dielectric constant of ZnO and equals 8.5 [8], ϵ_0 is the permittivity of vacuum $8.85 \cdot 10^{-14}$ F cm^{-1}, e is the electronic charge, k is the Boltzmann constant, n_∞^- the bulk density of free electrons, $\varphi_0 - \varphi_\infty$ is the surface potential. One finds $\varphi_0 - \varphi_\infty$ in oxygen to be +150 mV; consequently in vacuum the surface potential should be about +50 mV.

The proposed mechanism corresponding to a negative R change when oxygen is admitted on a (0001) face can be summarised as follows. Starting in vacuum the surface is surface state compensated: to a first approximation the surface is neutral. When oxygen is admitted on such a surface it chemisorbs and traps conduction band electrons thereby raising the bands by a maximum of 500 mV [7], the charged surface states are pushed over the Fermi level (at 180 mV from the conduction band in the bulk) and emit their charge. Further one has to assume that oxygen obliterates the surface states and their emitted charge finally makes the surface more enriched. The measured ΔN thus corresponds to the charge emitted out of the surface states N_{ss} minus the charge trapped in chemisorbed oxygen. This last one is at most $1.7 \cdot 10^{11} cm^{-2}$; a minimum density of the charged surface states therefore is given by:

$$N_{ss} = \Delta N - 1.7 \cdot 10^{11} \cong 6 \cdot 10^{11} cm^{-2}$$

and the surface states are situated by 180–50 = = 130 mV below the conduction band.

5. Conclusion

Nothing was said about the (000$\bar{1}$) face. It only shows a negative ΔR when completely faceted; on such a surface the (000$\bar{1}$) plane actually is absent, therefore any correlation with LEED results is meaningless.

On the (0001) surfaces an interplay between two surface compensation mechanisms was clearly demonstrated. The main reason for this effect to be measurable is the great difference in time constant for the two processes: oxygen can make the surface states to emit in about 15 min., depending on the temperature, oxygen takes over the surface compensation definitively only after at least 24 hours. The time- and temperature dependence of these two processes is now under study.

Acknowledgements

The authors are much indebted to R. Deroost and G. Engelbosch for their skilful technical assistance. Many helpful and stimulating discussions with A. Neyens and R. van Overstraeten are gratefully acknowledged. In the early stages of this work, some parts were discussed with G. Heiland and S. Morrison, R. Helbig of the Erlangen University provided kindly the ZnO crystals and the SEM pictures were taken at the Centre of Electron Microscopy at the Leuven University.

References

[1] A. Mariano and R. Haneman, J. Appl. Phys. **34**, 384 (1963); G. Heiland, P. Kunstmann and H. Pfister, Z. Phys. **176**, 485 (!963).
[2] R. Nosker, P. Mark and J. Levine, Surf. Sci. **19**, 291 (1970).
[3] M. Chung and H. Farnsworth, Surf. Sci. **22**, 93 (1970).
[4] H. van Hove and R. Leysen, submitted to Phys. Stat. Sol. *(a)*.
[5] S. Davison and J. Levine, Solid State Phys. **25**, 2 (1970).
[6] J. Marien, R. Leysen and H. van Hove, Phys. Stat. Sol. *(a)* **5**, 121 (1971).
[7] R. Collins and D. Thomas, Phys. Rev. **112**, 388 (1958).
[8] A. Hutson, Phys. Rev. **108**, 222 (1957).
[9] H. Krusemeyer and D. Thomas, J. Phys. Chem. Solids. **4**, 78 (1958).

Discussion

G. Heiland: I would like to mention that the influence of oxygen on the surface conductivity of the polar surfaces has been studied by Heiland and Kunstmann [*Surf. Sci.*, **13**, 72 (1969)] only after previous treatment of the surfaces with atomic hydrogen. The high surface conductivity created by the hydrogen is decreased by oxygen with different speed on both surfaces. In the same paper, measurements with a mercury contact are reported. It is shown that a depletion layer appears after exposure to air on both surfaces, but stronger on the (0001) Zn surface.

H. Meyer: Have you made any measurements on the photoelectric work functions of Zn-surfaces and O-surfaces? Are there any differences?

H. van Hove: Work function differences for the $(000\bar{1})$ and (0001) faces were measured by Field Emission Techniques. [*Phys. Stat. Sol (a)* **5**, 121 (1971)] and found to be about 4 eV for the $(000\bar{1})$ and 5 eV for the (0001) oriented tips.

On the Dark Decay of Negatively Charged ZnO Single Crystals

H. Kiess

Abstract

A discussion is first given of various possible processes for the dark decay of electrostatically charged materials. Then, charging and discharging experiments with conducting and insulating ZnO crystals are described. The dark decay of the surface voltage was found to obey logarithmic time laws for both types of crystals. It is shown in the discussion that during the charging a photocurrent is observed which is caused by the light emitted from the corona. The dissipation of the charge in the dark after the switching off of the corona was found to be due to the not yet completely decayed photocurrent. Since the time law is logarithmic the decay of the charge, i.e. also of the photocurrent, is due to a continuum of states whose density is constant in energy between the conduction and valence band.

1. Introduction

The dark decay of negatively charged ZnO-binder layers was investigated by Hauffe and coworkers [1]. Their experimental surface voltage versus time curves can be best fitted by a logarithmic time law which was interpreted by the authors by a field dependent hopping conduction of injected electrons through the layers. Other authors [2, 3] explained the dark decay of charged Electrofax layers by a tunnelling process of charge carriers through the surface barrier.

In order to avoid the complication of the complex ZnO grains-organic binder system we decided to study the dark decay of charged ZnO single crystals. The simplicity of this system should alleviate the difficulties of experimental interpretation. The rather extended work by Morrison [4] was devoted to the determination of the surface states associated with chemisorbed species on ZnO single crystals by evaluating the dark decay of electrostatically charged ZnO crystals. The interpretation of his results, of course, implies that the details of the mechanism for the dark decay are known. Otherwise, a determination of the energy levels of the surface states would not be possible. In fact, Morrison observed a variety of time laws for the decay of the surface voltage from which he selected the one with a linear relationship between the square root of the voltage and time as easily interpretable and therefore suited for the evaluation of his results.

In order to show the complexity of the problem, it seems to be best to give in the following section first a description of the different physical processes which might lead to a dissipation of the surface charge and to state qualitatively the time laws to be expected without going into the details of a mathematical derivation. Thereafter, our experiments will be described and the results will be compared with the anticipated discharge processes.

2. Physical Processes for the Dissipation of the Surface Charge in the Dark

The dissipation of the surface charge may take place from the surface itself, from the back electrode or from the bulk. Obviously the processes which might occur on the surface may also take place at the back electrode, and the answer to the question from where the charge is dissipated can only be given if certain experimental measures are taken to exclude carrier injection from the back electrode. This will be assumed to be true throughout these considerations for the sake of simplicity and clarity. For ZnO single crystals this assumption is fulfilled if the crystals are conducting and if a Schottky barrier is formed during the charging process.

2.1. Injection of Charge Carriers from the Surface

Let us assume that the band model shown in Figure 1 holds after the ZnO single crystal was electrostatically charged. The negative surface potential is indicated by $-V$, surface states associated with the negative ions used for the charging are indicated by I^- at an energy E_I. In addition, neutral species might be present on the crystal surface which become positively charged upon injecting an electron into the crystal. These levels are indicated by N^x at an

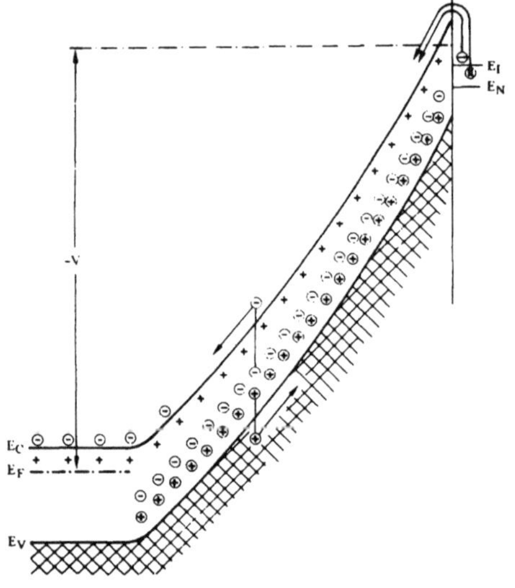

Fig. 1. Band model of conducting ZnO after charging in a corona discharge. Since a highly non-equilibrium situation prevails, the occupation of states in the depletion layer and on the surface cannot be described by a Fermi level, but rather it is given by the generation and recombination processes during charging.

energy E_N. In both cases, whether the electron is injected from the negative ion or from the neutral species, the surface charge is decreased by a unit charge. The negative surface charge is compensated by a positive space charge which is assumed to be equal to it. Due to the band bending $\left(V \gg \frac{kT}{e}\right)$ no electrons will be able to move from the bulk to the surface. Thus changes of surface potential with time are expected to occur due to irreversible electron injection from surface states into the conduction band.

Let us assume that the negative ion surface density is I^-. Then the rate of change of I^- is given by

$$\frac{dI^-}{dt} = -p_I I^- \tag{1}$$

where p_I is the probability per unit time for the emission of an electron into the conduction band. If however the electron is emitted from neutral species on the surface involving a charge transfer reaction such that $N^x \rightarrow N^+ + e$, we have for the rate of change of the neutral species:

$$-\frac{dN^x}{dt} = \frac{dN^+}{dt} = p_N(N_0^x - N^+) \tag{2}$$

where N_0^x is the initial density of neutral species at time zero and p_N the probability of the charge transfer. The rate of change in the surface charge density Q_s is then given by

$$\frac{1}{e}\frac{dQ_s}{dt} = \frac{dI^-}{dt} - \frac{dN^+}{dt}$$

In the following two sections 2.1.1) and 2.1.2. we will discuss cases in which (dQ_s/dt) is either due to the rate of change in ion density I^- or of the neutral species N^x, i.e. we assume that one of the two processes dominates. If both are of similar magnitude, then we have simply to add both rates, since no interaction between the two species I^- and N^x is assumed to take place.

2.1.1. Thermal activation of Electrons from Surface States

The decay mechanisms of the charge under these conditions are summarized in Table 1a. The simplest case is the one in which the probability p_I and p_N in equation (1) and (2) is constant in time. This is true if the electron is thermally activated from surface states of a well defined energy. Integration of equations (1) or (2) then gives an exponential time law for the decay of the surface charge. A linear time law will be obtained if in equation (2) the density N_0^x is much larger than N^+ during the whole decay process. This may happen if the deposited ion density I_0^- is small compared to N_0^x. Since in the discharged state $Q_s = 0$, the maximum value which N^+ may attain is equal to I_0^-, which was supposed to be small compared to N_0^x. If the surface states are continuously distributed in energy, the probability for the emission decreases exponentially with increasing separation of the energy level from the conduction band edge. In this case we have to evaluate (1) and (2) for all energies E between conduction and valence band. If the density of surface states per unit energy is assumed to be constant, we obtain upon integration a logarithmic time law. Mathematical details of this case are presented in the appendix.

2.1.2. Field Dependent Injection of Electrons into the Conduction Band

We assume now that the probability for the charge injection into the conduction band depends on the electric field E_s in the ZnO immediately beneath the surface and therefore on the surface charge density eI^-, since $|\vec{E}_s| = |(e/\epsilon\epsilon_0 I^-|$. The decay of the surface charge is now mathematically more complicated and the integration of equation (1) is not straightforward. We will not go into details of a derivation of the time laws, but only quote the physical processes and the resulting implications we expect. A summary is given in Table 1b. There are three processes which are strongly field dependent, namely the tunnelling process, the lowering of the surface barrier by the Schottky effect which is due to the image force, and the lowering of the surface barrier by the mutually repulsive electrostatic forces of the deposited ions. Under certain assumptions logarithmic time laws can be derived for all three processes. It is important, in this context, to note that the decay rate depends exponentially on the charge density, i.e. it depends upon the magnitude of the surface voltage to which the crystals were charged. Thus the decay rate drops drastically with decreasing initial voltage.

2.2. Discharge by Bulk Processes

During charging a non-equilibrium situation exists in the depletion layer, and the occupation of the states between the conduction and valence band cannot be described by a Fermi level. The reason for this will be discussed later in section 5. Therefore, after the switching off of the corona, deep-lying donors and acceptors may release electrons and holes into the conduction and valence band respectively. An emission of holes into the valence band will change the surface charge and—if the trapped electrons are immobile—also the space charge density. The emission of holes from the acceptor states A^+ into the valence band is given by

$$\frac{dA^+}{dt} = -p_A \cdot A^+. \tag{4}$$

Let us assume that the free holes are swept to the surface without retrapping and recombination losses then we find for the rate of change of the surface charge density

$$\frac{1}{e}\frac{dQ_s}{dt} = \frac{d}{dt}\int_0^d A^+ dx = p_A \cdot \int_0^d A^+ dx \cdot \tag{5}$$

If a thermal emission process is dominant, and if the acceptors can be assigned to only one energy level, then the integration will give an exponential time law for the decay of the surface charge. Similarly to 2.1.1, we will find a logarithmic time law in the case of acceptors being continuously and uniformly distributed in

Table 1a. Discharge by Thermal Processes Occuring at the Surface

Physical Process and Emission Probability	Rate of Change of Surface Charge	Decay Law for the Surface Charge
Thermal activation of electron from the deposited ion $P/_I = a_I \cdot e^{-E_I/kT}$	$\frac{1}{e}\frac{dQ_s}{dt} = \frac{dI^-}{dt} = -a_I \cdot e^{-E_I/kT} \cdot I^-$	$\frac{1}{e} Q_s = I^-(t) = I_0^- \cdot e^{-p_I \cdot t}$
Thermal activation of electron from neutral species $p_N = a_N \cdot e^{-E_N/kT}$	$\frac{1}{e}\frac{dQ_s}{dt} = -p_N \cdot e^{-p_N \cdot t} \cdot N_0^x$	$\frac{1}{e} Q_s = I_0^- - N_0^x (1 - e^{-p_N t})$
Thermal activation of electron from neutral species with assumption $N_0^x \gg N^+$ $p_N = a_N \cdot e^{-E_N/kT}$	$\frac{1}{e} \cdot \frac{dQ_s}{dt} = -p_N \cdot N_0^x$	$\frac{1}{e} Q_s = I_0^- - N_0^x p_N \cdot t$
Thermal activation from continuum of surface states $p(E)\,d(E) = a \cdot e^{-E/kT}\,dE$	$\frac{1}{e} \cdot \frac{dQ_s}{dt} = \frac{dI^-}{dt} = \frac{I_0^- kT \cdot N_c v_n \sigma_n}{(E_{D(0)} - E_v)\, e^{E_{D(0)}/kT}} \cdot \frac{1}{1 + \frac{N_c v_n \sigma_n}{e^{E_D/kT}} \cdot t}$	$\frac{1}{e} Q_s = I_0^- - \frac{I_0^- \cdot kT}{E_{D(0)} - E_v} \cdot \ln\left(1 + \frac{N_c v_n \sigma_n}{e^{E_{D(0)}/kT}} \cdot t\right)$

energy between conduction and valence band. If only electrons are emitted from donor states into the conduction band then no real change in surface charge occurs but only a redistribution of the bulk charge i.e. an increase in space charge density. This may, of course, also reduce the surface voltage; therefore it has to be established in all cases whether the charge on the surface is really dissipated. If both, electrons and holes are liberated from their traps, then the surface charge drops and the space charge density stays unaffected.

2.3. Relation Between Decay of Surface Charge and of Surface Potential

Since the surface potential rather than the surface charge is measured the decay laws for the surface charge deduced for the various models have to be transformed into voltage decay laws, so that experiment and theory can be compared. The surface voltage depends on the detailed distribution of the charge in the bulk, the latter being in general a function of space and time. Usually the following simplifying assumptions are made for calculating the voltage: The charge due to injected carriers is neglected in calculating the voltage, though trapping is taken into account for calculating the number of charge carriers which drift through the space charge layer [5]. This simplification is certainly only self-consistent if the ratio of the trapped to the number of arriving charge carriers is small. In many cases trapping is neglected, and it is assumed that all charge carriers drift completely through the space charge layer. This is allowed if the transit time is small relative to the mean

Table 1b. Discharge by Field Dependent Processes Occuring at the Surface

Physical Process and Emission Probability	Rate of Change of Surface Charge	Decay Law for the Surface Charge
Tunnel effect $p(\vec{E}) = K \cdot e^{-\frac{4}{3}\frac{(2m^*)^{1/2}}{h\,e}\frac{(\Delta E)^{3/2}}{\vec{E}}}$	$\frac{1}{e}\frac{dQ_s}{dt} = \frac{dI^-}{dt} = -K I^- \cdot e^{-\frac{\beta}{I^-}}$ $\beta = -\frac{4}{3}\frac{(2m^*)^{1/2}}{h\,e} \cdot \frac{\epsilon\epsilon_0}{e}(\Delta E)^{3/2};$ $\lvert\vec{E}\rvert = \frac{e}{\epsilon\epsilon_0}\lvert I^-\rvert$	$\frac{1}{e}Q_s = I_0^- - \frac{I_p^{-2}}{\beta}$ $\cdot \ln\left(1 + \frac{K\beta}{I^- e^{\beta/I_0^-}} \cdot t\right)$ for $1 \gg \frac{I_0^-}{\beta}$
Schottky emission $p(\vec{E}) = K e^{\left(\frac{e^3}{4\pi\epsilon\epsilon_0}\right)^{1/2}(\vec{E})^{1/2}}$	$\frac{1}{e}\frac{dQ_s}{dt} = \frac{dI^-}{dt} = -K I^- e^{\beta(I^-)^{1/2}}$ $\beta = \left(\frac{e^3}{4\pi\epsilon\epsilon_0}\right)^{1/2} \cdot \frac{\epsilon\epsilon_0}{e};$ $\lvert\vec{E}\rvert = \left\lvert\frac{e}{\epsilon\epsilon_0} I^-\right\rvert$	$\left(\frac{Q_s}{e}\right)^{1/2} = (I_0^-)^{1/2} - \frac{1}{\beta}$ $\cdot \ln\left(1 + \frac{(I^-)^{1/2} \cdot \beta \cdot K}{2\,e^{-\beta(I_0^-)^{1/2}}} \cdot t\right)$
Electrostatic interaction of deposited ions $p(I^-) = a\, e^{-E/kT} e^{\beta^*(I^-)^{1/2}}$ $= p_0 \cdot e^{\beta^*(I^-)^{1/2}}$	$\frac{1}{e}\frac{dQ_s}{dt} = \frac{dI^-}{dt} = p_0 I^- e^{\beta^*(I^-)^{1/2}}$	$\left(\frac{Q_s}{e}\right)^{1/2} = (I_0^-)^{1/2} - \frac{1}{\beta^*}$ $\cdot \ln\left(1 + \frac{(I^-)^{1/2}\beta^* p_0}{2\,e^{-\beta^*(I_0^-)^{1/2}}} \cdot t\right)$

free time before deep trapping. Dealing with high mobility material like ZnO single crystals, this condition is most likely fulfilled. For a conducting ZnO crystal with a Schottky-barrier at the surface we have under these conditions

$$V - V_0 = \frac{(e\,Q_s)^2}{2\,\epsilon\epsilon_0\,\rho} \qquad (6)$$

where ρ is the space charge density and V_0 the surface barrier before charging. The other symbols have their usual meaning. Insertion of the time dependences of Q_s and ρ gives the appropriate decay law for the surface voltage for the different processes. Clearly ρ is time dependent only for a bulk process. We expect therefore from equation (6) that the square root of the surface voltage is proportional to the decay law for the charge if surface processes alone are effective. If bulk processes dominate and only holes are transported then the surface voltage may be directly proportional to the decay law for the surface charge. We find this result realizing that ρ is proportional to the density of acceptors A^+; with equation (5) we obtain that $\rho(t) \sim A^+(t) \sim Q_s(t)$ and therefore equation (6) gives a linear relation between surface charge and voltage.

Table 1c. Discharge by Thermal Processes Occurring in the Bulk

Physical Process and Emission Probability	Rate of Change of Surface Charge	Decay Law for the Surface Charge
Thermal activation of holes from well defined energy level A^+ in the bulk $p_A = a^* \cdot e^{-E_A/kT}$	$\frac{1}{c}\frac{dQ_s}{dt} = -p_A \int_0^d A^+ dx$	$\frac{1}{e} Q_s = \bar{I_0}\, e^{-p_A \cdot t}$
Thermal activation of holes from continuous distribution of energy levels in the bulk $p_A(E)\, dE = a^* \cdot e^{-E/kT}\, dE$	$\frac{1}{e}\frac{dQ_s}{dt} = \frac{A_0 \cdot kT}{(E_{D(0)} - E_v)} \cdot \frac{N_c v_n \sigma_n}{e^{E_{D(0)}/kT}} \cdot \frac{1}{1 + \frac{N_c v_n \sigma_n}{e^{E_{D(0)}/kT}} \cdot t}$	$\frac{1}{e} Q_s = \bar{I_0} - \frac{A_0 \cdot kT}{E_{D(0)} - E_v} \cdot \ln\left(1 + \frac{N_c v_n \sigma_n}{e^{E_{D(0)}/kT}} \cdot t\right)$

If electrons and holes are freed from the traps in the bulk at about equal rates, then the space charge density is again unaffected and the square root of the voltage should be proportional to the time law of the dark decay.

4. Measurements

The ZnO single crystals were obtained from different sources. Conducting crystals with a resistivity of approx. 10 Ωcm were obtained from 3M and crystals with a resistivity of about $10^{11}-10^{12}$ Ωcm from Airtron.

They were cut into discs of about 1–2 mm in thickness with the hexagonal c-axis perpendicular to the broad face. The diameter of the discs was about 4–5 mm. Both faces, the (0001) and $(000\bar{1})$ surface were used for the measurements. However, no significant difference in the charging and discharging behaviour between the two surfaces was found. The surfaces were polished in successive steps with 3, 1 and .25 μm diamond paste and then etch-polished in HCl or H_3PO_4 using a technique described by several authors [6]. Ohmic contact to the conduction band was made either by In or silver paste electrodes. The crystals were charged in a corona discharge. The charge on the crystals was determined by integrating the current during the charging process (see Fig. 2). A screen was used between corona wire and crystal to control the charging of the crystal. The screen could be rotated so that the charge deposition was homogeneous on the crystal surface. After charging, the crystal was moved to the measuring electrode and the induced voltage was measured with a Cary Model 31 V or a Keithley No. 640 electrometer. The air was kept dry in the system by phosphorus-pentoxide.

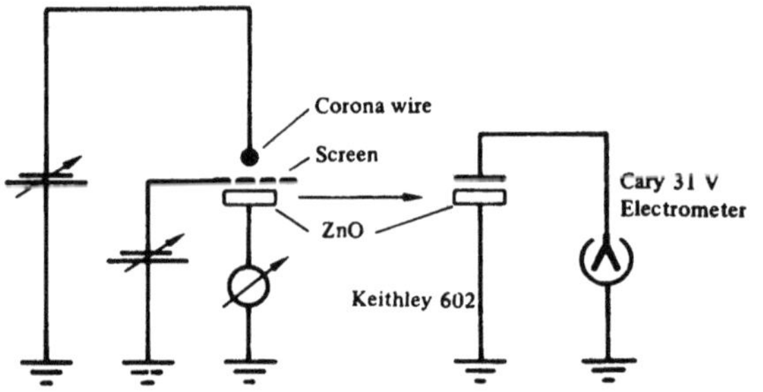

Fig. 2. Schematic diagram of the measuring apparatus. The Keithley electrometer is either used to measure the current during charging or as a Coulombmeter. After charging the crystal is moved opposite to a transparent electrode in order to measure the surface voltage.

5. Results

5.1. Insulating Crystals

These crystals were Li-doped, and the resistivity was found to be about 10^{11}–10^{12} Ωcm. A typical curve of the charging current versus time and of the time integral of the current as a function of time is shown in Figures 3a and 3b. The curves indicate a capacitor-like behaviour of the ZnO crystal during the charging process with a leakage current smaller than 10% of the peak current. The surface voltage was measured about 1–2 sec after the charge measurement. The charge depends linearly on the surface voltage as deduced from these measurements and the capacitance corresponds to the one calculated from the geometrical data and the dielectric constant of ZnO. The charge density amounts to about $1.7 \cdot 10^{-9}$ Cb/cm^2 or about 1×10^{10} charges/cm^2 at 300 V surface voltage and a crystal thickness of $5 \cdot 10^{-2}$ cm. The leakage current as a function of surface voltage is plotted in Figure 4. It increases slightly super-

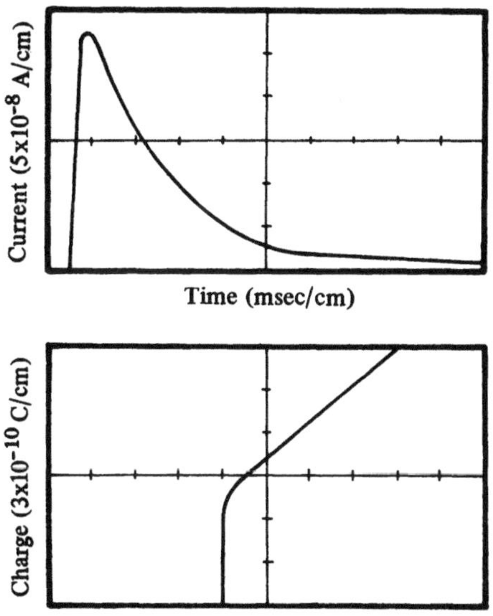

Fig. 3a and 3b. Current and charge as a function of time during charging of an insulating ZnO crystal. The surface voltage was 255 V after charging, the leakage current 1.35×10^{-9} A.

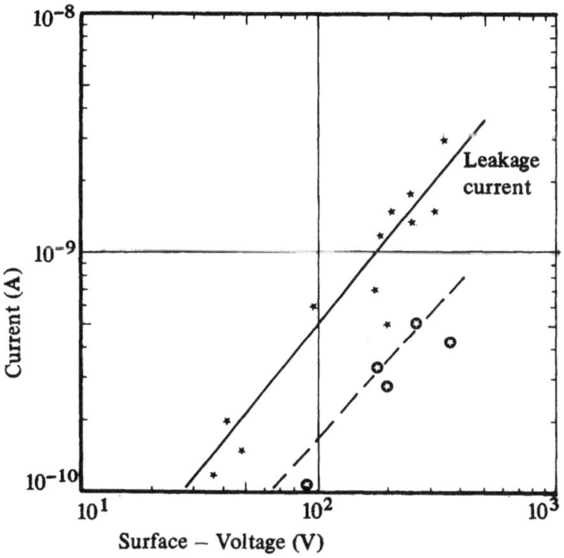

Fig. 4. The upper curve represents the leakage current through the crystal during charging as a function of the surface voltage which was measured immediately after charging. The lower curve gives the current through the crystal at the beginning of the dark decay measurement.

linearly with surface voltage. This indicates that no strongly field-dependent processes such as tunnelling cause the leakage current in this voltage range. A curve with the same slope but shifted to lower values in current by a factor of 2 to 4 is found for the current measurements immediately after the beginning of the dark decay measurement.* Since we have a delay of about 1–2 sec between the switching off of the corona and the beginning of the voltage measure-

* The current was obtained from the dark decay rate of the voltage, using the formula $J = (dQ)/(dt) = C\,(dV)/(dt)$. The time derivative of the voltage was given by the logarithmic time law for the decay as $(dV)/(dt) = -K/\tau$. Using these formulas it was implicitly assumed that the decay of the surface voltage is due to a decay of surface charge and not due to an increase in the space charge density in the bulk by electron emission from traps. This was checked and found to be true by recharging the crystal after a partial decay of the voltage; the same surface voltage and the same decay rate of the voltage were observed, which one would not expect if a volume charge had built up during the first measurement.

ment, we have to assume that the leakage current and the decay current would have been identical if the surface voltage could have been measured immediately after the interruption of the charging process. Consequently, we assume that the physical processes leading to the dis-

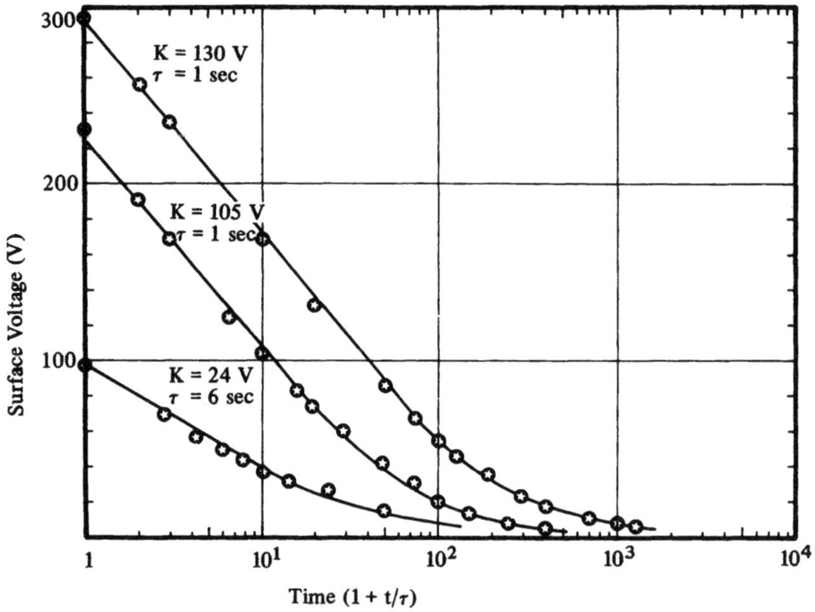

Fig. 5. The dark decay of the surface voltage of Li doped ZnO single crystals. The time laws are logarithmic and given by $V = V_i - K \log(1 + t/\tau)$. The constants K and τ for the individual curves are given in the Figure.

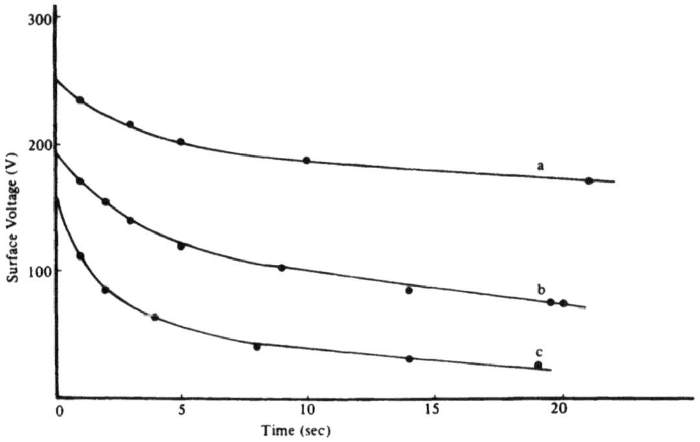

Fig. 6. Dark decay of the surface voltage of Li doped ZnO crystals. Voltage and time are represented in linear scales,

a) the crystal was dark adapted for 60 hours,
b) the crystal was exposed to white light and dark adapted for 8 minutes,
c) the crystal was exposed to white light; no dark adaption. All three curves give straight lines in logarithmic plots $V \sim \log(1 + t/\tau)$.

charge and the leakage current during the charging are the same. Hence thermal processes rather than field-dependent processes determine the rate of the dark decay.

The dark decay of the surface voltage is plotted in Figure 5. The crystals were kept in the dark for about 10 hours before measurement. The decay laws are logarithmic and can be represented in the following analytical form in a time range of about 2 orders of magnitude:

$$V = V_i - K \log \left(1 + \frac{t}{\tau}\right) .$$

It is obvious from the curves that the decay rate of the voltage does not drastically increase with increasing initial voltage as already evidenced before by the nearly linear relationship between current and voltage. If, however, the samples are exposed to light before charging, then the decay of the charge is faster and the initial voltage is lower under otherwise identical charging conditions. This is shown in Figure 6. Curve a was measured 60 hours, curve b 8 minutes after exposure to light. Curve c was obtained without having the crystal dark adapted after illumination.

5.2. Conducting Crystals

The resistivity of the crystals was between 1–10 Ωcm. Typical current vs. time curves dur-

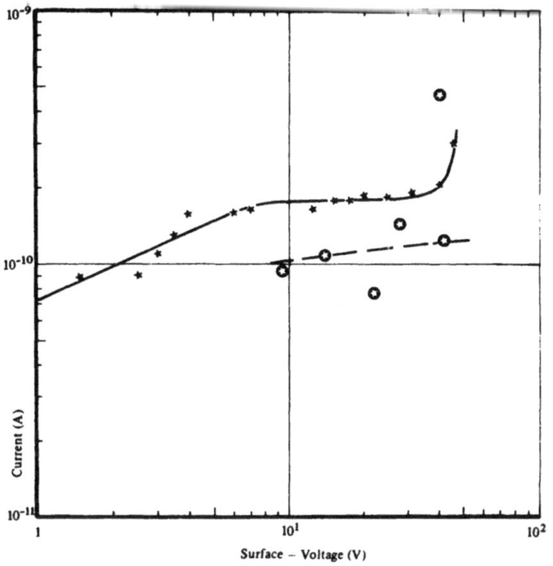

Fig. 8. The leakage current as a function of surface voltage. The lower curve gives the current through the crystal at the beginning of the dark decay. The point above the leakage current at 40 V surface voltage which was also gained from the voltage decay measurement, was obtained after having exposed the crystal to light before charging.

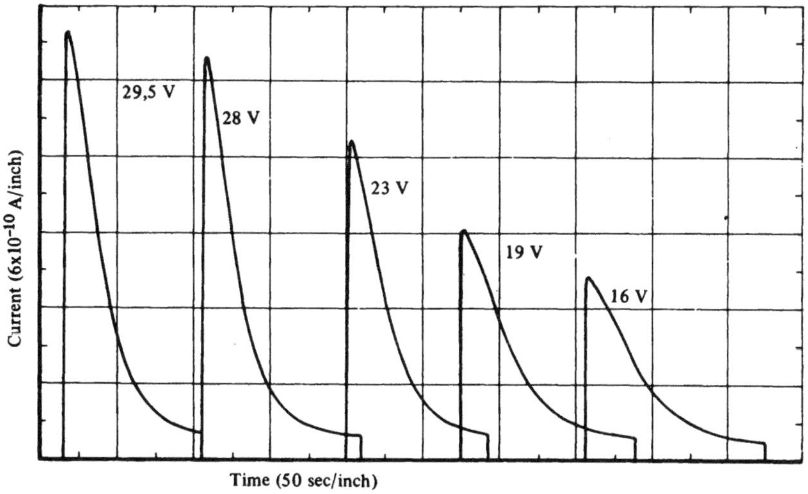

Fig. 7. Current as a function of time during charging of a conducting ZnO crystal. The surface voltage, measured immediately after charging, is noted at each curve. Note that the leakage current did not increase substantially, although the surface voltage increased from 16 V to 30 V.

ing the charging are shown in Figure 7. The crystal shows a capacitor-like behaviour upon charging, however, in this case of conducting crystals the charge stored on the surface is proportional to the square root of the surface voltage. This indicates the presence of a Schottky barrier. The surface charge density is at 30 volts about 6.4 x 10^{-7} Cb/cm^2 which corresponds to approximately 3.2 x 10^{12} charges/cm^2. This density is greater by up to two orders of magnitude than the one to which insulating crystals could be charged. The leakage current (Fig. 8) increased as the square of the voltage up to 10 volts. The current was saturated in the voltage range between 10 and 30 volts and increased steeply above 35 volts. In some cases the plateau was not found and the square root dependence merged into the sharp rise of the current which then mostly occurred at voltages below 35 volts. The lower curve in Figure 8

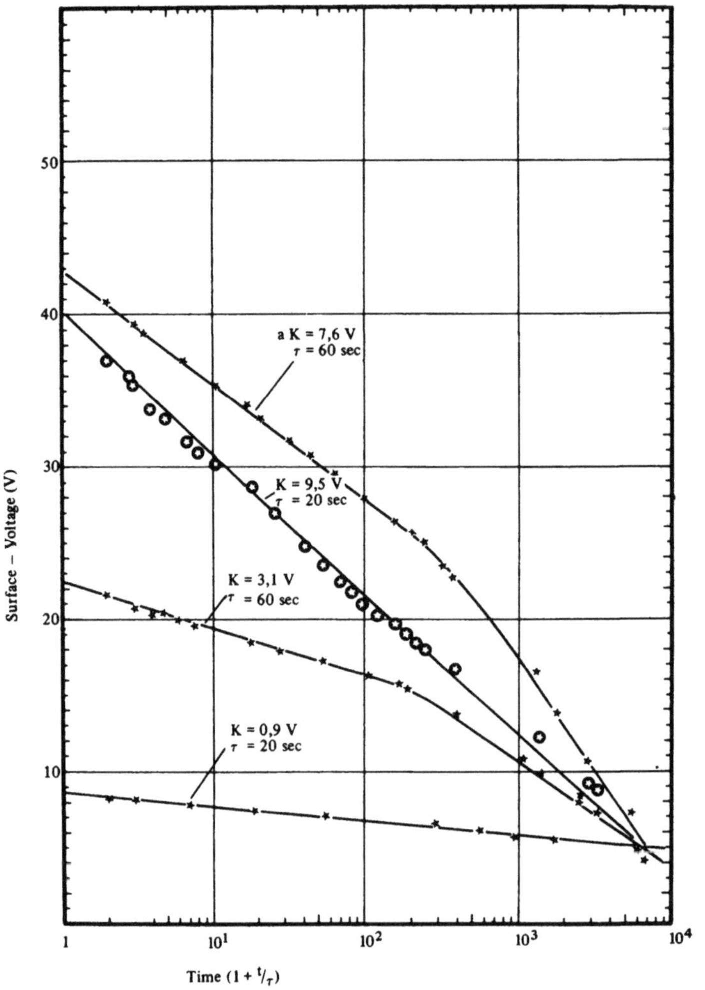

Fig. 9. The dark decay of the surface voltage of conducting ZnO crystals. The time laws are logarithmic and given by $V = V_i - K \log (1 + t/\tau) - \kappa \log \left(1 + \frac{t}{\tau} \cdot \frac{\tau}{\tau_K}\right)$. K and τ are indicated at each individual curve. Curves a und b are measured after having not and having exposed the crystal to light respectively.

gives the current found at the beginning of the dark decay. Though a considerable scatter of the values is observed, they are all within a factor of two below the curve for the leakage current. This is considered to be within the limits of error to show that the leakage current and the dark discharge must be traced back to the same physical cause.

In order to give further evidence for this idea, the dark decay of the surface voltage is shown in Figure 9. The curves in this semi-logarithmic plot are straight lines which indicate that the dark decay follows logarithmic time laws. In the cases with initial voltages higher than about 10 volts the curves were composed of two straight lines, the slope of the straight line observed at the beginning of the dark decay being always smaller. If the sample is exposed to light before charging, then–under identical charging conditions–the initial voltage is found to be lower, the rate of the voltage decay faster and the decay curve can be represented by just one straight line instead of two. This is shown by curve a and b, being measured without and with having the sample exposed to light before charging.

6. Discussion

Our measurements have shown that in the cases of both insulating and conducting crystals the dark discharge obeys logarithmic time laws and that the dark discharge is sensitive to previous illumination of the samples. Since there is evidence that the physical cause for the leakage current during charging and the discharge current in the dark are the same, a discussion of the leakage current seems to be appropriate.

Our measurements confirmed the previous finding of other authors [7] that the light emission of the negative corona discharge is due to the second positive system of nitrogen and that strong emission lines exist, the light of which is strongly absorbed by the ZnO. The spectral sensitivity of the insulating and conducting ZnO

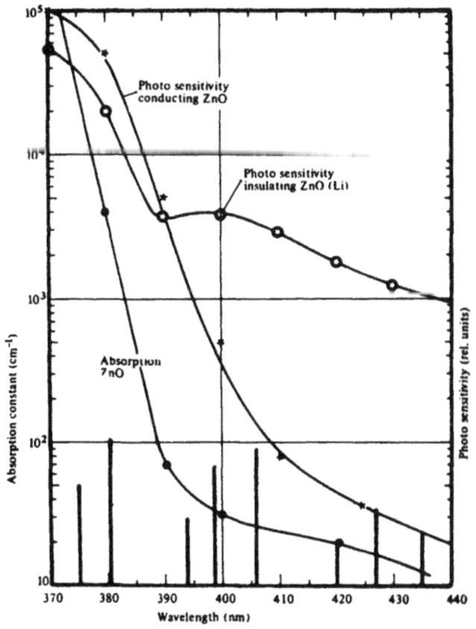

Fig. 10. Photo-sensitivity of conducting and insulating Li doped ZnO crystals. Emission lines observed in the corona discharge in air are indicated by bars, the length of which is a relative measure of intensity, not taking into account the spectral sensitivity of the photomultiplier.

are plotted in Figure 10 together with the emission lines in the wavelength region of interest. These lines coincide with the region of high photosensitivity of the crystals, and we found that under our experimental conditions the light intensity was sufficient to discharge the samples. Therefore we suggest that the leakage current observed during charging is in fact a photocurrent. In the case of the conducting crystals it is assumed that the photocurrent is saturated. Since the width of the Schottky depletion layer increases with the square root of the voltage we expect also an increase in photocurrent proportional to $V^{1/2}$ as long as the penetration depth of the light is larger than the thickness of the depletion layer. At high voltages and with sufficiently pure ZnO we may find a situation in which all the light is absorbed so that the photocurrent becomes independent on voltage. At still higher voltages breakdown occurs. For the insulating crystals the electrical fields across the whole crystal are too low so that the photocurrent is still in the regime of Ohm's law.

Immediately after switching off the corona the photocurrent will start to decay and the occupation of the states in the forbidden gap will correspond to the non-equilibrium situation during charging. This situation is characterized by the recombination and trapping processes of the photogenerated charge carriers. The rate of the thermal release of the electrons and holes from these states will determine the decay law of the photocurrent. These charge carriers move to the surface and the charge there will be dissipated. Hence the discharge in the dark is not a real dark decay but rather a dissipation of the surface charge by carriers generated before by the light emitted from the corona. This is substantiated by the fact that a previous exposure to light gives a faster charge dissipation.

Logarithmic time laws for the decay of the photocurrent have been reported repeatedly and Medved [8] has proposed a model wherein the basic process is the penetration of the surface barrier by the photoconduction electrons and the consequent formation of adsorbed ions. This cannot be true in our case since no electrons are able to move from the bulk to the free, negatively charged surface. The reason for the logarithmic decay law of the voltage must therefore be due to electrons and holes which are distributed in a continuum of traps (Table 1), their numbers being approximately equal. The observation of a continuum of traps in connection with photo-conductivity measurements was also reported by Heiland [9]. However, in contrast to his statement, we have to assume that the traps are located in the bulk and not on the surface. Otherwise in the case of conducting crystals the square root of the surface voltage would be proportional to the logarithm of time which was found not to be true. The nature of the traps is unknown. At the present state it is not known, which process would dissipate the surface charge and finally limit the storage time of the charge if the crystals were charged without simultaneous exposure to light.

Acknowledgement

The author wishes to thank Dr. A. M. Goodman and Dr. W. J. Merz for helpful discussion and careful reading of the manuscript.

Appendix

For the calculation of the logarithmic time law for the charge decay in the dark we have to assume that a continuum of surface states exists between the valence band and conduction band, the density of which is constant per unit energy. These states shall become occupied by the arriving negative ions. The occupation at the beginning is determined by the kinetic processes during the charging rather than by thermodynamics; therefore states may be only partially occupied e.g. 1 per cent of the available states per unit energy. We may now define a demarcation level at whose energy the probability per unit time for the thermal activation of an electron from the ion into the conduction band is equal to the probability per unit time for the deposition of an ion. This energy shall be denoted by E_D. For energies $E < E_c - E_D$ the surface states shall not be occupied, for greater energies $E \geqslant E_c - E_D$ the occupation shall be constant and equal to 1. After switching off the charging process the surface states will be thermally emptied, and the demarcation level will move to lower energies, since the probability for thermal activation depends exponentially on energy, and therefore the levels lying close to the conduction band will be emptied first. Hence the following equation for the discharge or the surface states is valid:

$$\frac{d}{dt}\left[H_0 \cdot f(E)\,dE\right] = p(E) \cdot f(E) \cdot H_0 \cdot dE \qquad (1)$$

H_0 is the surface state density per unit energy. For the occupation probability f(E) we have according to our assumption:

$f(E) = 0$ for $E < E_c - E_D$

$f(E) = 1$ for $E \geqslant E_c - E_D$.

p(E) is the probability per unit time for thermal activation from the surface state at energy E into the conduction band. It is

$$p(E) = N_c v_n \sigma_n \cdot e^{-E/kT}$$

N_c effective density of states, v_n = thermal velocity, σ_n = capture cross section.
From equation (1) we find, integrating between demarcation level and valence band:

$$H_0 l \frac{d}{dt} \int_{E_v}^{E_D} dE = H_0 l N_c v_n \sigma_n \int_{E_v}^{E_D} e^{-E/kT} \, dE$$

Assuming that $e^{-E_D/kT} \ll e^{-E_v/kT}$ we find for the time dependence of the demarcation level:

$$E_D(t) = E_D(0) + kT \ln \left(1 + \frac{N_c v_n \sigma_n \cdot t}{e^{E_D(0)/kT}}\right)$$

and for the ion density per unit energy on the surface:

$$H_0 l(E_v - E_D) = H_0 l [E_v - E_D(0)] - kT H_0 l \times \ln \left(1 + \frac{N_c v_n \sigma_n}{e^{E_D(0)/kT}} \cdot t\right)$$

Since $H_0 l(E_v - E_D)$ is identical with the ion density $I^-(t)$ and $H_0 l = \frac{I_0^-}{E_v - E_D(0)}$ we have

$$I^-(t) = I_0^- - \frac{kT}{E_v - E_D(0)} \cdot I_0^- \ln \left(1 + \frac{N_c v_n \sigma_n}{e^{E_D(0)/kT}} \cdot t\right)$$

A similar logarithmic time law will also be obtained if a continuum of states exists between conduction and valence band in the bulk of the material, the calculation being the same except that we have in addition to integrate over the depletion layer.

References

[1] K. Hauffe, R. Stechemesser, Phot. Scienee and Engin. **11**, 145 (1967).
[2] J. A. Amick, RCA Review **20**, 753 and 770 (1962).
[3] W. Ruppel, H. J. Gerritsen and A. Rose, Helv. Phys. Acta **30**, 495 and 504 (1957).
[4] S. R. Morrison, Surface Sci. **13**, 85 (1969).
[5] R. M. Schaffert, Electrophotography, Focal Press, London 1965, p. 266.
[6] See e.g. W. H. Strehlow, J. Appl. Phys. **40**, 2928 (1969).
[7] W. B. Pearse and A. G. Gaydon, Identification of Molecular Spectra, John Wiley (1950); C. F. Gallo, A. G. Leiga, J. A. McInally, Phot. Sc. and Engin. **11**, 11 (1967).
[8] D. B. Medved, J. Phys. Chem. Sol. **20**, 255 (1961).
[9] G. Heiland, J. Phys. Chem. *Sol.* **22**, 227 (1961).

Analysis of Electrophotographic Discharge by Independent Measurement of Surface Charge and Surface Voltage

R. B. Comizzoli and H. Kiess

Abstract

The electrophotographic discharge processes for three materials are studied by independent measurements of the voltage decay with light as a function of time *and* of the surface charge remaining on the layer after the voltage is partially reduced by light.

The surface charge is measured by determining the amount of charge of opposite sign necessary to reduce the surface voltage of the previously charged layer to zero.

The materials studied are Rhodamine B dye-sensitized poly-N-vinylcarbazole (PVK), fluorescein dye-sensitized ZnO-resin binder layers, and sputtered ZnO layers. For each layer various limitations on the discharge process are found involving such phenomena as light penetration, trapping, and field dependence of photogeneration.

The discharge of the dyed PVK with light weakly absorbed by the dye is shown to be limited by the fact that only positive carriers are mobile. Thus, a negative trapped space charge region forms within the layer during exposure. Under our experimental conditions, voltage discharge by UV radiation takes place without the formation of a space charge region.

The ZnO-resin layer discharge with light absorbed by the dye involves a process similar to that of the PVK, while our results on the decay by UV radiation are not contradictory to previous models which explain the S-shaped curve by trapping of charge carriers.

The sputtered ZnO layers are of two types: A "poor" layer with UV discharge limited apparently by trapping and a "good" layer with no trapping limits. The discharge of the "good" layer with weakly absorbed light just below the band edge is similar to the discharge of the PVK and ZnO-resin layers due to light absorbed by the dye.

1. Introduction

Electrophotographic materials may be classified in several ways, one among them being a classification according to the optical generation and carrier transport properties. The nature of the generation and transport properties strongly affect the shape of the light induced voltage discharge curve [1–4]. Often the voltage as a function of time under illumination is fitted according to some generation and transport model. The interpretation is, however, quite difficult since the curve of voltage versus time may be explained by different models. Additional informations may be gained by considering the field dependence of the photocurrent [5–6]. Another possibility is to measure not only the surface voltage but also the surface charge corresponding to that voltage [7]. From the relation between the measured voltage and surface charge, one may infer the presence or absence of space charge within the layer and this can be an important clue in understanding the discharge process.

In this paper we consider several types of electrophotographic materials which exhibit quite different discharge kinetics, and show how a measurement of surface charge can identify the particular generation and transport processes. In the three materials–poly-N-vinylcarbazole, ZnO-resin binder, and sputtered ZnO layers–the relative importance of generation and trapping limits on the discharge processes are considered.

2. Experimental

2.1 Samples: Three kinds of layers were examined.

2.1.1. Poly-N-vinylcarbazole (obtained from Polysciences, Inc., Rydal, Pa.) was dissolved in 4 : 1 weight ratio of toluence/cyclohexanone; the solution contained about 1 g PVK per 10 ml. Rhodamine B dye was added to the polymer solution before coating in concentrations ranging from 10^{-4} to 10^{-2} g dye/g polymer. The substrates were made by evaporating gold on aluminum sheets, and the PVK solution was doctor bladed onto these. The layers were air dried, baked for about 30 minutes at 70° C, and stored in the dark.

2.1.2. ZnO-resin binder layers were made dispersing ZnO powder (N. J. Zinc Photox 801) in a polyester resin with toluene as solvent. The particle size was .3 x 10^{-4} cm. The ZnO: resin ratio was 6 : 1 by weight. The ZnO was sensitized with 2 x 10^{-4} g of fluorescein dye per g of ZnO. The mix was de-aerated and coated on aluminum sheets with a doctor blade. After air drying, the plates were baked in an oven at 100° C for 10 minutes, and then stored in the dark.

2.1.3. The sputtered ZnO plates were produced by sputtering Li doped zinc (188 ppm) in argon/oxygen atmosphere. The substrates were either chrome on steel or tin oxide coated quartz. The thickness of the layers used varied from 3 to 5 x 10^{-4} cm.

2.2. Measurements

Two kinds of measurements were performed: the decay of surface voltage under illumination, and measurements of the surface charge as a function of surface voltage.

2.2.1. Decay of Surface Voltage under Illumination

The samples (1.5 x 2 in) were charged in the dark with appropriate polarity by a corona discharge and transported to a measuring station. The surface voltage was measured with a vibrating (10 Hz), transparent electrode which was connected to an amplifier. A tungsten lamp was the light source, and a *Bausch and Lomb* 250 mm monochromator was used for varying the wavelength of the light. The electrode-sample spacing and relative areas were such that only the uniformly charged central area of the sample was probed.

2.2.2. Surface Charge Measurements

The surface charge of the sample as a function of surface voltage was measured by a technique described by Comizzoli [7]. Essentially, the surface charge is determined by measuring the amount of charge Q_c of opposite sign deposited on the sample required to reduce the surface voltage to zero.

Useful data are obtained from these measurements only if certain requirements are met. These involve the dark decay rate, the contact properties of the substrate-layer interface, and the surface charge uniformity.

The dark decay rate must be sufficiently low so that during measurement the change in surface voltage due to dark decay is negligible compared to the change in voltage due to the deliberate neutralization of the surface charge by the deposition of charge Q_c of opposite sign. In the work discussed here, the change in voltage due to dark decay during the measurement was less than 10% of the initial voltage.

The contact properties of the substrate-layer interface are important if part or all of the compensating charge is in the form of a space charge within the layer. Consider a negatively charged electrophotographic layer, the surface charge of which is compensated by a positive space charge region, as shown in Figure 1. If the substrate does *not* inject negative charge into the layer then a unit positive charge deposited on the negatively charged surface induces a unit negative charge at the substrate interface. Therefore the positive space charge is not reduced and hence the surface voltage changes by a factor proportional to the geometrical sample capacitance. The surface voltage will be reduced to zero when a quantity of positive charge Q_c given by

$$Q_c = CV_0,$$

where C is the geometric capacitance and V_0 the initial surface voltage, has been deposited. But this charge Q_c is *less* then the actual surface charge since the compensating charge is in the form of a space charge region within the sample, resulting in a higher effective capacitance. In order to measure the actual surface charge, the substrate must inject a negative charge into the layer for each positive charge deposited, so that the space charge layer is neutralized. A necessary but not sufficient condition for the charge injection is the inability to charge up the layer with charge of sign Q_c. This criterion was met for all samples with one exception.

Charge uniformity is assured by surrounding the sample edges with a plate which is set at the same voltage as the sample. This was done for the corona charging and discharging steps. Both the charging and the discharging are done in increments so that the bias of the plate can be adjusted to the corresponding surface voltage. With this "pseudo guard ring" technique, the surface voltage proved to be within 10% of its average value over 90% of the sample area as measured with a high-resolution electrometer. The sample holder and measurement circuit are shown schematically in Figure 2.

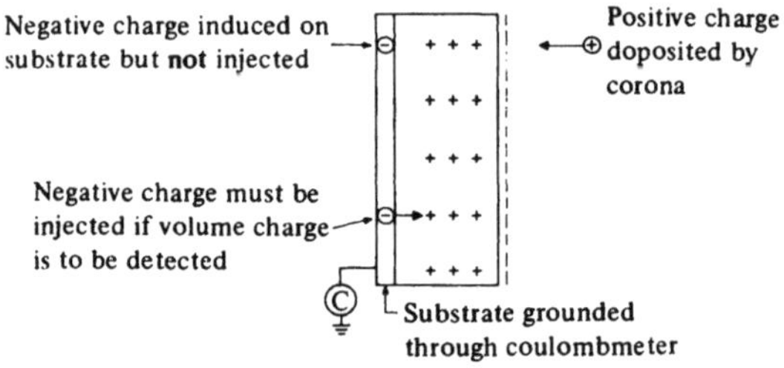

Fig. 1. The substrate must be capable of injecting charge of sign opposite to that of the space charge in order to permit detection of the space charge.

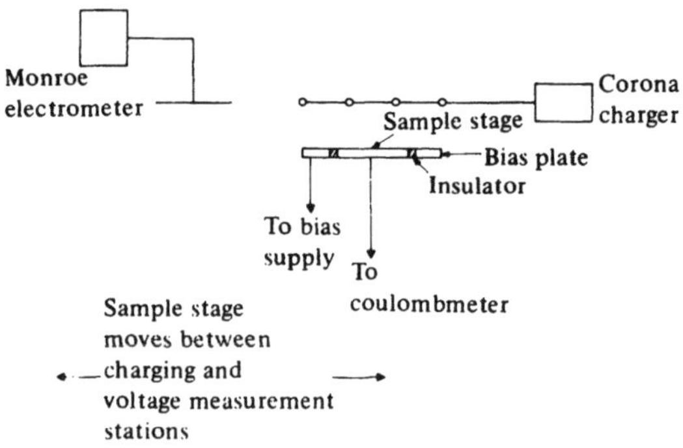

Fig. 2. Schematic of experimental apparatus.

3. Results and Discussion

First we consider in a general way the conclusions reached for the three kinds of layer. Table 1 shows for each material the characteristic decay curve of the voltage under illumination, lists the results of the surface charge measurements, and gives for the various cases the generation and transport model deduced.

We next consider each of the three layers in detail showing how the decay curves of the voltage and the surface charge measurements lead to the conclusions of Table 1.

3.1. Poly-N-Vinylcarbazole (PVK)

3.1.1. Discharge by Strongly Absorbed Light

Regensburger [6] has shown that the quantum efficiency follows the spectral dependence of the absorption constant of PVK. At 340 nm

MATERIAL	LIGHT INDUCED VOLTAGE DECAY	SURFACE CHARGE MEASUREMENT AFTER LIGHT EXPOSURE	TRANSPORT MODEL
PVK	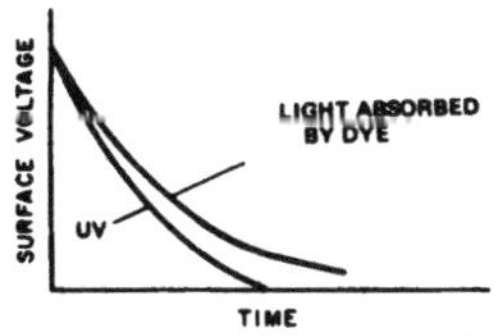	UV– SURFACE CHARGE REDUCED BY LIGHT. LITTLE OR NO SPACE CHARGE.	FIELD DEPENDENT GENERATION RATE.
		DYE–SURFACE CHARGE ESSENTIALLY UNCHANGED BY LIGHT.	NEGATIVE CHARGE CARRIERS IMMOBILE. VOLTAGE DECAY BY FORMATION OF NEGATIVE SPACE CHARGE.
ZnO POWDER-RESIN BINDER	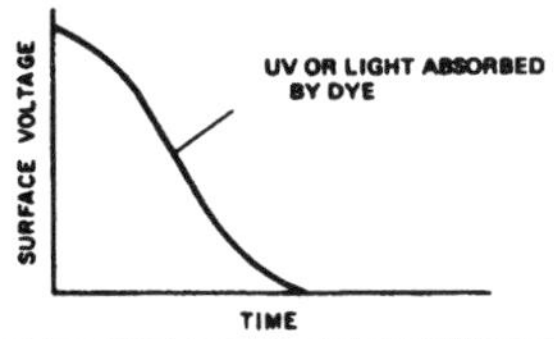	UV–SURFACE CHARGE REDUCED BY LIGHT. SLIGHT SPACE CHARGE INTRODUCED BY LIGHT EXPOSURE.	CARRIER TRANSPORT WITHOUT RECOMBINATION. CARRIER TRAPPING AT LOW FIELDS.
		DYE–SURFACE CHARGE UNCHANGED BY LIGHT. (EXCEPT AT LOW VOLTAGES).	POSITIVE CHARGE CARRIERS IMMOBILE. VOLTAGE DECAY BY FORMATION OF POSITIVE SPACE CHARGE.
SPUTTERED ZnO A	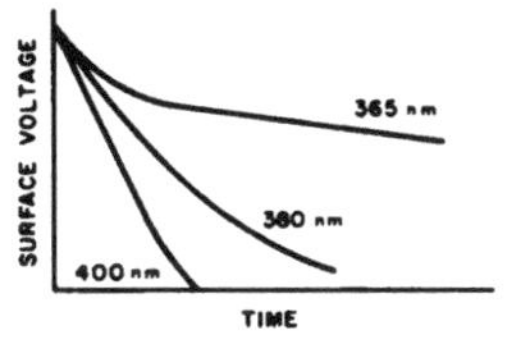	THESE SAMPLES CHARGE POSITIVELY; SURFACE CHARGE CANNOT BE DETERMINED BY THIS TECHNIQUE.	3650A – TRAPPING OF ELECTRONS AND SPACE CHARGE LIMITED CURRENT FLOW. 4000A – TRAPPING OF ELECTRONS AND HOLES AT APPROXIMATELY EQUAL RATES. TRANSPORT WITHOUT RECOMBINATION.
SPUTTERED ZnO B	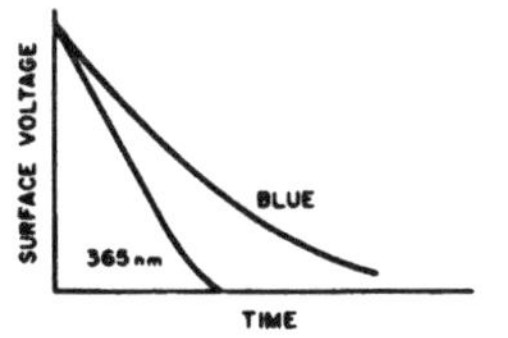	3650A –SURFACE CHARGE REDUCED BY LIGHT. LITTLE OR NO SPACE CHARGE.	TRAPPING OF ELECTRONS INSIGNIFICANT. SATURATED CURRENT FLOW.
		BLUE–SURFACE CHARGE ESSENTIALLY UNCHANGED BY LIGHT.	POSITIVE CHARGE CARRIERS IMMOBILE. VOLTAGE DECAY BY FORMATION OF POSITIVE SPACE CHARGE.

Table 1. Generation and Transport Models Deduced from Light Induced Voltage Decay and Surface Charge Measurements for PVK, Powder-Binder, and Sputtered Layers.

Regensburger reports the absorption constant as $2.5 \times 10^4 \text{ cm}^{-1}$ and the quantum efficiency as 2% at a field of 83 V/micron. Our samples had the same quantum efficiencies. PVK layers were charged positively and exposed to strongly absorbed light. The shape of the decay is shown in Figure 3 for 340 nm illumination. The general shape is independent of wavelength from 350 to 290 nm and independent of thickness for layers charged to maximum voltage from approximately 2 to 20 microns. Attempts to fit the curve with a hyperbolic or exponential time law were unsuccessful.

The strong decrease in the slope of the discharge curve with time suggests that either trapping, a decrease in carrier generation rate, or recombination limit the discharge. The measurement of the positive surface charge remaining on the sample after having reduced the voltage by light to some fraction of its initial value is shown in Figure 4. In this Figure the positive surface voltage is plotted against the negative charge Q_c deposited on the surface. Note that the capacitance as evidenced by the slope of the charge-voltage curve is, to a good approximation, unchanged after light exposure. Thus, no space charge is created (or neutralized) by trapping of light-generated carriers. Tentatively we conclude on this basis that the discharge curve is not limited by trapping but must be limited by the generation process; i.e., the optical generation rate decreases with the decrease in field, as found by Regensburger [6]. This was confirmed by measuring the initial slope of the decay curve for various thicknesses and voltages. The result is shown in Figure 5.

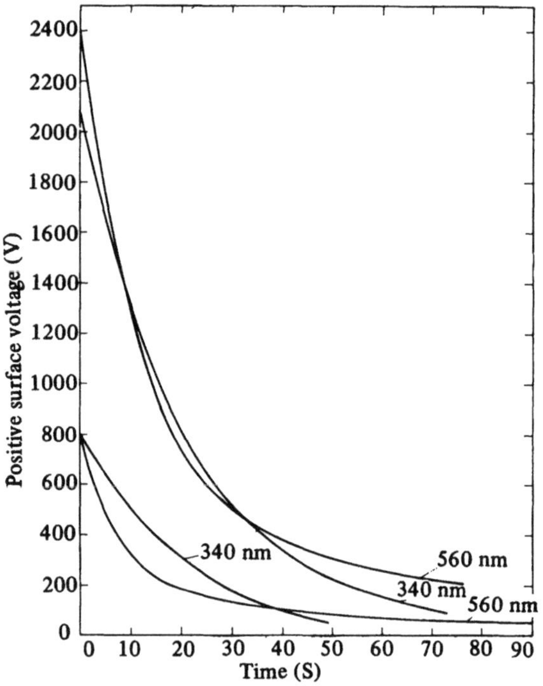

Fig. 3. Positive surface voltage vs. time for exposure of dye sensitized PVK layers to 340 nm and 560 nm light. Two layer thicknesses are shown. Dye concentration is 10^{-4} g dye/g PVK.

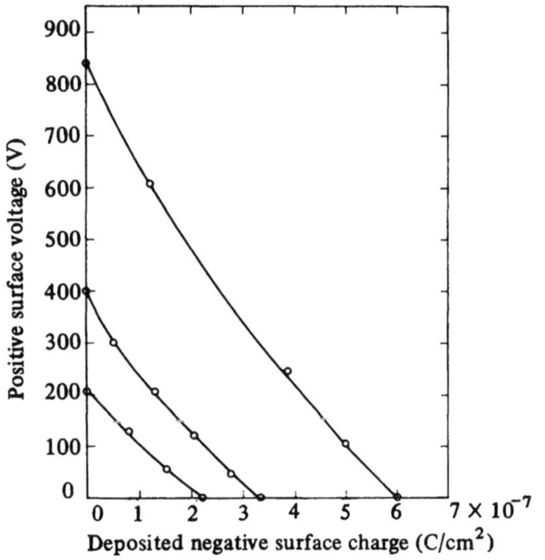

Fig. 4. Positive surface voltage vs. negative charge deposited on PVK layer showing reduction of surface charge by UV radiation. Uppermost curve represents sample charged positively to maximum voltage and the lower two curves represent samples exposed to 340 nm light to reduce the voltage to a value below V_{max}.

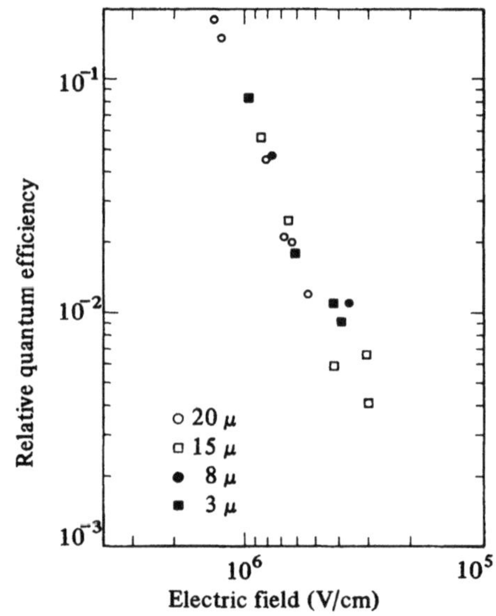

Fig. 5. Plot of relative quantum efficiency at 340 nm vs. electric field for PVK layers of various thicknesses.

3.1.2. Discharge by Weakly Absorbed Light

PVK layers sensitized with Rhodamine B dye can be discharged with visible light in the region of 500–600 nm. A typical discharge curve is shown in Figure 3. Note that even greater changes in slope occur than in the case of discharge by UV. This wavelength region we refer to as "yellow" in the following.

In Figure 6 the results of surface charge measurements after exposure to yellow light are shown. Note that there is little or no decrease in the charge Q_c compared proportionately to the decrease in surface voltage due to the light. Thus, the decrease in surface voltage by yellow light occurs principally by a change in the internal charge distribution and not by a decrease in the surface charge. This is expected for a discharge process in which weakly absorbed light generates only one sign of mobile carrier, as illustrated in Figure 7. In effect, the substrate charge is transferred to the bulk. As discussed by Comizzoli and Ross [4], the long tail arises from two effects. As illumination continues the space charge region breaks away from the substrate and shrinks towards the surface. Thus, a

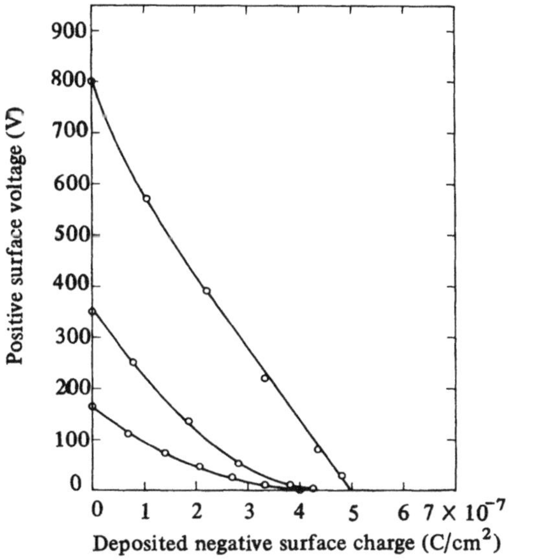

Fig. 6. Positive surface voltage vs. negative charge deposited on PVK layer showing formation of space charge by exposure to light absorbed by dye. Upper curve represents sample charged positively to maximum voltage and the lower two curves represent samples exposed to 560 nm light to reduce the voltage to a value below V_{max}.

progressively smaller fraction of the incident light is effective in generating carriers. Further, the effective capacitance increases with decreasing thickness of the space charge layer. Therefore dissipation of the same amount of charge gives a smaller change in voltage at lower surface voltages.

The dye concentrations used were sufficiently low so that the yellow light was only weakly

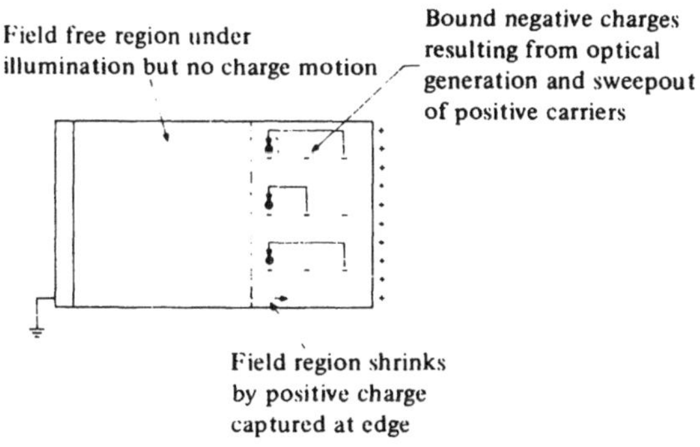

Fig. 7. Model for discharge by uniformly absorbed light for the case of a dye-sensitized PVK layer.

absorbed. This was checked by transmission measurements which showed that only a few percent of the light in the 500–600 nm regions is absorbed.

The apparent decrease in the charge Q_c at greater exposures may be due to several causes. There may be a small mobility of the negative charges, or the decrease may indicate that the dark decay of the charge is not negligible. Or, the surface voltage measurement at low voltage may not be sensitive enough to pick out the exact amount of charge needed to reduce the surface voltage to zero.

In contrast to the case with strongly absorbed light, the change in the slope of the decay curve is primarily a space charge effect and not determined by the field dependence of the optical generation rate. This is a rather surprising result.

3.2. ZnO Powder-Resin Binder Layers

3.2.1. Discharge by UV-radiation ($\lambda \leqslant 380$ nm)

Light of wavelengths shorter than 380 nm is strongly absorbed by ZnO, however not by a ZnO-powder binder layer [4]. This is connected with the fact that the binder usually transmits UV above approximately 300 nm and therefore by multiple reflection on the ZnO grains a substantial fraction reaches the back electrode.

A typical discharge curve at 380 nm is shown in Figure 8. Note that a large initial "induction period" exists, followed by a linear discharge to about 50 volts, after which there is a slight decrease in slope. The induction period has been attributed to trapping of carriers during the initial period of photoexcitation [3].

The linear discharge region has been described as a primary saturated photocurrent, i.e., electrons generated near the negatively charged surface travel completely through the layer to the substrate [4, 8, 9]. Positive carriers generated near the surface reduce the negative surface charge. The linear process has a quantum efficiency near unity [4]. The lower portion of

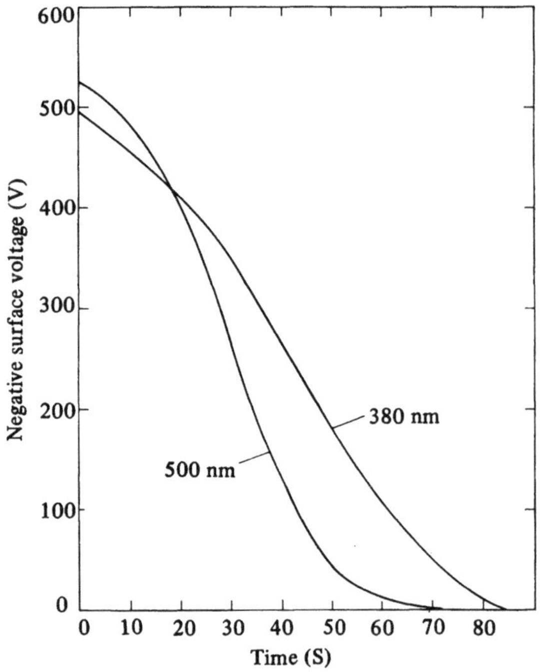

Fig. 8. Negative surface voltage vs. time for dye-sensitized ZnO-resin binder layers. Wavelength of light 380 and 500 nm.

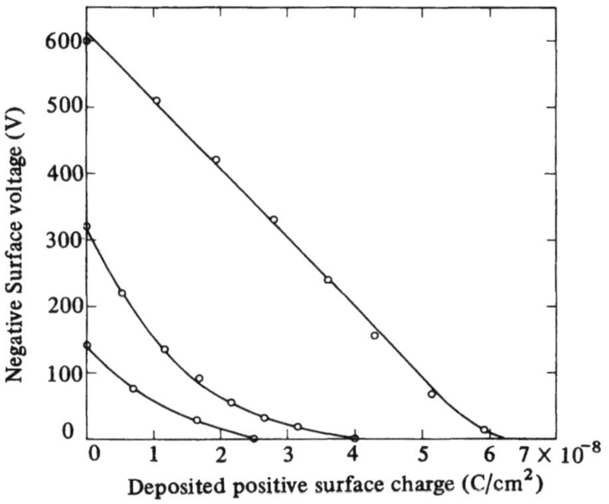

Fig. 9. Negative surface voltage vs. positive charge deposited on ZnO-resin binder layers. Reduction of surface charge by 380 nm light is clearly indicated. Uppermost curve represents sample charged negatively to maximum voltage and the lower two curves represent samples exposed to 380 nm to reduce the voltage to a value below V_{max}.

the discharge curve with decreasing slope has been related to the presence of a space charge within the layer [9].

This interpretation of the linear region is supported by surface charge measurements after exposure to UV. Figure 9 shows the surface charge on the ZnO-resin layer after exposure. Note that for all exposures there is a decrease in surface charge.

At the lower voltages, increasing curvature of the Q–V curve indicates the presence of space charge *due to the light exposure.* The space charge is generated during the light-induced discharge and is present even for exposures for which the voltage after exposure is still within the linear region of discharge. The Q-V curve without exposure compared with that after exposure demonstrates that the major portion of the space charge was *not* present before exposure, but was generated by the light discharge. One interpretation of this result is that carrier trapping during the initial slow "induction period" is responsible for the space charge. Since even UV radiation is not very strongly absorbed in powder-binder layers [4], it is possible that either electrons or holes may be the trapped carriers.

3.2.2. Discharge with Light Absorbed by the Dye ($\lambda > 380$ nm)

The dye concentration was sufficiently low so that the light absorption was essentially uniform throughout the layer thickness as evidenced by light transmission measurements. The typical discharge curve is shown in Figure 8. Disregarding a scaling factor depending on light intensity and spectral sensitivity of the layer, this curve is quite similar in shape to that obtained by illumination with UV, except for a slightly greater decrease in slope and a more pronounced tailing.

The surface charge measurements after exposure to light are shown in Figure 10. For an exposure which reduces the voltage by more than 50%, there is no change in surface charge, while for exposures greater than this there is some

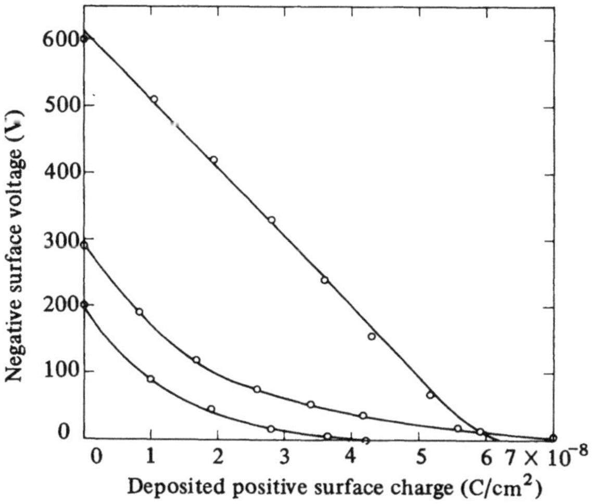

Fig. 10. Negative surface voltage vs. positive charge deposited on ZnO-resin binder layer. Uppermost curve represents sample charged negatively to maximum voltage and the lower two curves represent samples exposed to 500 nm to reduce the voltage to a value below V_{max}.

decrease. The behavior here is quite similar to that of the PVK layer dyed with Rhodamin B and reflects similar processes: a uniform generation of mobile carriers of only one sign. What is *not* understood for the ZnO-binder layer at the present time is the absence of a more pronounced tail in the voltage discharge curve with dye exciting light, as found for the PVK layer.

3.3. Sputtered ZnO Layers

The sputtered ZnO layers were found to be of two characteristic types, which we designate A and B. The important differences were twofold: Type A showed only partial voltage decay with strongly absorbed light, and could be charged positively. Type B could be discharged completely with strongly absorbed light and could not be charged positively. The reasons for these differences are at present unknown.
Since the type A charges positively, our technique of surface charge measurement cannot be used to unambiguously determine the surface charge. For layer A we consider only the voltage discharge curves, as shown in Figure 11 for 3 wavelengths. With strongly absorbed light (365 nm) there is a fast decrease in voltage to about 80% of the maximum voltage followed by a long tail. As the light penetration depth increases, the tail effect decreases, until for weakly absorbed light (400 nm) there is essentially a linear fall in voltage almost to zero.

Carrier trapping can account for the abrupt change in slope and long tail at 365 nm. We have here a problem similar to that encountered in Vidicon tubes, which has been discussed in detail by du Chatenier [11] and by Goodman and Rose [12]. The light is essentially absorbed within about .1–.2 μm of the surface. The electrons generated in this region either all move to the back electrode or else a concentration higher than that in the bulk builds up. In the first case—at high voltages—the current is saturated and the voltage drops linearly with time. At lower voltage the electron density in the surface region increases and the current through the layer becomes space-charge limited. The voltage dependence of space charge limited currents is a strong function of the detailed trap

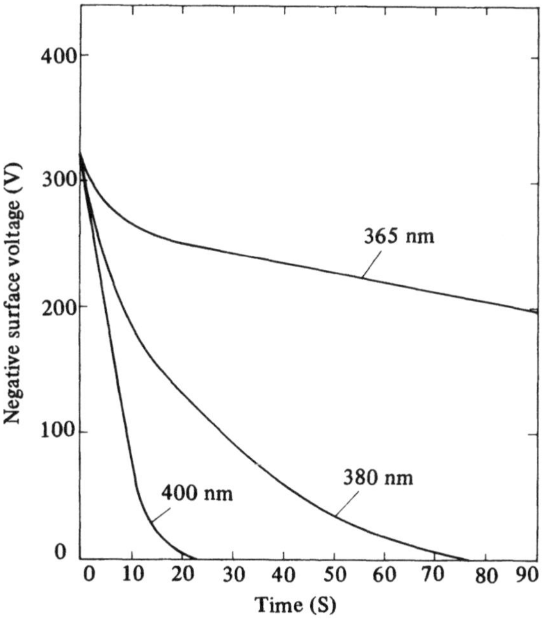

Fig. 11. Negative surface voltage vs. time under illumination. The shape of decay curves of Type A sputtered ZnO layers is strongly affected by the wavelength of the incident light. Layer thickness is 3.75 x 10^{-4} cm.

density and distribution. The voltage decay is therefore expected to be correspondingly sensitive to traps and very slow since the current through the layer may drop as a high power of the voltage. It is essential to our argument that the case with strongly absorbed light be a one-carrier process, though electrons and holes are generated in the surface region.

Illumination of the sample with weakly absorbed light (λ = 400 nm) results in a different situation: Electrons and holes are generated throughout the layer and we have to deal with a two-carrier problem. Assuming an approximately equal trapping rate of charge carriers of both signs, space charge will not form. Therefore the current stays saturated as long as the condition is fulfilled that the transit time of the minority carriers is small compared to the recombination lifetime. This condition may be fulfilled down to much lower voltages and therefore a voltage drop linear with time to relatively low voltages may be anticipated.

The discharge curves of type B layers are shown in Figure 12. The 365 nm discharge is linear for 75% of the decay with quantum efficiency of about 0.5. Surface charge measurements are shown in Figure 13. At 365 nm the surface charge is reduced in proportion to the surface voltage. We conclude for this layer that the electrons optically generated at the negative surface travel through the layer as a saturated photocurrent, which is in agreement with the linear time decay of the voltage.

Exposure with weakly absorbed light gives a discharge curve with a long tail, as in Figure 12. As shown in Figure 13, the surface charge is unchanged by exposure to this light. The discharge process here, then, involves the same type of model for weakly absorbed light discussed for the PVK, shown in Figure 7. Carriers are generated uniformly, but essentially only one carrier is mobile. Thus, the surface voltage is reduced by a transfer of substrate charge into the bulk, with essentially no change in surface charge density.

In this paper we have considered several electrophotographic materials which exhibit quite different light induced voltage discharge curves. By determining changes in surface charge density in addition to measuring the surface voltage photodecay, we have been able in most cases to

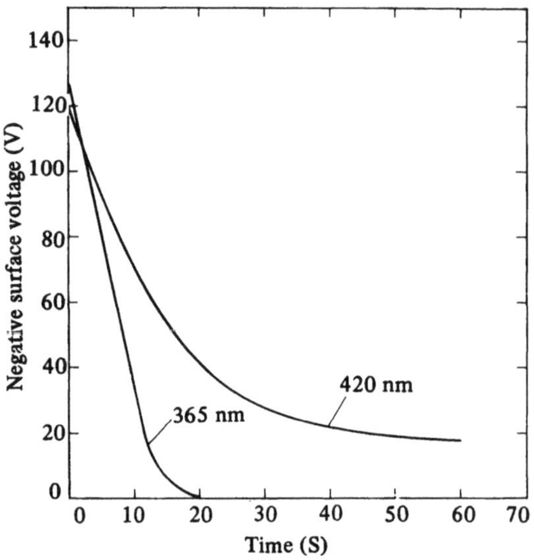

Fig. 12. Decay of negative surface voltage vs. time under illumination for Type B sputtered ZnO layer.

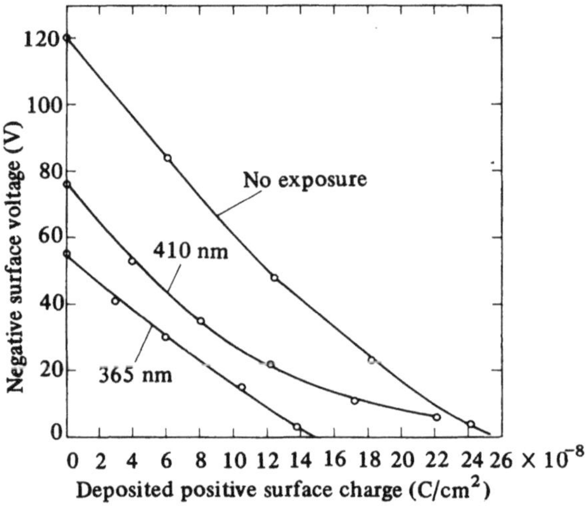

Fig. 13. Negative surface voltage vs. positive charge deposited on Type B ZnO sputtered layer. Uppermost curve represents sample charged negatively to maximum surface voltage and the lower two curves represent samples exposed to light to reduce their surface voltage to values below V_{max}.

identify the physical transport process involved in each layer without attempting to fit detailed mathematical models to the voltage decay time dependence.

Acknowledgement

The authors wish to thank Dr. J. J. Hanak for producing the sputtered layers, and Mr. H. M. Azarian for performing some of the measurements. We also thank Dr. A. M. Goodman for drawing our attention to the Vidicon tubes, for which analogous problems have been discussed before in detail.

References

[1] P. J. Warter, Jr., Applied Optics: Suppl. 3 on Electrophotography, p. 65 (1969).
[2] H. Kiess, Applied Optics: Suppl. 3 on Electrophotography, p. 100 (1969).
[3] R. Arneth, B. Lorenz, Reprographie **3**, 199 (1963).
[4] R. B. Comizzoli, D. A. Ross, Phot. Sci. and Eng. **13**, 265 (1969).
[5] M. D. Tabak, P. J. Warter, Jr., Phys. Rev. **173**, 899 (1968).
[6] P. J. Regensburger, Photochem. Photobiol. **8**, 429 (1968).
[7] R. B. Comizzoli, Phot. Sci. and Eng. **14**, 210 (1970).
[8] H. J. Gerritsen et. al., Helv. Phys. Acta **30**, 504 (1957).
[9] J. A. Amick, RCA Review **20**, 770 (1959).
[10] E. Mollwo in Photoconductivity Conference, ed. by Breckenridge et al., Wiley (New York) 1956, p. 509.
[11] F. J. du Chatenier, Philips Res. Reports. **23**, 142 (1968).
[12] A. M. Goodman, A. Rose, to be published in Journal of Applied Physics.

Electrophotographic Properties of Amorphous Selenium-Sulfur Layers

K. H. Kassel

Abstract

The electrophotographic properties of evaporated layers of amorphous selenium containing up to 30 atomic percent of sulfur were investigated. The light sensitivity decreases with increasing sulfur content. The spectral sensitivity is shifted to shorter wavelengths. By additional doping with iodine the light sensitivity decreases slightly for small concentrations until a relatively abrupt rise occurs at high concentrations.

Investigations of selenium-sulfur alloys were made with the differential-calorimeter apparatus. When using selenium-sulfur alloys which were additionally doped with iodine, the crystallization peak shifts to lower temperatures with increasing iodine concentration.

For electrophotographic layers amorphous selenium is an appropriate substance, as is well known. This modification consists of two types of molecules, namely of relatively long disordered chains and of 8-membered rings. The electrophotographic properties of amorphous selenium can be considerably influenced by addition of certain inpurities. The influence is qualitatively different for certain groups of elements, and this difference is due to the different location of the added atoms in the rings or chains. These facts have recently been pointed out in a report on atomic structure and charge carrier transport properties of several vitreous alloys of selenium by Schottmiller et al. [1]. The main results of this report were the following: Univalent additives such as chlorine are able to break the chains and terminate the chain ends. Isoelectronic additives such as sulfur induce the formation of mixed 8-membered selenium rings and mixed chains. A larger number of sulfur atoms will go into the rings than into the chains, and therefore the density of 8-membered pure selenium rings decreases with increasing sulfur content. Branching of chains occurs by additions such as arsenic. Moreover the ring concentration decreases rapidly with increasing amounts of arsenic.

First results will be given on the influence of sulfur on the electrophotographic properties of amorphous selenium. The evaporated layers we investigated contained up to 30%* of sulfur.

The decay in surface potential through exposure after positive charging of the surface was measured, and the light intensity was held constant for all measurements. The results are shown in Figure 1. In the diagram 1a the light decay in % is logarithmically plotted against photon energy for pure amorphous selenium and for three selenium-sulfur alloys. It can easily be

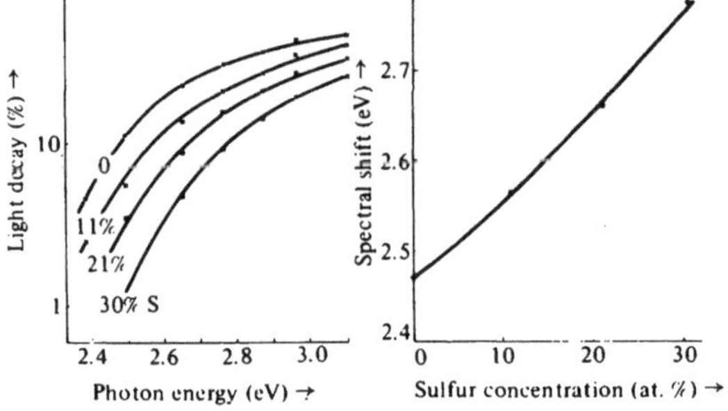

Fig. 1. Light decay and its spectral shift in the Se-S-system

* All concentrations are here expressed in atom percent.

seen that with increasing concentration of sulfur the curve shifts to higher energy, i.e. to shorter wavelengths. This shift is shown in the diagram 1b in more detail taken for a light decay of 10%. It is in agreement with the corresponding shift of the absorption edge [2].

Some other electrophotographic properties we investigated for selenium-sulfur alloys are shown in Figure 2. These results were in contrast to

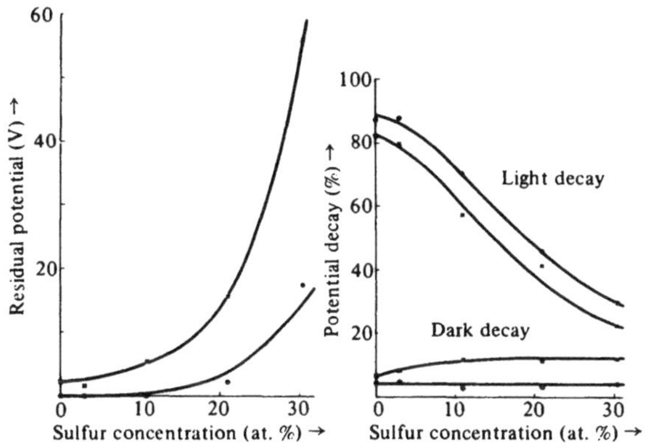

Fig. 2. Electrophotographic properties of Se-S alloys
a) Residual potential
b) Potential decay
○ – ○ after dark storage
x – x after prior illumination

those of Figure 1 obtained using the continuous spectrum of an incandescent lamp. In the diagram 2b is shown the dependence of light decay and dark decay on sulfur concentration. The light decay decreases rapidly with increasing sulfur concentration. The reason for this behaviour is on the one hand the spectral shift of photoconduction to shorter wavelengths and also the drop in light-intensity of the incandescent lamp towards shorter wavelengths. On the other hand the change of the hole mobility is of influence, which decreases with increasing sulfur content [1]. The upper curve is found for layers, which were stored in the dark for a long time, whereas the lower curve is obtained after exposure to light. The dark decay is almost unchanged by the sulfur concentration when the sample has been stored in the dark for a long time. After prior exposure to light it increases slightly. The light fatigue, i.e. the difference between the dark decay curves, therefore rises somewhat. In the diagram 2a is shown the influence of sulfur addition on the residual potential. This quantity is considerably enhanced with increasing sulfur concentration, particularly for layers which are fatigued by prior illumination.

To get some more insight into the doping mechanism we have also studied selenium layers which, besides the doping with sulfur, were additionally doped with iodine. Figure 3 shows the results for layers which contained 26% sulfur and a different content of iodine over several orders of magnitude. In the diagram 3a is

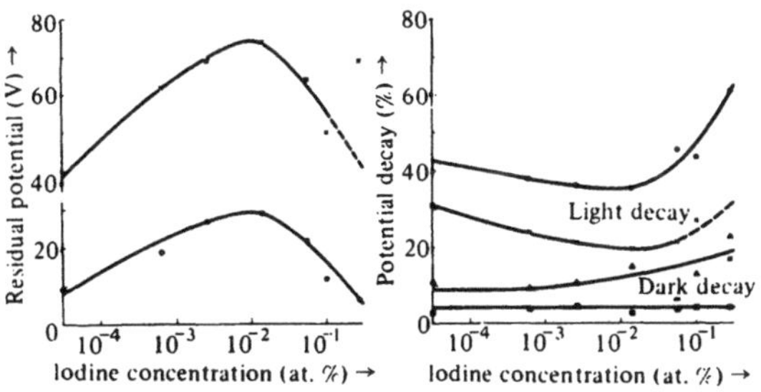

Fig. 3. Electrophotographic properties of a Se-S alloy (26% S) doped with iodine
a) Residual potential
b) Potential decay
○ – ○, □ – □ after dark storage
x – x, △ – △ after prior illumination

shown the influence of iodine on the residual potential and in 3b its influence on potential decay. For small concentrations the light decay decreases slightly until a relatively abrupt rise occurs at high concentrations. This is found not only for the samples which stayed in the dark, but also after exposure to light. The dark decay, on the other hand, is again not much affected by addition of iodine, except that at high concentrations it increases somewhat after exposure to light. The residual potential in the diagram 3a shows the reverse behaviour as the light decay, since it increases at first and after a maximum at 10^{-2}% iodine content it decreases. The character of the curve is similar again for storage in the dark and after exposure.

One important result of these investigations is that by addition of sulfur the residual potential increases considerably. This indicates that positive charge is stored within the selenium layers; therefore, hole traps are present in agreement with the results of Schottmiller et al. [1], who found a decrease of hole mobility and also of hole lifetime.

Remarkable is also the behaviour when both sulfur and iodine are added. The existence of a minimum in the light decay curves or a maximum in the curve for the residual potential indicates that two different processes exist, which influence these quantities in opposite ways. The increase in residual potential and the corresponding slight decrease in light decay at low iodine concentrations seems to indicate that the density of hole traps increases with increasing iodine content. Schottmiller et al. [1], however, found that addition of halogen impaired only the transport of electrons, indicating that electron traps are created. In our results the presence of electron traps would be in agreement with the increasing light fatigue in dark decay. The rapid increase in light decay at high iodine concentrations and the strong decrease of residual potential, however, are not yet understood.

In conclusion I would like to show an interesting result of a differential-thermal-analysis on the selenium-sulfur alloys. In Figure 4a differential-calorimetric curves are shown. The solid curve has been found for amorphous selenium with 11.5% sulfur, the dashed curve for material with additional iodine content of 10^{-2}%. Maxima are related to exothermic, minima to endothermic processes. In contrast to pure selenium, where only one exothermic maximum is found when crystallization occurs, these

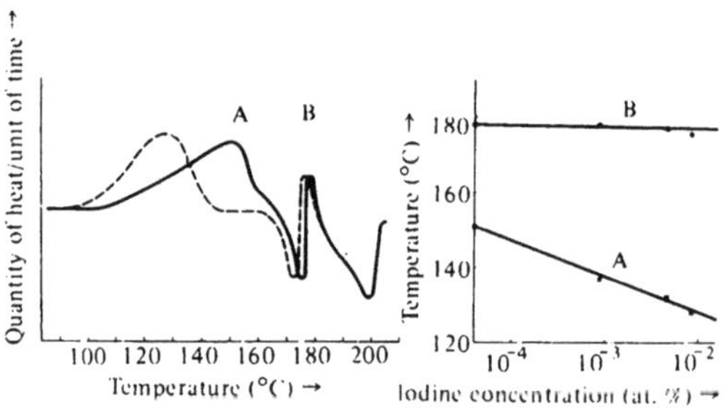

Fig. 4. Differential-calorimeter curve of a Se-S alloy (11,5% S) doped with iodine

materials with sulfur content show the two exothermic maxima A and B. The maximum A is found for samples without additional iodine content at 150° C. By addition of 10^{-2}% iodine it is shifted to about 130° C. The second maximum B, however, is only affected to a small extent. This is shown in the diagram 4b in more detail. The shift of A to lower temperatures can be attributed to the increasing tendency to crystallize of the samples containing iodine, because iodine can break up the chains and, therefore, enhances crystallization. The remaining part of the curve at higher temperatures is probably due to phase transitions. These results indicate that the iodine atoms preferably influence the chain part of amorphous selenium and that, on the other hand, the sulfur preferably goes into the 8-membered rings.

The work on which this report is based is subsidized by funds of the Federal German Ministry of Education and Science (File No. NT 209). The Federal Minister for Education and Science does not warrant the correctness, accuracy and completeness of the work as well as observance of private rights of third parties.

Reference

[1] J. Schottmiller, M. Tabak, G. Lucovsky, A. Ward, J. Non-Crystalline Solids **4**, 80 (1970).

[2] O. Watanabe, S. Tamaki, Electrochimica Acta **13**, 11 (1968).

Model of Electrophotographic Binder Layers

V. Gaidelis, E. Meskuotiene, D. Jurevicius and Z. Pocius

Abstract

A short review of photoelectric properties of ZnO electrophotographic layers and some models, as well as some new data concerning ZnO and CdS layers with a great amount of binder are presented in this paper. Charge carriers are shown to be injected into a binder and move in it. Some properties of PbO layers involving silicone polymer and CdS layers incorporating inorganic glass are described.

A model is presented which qualitatively takes into account electrostatic forces of barriers within the microcrystals. This model is able to account for major properties of ZnO and other binder layers.

1. Introduction

Electrophotographic layers can be prepared by using various semiconducting powders and various binding materials. ZnO layers involving organic and silicone polymers are widely employed in practice and have been investigated in detail. Their basic properties have been first described in [1].

In addition, PbO [2] and CdS [3] layers with polymers, CdS, CdSe, CdS_xSe_{1-x} layers involving an inorganic glass, as well as binder layers involving various inorganic semiconductors and organic semiconducting substances (for example, poly-N-vinylcarbazole) are promising and of great interest.

Although ZnO layers are extensively investigated, their basic characteristic, i.e., the slow potential decay at the exposure beginning or the so-called S-shaped photodischarge curve has remained one of the least understood phenomena in electrophotography for a long time. In order to understand the reasons of such a photodischarge, basic properties of ZnO layers should be considered and compared with those of other binder layers, single crystals and sintered pellets.

Although characteristics of various binder layers are different, they can be divided into two groups. ZnO layers with a normal amount of a binder, i.e., when there are direct contacts between semiconductor microcrystals, as well as CdS_xSe_{1-x} layers involving an inorganic glass belong to the first group. Apart from the S-shaped photodischarge, the space screening charge formed during charging, the small residual potential, as well as the ability to retain charges of one sign are characteristic of these layers. Features typical for the second group of layers, such as the considerable residual potential, the ability to be equally effective when charged negatively or positively, the absence of the S-shape, arise when the binder amount is increased. Apart from layers involving a large amount of binder, some layers having a normal binder quantity also belong to this group, for example, PbO layers or layers of CdS activated by acceptor impurities.

Now we shall briefly discuss basic properties of ZnO layers and models describing their behaviour. Later we shall present some data concerning layers with a large amount of binding material and show that charge carriers can be injected into the binder and move within it. We shall also present some results concerning CdS layers with an inorganic glass and PbO layers involving a silicone polymer. Finally, we shall present a model taking into consideration the electrostatic field of microcrystals. The model accounts for the most important characteristics of the layers.

2. Review of Properties of ZnO Layers Involving a Normal Amount of Binder. Models

Since properties of such ZnO layers have been investigated and described in literature in detail, we shall not present new results and shall only briefly deal with known facts which are important for the understanding of physical processes occurring in the layers.

The photodischarge process has been studied in [1, 5–8]. The following regularities have been observed:

(a) The photodischarge rate at the exposure beginning is small, then increases, reaches a maximum value and again drops to zero as the potential approaches zero, i.e., the photodischarge follows an S-shaped curve.

(b) The photodischarge rate at the exposure beginning increases faster when the initial potential is small, as well as when the layer was illuminated prior to or after the charging.

(c) The S-shaped photodischarge is observed over a wide range of light intensities which corresponds to potential half-decay times from several tens of seconds to fractions of miliseconds [5, 6].

(d) A significant drop of the potential is observed when the illumination is stopped.

(e) The S-shape is more pronounced in the spectral sensitization region than in the intrinsic region.

(f) The S-shape is observed in ZnO layers both negatively and positively charged [6].

The above mentioned regularities are valid when the layer is exposed to continuous light. The results obtained when exposing to short light pulses (10^{-7} sec) are very important for the understanding of processes occuring in the layers. This is carried out in such a way that the potential after illuminating with a single pulse would be significantly smaller than the initial potential. These investigations have disclosed the following regularities [6, 7]:

(a) When a ZnO layer is illuminated with a short light pulse, a fast decay of the potential is observed during exposure and a much slower but greater decay after exposure. The magnitude of the fast decay is only 2 to 3% of the total decrease.

(b) The magnitude of the fast decay of the potential increases with potential increase in the region of small potentials and is independent of the potential in the region of high potentials.

(c) The slow decay rate reaches a maximum at some potential value and then considerably decreases as the potential is further increased.

The density of the space screening charge in ZnO layers has been first determined by Amick [9]. This charge density varies in the range 10^{14} to $10^{16}\,e/cm^3$ for various ZnO samples. The space charge density of the sample strongly depends upon its storage time in the dark. After long dark storage the space charge density is about $10^{15}\,e/cm^3$ and increases by an order of magnitude when the sample is illuminated prior to measurement [6].

ZnO layers can be charged negatively much better than positively. However, layers kept in the dark for a long period of time and involving electron acceptor substances can be charged to several hundred volts. The rate of potential decay in the dark is greater and the photosensitivity is smaller for positively charged layers than for negatively charged ones.

In order to account for the S-shaped photodischarge of ZnO layers, Gerritsen, Ruppel and Rose [1] have proposed a model in which the major role is ascribed to the trapping of charge carriers. This model was also used by Arneth and Lorenz [5]. Later a model involving trapping levels has been investigated theoretically [7, 10]. It has been shown in these publications that trapping alone is unable to account for the principal phenomena observed in ZnO layers.

Some authors identify the trapping levels with positively charged ionised donors. For example, Schaffert [12] described the photodischarge process in ZnO layers making use of concepts of "direct" and "indirect" recombination. The "direct" recombination is the recombination of free pairs, and the "indirect" one the electron trapping on levels of ionised donors. Thus, this theory presents, in general, nothing new comparing with the model described in [1].

Hauffe [13] contends that the S-shape is caused by oxygen desorption during exposure. When oxygen is desorbed from ZnO microcrystals, the density of electron trapping levels is decreased and, therefore, the photodischarge rate is enhanced.

Photodesorption seems to play an important part in ZnO layers. It might explain the light fatigue observed in layers. However, it is unable to account for the occurrence of an S-shape over a wide range of light intensities.

A barrier model has been advanced in [6, 7]. It is known that a rather high density of donor levels exists in ZnO microcrystals and a high density of acceptor levels on their surface. By microchemical analysis it has been found [11] that there are 10^{-5} weight parts of excessive Zn in ZnO powder which corresponds to a donor density of $10^{18}\,cm^{-3}$. The surface acceptor levels are created by adsorbed oxygen or by other electron-acceptor substances. It follows from the temperature dependence of conductivity that the barrier height in ZnO binder layers may amount to 1 eV [6]. A model composed of layers parallel to the surface, involving potential barriers, and ideally contacting each other, has been studied in [6, 7]. Such a model is able to explain the S-shape, it fails, however, to account for the S-shape increase as the initial potential increases.

The barriers in the system made up of layers with parallel, ideally contacting walls exist until the outer field is weaker than the barrier field. When the outer field gets stronger, barriers are strengthened and cannot exert any appreciable influence upon the motion of charge carriers. The effective mobility of charge carriers in such a system is monotonously enhanced with the field strength, and it follows from considering ZnO properties that mobility will drop with field strength in the region of strong fields. This behaviour of ZnO layers is determined by a heterogeneous structure, i.e. by the presence of binder interlayers between semiconductor grains and by Schottky-type barriers in the grains themselbes. An essential part is also played by small-area contacts between semiconductor microcrystals. In order to understand this better it is necessary to consider some new experimental data.

3. Layers Involving a Large Amount of Binder. Charge Carrier Injection into Binder

There are few data in the literature dealing with layers involving a large amount of binder. CdS polystyrene layers have been investigated in [15] in which the ratio of CdS and the binder was 1 : 2. Upon illumination the potential of such layers drops by ~10%.

Below we present some data relating to ZnO and CdS layers involving polyvinylbutyral, polybutylmethacrylate and poly-N-vinylcarbazole as binders. The ZnO "Cda" used was sensitized in an alcohol eosine solution with subsequent evaporation of the solvent. CdS of "luminaphor" grade was used.

Photodischarge curves of ZnO layers with various amounts of semiconductor are given in Figure 1. The concentration of eosine is 3 mg per g of ZnO. It is seen from the figure that the S-shape disappears and the residual potential

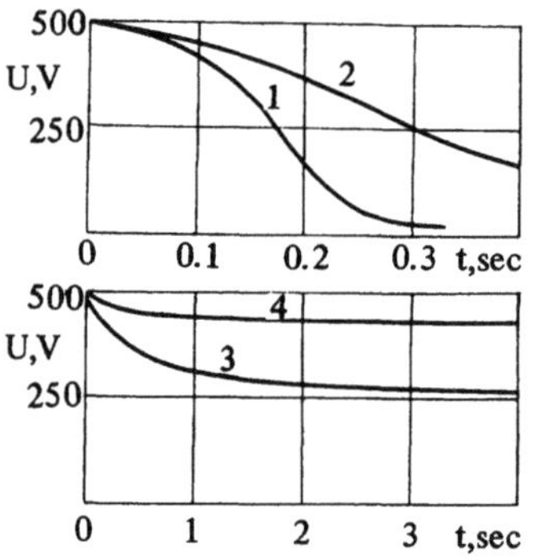

Fig. 1. Photodischarge curves of ZnO layers involving various amounts of polyvinylbutyral.
1–64 vol. % of ZnO; 2–32%; 3–16%; 4–4%.

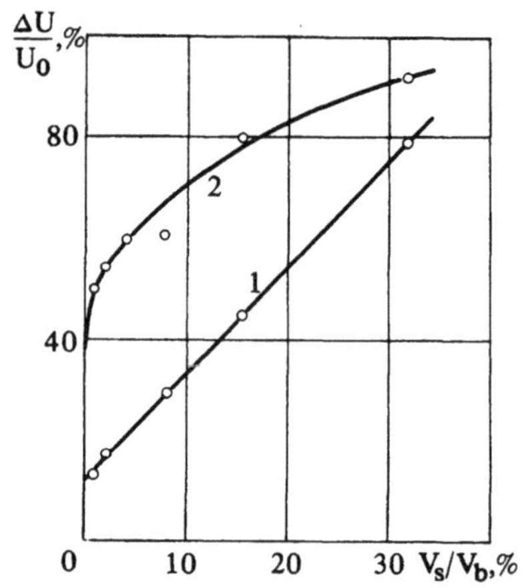

Fig. 2. Dependence of $\Delta U/U_0$ upon V_s/V_b. 1–CdS involving polybuthylmethacrylate; 2 – CdS involving poly-N-vinylcarbazole.

becomes significant as the amount of binder is increased. The initial rate of potential decay is considerably decreased. Thus, the presence of contacts between semiconductor microcrystals is essential for the appearance of the S-shape.

An important feature of layers involving a small amount of semiconductor is a rather strong decay of the potential ΔU upon illuminating which leads to the ratio $\Delta U/U_0$ markedly exceeding that of the semiconductor volume V_s to the binder volume V_b (Fig. 2). If charge carriers could move only within each microcrystal then the potential drop could not exceed the volume ratio V_s/V_b. This can be demonstrated in the following way.

Let us assume that a whole semiconductor makes up a continuous layer whose thickness is l_s, and a binder makes up the remaining portion of a layer whose thickness is l_b. Then it can be easily shown that the following formula is valid.

$$\frac{\Delta U}{U_0} = \frac{l_s\epsilon_b}{l_b\epsilon_s + l_s\epsilon_b} = \frac{V_s\epsilon_b}{V_b\epsilon_s + V_s\epsilon_b} \approx \frac{V_s}{V_b} \cdot \frac{\epsilon_b}{\epsilon_s} \cdot (1)$$

where ϵ_s and ϵ_b are the dielectric constants of semiconductor and binder, respectively.

Since $\epsilon_b < \epsilon_s$, then $\Delta U/U_0$ should also be less than V_s/V_b.

One may think that formula (1) also holds in some approximation for the real distribution of a semiconductor in a layer. The large difference between the ratios $\Delta U/U_0$ and V_s/V_b cannot be accounted for by the grain distribution, and one should assume that charge carriers are injected from a semiconductor into the binder and can move in it.

The effective range of charge carriers Δl, can be found from the potential decay rate when illuminating a layer with light of known intensity. Let us assume that the photogeneration quantum

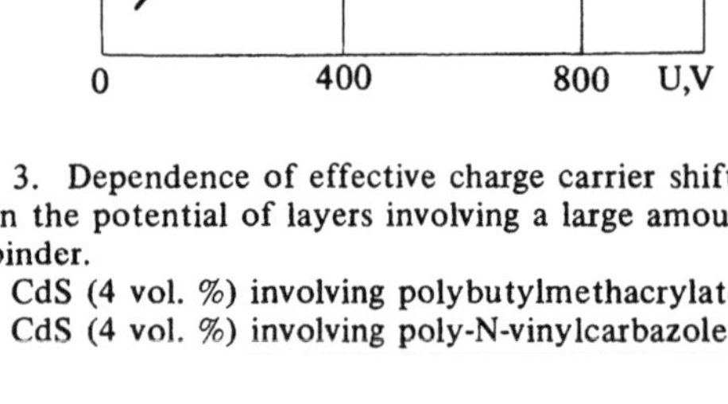

Fig. 3. Dependence of effective charge carrier shift upon the potential of layers involving a large amount of binder.
1 – CdS (4 vol. %) involving polybutylmethacrylate;
2 – CdS (4 vol. %) involving poly-N-vinylcarbazole.

yield is β, the diffuse reflection coefficient of the layer is R and the quantity of light photons incident upon a unit surface area is N. Then the rate of the potential decay can be expressed as follows

$$\frac{dU}{dt} = eN(1 - R)\beta \Delta l \, . \tag{2}$$

Figure 3 shows the dependence of Δl upon the initial potential of a layer. The dependence was calculated from the initial potential decay rate upon illuminating with 510 nm-light and assuming $\beta = 1$. Δl linearly increases with potential; however, these straight lines do not always pass through the origin. The finite range of charge carriers in the region of low potentials can be explained by the shift in semiconductor microcrystals, and its increase with potential, by the injection of charge carriers into the binder. The charge carrier shift is greater in poly-N-vinylcarbazole than in dielectrics, such as poly-butyl-methacrylate. It is quite understandable since charge carriers in poly-N-vinylcarbazole, especially holes, are rather mobile.

Charge carriers injected into the binder are captured; however, they may be freed and thus may drift slowly. Figure 4 demonstrates this case where the dark decay of the potential for a ZnO layer is given. In one case (curve 1) the layer after charging was not exposed, and in the other it was illuminated, so that the potential dropped down to the residual one (curve 2). As seen from Figure 4 the potential decay is considerably faster in the second case which can be explained by the drift of charge carriers injected into the binder.

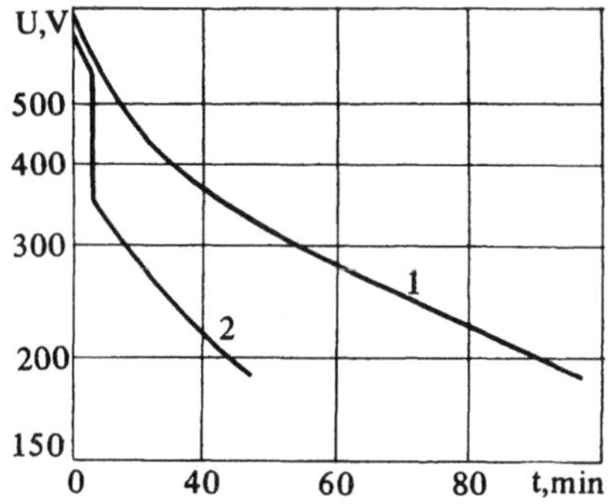

Fig. 4. Dark decay of potential (curve 1) and the decay of initial potential (curve 2) for ZnO layers involving polybutylmethacrylate (8 vol. % of ZnO, 92% of binder).

The injection of charge carriers into the binder and their slow drift in it are essential phenomena also in layers with a normal amount of binder. These phenomena will be taken into consideration later when describing a model of heterogeneous layers.

4. PbO Layers Involving a Normal Amount of Binder

Making use of PbO layers we shall show that the monotonous decrease in the photodischarge rate is characteristic not only of layers involving a large amount of binder. PbO layers display no residual potential; nevertheless the photodischarge rate always diminishes during exposure.

The PbO layers studied were prepared according to [16]. The red tetragonal PbO was dispersed in KO-815 silicone laquer. 25 g of PbO contained 6.8 g of the lacquer (dry residue). The initial thickness of a layer after coating on a duralumin plate was 250 μm. The upper part of the layer was mechanically removed so that the thickness of the remaining layer was 90 μm. Then the layer was heated at 220° C for 10 min. Layers prepared in this manner could be charged

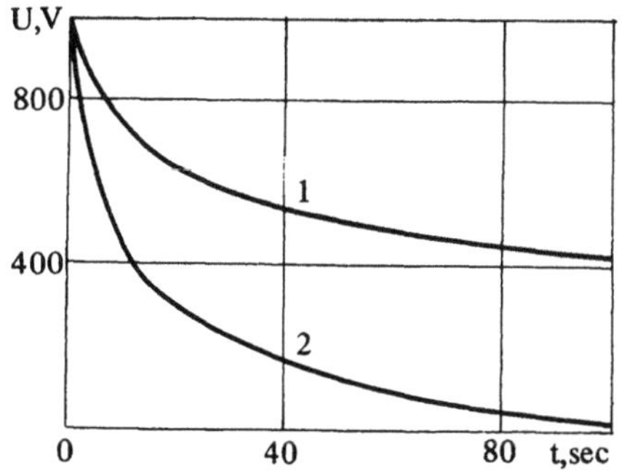

Fig. 5. Photodischarge of PbO layers. 1 – positive potential; 2 – negative potential.

by both charges up to 700–1000 V. The photoresponse is higher for positive potentials. Photodischarge curves are presented in Figure 5. The positive residual potential is insignificant which suggests that there are many contacts between PbO microcrystals. Contacts in these layers arise owing to the fact that upon coating the suspension, the heavy PbO will sediment, so that a semiconductor-rich layer is formed at the base and a layer of nearly pure binder at the top. Upon removing the top, a layer containing the optimal amount of binder is left.

5. CdS Layers Involving Inorganic Glass Binder

These layers are of interest since their photoelectric properties can be easily changed by adding acceptor impurities. Properties of layers without specially introduced impurities are similar to those of ZnO layers involving a normal amount of binder. A small amount of Cu strongly diminishes the S-shape or even completely removes it.

Samples were prepared according to [4]. CdS powder was mixed with finely ground glass of the following composition: B_2O_3 – 15 weight %, ZnO – 6% and PbO – 79%. The weight ratio of the glass and CdS was varied from 4 : 1 to 2 : 1. The powder mixture was heated at 700° C for 5 min. The obtained sintered mass was then ground, mixed with ethyl alcohol and evenly coated on glass plates having a conductive SnO_2 layer. After the solvent was evaporated layers were first heated at 500–550° C for 2 to 5 min and then slowly cooled to room temperature. $CuCl_2$ impurity (~0.05% by weight of the CdS) was added when preparing the suspension of the glass and sintered CdS.

Layers without $CuCl_2$ impurity either cannot be charged positively at all, or the dark decay of the positive potential of these layers is much faster than that of the negative potential. Layers containing the impurity can always be charged both positively and negatively and rates of potential decay differ insignificantly.

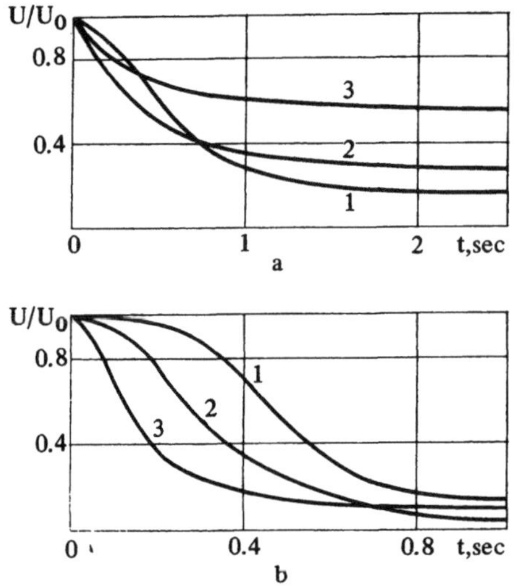

Fig. 6. Photodischarge of CdS layers involving a glass binder; a – layers without any impurity; b – layers with $CuCl_2$ impurity (0,05% of CdS by weight).

Photodischarge curves are presented in Figure 6. Layers without Cu impurity display clearly pronounced S-shape when the initial potential is high. It disappears in the region of low potentials, and can hardly be observed in layers involving a small amount of Cu.

Figure 7 shows the dependence of surface charge density upon potential for CdS layers. It demonstrates that $CuCl_2$-free layers have a significant space screening charge. The layers with the impurity have no such charge. This can be accounted for by the fact that the acceptor Cu impurity compensates donor levels in CdS microcrystals.

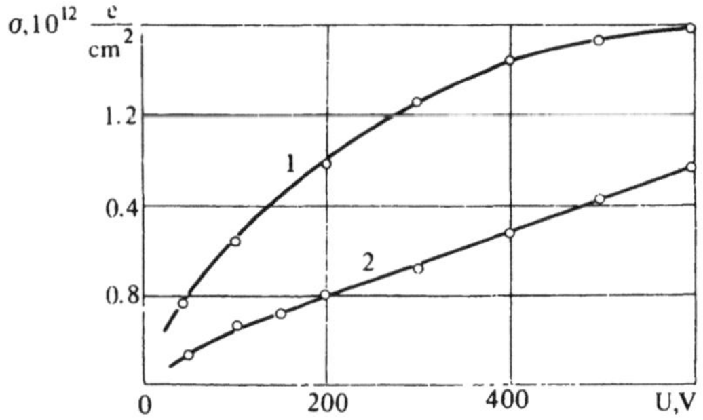

Fig. 7. Dependence of surface charge upon potential for CdS layers.

A significant space screening charge has been found in CdS_xSe_{1-x} layers involving a glass binder [8].

6. Single Crystals and Sintered Pellets

In order to understand better the properties of binder layers it is useful to compare them with properties of single crystals and sintered pellets of the same substances. Undoped ZnO single crystals can be charged negatively up to 50 to 80 V and cannot be charged positively. The potential decay is monotonously slowed down during exposure [17, 18]. Li-doped samples can be charged negatively up to 1000 to 1300 V, but again cannot be charged positively. In most cases the photodischarge rate is slowed down with time, but there are samples in which the S-shape can be observed. This S-shape disappears in the region of low light intensities (Fig. 8), and this can be accounted for by electron trapping in shallow levels [18]. Properties of ZnO pellets sintered at temperatures between 800 to 1000° C differ little from those of single crystals [19]. The difference in essential characteristics of ZnO pellets and single crystals on the one hand and binder layers on the other suggest that the dielectric interlayers play an important part in binder layers.

Sintered pellets of CdS have been prepared by somewhat different methods [20]. CdS powder was pressed, then heated in air at a temperature of 460° C for 12 to 24 hours. Undoped CdS powder as well as CdS containing some amounts of Cu and Cl have been used. The pellets pre-

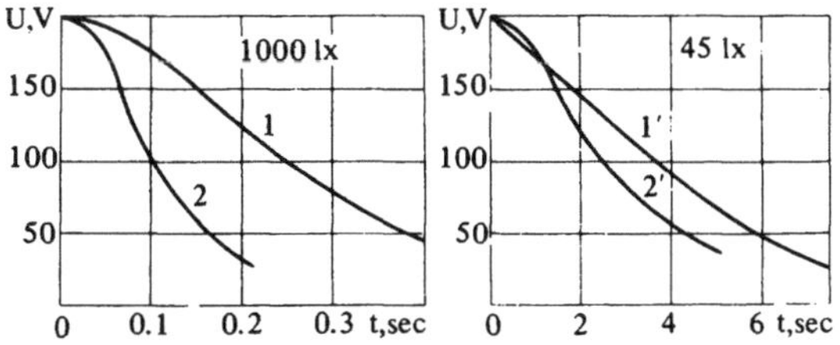

Fig. 8. Photodischarge curves of ZnO single crystals (1,1') and ZnO layers involving polyvinylbutyral (2,2'). Left – high light intensity; right – lower light intensity.

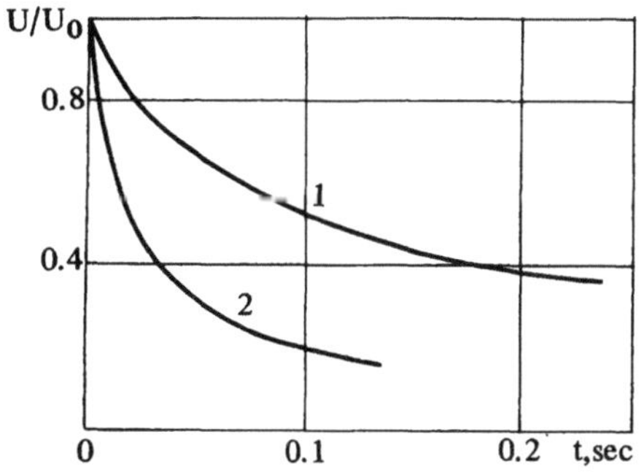

Fig. 9. Photodischarge curves of CdS pellets. 1 – undoped CdS pellet, U_0 = 70 V; 2 – Cu and Cl-doped CdS pellet, U_0 = 800 V.

pared in this way cannot be charged. This fact can be accounted for by the presence of micropores in the pellets. When pellets are impregnated with K-55 silicone lacquer and dried, they can be charged and are photosensitive. Figure 9 shows phtodischarge curves of such pellets. It is seen from the figure that the properties of these pellets significantly differs from those of CdS layers involving a glass. This can be explained by different amounts of binding material; in pellets there is little binder which fills only narrow micropores, and the contact area between microcrystals is large. In layers involving a glass binder they are contacting through small areas (separate points) due to a much greater amount of binder. The small value of the maximum potential of both ZnO and CdS pellets without impurities is explained by the presence of a space screening charge. It follows from this that donors present in microcrystals are not effectively compensated by surface levels.

7. Model of Heterogeneous Layers

The structure of layers involving various amounts of binder is given in Figure 10. Figure 10a shows the cross-section of a layer incorporating a large amount of binder; the semiconductor microcrystals are separated by thick interlayers of a binding material as shown earlier; the potential decay upon illuminating the layer is due to the shift

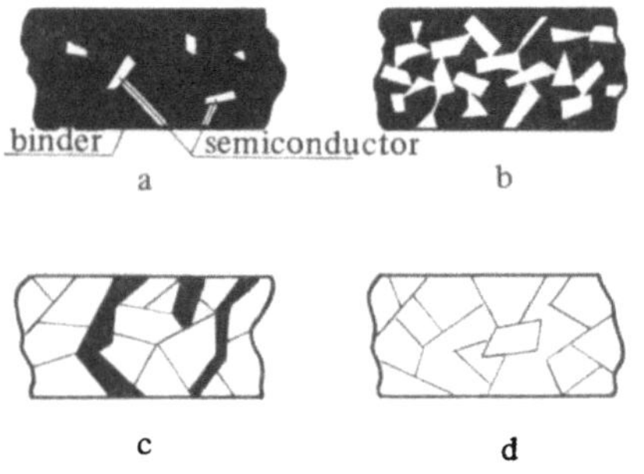

Fig. 10. Structure of electrophotographic layers involving various amounts of binder.
a – layer with large amount of binder;
b – layer with normal amount of binder;
c – layer with a small amount of dielectric interlayers;
d – polycrystalline layer without any dieletric interlayers.

of charge carriers within microcrystals as well as to their injection into the binder and their motion in it.

Figure 10b shows the structure of layers with a normal amount of binder. Microcrystals are contacting each other by separate points and the space between them is filled with a binder. As shown earlier, the S-shaped photodischarge can occur in such layers.

Figure 10c demonstrates layers similar to the above-described CdS sintered pellets impregnated with lacquer. In these layers microcrystals are contacting each other by nearly all their surface except for areas of small dielectric interlayers which fill micropores.

Finally, Figure 10d shows a polycrystalline layer without dielectric interlayers, for example, a well sintered ZnO pellet.

In layers as presented in Figures 10c and 10d charge carriers can freely pass from one microcrystal into another. Barriers on the microcrystal surface are much smaller in this case than in binder layers, and besides, there is little or no dielectric in the form of interlayers.

Now we shall consider the model of layers presented in Figure 10b. In layers of such a kind with S-shaped discharge curves a space screening charge is involved. This is not mere coincidence since the presence of the S–shape can be foreseen for layers in which semiconductor microcrystals have a sufficiently high space charge and high potential barriers on their surface.

It has been found earlier that the layer model composed of ideally contacting flat sheets involving potential barriers fails to account for all experimentally observed phenomena. The effective mobility of charge carriers in such a system monotonously rises with field strength. However, in a system composed of semiconductor crystals contacting by separate points and when there are high Schottky-type potential barriers and dielectric interlayers on the crystal surface, the effective mobility can be very low in the region of strong fields and increases as the outer field decreases.

Let us consider the system presented in Figure 11 and composed of four freely oriented microcrystals. We shall try to evaluate conditions of charge carrier drift which could be found in a real layer.

The electric field barrier exists inside microcrystals. In the lower part of microcrystals this field is directed against the outer field created during the layer charging. The conditions of charge carrier motion in a microcrystal are, of course, greatly dependent upon the resultant direction of the field at the semiconductor-binder boundary. When the outer field is strong, i.e., when the layer surface potential is high, the resultant force presses

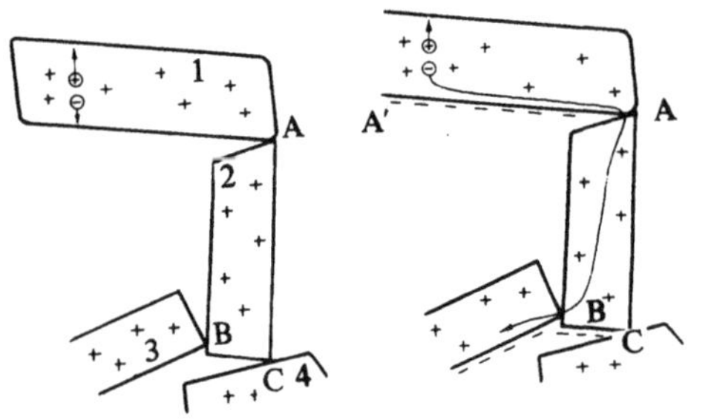

Fig. 11. System composed of four contacting microcrystals:
a – strong outer field;
b – weak outer field.

majority charge carriers (electrons in ZnO or CdS) towards the binder. Charge carriers cannot drift down the inclined lower wall of the 1st crystal (Fig. 11a) since under the action of the strong field they are injected into the binder and are quickly trapped in it. Therefore, the free shift of charge carriers is small and the photodischarge is slow in the case of a strong field.

When many localized electrons are accumulated at the lower wall of a microcrystal (Fig. 11b), the field is weakened and the component perpendicular to the wall becomes zero or even changes its direction. Then photogenerated charge carriers are no longer pressed towards the binder and under the action of parallel components of the field can drift freely until they reach another microcrystal and enter it. Thus, transition of charge carriers from one microcrystal into another becomes possible and the free range of charge carriers is considerably enhanced.

The greater the inclination of microcrystal walls with respect to the layer surface the sooner charge carriers will begin to move freely. For example, the 2nd crystal (Fig. 11) is nearly perpendicularly oriented and charge carriers can freely move down it, independent of the field strength.

Apart from the movement down the inclined walls of microcrystals, the transition of charge carriers over contacts into neighbouring microcrystals is essential for the appearance of the S-shaped photodischarge.

Contact A in Figure 11 finds itself on a nearly horizontal wall, therefore, the strong field reduces the barrier and charge carriers are able to pass through it from the 1st microcrystal into the 2nd one if they reach the contact at all, i.e., if they are generated in the contact region. The barrier is not recovered during exposure and when charge carriers start moving down an inclined wall towards to contact, conditions of their passage across the contact are not appreciably deteriorated. Thus, A-type contacts remain open for charge carriers during the whole exposure.

In contrast to A-type contacts, B-type contacts (Fig. 11) which are on nearly vertical walls are closed at the exposure beginning due to high barriers, their height being independent of the outer field strength. During exposure the barrier height diminishes due to the drift of electrons and holes generated in a microcrystal, and the probability of passage across the contact increases.

There are intermediate types of contacts on strongly inclined walls in real layers. Such contacts can work either as A-type or B-type ones depending upon the outer field strength.

One more circumstance can be essential for the appearance of the S-shape in real layers, namely the presence of deadlocks in which charge carriers can be caught. The lower part of the 2nd microcrystal (Fig. 11) might serve as such a deadlock, provided there is no contact with the 4th microcrystal. At the exposure beginning electrons are accumulating in it. When there is a sufficient amount of electrons, some of them will stay no longer in the lower part of a microcrystal due to the mutual repulsion and remain free being able to reach contact B. Only after this can charge carriers pass from the 2nd microcrystal into the 3rd one.

Thus, the motion of photogenerated charge carriers within a microcrystal at the exposure beginning creates conditions for charge carriers to pass from one microcrystal into another. In order to improve the drift conditions in the layer bulk it is necessary for the light to penetrate deeply into the layer.

The light penetrates into a layer owing to the fact that the binding material is transparent and some portion of the light is reflected from the microcrystal surface. In some cases, e.g. in ZnO polyvinylbutyral layers sensitized with a large amount of dye (5–10 mg of eosine per 1 g of ZnO), the S-shaped photodischarge is well pronounced near the limits of the sensitization region and is absent at the dye absorption maximum. Preliminary investigations have diclosed that the light does not get through the layer at all at the absorption maximum. The intrinsic light can create both electrons and holes. Since holes effectively modulate barriers, then upon illuminating with the intrinsic light B-type contacts are opened faster and the S-shape is less pronounced in contrast to the light from the sensitization region where charge carriers of one sign only are generated.

Thus, the following points should be fulfilled for the S-shaped photodischarge to occur:

1 – high potential barriers in microcrystals,
2 – small-area contacts in microcrystals and
3 – light penetration into the layer bulk.

The first condition is not fulfilled in PbO layers. It is not also fulfilled in Cu-doped CdS layers involving a glass binder. Since Cu as an acceptor impurity is in the bulk and compensates donor levels, the barriers on the microcrystal surface are lowered. Layers of this type does not exhibit the S-shape.

The second condition is not fulfilled in layers with a large amount of binder. Microcrystals may not contact each other in these layers. The conditions is also unfulfilled in CdS sintered pellets: here microcrystals will contact over large areas and the small dielectric interlayers have an insignificant effect upon the charge carrier drift through the layer.

In layers without any dielectric interlayers but which have high barriers on the microcrystal surface, the S-shape is possible only in the region of low potentials when the outer field is unable to smooth down the barriers. Generally speaking the model described in [6, 7] might be valid for such layers.

The third condition is not fulfilled in ZnO layers incorporating a large amount of a sensitizer in the spectral region of sensitization, provided the sensitizer also dyes the binder.

Some words should be said about ZnO and CdS layers with a glass which can be charged positively. According to the model described above, the slowly relaxing negative screening charge can be built up in microcrystals due to the features of B-type contacts (Fig. 11), as well as in deadlocks. If the layer is illuminated prior to charging, the barriers are diminished and the transition conditions through B-type contacts are improved. As a result, layers can be worse when charged positively.

A purely qualitative model has been described here which takes into account electrostatic forces inside microcrystals. In real layers, a significant role may be played by phenomena not discussed here, such as the photodesorption, recombination and trapping of charge carriers. However, the feature distinguishing binder layers from single crystals, pellets or vacuum-evaporated layers is the heterogeneous structure of binder layers.

The authors wish to acknowledge the kind assistance of Mr. J. Rakauskas in preparing the PbO layers.

Reference

[1] W. Ruppel, H. J. Gerritsen, A. Rose, Helv. Phys. Acta, **30**, 495 (1957); **30**, 504 (1957).
[2] Van Dorn, US Patent, No. 3008825, 1961; V. Baltrusaiticne, B. Kalinauskiene, E. Montrimas, Patent USSR Nr. 168127, 1965.
[3] M. Smith, A. J. Behringer, J. Appl. Phys. **36**(11), 3475 (1963).
[4] US Patent, No. 3288604, 1966.
[5] R. Arneth, B. Lorenz, Reprographie, **3.** Jg. No. 9, 199 (1963).
[6] V. Gaidelis, N. Markevic, E. Montrimas, Fizitcheskoje procesi v elektrofotografituheskich slojach ZnO (in Russian), Vilnius, 1968.
[7] J. Viscakas, V. Gaidelis, Reprographie,II, Köln 1967, p. 79.
[8] V. Gaidelis, Zh. nautch. i prikl. foto i kinematografii, **13**(2), 89 (1968).
[9] J. A. Amick, RCA Rev., **20**.(4), 770 (1959).
[10] V. Gaidelis, J. Kalade, Lietuvos fizikos rinkinys, **8** (1–2), 197)1967); V. Gaidelis, J. Kalade, E. Montrimas, Lietuvos fizikos rinkinys, **8** (3), 429 (1968); **8** (5–6), 819 (1968).
[11] G. Korsunovskij, Fizika tverdovo tela (in Russian) **4** (4), 965 (1962).
[12] R. M. Schaffert, Electrophotography, Focal Press, London and New York, 1965.

[13] K. Hauffe, J. Photogr. Sci., **10**, 321 (1962).
[14] J. Viscacas, V. Gaidelis, Z. Pocius, Zh. nautch. i prikl. foto. i kin., **14** (4), 273 (1969).
[15] S. Aftergut, J. J. Bartfai, B. C. Wagner, Applied Optics, Suppelement 3, 161 (1969).
[16] V. I. Bogolovskij et al., Fizitcheskie osnovi elektrofotografii, Vilnius, 1969, pp. 36, 40.
[17] H. Kiess, Applied Optics, Suppl. 3, 100 (1969).
[18] J. Viscakas et al., Fizitcheskie osnovi elektrofotografii, Vilnius, 1969, p. 82.
[19] V. Gaidelis, Z. Pocius, Fizitcheskie osnovi elektrofotografii, Vilnius, 1969, p. 119.
[20] V. Gaidelis, D. Jurevicius, Fizitcheskie osnovi elektrografii, Vilnius, 1969, p. 123.

Discussion

J. S. Boroky: Were EP binders other than PVB tried for the experimental conclusion of discharge models of this paper? The reason behind this question is that polymeric binders are chemically ill defined materials; they are polydisperse in their UW, and dependent upon the polymerisation process are the surface radicals on segments of polymeric chains. Therefore one cannot exclude the possibility that the shape of the discharge curves upon illumination is strongly dependent on what type of polymeric binder is used. If other binders were used, have the authors found any exceptions or deviations from the described model at one particular ratio of ZnO to binder?

A. Matulionis: Many other binders such as enamels, lacquers, resins etc. were used. (Please, consult the main text of the written report).

J. W. Weigl: The complications due to polymeric binders having a poorly defined composition and electrical properties could be avoided by using a model system comprising pure crystalline organic compounds as binders. Such a system was first proposed by a group from the Research Laboratories of Australia.* To the best of my recollection is exhibits no residual, and little or no induction phenomenon.

K. Verhille: How do you incorporate into your model the transient nature of the concentration of charge carriers when light is switched on? As published in the Rochester conference this effect, together with the continually decreasing electric field over the layer, explains experimental results to a certain degree.

A. Matulionis: The model emphasises electrostatic forces inside the micro crystal that are very important for understanding of specific properties of binder layers in general. Such phenomena as photogeneration, recombination, carrier trapping, photodesorption are also quite important, but they cannot explain the disappearence of the S-shaped photodischarge in the layers charged to lower voltages. An accurate treatment of a trapping model is given by Viščakas, Gaidelis at Köln in 1967 [1].

[1] J. Viščakas, V. Gaidelis, Reprographie II, Köln 1967, p. 79.

J. Rochlitz: I would like to make a comment in relation to the S-shape light-decay: We have investigated *homogeneous* Polyvinyl Carbozole binder layers and found a S-shaped light-decay. Low amounts of electron acceptors change the S-shape type. If the concentration of the electron-acceptors is enlarged, the normal exponential light decay can be observed. Is it possible to give an explanation for their behaviour by the given model?

A. Matulionis: Thank you for the contribution. I hope that the properties of that homogeneous layers should be explained by one of the models developed for homogeneous layers.

* I. Matkan, R. S. Wright, U. S. Pat. 3, 406, 063 (1968)

Investigation of Space Charge Formation and its Distribution in Electrophotographic Layers

V. Gaidelis, E. Montrimas, A. Pazera and J. Viščakas

Abstract

The kinetics of space charge formation and its geometric distribution is investigated in Se and As-Se electrophotographic layers.

The suggested technique to determine the space charge density and distribution is based on gradual neutralisation of space charge by injected carriers when the applied voltage is reduced linearly with time.

It was found that negative and positive space charges in Se electrophotographic layers are formed as a result of carrier injection from the electrodes and are localized near the injecting electrode in a narrow region of the layer. The density of this charge near the electrode is 10^{14} to $10^{15} cm^{-3}$ and drops sharply with distance from the electrode. The density of the space charge can be substantially increased when injection is increased, e.g. by illuminating the electrode with strongly absorbed light.

In As-Se electrophotographic layers a negative space charge is formed when the voltage is applied, due to the sweep out of free and ionized holes from the layer. The space charge density is low and approximately constant if the layer is kept under voltage for less than 0.1 s.

Later on space charge redistribution takes place and its density increases at the anode up to 10^{17} to $10^{18} cm^{-3}$. The redistribution time is shorter at higher voltages. It also depends on temperature and pre-exposure.

1. Introduction

In many semiconductor devices a space charge is formed which has a substantial effect upon their characteristics and operating conditions. In some devices it is formed at the blocking contact for majority charge carriers, in others, for example, in vidicon targets or electrophotographic layers, it is formed upon charging and during the potential decay in the dark or under illumination. The presence of the space charge in electrophotographic layers and vidicon targets can lead to an undesirable electrostatic memory which manifests itself by developing the previous electrostatic image, by diminishing the photosensitivity, and by changing other electrostatic properties of layers. The space charge not only substantially alters electric and photoelectric properties of layers but also makes it difficult to determine such parameters as the drift mobility [1, 2], the effective lifetime of charge carriers [3] etc.

Notwithstanding the significance of a space charge, the known methods of its determination are not sufficient. The determination of a space charge density in ZnO electrophotographic layers using the dependence of maximum potential of charging upon layer thickness [5, 6] or from the dependence of surface potential upon surface charge density [4, 7] is possible in the layer only when the density is independent of the coordinate. Otherwise, only the mere presence of a space charge in the layer can be stated, and to find its density is impossible. Below we shall describe a way by which it is possible to measure the space charge distribution in a semiconductor layer. This technique can be employed in the case of semiconducting materials involving a significant density of free charge carriers, and allows to determine their density or the space charge density in a depletion layer. This technique can under certain conditions also be used for high-resistance semiconductors.

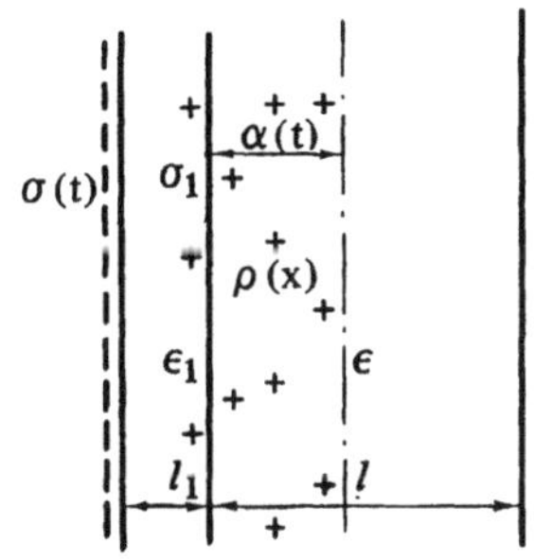

Fig. 1. Model of semiconductor involving a space charge.

2. The Investigation Technique

The following method may be employed for investigating the formation and the spatial distribution of a space charge in plane-parallel samples of semiconductors and photodielectrics [8].

The flat semiconductor or photodielectric is provided with electrical contacts on both sides. One contact is blocking, the other one will inject charge carriers, their polarity being opposite to the space charge. The injection of neutralizing charge carriers can also be performed by strongly absorbed light through a semi-transparent electrode.

After some time, the outer voltage applied to the sample is made to decrease according to a definite law (preferably linearly), and as a result, charge carriers are injected from one contact and neutralize the space charge. The neutralization of the space charge begins at the injecting electrode and advances into the sample following the plane of electrical field inversion (plane M, Fig. 1).

The neutralization of the space charge will be complete if the neutralizing charge carriers are not captured before reaching the boundaries of the neutralization region M, and if their amount is sufficient. The voltage on the neutral part of a layer should be insignificant, and the time of the voltage drop to zero much less than the relaxation time of a space charge. In the general case (Fig. 1), one of the semiconductor surfaces may be coated with a dielectric layer whose thickness is l_1 and electrical permittivity ϵ_1. The surface charge having density σ_1 may be present on the semiconductor-dielectric interface. Let us assume that this charge is constant. The charge density on the electrode $\sigma(t)$ and the screening depth $L(t)$ vary with voltage. Let us designate the layer thickness, the electrical permittivity of a semiconductor and the electrical permittivity of vacuum by l, ϵ and ϵ_0, respectively. By integrating the Poisson equation one can obtain the following expression for the surface potential:

$$U(t) = \frac{\sigma(t)\, l_1}{\epsilon_0 \epsilon_1} + \frac{\sigma(t) + \sigma_1}{\epsilon_0 \epsilon} L(t) + \frac{1}{\epsilon_0 \epsilon} \int_0^{L(t)} \int_0^{x} \rho(x)\, dx\, dx \qquad (1)$$

Differentiating this equation with respect to t and taking into account that

$$\int_0^{L(t)} \rho(x)\, dx = -[\sigma(t) + \sigma_1],$$

and that $$\frac{dL(t)}{dt}\, \rho(L) = -j = -\frac{d\sigma(t)}{dt} \qquad (2)$$

we obtain

$$L(t) = \frac{\epsilon_0 \epsilon}{j} \frac{dU}{dt} - \frac{l_1 \epsilon}{\epsilon_1}. \qquad (3)$$

If the current density j is known at time moment t, one can determine the screening depth in a semiconductor using the latter formula. Differentiating equation (3) with respect to t and taking into account equation (2), we can get a formula for the determination of ρ (L) at the boundary of the space charge region.

$$\rho(L) = \frac{1}{\left(\frac{\epsilon_0 \epsilon}{j^3} \cdot \frac{dU}{dt} \cdot \frac{dj}{dt} - \frac{\epsilon_0 \epsilon}{j^2} \cdot \frac{d^2U}{dt^2}\right)} \qquad (4)$$

If the voltage drops according to a linear law with the rate (dU/dT) = const. then formula (4) can be simplified.

If the space charge is formed due to the removal of free charge carriers then its density may be determined by a similar technique also when the voltage is increased according to some definite law. If, however, the space charge formation is caused by other reasons, for example, by generation of charge carriers of one kind from local levels or by trapping of charge carriers injected from contacts, then its distribution may be determined only when the voltage is decreased. Even in this case, the analysis of current regularities when the voltage is increased may provide valuable information regarding space charge formation.

3. Experimental Results and Discussion

It follows from the above that the technique described may be employed most advantageously for investigating the density of space charges or the concentration of free charge carries in classical low-resistance semiconductors. In these materials there are no problems associated with incomplete neutralization of space charge during measurement. Single crystals of B-doped silicon having p-type conductivity well illustrate the above.

Silicon plates were first coated with a 0.5 μm-thick SiO_2 film on one side; then Al contacts were evaporated on to both surfaces. When negative voltage is applied to the electrode on SiO_2 film, the holes are attracted to the Si-SiO_2 interface. Some these holes are captured by

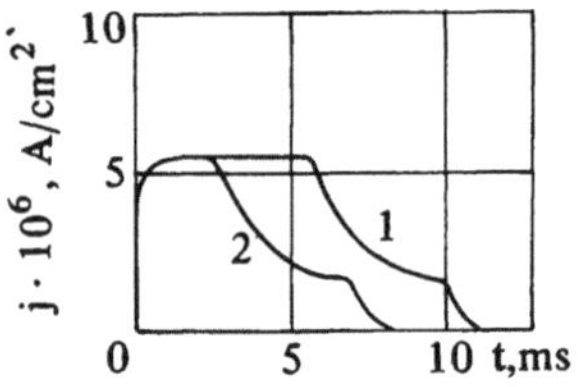

Fig. 2. Time dependence of the discharge current for p-type Si sample.
SiO_2 film thickness 0.52 μm.
1 – U_0 = 120 V, t = 7.8 · 10^{-3} sec;
2 – U_0 = 84 V, t = 5.4 · 10^{-3} sec (dU/dt is equal for both curves).

deep levels, the others remain free. At the beginning of the voltage decrease (Fig. 2), the current corresponds to the geometrical capacitance of the SiO_2 film. This is true until there are free holes at the interface. At some moment when the reserve of free holes at the interface diminishes, the current decreases which is associated with the formation of a depleted region. Figure 3 shows a curve of space charge density distribution in the depleted region when the voltage dropped to zero. The curve was calculated making use of the results presented in Figure 2. It is clearly seen from Figure 3 that there a thin channel (~1 μm) of enhanced conductivity at the semiconductor surface. In the bulk, the space charge density (~$1.5 \cdot 10^{15}$ e/cm^3) corresponds to the density of free charge carriers determined by alternative methods ($1.4 \cdot 10^{15}$ e/cm^3).

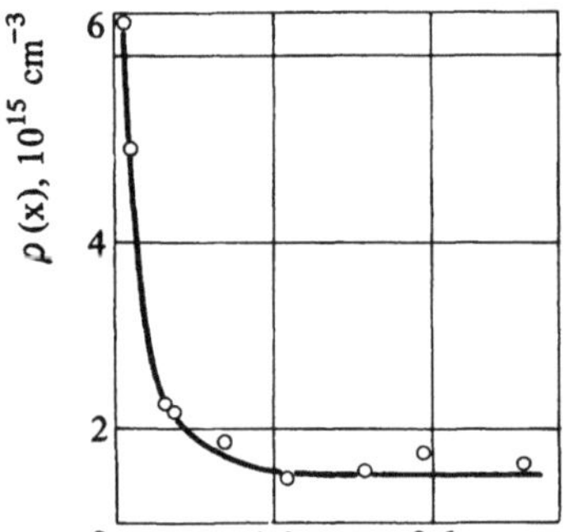

Fig. 3 Distribution of space charge density in the depleted region of p-type silicon formed due to hole trapping at Si-SiO_2 interface.

It is quite easy to calculate the surface density of holes captured at the Si-SiO_2 interface making use of values of the voltage at which the current decrease begins (Fig. 2). Values ranging from $1.7 \cdot 10^{12}$ to $9.2 \cdot 10^{12} cm^{-2}$ have been obtained for various samples. As seen from Figure 2, the density of captured holes is voltage–independent in the region of high voltages.

Using the above-described technique an attempt was made to study the space charge distribution in Se and As-Se electrophotographic layers.

Curves of the charging current in Se and As_2Se_3 electrophotographic layers are presented in Figure 4. For Se layers, the charging current in the dark usually corresponds to that of the sample geometrical capacitance (Fig. 4, curve 1), and in some cases a through current of long enough duration is observed. Curves of the first type are observed with samples having Al contacts which are blocking for holes and electrons in the case of amorphous selenium. In this case, no space charge is formed in the layer. Curves of the second type are characteristic of selenium samples when a Pt or Au anode is applied to the crystalline interlayer or when the contact is illuminated with strongly absorbed light. These contacts inject holes into the layer. Some of the charge carriers injected are trapped by deep levels which leads to the formation of a space charge in selenium electrophotographic layers. Quite different regularities govern the charging current kinetics in As_2Se_3 electrophotographic layers. In layers previously illuminated with weakly absorbed light, the charging current after switching on the gradually increasing voltage first decreases, then reaches a maximum value and finally drops to zero or to some definite value (Fig. 4, curve 3). For layers kept in the dark for several minutes, the charging current first decreases, then, at some moment increases up to the inflection point and finally diminishes to zero (Fig. 4, curve 4). The appearance of the maximum on the charging curve (shaded area in Fig. 4) is caused by the formation of a space charge in the layer. The latter is associated with the electrical field strength. As long as the field strength does not exceed ~10^4 volts/cm, the charging current is conditioned by the geometrical capacitance of the sample. When this value is exceeded, a space charge begins to be formed. Decreasing the sample voltage to zero and its repeated switching on within a time interval not exceeding 0.4 sec results in the charging current curve which coincides with that of the sample geometrical capacitance (Fig. 4, curve 7). This can be accounted for by the fact that neutralization of negative space charge does not occur upon decreasing the voltage and during a time of 0.4 sec after the voltage has dropped to zero. If the time interval is greater a charging current is observed which already differs from that of the sample geometrical capacity (Fig. 4, curves 5, 6). The negative space charge is completely neutralized only in 3 to 4 minutes and the charging current can again be described by curve 4 (Fig. 4). Such a slow neutralization process of the space charge seems to be caused by thermal generation of holes and their redistribution on the region of negative space charge. It has been observed that charging current curves for As_2Se_3 electrophotographic layers presented in Figure 4 are independent of the contact nature, provided it is blocking for holes. If the contact is nonblocking for holes then the regularities described disappear, since a through current of considerable magnitude flows through the layer. On the basis of the regularities des-

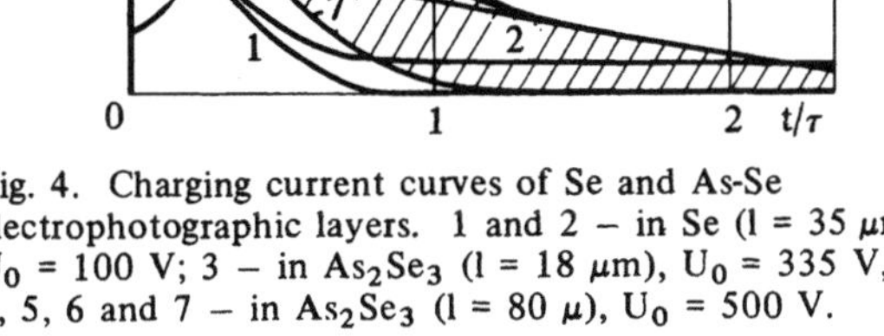

Fig. 4. Charging current curves of Se and As-Se electrophotographic layers. 1 and 2 – in Se (l = 35 μm), U_0 = 100 V; 3 – in As_2Se_3 (l = 18 μm), U_0 = 335 V, 4, 5, 6 and 7 – in As_2Se_3 (l = 80 μ), U_0 = 500 V.

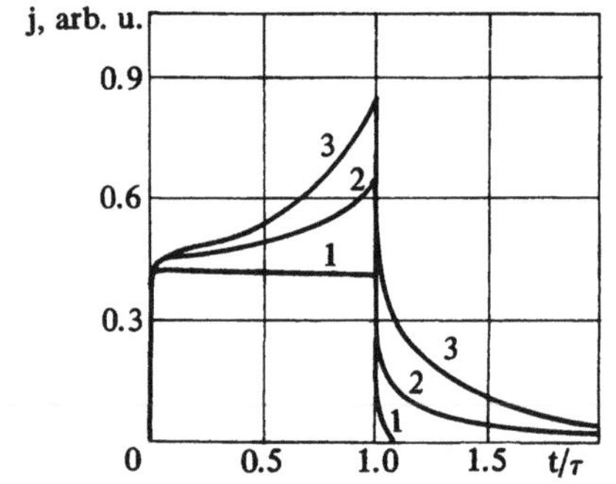

Fig. 5. Current variation in Se samples with linearly decreasing voltage. 1 and 2 – l = 39 μm, U_0 = 265 V, t = 37 · 10^{-3} sec; 3 – l = 35 μm, U_0 = 260 V, t = = 37 · 10^{-3} sec.

cribed concerning the charging of As_2Se_3 samples, one would conclude that the negative space charge is formed due to the field-induced removal of generated holes from the layer bulk.

Typical discharge curves for Se and As-Se layers are presented in Figures 5 and 6. Curve 1 in Figure 5 was obtained for an Se layer upon decreasing the voltage in the dark from 265 V to zero and curve 2–with the simultaneous illumination of the anode with strongly absorbed light. Curve 3 was obtained upon decreasing the voltage and illuminating the layer cathode. As seen from curves 2 and 3, the current density reaches zero not simultaneously with the decreasing voltage but gradually after some time. This can be accounted for by the insufficient photoinjection intensity of charge carriers which neutralize the space charge. This becomes apparent from the following experiment. A sample is illuminated with modulated light of frequency 100 Hz. At those moments in time when the light intensity is zero the current reaches its dark value both upon decreasing the voltage and after its drop to zero (see Fig. 7). This means that injected charge carriers are following the field inversion plane. In this case, after the voltage decrease to zero the current is caused by the numbers of charge carriers being insufficient to neutralize the space charge.

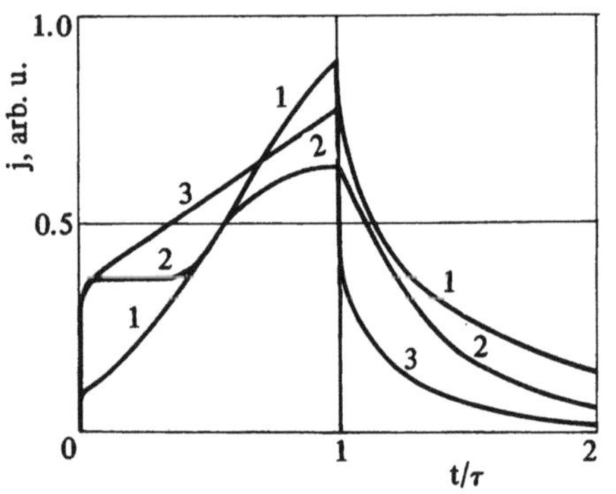

Fig. 6. Current variation in As-Se samples with linearly decreasing voltage.
1 – in As_2Se_3 (l = 14 μm), U_0 = 263 V, t = 26 · 10^{-3}sec;
2 – in $AsSe_4$ (l = 18 μm), U_0 = 220 V, t = 10 · 10^{-3}sec;
3 – in As-Se (l = 17 μm), U_0 = 200 V, t = 10 · 10^{-3}sec.

Figure 6 shows the characteristic change in current in the dark and upon illuminating with strongly absorbed light for As-Se layers when the voltage is linearly decreased. Curve 1 was obtained for an As_2Se_3 layer when hole injection occurs in the dark. As seen from this curve, before voltage has dropped to zero, the current first increases and then gradually decreases to zero. Similarly, curve 2 was obtained for an $AsSe_4$ layer. Here the current is constant for some time and corresponds to the sample geometrical capacitance. This means that at the beginning the charge is removed from an electrode when the voltage is decreased. Later the current starts increasing and then gradually drops to zero when the voltage reaches zero. This is characteristic of all As-Se layers. The experiment involving modulated light has shown that the discharge current of As-Se layers at zero voltage is caused by insufficient mobility

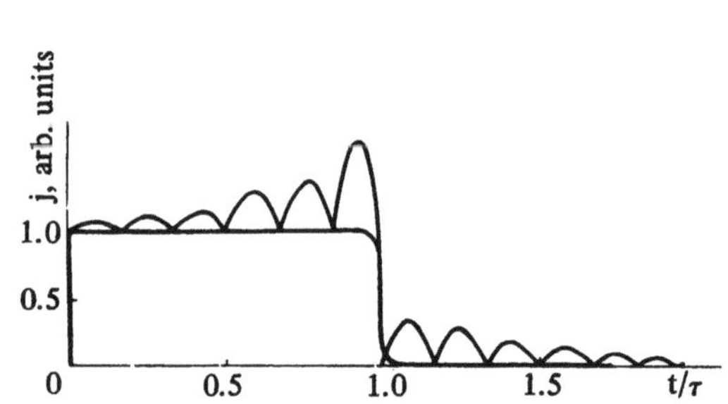

Fig. 7. Current variation in Se electrophotographic layers with linearly decreasing voltage and simultaneous illumination with modulated light.

of charge carriers, the latter being approximately 10^{-5} cm^2/volt·sec in many compositions [9]. This conclusion may be reached on the basis that upon illuminating the sample with modulated strongly absorbed light, at those moments in time when the light intensity is zero, the current value differs from that in the dark both upon decreasing the voltage and after it has reached zero. It follows that charge carriers are unable to follow the field inversion region, especially when the decreasing voltage approaches zero. Incomplete neutralization of the space charge leads, of course, to some errors in determining $\rho(x)$. However, one may suppose that under real conditions the space charge density and its bedding depth will be not less than the calculated value.

Figure 8 illustrates the dependence of space charge density upon the coordinate in Se and As-Se layers. The density of the positive space charge of a Se layer calculated from curve 2 in Figure 5 increases towards the cathode (Fig. 8, curve 6). A similar distribution has also been found in Se layers for a negative space charge.

The density of a negative space charge in As-Se electrophotographic layers calculated from curves 1 and 2 (Fig. 6) increases towards the anode (curve 5 in Fig. 8a and curve 1 in Fig. 8b, respectively). The density and the geometrical distribution of the space charge have been found to be substantially dependent upon the action time of the electrical field and its magnitude: when the time during which the sample is kept under voltage is increased, the space charge density is reduced at the cathode and enhanced at the anode (curves 1 and 2, Fig. 8a). This redistribution is markedly speeded up upon increasing the voltage (curves 3, 4 and 5). If the field strength and its action time are small, then the density of the negative space charge in the layer is weakly dependent upon the coordinate (Fig. 8a, curve 1). But upon increasing the field strength and its action time, the space charge in the layer is redistributed so that its density abruptly rises towards the anode.

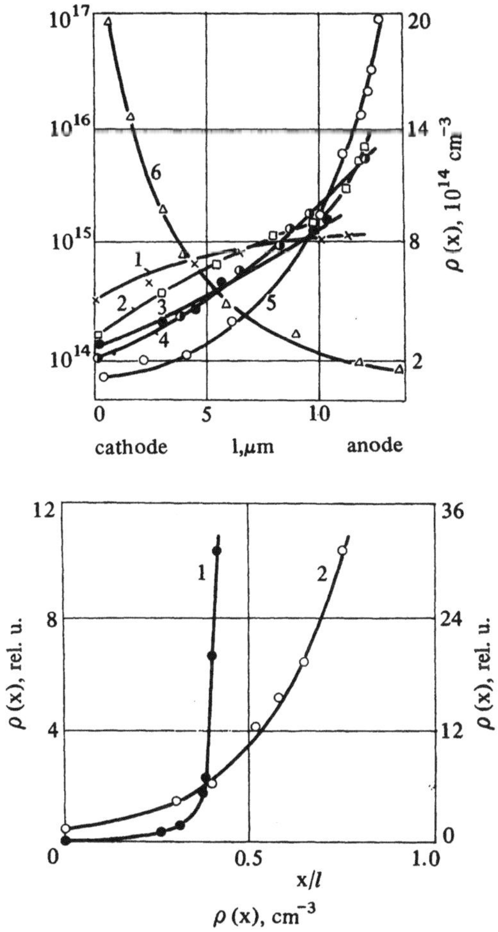

Fig. 8. Space charge distribution in Se and As-Se electrophotographic layers.
a) in As_2Se_3 (l = 14 μm) (scale on the left).
1 – $\epsilon = 7.2 \cdot 10^4$ V/cm, t_E = 5 sec; 2 – $\epsilon = 7.3 \cdot 10^4$ V/cm, t_E = 7 min; 3 – $7.6 \cdot 10^4$ V/cm, t_E = 10 sec; 4 – $\epsilon = 11.1 \cdot 10^4$ V/cm, t_E = 10 sec; 5 – $\epsilon = 18.8 \cdot 10^4$ V/cm, t_E = 10 sec; 6 – in a Se layer (l = 39 μ), (scale on the right).
b) 1 – in $AsSe_4$ layer (scale on the right), U_0 = 220 V, T = $10 \cdot 10^{-3}$ sec, 2 – in AsSe layer (1 scale on the left), U_0 = 400 V.

References

[1] E. A. Montrimas, A. A. Pazera, Lietuvos fiz. rink. 8, 853 (1968).
[2] J. K. Viscakas, E. A. Montrimas, A. A. Pazera, Appl. Optics, Suppl. 3, 79 (1969).
[3] R. J. Dienys, J. A. Kalade, E. A. Montrimas, A. A. Pazera, Lietuvos fiz. rink. 11, (4), (1971).

[4] H. J. Gerritsen, W. Ruppel, A. Rose, Helv. Phys. Acta **30,** 504 (1957).
[5] D. C. Hoesterey, J. Appl. Phys. **33,** 992 (1962).
[6] K. Hauffe, J. Photogr. Sci. **10,** 321 (1962).
[7] V. J. Gaidelis, N. N. Markevicius, E. A. Montrimas. "Fizitcheskie procesi v elektrofotographitcheskich slojach ZnO", Vilnius 1968.
[8] J. K. Viscakas, V. J. Gaidelis, E. A. Montrimas, T. Lasovskis, A. A. Pazera, Lietuvos fiz. rink. **9,** 1093 (1969).
[9] E. A. Montrimas, A. A. Pazera, J. K. Viscakas, Phys. stat. sol. *(a)* **3,** K 199 (1970).
[10] J. J. Lander, J. Phys. Chem. Solids **15,** 324 (1960).

Discussion

W. Eckenbach: How large are the loss currents in your experiment?

A. Matulionis: Dark discharge takes as long as 1000 s in As_2Se_3. The forced discharge (reduction of voltage according to the linear law) usually takes 20–100 ms, therefore the current in the outer circuit exceeds sufficiently the leakage current.

H. Meier: I want only to comment on your paper that we also have measured quantum efficiencies larger than 1 on dye-photoconductors and that we agree with the mechanism which you had given. We think that the reason is the same in the photo-conductors measured by us.

Dielectric Properties and Quantum Efficiency of ZnO-Binder-Layers

W. Eckenbach

Abstract

Electrophotographic ZnO binder layers are negatively charged to about 10^{-7} C · cm^{-2} by means of a corona discharge. The evaluation of this surface charge involves some complications. The usual potential measurement with an electrometer probe ist not applicable to this problem, because either the probe distance has to be comparable with the layer thickness or the dielectric constant of the layer must be known.

Differential compensation of the negative surface charge by positive ions from a corona discharge proved to be a suitable method for direct determination of surface charge. In addition to this method the capacitance between metallic contacts was studied in the frequency range from 2 Hz to 250 kHz.

The quantum efficiency of different sensitized and unsensitized layers has been calculated as a function of wavelength; the surface charge density was evaluated by the corona compensation method and the number of absorbed photons by measurements of incident, reflected and transmitted light. In the dye absorption region a remarkable maximum of the quantum efficiency is found with values sometimes greater than in the intrinsic region of ZnO.

1. Evaluation of Surface Charge

Electrophotographic ZnO binder layers are negatively charged to about 10^{-7} C · cm^{-2} by means of a corona discharge. An exact knowledge of the surface charge density is desirable for two reasons: firstly it is related to the kinetics of the development process and the maximum obtainable optical density, secondly it is an essential factor in the determination of the quantum efficiency.

One method of determining the surface charge would be to use the potential distribution normal to the layer, as recorded by an electrometer probe [1]. For this method, however, the probe distance must be comparable with the layer thickness. This is practically impossible for layer thicknesses of about 10 μm.

If on the other hand the surface charge is to be determined from the usual potential measurement above the layer, it is necessary to know the layer capacitance. This can be achieved simply by contacting the layer with metallic electrodes and measuring the capacitance of this system. In so doing, however, the boundary conditions of the layer are changed as compared with the electrophotographic process.

Theoretical values of the capacitance of a heterogeneous dielectric can be calculated from the simple capacitor model using the layer thickness and dielectric constant. For the derivation of the dielectric constant from the dielectric properties of the components many approximation formulae exist [2] which in the case of ZnO-binder layers, however, only roughly hold because of the irregular particle shape and the low binder concentration.

Another method which simulates the conditions of the electrophotographic process makes use of the fact that the ZnO binder layer as a whole appears to be neutral, and thus an equal amount of countercharge has accumulated on the rear side of a charged layer. By integrating the current which flows during corona charging to the rear side of the layer a measure of the actual surface charge could be obtained. But because of a superposition of charging current and loss current even this method is not applicable in this case. The loss current resulting from the poor insulating ability of the ZnO binder layer depends on the layer potential in an unknown way and can reach a multiple of the real charging current.

A method avoiding this difficulty was suggested by Comizzoli [3]. A negatively charged layer is discharged, in stages, by means of a positive corona source. Under these circumstances the loss current remains small compared with the discharging current, because of the electrical field distributions above and within the layer.

A typical measurement of this kind is presented in Figure 1. It indicates the charge which is gradually compensated versus the layer potential. Over the whole voltage region a good proportionality between charge and voltage is observed. This constancy of capacitance allows ZnO binder layers of the thickness used in these investigations to be described as plate capacitors. Deviation from linearity occurs at low potentials; this indicates a small loss current.

In addition to the corona compensation method we have tried to come to a better understanding of the capacitance by means of capacitance measurements between metallic electrodes. Because of the high DC conductivity only low-frequency AC methods could be used. The frequency of 1 Hz is a good approximation to the process times in electrophotography. Figure 2 shows some results of these investigations, in this case the capacitance and loss tangent as a function of frequency from 2 Hz to 250 kHz. Measurements were made on two samples, one on conductive paper, the other on aluminium substrate. The capacitance increases monotonically with decreasing frequency. The increase is much stronger on the conductive paper substrate; this is apparently due to interfacial polarization between the paper and the binder layer. The loss tangent is rather high, reaching values of more than 1. Its maximum can be correlated to a paper volume resistance of about $10^{10}\,\Omega$cm.

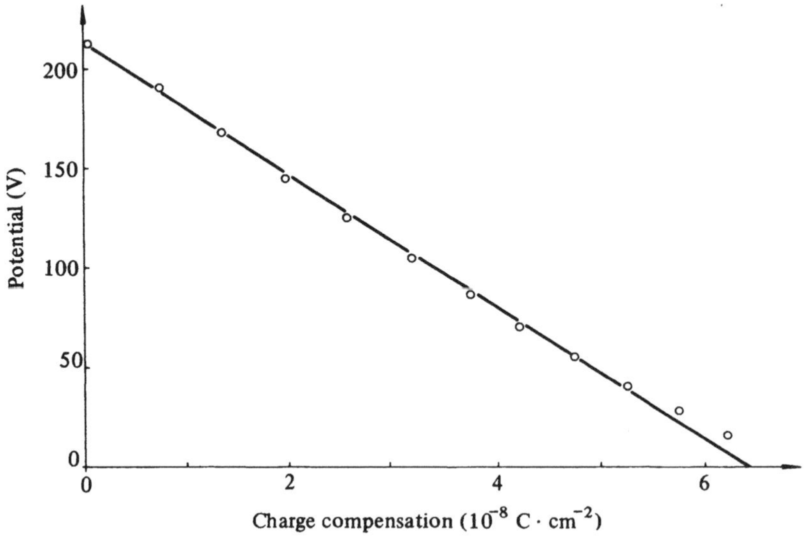

Fig. 1. Corona compensation method: compensated charge a a function of the layer potential.

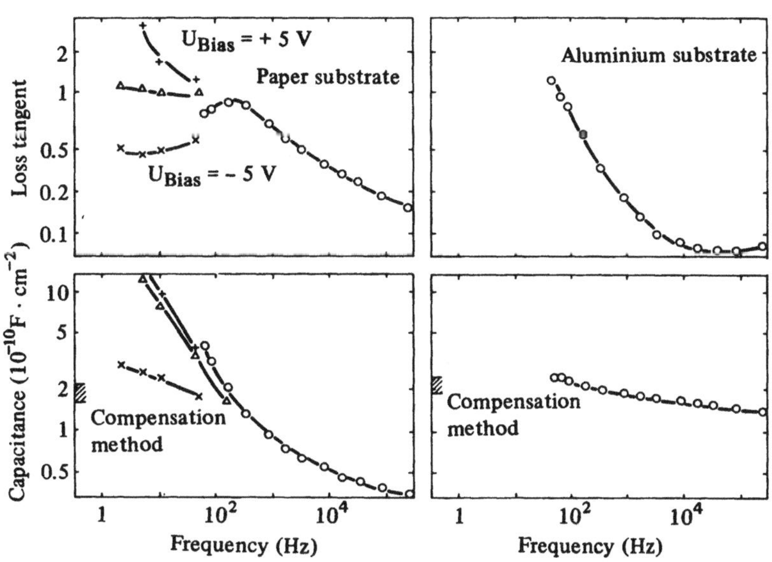

Fig. 2. Frequency dependence of capacitance and loss tangent of unsensitized ZnO binder layers on paper and aluminium substrates (parallel equivalent circuit).

There is no straight-forward way of extrapolating the capacitance to the DC value obtained by the compensation method; this is especially true in the paper case. Only by applying a reverse bias voltage can a displacement in the required direction be observed. Further experiments into this behaviour are necessary.

2. Determination of Quantum Efficiency

The quantum efficiency is determined with the capacitance values from the corona compensation method. Q. E. is defined as

$$\eta = \frac{N_Q}{N_P} = \frac{\text{Number of Charges transported through the Layer}}{\text{Number of Absorbed Photons}}$$

$$= \frac{\Delta Q/e}{E_{abs}/h\nu} = \frac{C \cdot \Delta U/e}{E_{inc} \cdot \alpha/h\nu} \; ; \; \alpha = 1 - \rho - \tau$$

with E_{inc} = incident photon energy
E_{abs} = absorbed photon energy
α = absorption
ρ = reflectance
τ = transmittance

N_Q results from the potential drop ΔU across the layer as a result of light exposure and the above mentioned capacitance. In order to obtain N_P it is necessary to know the number of incident photons and the absorption in the layer. The incident photon energy was measured with a calibrated thermopile; a halogen lamp with interference filters served as a light source. The determination of the absorption was performed by measuring reflectance and transmittance in a setup with an Ulbricht sphere and a monochromator.

Figure 3 shows the results for a ZnO binder layer sensitized with Rose Bengal. It gives the transmittance and reflectance as a function of wavelength. The hatched area between the two curves represents absorption. At 370 nm the transmittance is very small, below 1% for a layer thickness of about 14 µm, and the reflectance amounts to about 5%. Between 380 and 400 nm transmittance and reflectance increase considerably, i.e. the absorption strongly decreases. In the dye absorption region both reflectance and transmittance closely follow the spectrum of Rose Bengal with its different absorption bands.

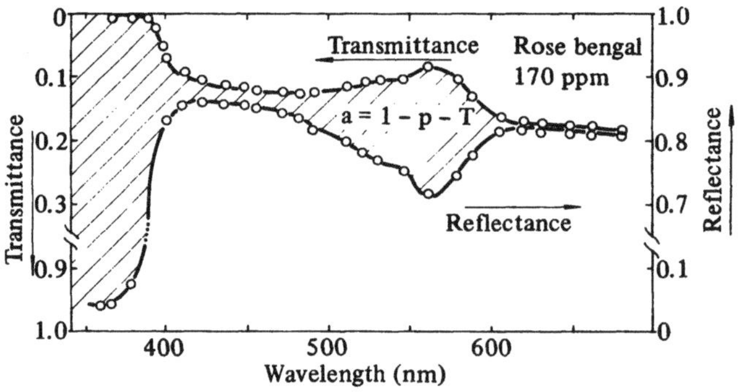

Fig. 3. Transmittance, reflectance, and absorption of a ZnO binder layer with Rose Bengal as a function of wavelength.

Hence, all necessary quantities for the determination of Q. E. are known. Figure 4 demonstrates the results for the Rose Bengal layer. The upper part shows again the absorption which was determined from the transmittance and reflectance shown in Figure 3. In the middle part the relative sensitivity can be seen; i.e. the efficiency normalized to the same incident photon energy. It has a similar character to the absorption curve. The lower part shows the Q. E. with regard to reflectance and transmittance. In the intrinsic region of ZnO it amounts to about

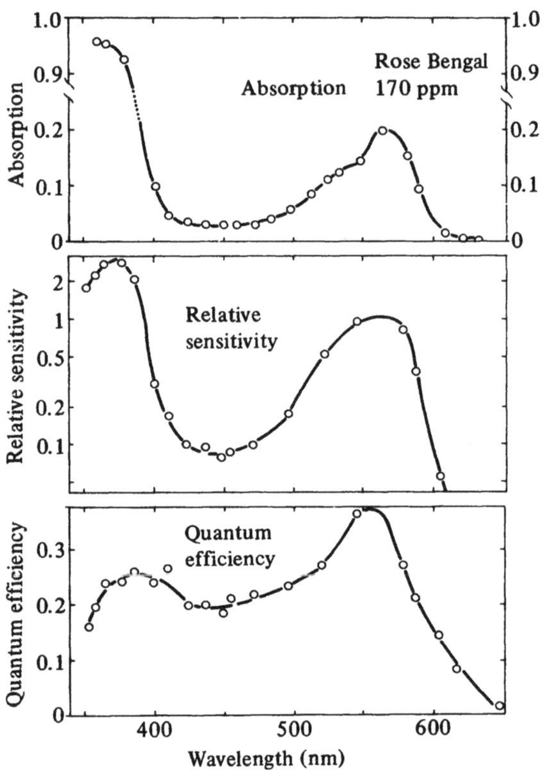

Fig. 4. Absorption, relative sensitivity, and quantum efficiency of a ZnO binder layer with Rose Bengal as a function of wavelength.

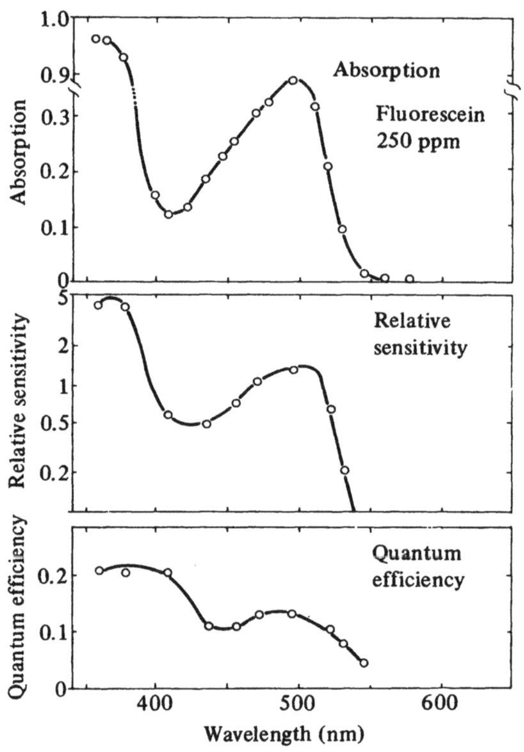

Fig. 5. Absorption, relative sensitivity, and quantum efficiency of a ZnO binder layer with Fluorescein as a function of wavelength.

0.25, in the dye absorption region is shows a pronounced maximum reaching values of about 0.4.

If absorption in the layer were wholly responsible for the maximum in the relative sensitivity, quantum efficiency should be constant with respect to wavelength. This remarkable maximum of the quantum efficiency with sensitization can be interpreted by means of transport phenomena, taking into account different penetration depths of the light and range limitation of the holes.

Figures 5 and 6 demonstrate the same measurements for ZnO binder layers sensitized with Fluorescein and Bromophenol Blue. The latter has a very broad absorption region. The maximum of relative sensitivity and quantum efficiency at 604 nm was studied as a function of dye concentration. This is shown in Figure 7. The concentration varied from 50 to 4,000 ppm.

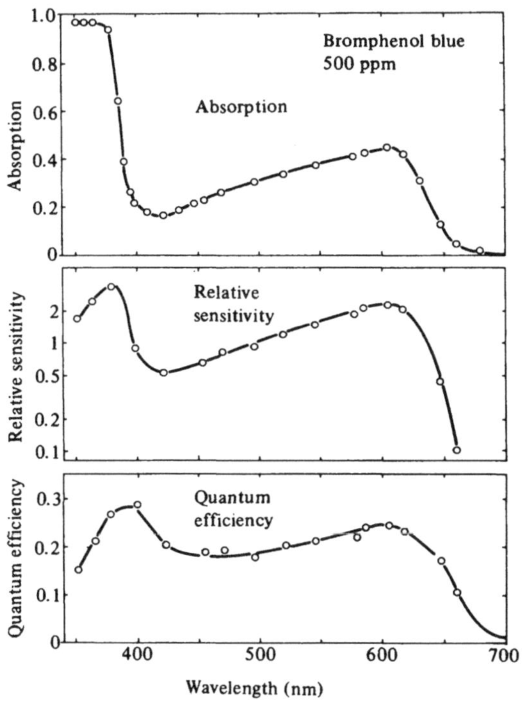

Fig. 6. Absorption, relative sensitivity and quantum efficiency of a ZnO binder layer with Bromophenol Blue as a function of wavelength.

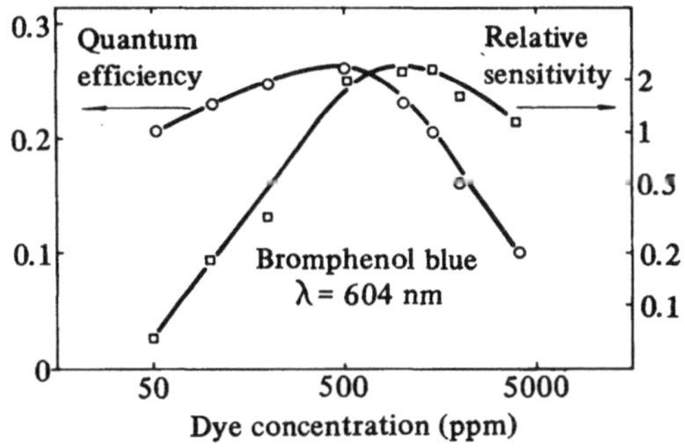

Fig. 7. Relative sensitivity and quantum efficiency at 604 nm of ZnO binder layers with Bromophenol Blue as a function of dye concentration.

At about 1,000 ppm the rel. sensitivity has a maximum. The Q. E., too, shows a maximum at a somewhat lower concentration at around 500 ppm. This means that the effective sensitization begins to fall off with increasing concentration even before a coverage of half a monolayer is reached. Possible causes are: 1) Clusters of adsorbed dye molecules are formed on the ZnO surface; 2) remaining dye molecules are caught in the binder matrix.

A separate investigation of the true Q. E. and light absorption offers the possibility of studying more exactly the discharging mechanism of sensitized layers and of performing more effectively the optimization of layers for electrophotography.

Acknowledgement

I wish to thank Mr. J. Eilts for his assistance in performing the experiments and Mr. E. Klein for the preparation of the samples. I am greatly indebted to Dr. N. Hansen for his continuous interest and helpful discussions.

References

[1] G. Heyl, Kunststoffe **60**, 45–52 (1970).
[2] L. K. H. van Beek, Progr. Dielectr. (London) **7**, 69–114 (1967).
[3] R. B. Comizzoli, Phot. Sci. Eng. **14**, 210–212 (1970).

Discussion

R. Schaffert: How do you measure sensitivity?

W. Eckenbach: Relative sensitivity is defined here, for convenience, as the reciprocal of the light energy (in $\mu Ws/cm^2$) which is required to reduce the surface potential to 50% of the initial value. The surface potential is measured simultaneously with light exposure by means of a transparent conductive probe.

K. Verhille: Could you elaborate somewhat on the Comizzoli method for determining the layer capacitance?

W. Eckenbach: The negative surface charge is compensated in stages by means of positive ions from a positive corona source. In each step the amount of positive charge is determined by integrating the discharging current and the potential drop is measured with an electrometer probe. This process is repeated until the layer is completely discharged.

R. Arneth: 1) Did you use always the same sample performing the measurement of spectral sensitivity and quantum efficiency? 2) If yes didn't you observe fatigue effects?

W. Eckenbach: Yes, all measurements were done on the same sample. We have always measured from red to blue with intersities as small as possible, so we did not observe severe fatigue effects.

H. Meier: Do you think that the observed decrease of quantum efficiency of ZnO by addition of some dyes (e.g. Rose Bengal) is the result of a real desensitization of ZnO?

W. Eckenbach: We did not observe a decrease of Q.E. in the intrinsic region of ZnO as a consequence of dye addition.

A. Meyer-Laack: I would like to know what kind of binder did you use for your ZnO binder layers?

W. Eckenbach: We tried several binders, e.g. silicone, pliolite, synolac. The results we have presented are obtained with synolac.

W. F. Berg: The optimum of Q. E. coinciding with the absorption maximum seems to be a general feature of dye sensitization, even on single crystals. This may be due to dye molecules having different situations on the crystal surfaces, the most efficient ones being those at the absorption optimum.

K. Verhille: Did you find an exponential decrease of voltage upon time when dark decaying the charged layer? If not does it not puzzle you that a constant value for the capacitance is given in your paper.

W. Eckenbach: The dark decay of our layers is not exponential with time. The relation between charge and voltage is almost linear; there is a slight parabolic dependence which might be due to a small space charge. But we believe that the non-exponential dark decay is related to the non-ohmic resistance of the ZnO binder layer. Several commercial Electrofax layers showed practically the same behaviour.

J. S. Boroky: You mentioned that in your investigation of the quantum efficiency and peak sensitivities of dye sensitized ZnO binder layers the concentration of dyes varied from a "half monolayer" thickness to multilayer coverages. How was this "half monolayer" thickness determined? You are undubitably aware that most dye sensitizers have plate like structures and it is not irrelevant whether they are positionned edgewise or parallel to the surface of the photoconductive crystals. Because of this relationship between the stacking arrangement of the dye sensitizers and the concentration of them naturally the quantum efficiency is also affected. It would be therefore very important to know just how exactly were the observed layer thicknesses established in your case.

W. Eckenbach: Our statement about coverage is only a rough estimate, based on the assumption that a single molecule of the dye needs an area of about 100 $Å^2$.

Latent Image Formation Mechanism in As_2Se_3 Electrophotographic Layers

E. Montrimas, S. Tauraitiene and A. Tauraitis

Abstract

The dark discharge and photodischarge mechanisms were investigated for As_2Se_3 layers, where the integral photosensitivity exceeds that of Se layers more than ten times.

It was found that the dark discharge of positively charged As_2Se_3 layers is determined by carrier generation from the levels distributed in bulk. The discharge rate is strongly dependent on the duration of dark storage after a pre-exposure. During charging, the free holes and those situated in shallow level are swept out and negative space charge is formed, the density of which substantially depends on temperature and pre-exposure. The discharge for layers stored for a long period of time in the dark is determined by the excitation of current carriers from a number of discrete levels, while for those pre-exposed, it is a result of generation from the continuous level spectra, where the most shallow ones are emptied. With field strength less than $3 \cdot 10^4$ V/cm, the discharge is determined by a constant rate of carrier generation, i.e. the potential decay rate does not depend on the potential.

High field effects and the resultant increase of dark discharge rate are observed at the higher fields.

The photodischarge is caused by drift of photogenerated holes; the photogenerated electrons are immobilized in the generation sites and later on can serve as recombination centres. The theoretical aspects of photo discharge in such a physical model are dealt with, and the regularities of discharge kinetics are determined for different wavelengths of light, the density and distribution of the negative space charge in the layer, and recombination intensity. A satisfactory fit to the experimental data is obtained, suggesting insignificant hole recombination.

The quantum efficiency of photodischarge essentially exceeds unity. Dependence of the quantum efficiency on temperature, wavelength of light and initial potential is investigated. It is presumed that during the photodischarge process a multiplication of charge carriers takes place due to the high field effect.

1. Introduction

During recent years arsenic triselenide layers have found wide use in electrophotography as information carriers in highspeed copiers and other photo-registering apparatus. These layers have fairly good light sensitivity, i.e., its integral photo-sensitivity is more than by one order of magnitude greater than that of the highest quality Se electrophotographic layers (EPL), while the spectral sensitivity of these EPLs covers the region from ultraviolet to 700 nm [1–3]. Besides that, As_2Se_3 EPLs, compared to selenium, have a higher thermostability in a wide temperature interval and are more durable [4, 5]. On the other hand, the potential dark decay rate of these layers is substantially greater than for Se layers and greatly depends on pre-illumination [5, 6]. All these features of

As_2Se_3 EPLs attract the attention of workers as well as research-physicists involved in this field. A large number of studies in which the technology of As_2Se_3 layer-preparation affect the electrophotographic parameters [1, 7], the charge transfer phenomena [5, 8], as well as other electric and photoelectric properties [8, 13], are available. But the physical processes taking place in As_2Se_3 EPLs in the dark as well as their photoinduced discharge have not been adequately investigated; that is why this paper is devoted to this problem.

2. Layer Preparation

The As_2Se_3 compound was made by vacuum heating of a corresponding mixture of As and Se at a pressure of 10^{-3} to 10^{-4} torr and at a temperature of 700° C during 6–7 hours. The composition of the layers was controlled by chemical methods [9], and it was found that the composition was uniform throughout the layer and did not differ from the stoichiometry of the As_2Se_3 compound. The amorphous condition of the layer was checked by means of X-ray and electron-diffraction methods.

The characteristic features of As_2Se_3 layers are as follows: electrons are practically immobile ($\mu_e < 10^{-8} cm^2/Vs$), the hole drift mobility is from 3 to $5 \cdot 10^{-5} cm^2/Vs$, [8] and the effective hole lifetime is from 1.2 to $3 \cdot 10^{-3}$s [10].

3. Dark Discharge

In order to clear up the As_2Se_3 EPL discharge mechanism, the dominating process was first of all determined: the generation of carriers in the layer volume or carrier injection from substrate or from surface. With this aim in mind two types of layers were prepared: (a) As_2Se_3 layers covered with a very thin (0.5 μm) film of selenium, and (b) Se layers evaporated on a thin (0.5 μm) underlayer of As_2Se_3. It follows first of all from the analysis of the dark decay discharge kinetic curves of the layers mentioned above (Fig. 1) that the As_2Se_3 underlayer (curve 1) forms a good barrier for electrons. This is possibly connected with the fact that the electrons are immobile in As_2Se_3. Curve 1 shows also that the intensity of hole injection from the surface of the selenium layer is very small. Therefore, a thin film of Se may serve as a barrier for holes. As the properties of As_2Se_3 layers with and without the film of selenium are identical (Fig. 1, Curves 2 and 3, 6 and 7), one may expect that the free charged surface of As_2Se_3 does not inject holes. Consequently, the dark discharge is a bulk property of As_2Se_3 layers. Hole generation in the dark from certain levels distributed in the layer bulk is the main reason of discharge. The electrons do not contribute to the discharge.

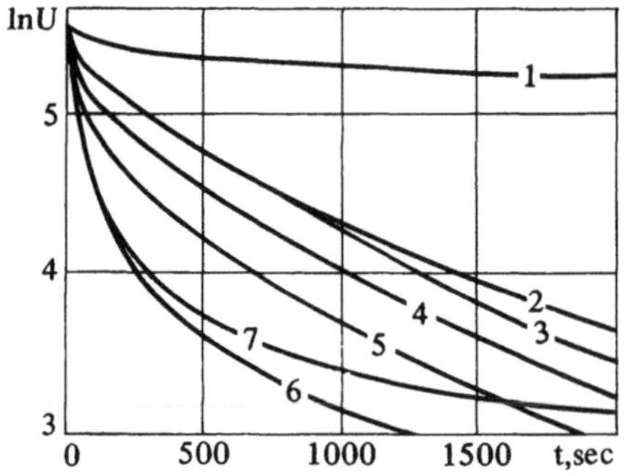

Fig. 1. Dark discharge of positively charged layers: 1 – a 22 μm thick Se layer on top of a 0.5 μm layer of As_2Se_3; 2, 7 – a 14 μm thick As_2Se_3 layer with a 0.5 μm thick film of Se on top; 3, 4, 5, 6 – a 14 μm thick layer of As_2Se_3. The dark storage times are: 2, 3 – few days, 4, 5, 6, 7 – 30 sec. Pre-exposure ($3.0 \cdot 10^5$ J cm^{-2} sec^{-1}, KS-17 filter): 4 – 3 sec; 5–6 sec; 6, 7 – 100 sec'

In As_2Se_3 layers a negative space charge forms as a result of charging and discharging. Its density grows with discharge time [11]. Curves 6 and 7 indicate that the discharge rate of pre-illuminated layers grows with the duration of pre-exposure.

More details of the main process of dark discharge (hole generation in the bulk) are determined through the study of the influence of electric fields (initial potential) on the initial discharge rate at various temperatures. In the relationship $dU/dt|_{t=0} = f(U_0)$ three parts are clearly defined (Fig. 2). The discharge rate increases with the initial potential at weak electric fields, then it saturates, and a steep increase of the discharge rate can be seen at higher potentials. The explanation is as follows. At low potentials the effective layer thickness and consequently the volume where hole generation takes place increase with the potential as a result of hole extraction. The discharge rate saturates when the extracted layer is as thick as the layer itself. At high potentials a high-field effect takes place, causing a steep increase in the rate of hole generation.

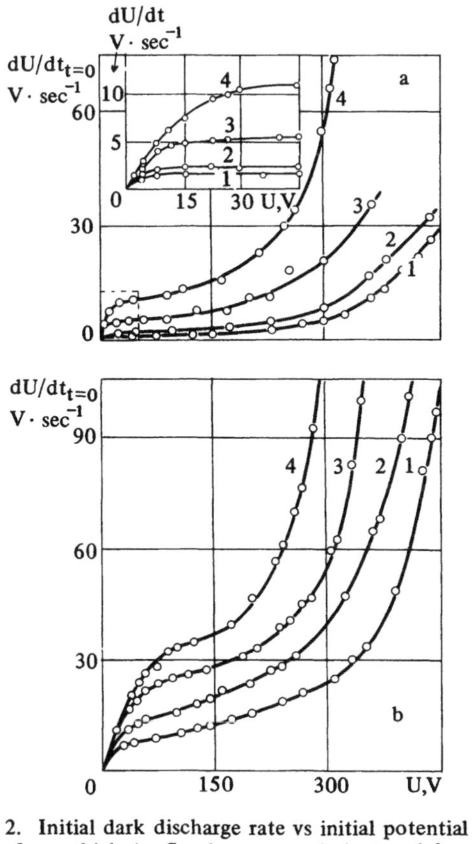

Fig. 2. Initial dark discharge rate vs initial potential for a 9 μm thick As_2Se_3 layer: a – dark stored for 3 days (enlargement in the left top corner – the initial part of the curves): b – pre-exposed by $1.8 \cdot 10^3$ J cm^{-2} through KS-17 and dark stored for 30 sec. Temperatures during the measurements: 1 – 20° C; 2 – 30° C; 3 – 40° C; 4 – 50° C.

The relationships $dU/dt|_{t=0} = f(U_0)$ obtained at various temperatures allow estimation of some important parameters of the layers. For example the saturation voltage can be used to calculate the density of the holes extracted by the field and the average density of the negative space charge. The density is dependent on temperature and on sample pre-illumination. It varies from $0.5 \cdot 10^{14}$ to $3 \cdot 10^{14} cm^{-3}$ as the dark storage temperature is varied from 20 to 50° C, and after pre-illumination from $2.9 \cdot 10^{14}$ to $8.5 \cdot 10^{14} cm^{-3}$.

Assuming that hole generation takes place uniformly throughout the bulk the hole generation rates was estimated from the saturation current. For dark-stored layers it is equal to: 0.82, 1.17, 3.05, 6.25 (in $10^{13} cm^{-3} sec^{-1}$) at temperatures of 20, 30, 40, 50° C, respectively.

When carrier generation in the bulk and space charge formation are taken into account the following analytical expressions obtained in [12] are valid for the dark discharge kinetics of an electrophotographic layer:

$$U(t) = U_0 - \frac{e l^2 m_0}{2 \epsilon\epsilon_0} [1 - \exp(-Rt)] , \qquad (1)$$

for $\sigma_0 \geqslant \rho l$, and

$$U(t) = \frac{2 \epsilon\epsilon_0 \rho_0}{\sigma_0^2} \left\{1 + \frac{e m_0}{\rho_0} [1 - \exp(-Rt)]\right\}^{-1}, \qquad (2)$$

for $\sigma_0 < \rho l$,
where U_0 is the initial potential, e – the elementary charge, l – the layer thickness, $\epsilon\epsilon_0$ – the dielectric constant, ρ_0 – the initial space charge density, m_0 – the initial density of the levels from which generation takes place, R – the probability of carrier thermal excitation from local levels ($R = \nu \exp(-E/kt)$, ν – the frequency factor, E – the activation energy, T – the temperature, k – the Boltzmann constant.

It was found that the discharge of As_2Se_3 layers in the dark cannot be described by equations (1) and (2) with constant parameters m_0

and R. However, when a layer was kept for a prolonged time in the dark a certain part of the $\ln(dU/dt) = f(t)$ relationship shows a linear part (Fig. 3, Curve 1). This, first of all, shows that during dark storage of the layer the more shallow levels will empty completely. Secondly, this allows the parameters of the deepest level to be estimated. The parameters determined from the initial discharge rates and from the discharge curves are [5]: $E = 0.91$ eV, $m_0 = 1.9 \cdot 10^{15} cm^{-3}$.

By substracting the contribution of the deepest levels, the presence and the contribution of the levels located above the deepest one can be determined if the curve $\ln(dU/dt)$ vs. t becomes straight. By using this technique more than four discrete levels were found in As_2Se_3. Unfortunately, this technique is not accurate enough to determine the parameters of these levels except for the deepest one.

The shallower levels are usually filled by the pre-illumination. In this case the linear part of the $\ln(dU/dt) = f(t)$ relationship is absent (Fig. 3, curve 2) and the level parameters cannot be determined.

Attention is to be paid to the characteristic bend-over point in Curve 1 (Fig. 3), which is present at the end of the linear part. The point mentioned is associated with the moment of time when the equality $\sigma = \rho l$ is fulfilled, i.e. no charge remains in the substratum. A fast

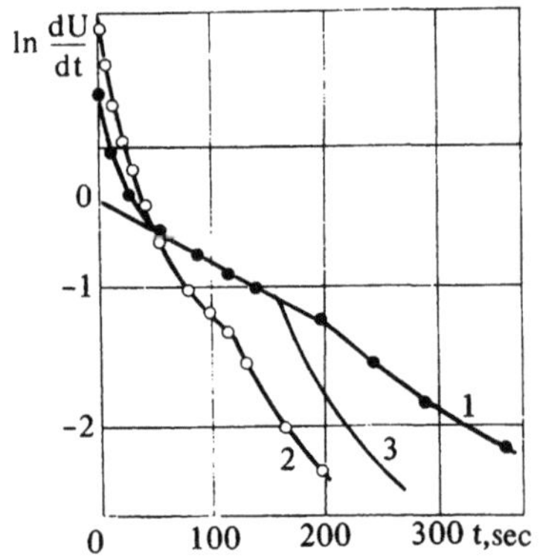

Fig. 3. ln (dU/dt) vs time for a 9 μm thick As_2Se_3 layer dark stored for 3 days.

decrease in the discharge rate results from the decrease in effective layer thickness. If the holes are generated uniformly throughout the layer the bend-over point is at $U = {}^1/_2 U_0$. The experimental bend-over point is, however, at $U < {}^1/_2 U_0$, consequently the generation is not uniform, and it is higher near the surface. This is also the conclusion from a number of other experiments. For example, under conditions of uniform hole generation the relationships $\frac{1}{l^2}\left(\frac{dU}{dt}\right)_{t=0} = f\,\frac{U}{l}$ obtained for layers of various thickness but of the same structure must coincide, which is not the case experimentally. A more detailed investigation into the properties of As_2Se_3 layers shows that hole generation being more intense at the surface results from atmospheric influence and from the presence of a larger density of centres.

4. Photodischarge

Electrons are not mobile in As_2Se_3 electrophotographic layers, and the photodischarge is exclusively due to the drift of holes [7]. It was also shown [11], that the partial photodischarge by weakly absorbed light results in a negative space charge of considerable density. The localized electrons can serve as recombination centres.

On the basis of the above findings the following system of equations can be used:

$$\frac{\partial j}{\partial x} + e\frac{\partial(p-m)}{\partial t} = 0\ ,\quad j = e\mu p\epsilon\ , \qquad (3)$$

$$\frac{\partial E}{\partial x} = \frac{e}{\epsilon\epsilon_0}\,(p-m)\ , \qquad (4)$$

$$\frac{\partial m}{\partial t} = -\gamma pm + \beta KL\exp(-Kx)\ , \qquad (5)$$

where m is the density of the levels filled by electrons, p – the hole density, x – the coordinate, j – the current density, μ – the hole mobility, E – the electric field, β – the photogeneration yield, K – the light absorption coefficient, L – the illumination intensity.

If the space charge of free holes is not taken into account, Eq. (3–5) can be transformed into:

$$\frac{\partial E}{\partial x} = -\,S\,m/[1 - \exp(-S)]\,, \qquad (6)$$

$$\frac{1}{m}\,\frac{\partial m}{\partial t} - \frac{\Gamma}{E}\,\frac{\partial E}{\partial t} = \frac{\exp(-Sx)}{m} \qquad (7)$$

where

$$\Gamma = \frac{\gamma\,\epsilon_0\epsilon}{e\mu}\,,\quad S = Kl$$

and the variables E, x, t, m are measured in E_0, l, T and M units respectively:

$$E_0 = \frac{\sigma_0}{\epsilon_0\epsilon}\,,\quad M = \beta K L T\,,$$

$$T = \frac{\sigma_0}{e\beta K L \int_0^1 \exp(-Kx)\,dx}\,, \qquad (8)$$

σ_0 – is the initial surface charge density.

The negative space charge formed on charging is postulated to be either uniform or exponentially distributed. Therefore the boundary and initial conditions are:

$$E(x,0) = \begin{cases} 1 - Bx \\ 1 - \frac{B}{r}\,[1 - \exp(-rx)], \end{cases} -E(0,t) = 1\,, \qquad (9)$$

$$m(x,0) = \begin{cases} \frac{B}{S}\,[1 - \exp(-S)]\,, \\ \frac{B}{S}\,[1 - \exp(-S)]\exp(-rs), \end{cases} -j(0,t) = 0,$$

where $B = \frac{e\,l}{\sigma}\,m(x,0)$, $r = \alpha l$, while $m(x,0)$ and α are determined by the initial value of m, i.e. either $m(x,0) = m_0 = \text{Const}$ or $m(x,0) = m(0,0)\exp(-\alpha x)$.

The analytical solutions of the system are available when B = 0 and Γ is equal either to zero or to unity [11]. Figure 4 gives the computed results of photodischarge rate against potential. Figure 5 represents the experimental data. Many similarities can be seen. Thus at the beginning of the process the photodischarge rate changes very little with decreasing of potential. The corresponding theoretical relationship is obtained for low intensity of carrier recombination ($\Gamma \leqslant 0.1$, Fig. 4, Curves 2, 6). A sharp drop begins when the number of localised electrons is equal to that of elementary charges on the surface. The negative space charge formed before exposure brings about a sharp decrease in discharge rate for higher potentials (Fig. 4, Curves 4, 8). Because the density of the negative space charge and its distribution depend on the potential [14], the shift of the sharp decrease towards higher potentials can be understood (Fig. 5, Curves 1, 2, 3). When the light is uniformly absorbed, the average shift of holes is halved. Therefore the maximum discharge rate for weakly absorbed light is half that for heavily absorbed light (Fig. 4, Curves 2, 6 and Fig. 5, Curves 2, 5). The decrease of the light absorption coefficient causes the sharp-decrease point to shift towards higher potentials (Fig. 5, Curves 2, 5).

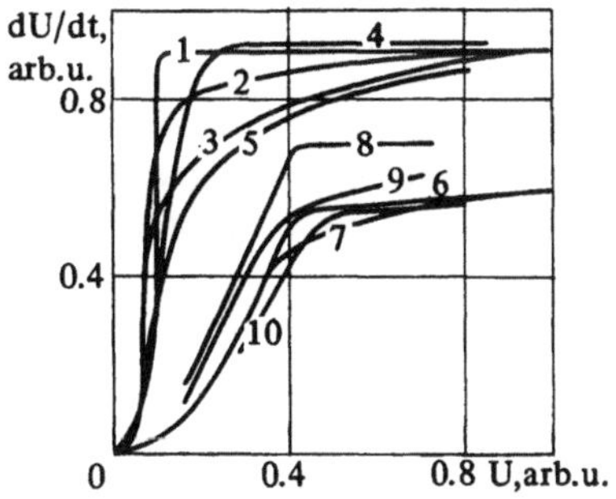

Fig. 4. Dependance of photodischarge rate on potential (calculated): 1 – no recombination (Γ = 0), no space charge before exposure (B = 0), strongly absorbed light (S = Kl = 10); 2 – Γ, B, S are 0.1, 0. 10); 3 – 0.3, 0, 10; 4 – 0, 10/19, 10; (the negative space charge is distributed exponentially), i.e. r = 1; 5 – 0.3, 10/19, 10. r = 1; 0.1, 0, 1 (weak absorption); 7 – 0.3, 0, 1; 8 – 0.1, 10/19, 1, r = 1; 9 – 0, 10/19, 2.5, r = 0; 10 – 0.3, 10/19, 2.5, r = 0.

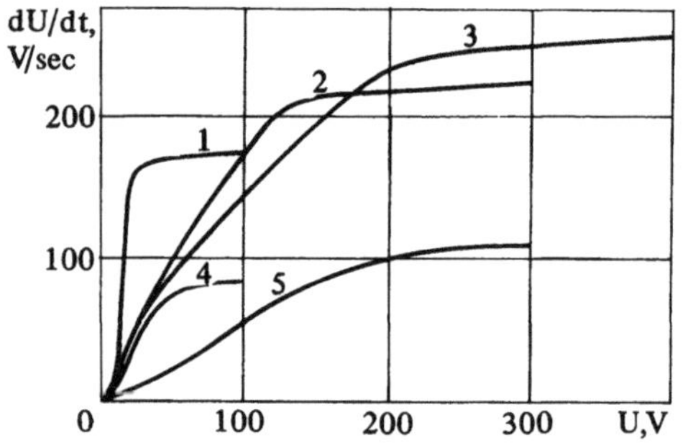

Fig. 5. Dependence of photodischarge rate on potential for a 15 μm thick As_2Se_3 layer (experimental): 1, 2, 3 – strongly absorbed light (400 nm); 4, 5 – weakly absorbed light (660 nm). The initial potential is: 1, 4 – 100 V; 2, 5 – 300 V; 3 – 400 V.

The increase in maximum discharge rate with increasing initial potential in the case of exposure to strongly absorbed light (Fig. 5, Curve 1, 3) as well as to weakly absorbed light (Curves 4, 5) cannot be explained in the framework of the model discussed. This feature of As_2Se_3 photodischarge is associated with the dependence of carrier photogeneration on the electric field [14]. Since the density of surface charge remains constant during photodischarge, the electric field at the surface and in the generation region decreases more slowly than in the case when both carriers are mobile. This is why the initial potential, but not the potential in general, governs photogeneration conditions during the significant time of photodischarge. A better understanding of the phenomena associated with the dependence of generation on electric field is gained by further investigation into quantum efficiency of electrophotographic layers of As_2Se_3.

5. Photodischarge Quantum Efficiency of As_2Se_3 Layers

The quantum efficiency was determined from the initial slope of the photodischarge curve by means of the following formula

$$\beta = \frac{\epsilon_0 \bar{\epsilon}\, dU/dt}{e L[1 - R_\lambda - \exp(-Kl)]\, l_{eff}},$$

where R_λ is the reflection coefficient, and l_{eff} is the effective thickness of the layer. R_λ and ϵ were taken from [15]. The estimation of l_{eff} presents some difficulties because it can be decreased by the space charge. Because of lack of reliable data on l_{eff}, it was taken to be equal to the geometrical thickness of the layer for the calculation of quantum efficiency. Therefore a methodical error was introduced which decreased the calculated quantum efficiency as compared to its true value. The relative error was estimated to be less than ± 25%.

With an increase in potential to its maximum value the quantum efficiency increases and reaches a value greater than unity. (Fig. 6, Curve 1). It increases further if the layer is illuminated by weakly absorbed light during its charging (Fig. 6, Curve 2), though the limiting potential of charging decreases. This is so because the negative space charge is formed by simultaneous charging and illumination. The increase in quantum efficiency is caused by the increase in surface charge and of the field at the surface, although the potential is reduced. As the exposure during the charging increases the quantum efficiency is reduced. The cause for this is the increase in layer capacity (a decrease in effective thickness) as well as a whole number of other reasons, whose explanation is, however, not the purpose of this investigation. Quantum efficiency exceeding unity can be caused by carrier multiplication in high electric

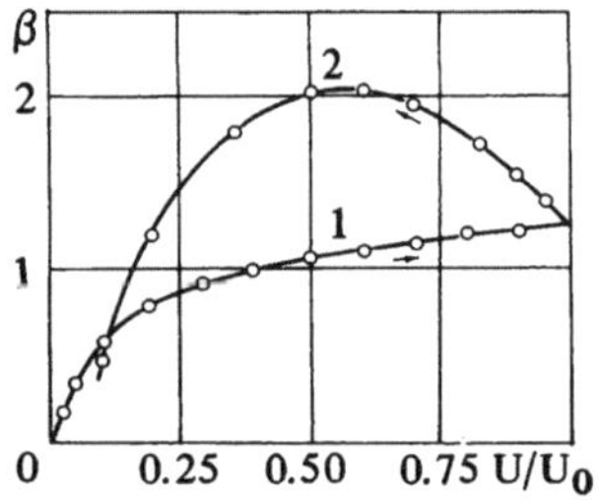

Fig. 6. Photodischarge quantum efficiency at λ = = 440 nm as a function of initial potential for a 20 μm thick As_2Se_3 layer: 1 – charged in the dark; 2 – simultaneously charged and exposed to weakly absorbed light of $2.5 \cdot 10^{12}\ cm^{-2}\ sec^{-1}$. The limiting initial potential is 600 V. The arrow shows the sequence of the measurements.

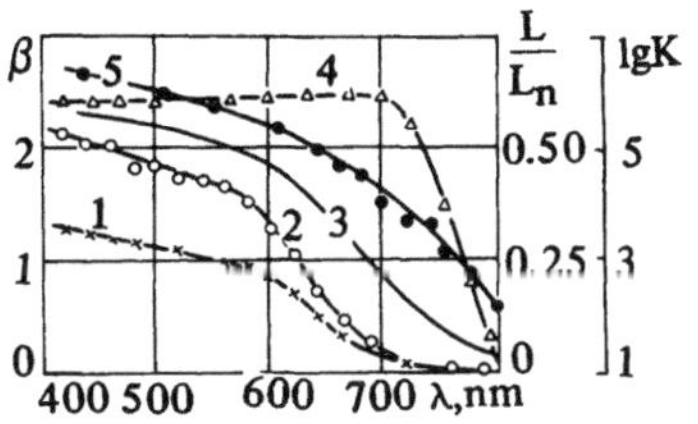

Fig. 7. Spectral dependance of quantum efficiency (1, 2), the absorption coefficient (3) and the ratio of absorbed to incident energy (4) for a 20 μm thick layer of As_2Se_3: 1 – at 1000 V after prolonged dark storage; 2 – at 600 V after pre-illumination to weakly absorbed light and simultaneous charging. 5 – Quantum efficiency for $As_2Se_3Sb_{0.1}$ layer at average field of 10 V/μm.

fields. An increase in quantum efficiency is favoured by the formation of a thin high-field region, although the optimal thickness depends on many other competing factors. This is also confirmed by the spectral characteristics. (Fig. 7). When the absorption coefficient decreases the quantum efficiency decreases too. This is so despite the fact that more energy is absorbed by the layer, but, possibly, outside the high-field region. The correlation of the quantum efficiency and the absorption coefficient applies for layers kept in the dark for a considerable time and for those illuminated while charging (Fig. 7, Curves 1, 2). Therefore in both cases a thin high-field region is formed. The negative space charge, even in the case of ordinary charging, is essentially non-uniform, its density being more than by four orders of magnitude higher at the surface than at the substratum [13]. Therefore the electric field at the surface is considerably higher than the average value ($5 \cdot 10^5$ V/cm) estimated from the limiting potential. It seems to fulfil the necessary conditions for carrier multiplication.

References

[1] A. Lukatskaya, S. Tauraitiene, A. Tauraitis. In collection "Fizitcheskie osnovy elektrofotografii", Vilnius, 1969, p. 213.

[2] E. J. Felty, Reprographie II, Verlag O. Helwich, 1967, p. 40.

[3] S. Tauraitiene, A. Tauraitis, E. Montrimas, "Int. Congress Photographic Sci., Section E", Moscow, 1970, p. 244.

[4] E. Montrimas, B. Petrietis, S. Tauraitiene, A. Tauraitis, In collection "Elektrofotografitcheskie kopirovalno-mnozhitelnye aparaty i materialy k nim", Vilnius, 1970, p. 153.

[5] S. Tauraitiene, Dissertation, Vilnius, 1971.

[6] E. Montrimas, A. Tauraitis, S. Tauraitiene, J. Balciunas, In collection "Fizitcheskie osnovy elektrofotografii", Vilnius, 1969, p. 258.

[7] B. Kolomietz, J. Kepeniene, E. Montrimas, S. Tauraitiene, A. Tauraitis, In collection "Elektrofotografitcheskie kopirovalno-mnozhitelnye aparaty i materialy k nim", Vilnius 1970, p. 108.

[8] E. Montrimas, A. Pazera, S. Tauraitiene, A. Tauraitis, Lietuvos Fizikos Rinkinys **9** (2), 355 (1969).

[9] V. Artamonova, S. Tauraitiene, In collection "Tonkie plenki i ich primenenie", Vilnius, 1969, p. 137.

[10] R. Dienys, J. Kalade, E. Montrimas, A. Pazera, Lietuvos Fizikos Rinkinys, **11** (4), (1971).

[11] V. Gaidelis, E. Montrimas, S. Tauraitiene, A. Tauraitis, Lietuvos Fizikos Rinkinys **9** (2), 355 (1969).

[12] V. Gaidelis, N. Markevich, E. Montrimas, "Fizitcheskie procesy v elektrofotografitcheskich slojach ZnO", Vilnius, Izdatelstvo "Mintis", 1967.

[13] E. Montrimas, A. Pazera, J. Viscakas, phys. stat. solidi *(a)*, **3**, 199 (1970).

[14] S. Grenishin, A. Tauraitis, V. Baliulis, In collection "Fizitcheskie osnovy elektrofotografii", Vilnius, 1969, p. 148.

[15] S. Vengris, V. Krisciunas, E. Montrimas, S. Tauraitiene, In collection "Fizitcheskie osnovy elektrofotografii", Vilnius, 1969, p. 56.

Dye Sensitization Problems in the Photoconduction of Zinc Oxide

Eiichi Inoue

Abstract

The sensitization of photoconduction in semiconductors has been one of the most attractive subjects for imaging scientists and engineers. To prove this, a number of papers have been devoted to the understanding of sensitized photoconduction. During the past ten years, we also added some experimental contributions to this problem, based on the concept that chemical factors should be important enough to determine the efficiency of the sensitization.
As for zinc oxide, there is well established evidence to be sure that chemisorbed oxygen plays an important role in photoconduction. Starting with a consideration on this point, some new chemical sensitizers were discovered, and the mechanism was examined. The chemical sensitizers studied were basically electron-affinitive molecules (EAM), such as phthalic anhydride. Though details still remain unsolved, there is some evidence to suppose that chemical sensitizers modify the surface states and serve to make photoelectrons long-lived. The existence of adsorbed oxygen was essential for chemical sensitization by EAM. When dyes were adsorbed on zinc oxide together with EAM, a marked increase in spectral sensitization efficiency was observed, and the term "hypersensitization" was given to this effect. This suggests that the chemical factors are decisive regarding efficiency, also in spectral sensitization by dyes. The photodesorption of oxygen from zinc oxide surfaces which has been known to occur in parallel with photoconduction, was observed again in the case of dye-sensitized excitation.
These findings consequently led us to study the adsorption states of oxygen on zinc oxide. It was generally accepted that at least two types of chemisorbed oxygen ions occur on zinc oxide surfaces. From a manometrical examination of the temperature effect of oxygen adsorption, it has been concluded that, at 25° C, most oxygen is chemisorbed as O_2^-, at 140° C as O^-, and at 280° C or higher temperatures as O^{2-}. At low temperatures, O_2^- accompanied another type of adsorbed species of a reversible and neutral character, presumably forming O_4^-. Photodesorption of oxygen was observed only in the presence of O_4^- and a small amount of water.
It may be readily supposed that some chemical interactions between dye and zinc oxide influence the sensitizing efficiency to a great extent. To examine this, a chemically active dye, the merocyanine form of 1, 3, 3'-trimethylspiro (indoline-2, 2'-benzopyran)-8' carboxylic acid was adsorbed on zinc oxide. A small portion of the adsorbed dye was found to be strongly adsorbed compared to the rest. The strongly adsorbed portion exhibited particularly efficient spectral sensitization. From extracted solution of the strongly adsorbed dye, zinc ions were detected. This is an indication of the relation between chemical factors and efficiency in spectral sensitization.
On the other hand, when another dye, formazan, existed on zinc oxide surface, the dye being photochemically formed from 2, 3, 5,-triphenyltetrazolium chloride on zinc oxide, distinct desentization was observed at wavelengths of intrinsic photoconduction of zinc oxide.
To have an insight into the nature of the chemical interactions, dye sensitization was further

investigated using several xanthene dyes. The rate of initial rise of sensitized photocurrent closely related to the density of surface-trapped electrons which was estimated assuming that zinc oxide was a perfect barrier-type semiconductor. Good sensitizers drew out more electrons from the bulk to the surface of zinc oxide, as a result of adsorption (in the dark).

Our most recent experiments showed that the energy transfer mechanism should not be overlooked. Single crystals of zinc oxide were used as an electrode in an electrochemical cell. Chemically-produced excited oxygen molecules ($^1\Delta g$ or $^1\Sigma^+g$) from H_2O_2 and NaOCl, increased the electric current flowing across the crystals. At the end of the reaction, the current started to decay down to the original level. One concludes that the excess energy of singlet oxygen must be transferred to surface traps to inject electrons from them into the volume of the crystals.

This paper is mainly concerned with the dye sensitization of photoconduction of zinc oxide powder for Electrofax. The object of the paper is to pick up the key knowledge of the primary sensitization act in zinc oxide and also to report our latest work. Before talking on dye sensitization, I would like to mention the intrinsic photoconduction properties of zinc oxide, because spectral sensitization behavior is closely related to the intrinsic photoconduction properties of zinc oxide.

1. Intrinsic Photoconductive Properties of Zinc oxide

Theoretically, the dark resistivity of zinc oxide is expected to be over 10^{25} ohm-cm, but the real crystal shows n-type dark resistivity below $10^6 \sim 10^{-2}$ ohm-cm in vacuum [1]. This is caused by a stoichiometric excess of zinc, either as interstitial zinc atoms or oxide-ion vacancies in the ionic crystal. The concentration of excess zinc atoms is about 10^{17} atoms/cm^3. The dark resistivity of zinc oxide is remarkably changed by adsorption of oxygen in air. The effect depends upon the specific surface area of zinc oxide, such as single crystals, thin films, sintered specimens, Electrofax coatings and others.

Pulverized zinc oxide exhibits a dark resistivity of 10^{17} ohm-cm in air. The specific surface area of the powder (mean particle size of 0.3 micron) is estimated approximately 4.5 m^2/g and the density of chemisorbed oxygen is about one oxygen per 1000 lattice atoms of zinc oxide crystal [2]. At thermal equlibrium, an insulating electron depletion layer under the surface is formed as shown in Figure 1 [1]. The increase of dark resistivity of zinc oxide powder in air was explained as follows [1]. By crushing up a large crystal into small grains, the zinc oxide becomes on the average more precisely stoichiometric in air and approaches more closely to intrinsic behavior. In zinc oxide powder with a grain size of about 10^{-5} cm, the thickness of the electron depletion layer becomes comparable with the grain size, as shown in Figure 2. In the powdered zinc oxide, oxygen adsorbed on the surface is one equivalent of oxygen uniformely diffused into the volume. A volume doping of 10^{15} atoms/cm^3 corresponds to a mean spacing of impurity atoms of 10^{-5} cm, i.e. with the dimension of zinc oxide powder grains used in Electrofax, as many as 100 oxygen atoms per grain still approximate uniform doping.

The kind of adsorbed oxygen, such as O_4^-, O_3^-, O_2^-, O^- and O^{2-}, has been discussed by many workers [3, 4, 5]. The surface barrier height of zinc oxide is estimated approximately 0.4 ~ ~ 0.8 eV, depending upon the adsorption states. The level of adsorbed oxygen on the surface is

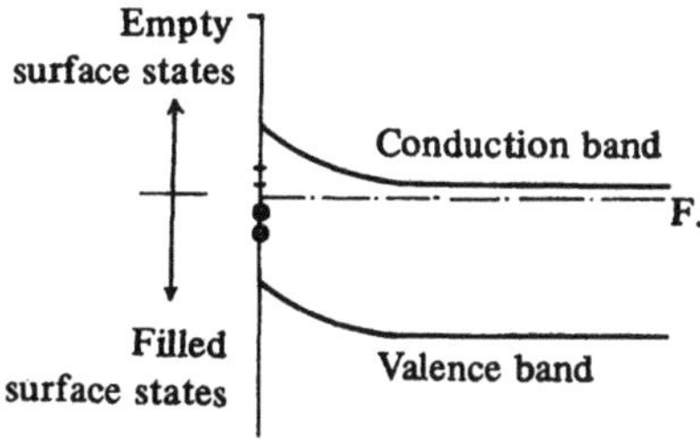

Fig. 1. The surface of a ZnO crystal at thermal equilibrium after Ruppel, Gerritsen & Rose [1]. F.L = Fermi energy.

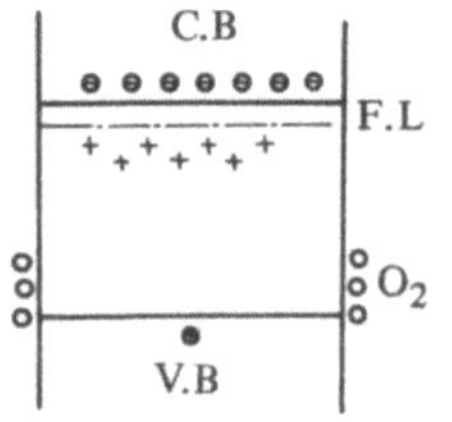

a) Before adsorption of Neg. Oxygen

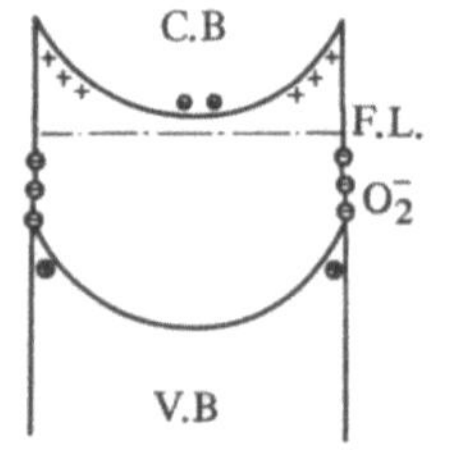

b) After dark storage thermal equilibrium Neg. Oxygen adsorbed

Fig. 2. After Ruppel, Gerritsen & Rose [1]

expected to be distributed at between 0.8 and 1.2 eV below the conduction electron level of zinc [6].

When the zinc oxide in air at room temperature is illuminated with light of wave length below the fundamental absorption edge (385 nm), photoconductivity is observed. The photoconductive response has been characterized by a slow process which builds up and decays over minutes or a fast process of millisecond-to second-duration, as shown in Figure 3. The slow process is related with the adsorption of oxygen in the dark and its desorption under

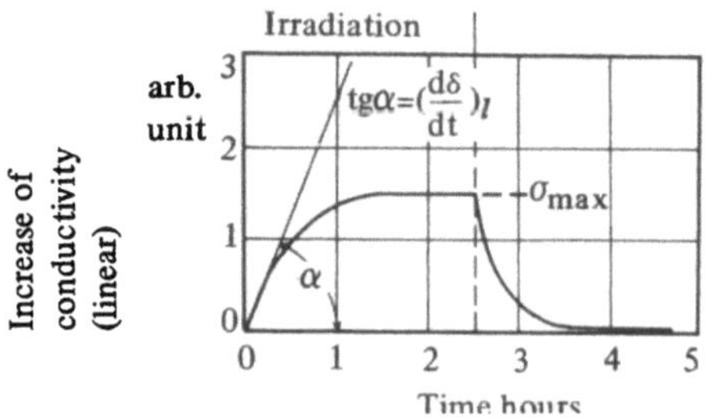

Fig. 3. Thin layer. Variation with time of conductivity during light irradiation in air. Slow process. After *H. Mahr*, Solid state phys. **8**, 279 (1959)

ultra violet irradiation [7]. The fast process is believed to be electronic in origin, and to be controlled by equilibrium between carriers and traps [8, 9].

The intrinsic light absorption spectrum of zinc oxide and the action spectrum of photoconduction are shown in Figure 4. The intrinsic photoconduction process, including the fast and slow processes, are expressed to be following an energy band model, as shown in Figure 5.

The interpretation of the slow rise and decay of photoconduction as an oxygen adsorption process has been supported by independent observations which showed that the conductivity increase under irradiation is accompanied by desorption of oxygen [8, 10, 11, 12]. Other

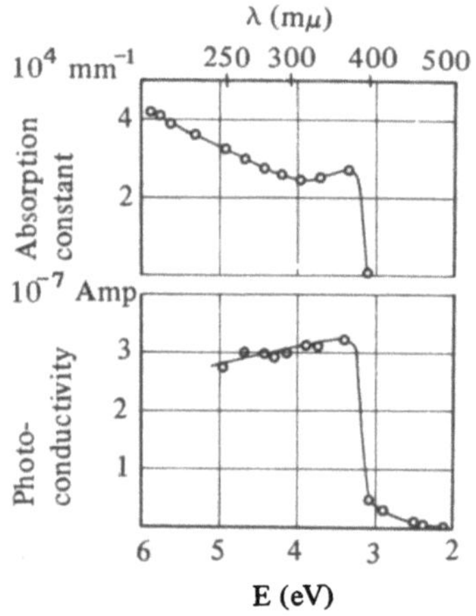

Fig. 4. Absorption spectrum and photoconduction spectrum of ZnO layer

After *E. Mollwo*: Z. Physik. Chem. **198**, 258 (1951)
H. Weiss: Z. Physik. **132**, 335 (1952)

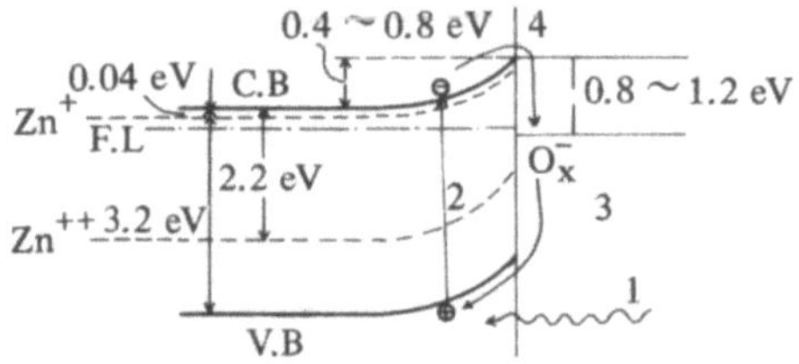

Fig. 5. Proposed energy-level diagram for zinc oxide

After *S. Roy Morrisson*: Advances in Catalysis, **2**, 259 (1955) investigations have shown that it is possible to observe a desorption or an adsorption of oxygen under irradiation depending on temperature, oxygen pressure, defect states of the surface and humidity [13]. The adsorption and desorption acts and the type of chemisorbed oxygen depend on the excess zinc content on the zinc oxide surface. There is a question about what type of chemisorbed oxygen is effective for photoconduction through photodesorption.

Molinari has demonstrated that there are two types of chemisorption, a weak and a strong one [14]. The weak chemisorption has been tentatively associated with the presence of O_2^- or O^- absorbed ions and the strong chemisorption with O^{2-} ions. The strong chemisorption is to be associated with excess zinc present on the surface [15]. At room temperature, it is assumed that weak chemisorption, which is largely affected by illumination, has only a very small effect on conductivity, while strong chemisorption should cause large variations in conduction even with comparatively small variations of surface coverage [16].

Kokes claimed that O^{2-} is stabilized above 200 to 300° C, and is less effective than O^- for conduction [17]. The adsorption and desorption characteristics are altered by water co-adsorption [18].

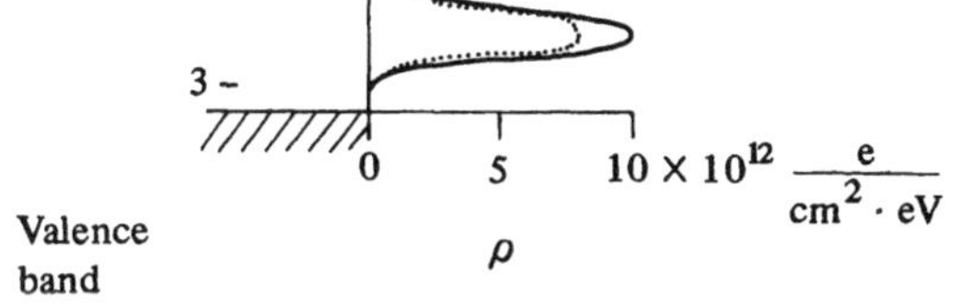

Fig. 6. Density of filled states vs energy measured downward from the conduction band edge on ZnO. After *Williams & Willis* [19]

It is clear that chemisorbed oxygen plays the principal role in the photoconduction of zinc oxide. Williams has observed the levels of chemisorbed oxygen at the surface of zinc oxide by infra-red illumination measurement. The energy levels are separated into two domains, depending on the type of chemisorbed oxygen, as shown in Figure 6 [19].

2. Chemical Sensitization of Intrinsic Photoconduction of Zinc Oxide by Electron-Affinitive Molecules

It has been found that electron-affinitive molecules other than oxygen are also chemisorbed on the zinc oxide surface and change both the dark and photocurrent, depending upon the character of adsorbant [20, 21, 22]. We have also observed the effect of electron accepting or donating molecules on zinc oxide conduction at different pressures [23]. The results obtained are shown in Table 1. It shows that adsorption of electronegative gases or vapors such as oxygen, nitrogen monoxide, nitrogen dioxide, or methyl-ethyl ketone lowers the dark current by several orders of magnitude compared with evacuated samples. However, the effect of

Table 1. Dark and Photocurrent Characteristics of ZnO Powder by Adsorption of Various Gases, after Kokado et al. [23]

Gas or vapor	Pressure mmHg	id/iv	ip/id	Decay
O_2	44	$\sim 10^{-2}$	$\sim 10^2$	rapid
NO	45	3×10^{-5}	$10^3 \sim 10^4$	rapid
NO_2	0.3	$<3 \times 10^{-5}$	$\sim 10^2$	very rapid
MEK	47	$10^{-2} \sim 10^{-3}$	2×10^2	slow
vacuum	–	–	$2 \sim 3$	very slow
CO_2	114	1	$2 \sim 3$	very slow
NH_3	42	~ 10	$2 \sim 3$	slow
Pyridine	7.5	$10^{-2} \sim 10^{-4}$	–	–

iv: Dark current in vacuum
id: Dark current in gas or vapor
ip: Photocurrent in gas vapor for 10^{12} photons/ $cm^2 \cdot sec$ of 365 nm light from an ultra-high pressure mercury lamp

After Kokado et al. [23].

ultraviolet irradiation was to restore approximately the vacuum conductivity in each case.

When organic acids or acid anhydrides such as maleic acid or phthalic anhydride were adsorbed on zinc oxide, the photocurrent was increased markedly, as shown in Figure 7 [24]. Normal organic acids and acid anhydrides behaved differently when adsorbed on zinc oxide [25]. Normal acids could react with excess zinc and increased the photocurrent. When their density exceeded the amount of excess zinc, the photocurrent was decreased. However, acid anhydrides sensitized the photocurrent of zinc oxide by co-adsorption with oxygen, at any density.

Kurita also confirmed that long-chain fatty acids, such as stearic acid, palmitic acid or behenic acid decreased the dark conductivity of zinc oxide, when they were adsorbed [26]. If a combination of electron-acceptor and donor molecule was co-adsorbed on zinc oxide, good sensitization was observed, as shown in Table 2 [27].

The photoresponse of zinc oxide is easily observed only if a surface barrier is formed to decrease the dark current.

Consequently, the maximum photocurrent in zinc oxide cannot exceed much the dark current in the absence of adsorbed molecules. The maximum photoresponse must correspond to the difference in dark current between chemisorbed and non-chemisorbed states. In general, the photoconduction of zinc oxide can be observed only if the zinc oxide is chemically sensitized. The response speed of photocurrent is thought to be altered with the characteristics of the chemisorbed species.

Table 2. Relative Sensitivity of ZnO Photoconduction by Co-adsorption of Acceptors and Donors, after E. Inoue et al. [27]

Second dopant \ First dopant	A 1	A 2	A 3	A 4	D 1	D 2	D 3	D 4
A 1	4.35 –							
A 2	– –	1.5 –						
A 3	7.0 –	160 –	2.3 6.6					
A 4	6.6 –	1.0 –	– –	1.12 60				
D 1	3.6 –	1.1 –	105 43	1.1 90	0.78 1.8			
D 2	37 –	3.6 –	1.5 7.6	0.56 8.0	– –	1.1 600		
D 3	88 –	0.78 –	– –	– –	260 90	2.2 450	0.6 16.5	
D 4	7.2 –	0.86 –	– –	– –	7 –	0.78 180	– –	0.54 3.65

upper. 0,01 mol%
lower. 0.1 mol%

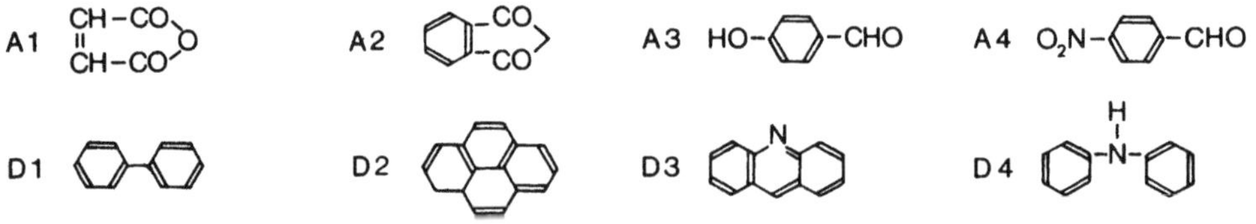

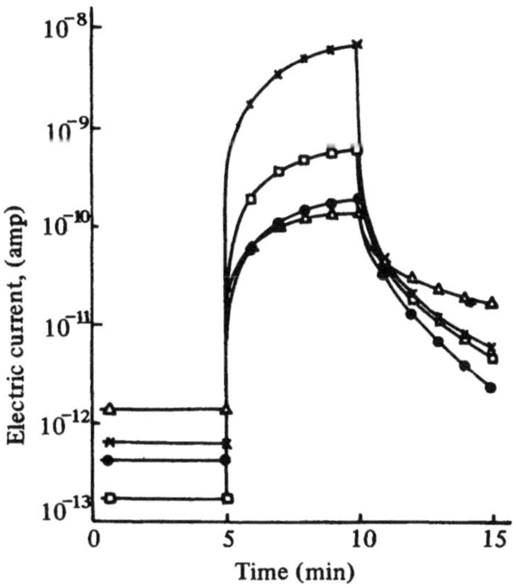

Fig. 7. Rise and decay of photocurrent in zinc oxide-oxygen-electron affinitive molecule systems

- • ZnO-O_2
- △ ZnO-O_2-p-chloranil (0.3 mole.%)
- □ ZnO-O_2-iodine (0.03 mole.%)
- x ZnO-O_2-phthalic anhydride (0.03 mole.%)

After *E. Inoue* et al. [24]

3. Spectral Sensitization of Zinc Oxide Photoconduction

Dye sensitization of the photoconduction of zinc oxide has been observed in various forms. The first report of this phenomenon was published by Putzeiko and Terenin [28]. They found that the Dember effect in the pressed powder was sensitized to visible light by adsorbed eosin, erythrosin, rhodamine B, or several cyanine dyes.

Since that time, other examples of zinc oxide spectral sensitization have been observed with dyes that are both p-type or n-type photoconductors. Also, Young and Greig have described spectral sensitization of Electrofax layers [29]. Recently, we have had many excellent reviews of this work [30].

Before discussing the mechanism of dye sensitization of zinc oxide photoconduction, I would like to point out the features of spectral sensitization, as follows.

(1). The sign of the charge carrier in all sensitization effects, including Becquerel effect, Dember effect, barrier-layer photoeffect, photocontact potential change and photoconduction, is that of the non-sensitized n-type zinc oxide [31].

(2). The primary sensitizing act requires less than 10^{-9} sec [32].

(3). The action spectra of the various photoeffects correspond to the absorption spectrum of the dyed substrates [33].

(4). The sensitizing dye can be used many times for absorption [34].

(5). The acidic dyes are chemisorbed strongly on the zinc oxide surface, but the basic dyes are not. The strongly adsorbed dyes are most effective in spectral sensitization [35, 36].

(6). The adsorption of electronegative molecules, including oxygen, is necessary and not avoidable for photoconduction in zinc oxide, as mentioned before. Also it is very difficult to eliminate chemisorbed oxygen from a zinc-oxide surface.

(7). Markievich and Putzeiko [37] observed that the chemisorbed particles containing oxygen act as centers responsible for the adsorption of acidic dyes and for the zinc oxide photo-effects, which are sensitized by these dyes. It appears that in air the adsorption bond of the dye with the zinc oxide needs the participation of an electron localized on the surface of the zinc oxide because of the influence of electronegative molecules of moist oxygen from the air, which have an affinity for electrons of the order of 1 eV.

(8). Putzeiko [38] found that the adsorbed electronegative molecules such as oxygen or quinone increase the efficiency of dye sensitization. We also followed up this fact using organic electron affinitive molecules and found a *hypersensitization* effect, which broadened the action spectrum, as shown in Figure 8 [39].

(9). Akimov made clear that fluorescence and phosphorescence of sensitizing dyes on zinc oxide are quenched entirely [40].

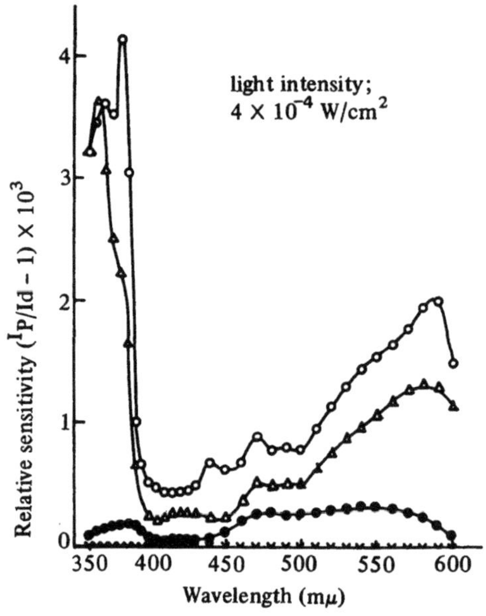

Fig. 8. Hypersensitization of zinc oxide photoconduction

- • basic system (dyed ZnO with 0.002 mole % rose bengal)
- o basic system + phthalic anhydride (0.02 mole %)
- △ basic system + iodine (0.03 mole %)
- x basic system + p-chloranil (0.03 mole %)

After *E. Inoue* and *T. Yamaguchi* [39]

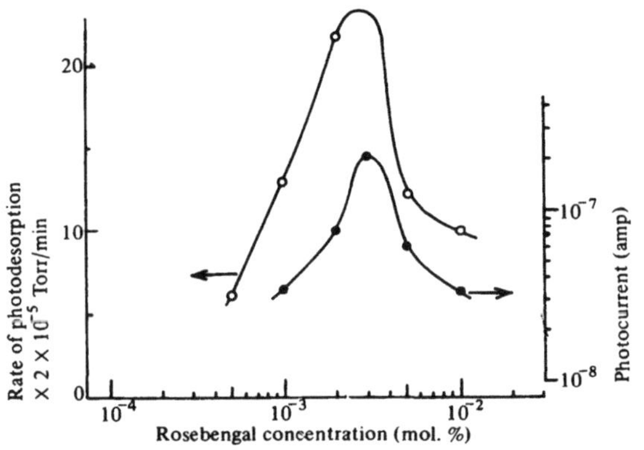

Fig. 9. Correspondence of oxygen photodesorption and photosensitivity of dyed ZnO with rose bengal; excitation: ~ 510–700 μm, 1.5 x 10^{-3} W./cm². After *E. Inoue* et al. [41]

(10). We found a photodesorption of oxygen from dyed zinc oxide, when the dyes are illuminated with light [41]. The desorption rate of oxygen corresponds to the photocurrent and depends upon the group of dyes, such as p-type or n-type, as shown in Figures 9 and 10.

(11). A *supersensitization* effect was found by Namba and Hishiki, when two kinds of dye were co-adsorbed on zinc oxide [42]. This suggests that one dye acts as the electronegative molecule to the other one on the zinc oxide surface.

(12). Dyes can sensitize spectrally the photoconduction of zinc oxide in submonolayer and aggregate states. The dark conductivity of a thin layer of dyed zinc oxide was changed with the adsorbed density of dye, as shown in Figure 11 [43]. When only a little dye is adsorbed the dark current is decreased [44], but when the dye density is increased much more the dark current is increased gradually. However, the characteristics of photocurrent do not correspond exactly to the dark-current change, because of the efficiency of light absorption and other factors.

(13). Reducing agents such as hydroquinone can increase the efficiency of dye sensitization

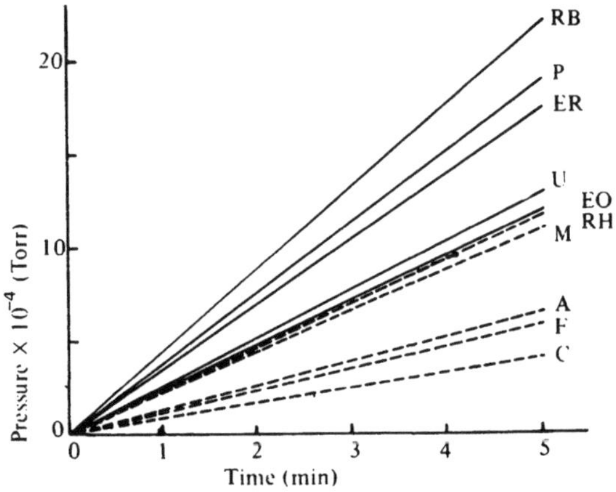

Fig. 10. Photodesorption of oxygen from ZnO dyed with p- or n-type dyes;

———— ; p-type

RB, rose bengal; P, phloxine; ER, erythrosine B; U, uranine; EO, eosine;

- - - - - - - - ; n-type

RH, rhodamine B; M, methylene blue; A, acridine orange; F, fuchsine; C, crystal violet.

After *E. Inoue* et al. [41]

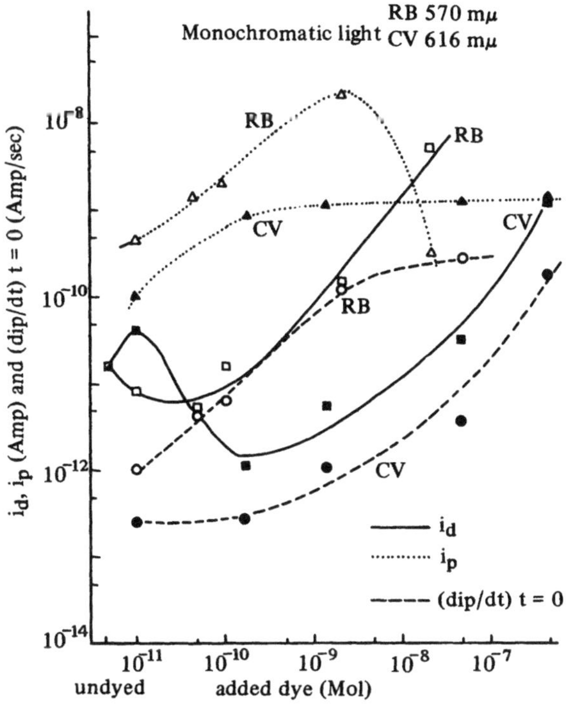

Fig. 11. Changes of i_d, i_p and (dip/dt) t = 0 as a function of amount of dye added.
After *Nakayama* et al. [43]

of zinc oxide crystals in electrolytic systems [45] and also in the normal dry measurement method [43].

(14). The photoconductive response decays gradually for long time illumination, although the dye is not destroyed.

4. Mechanism of Dye Sensitization

Two models of the mechanism of spectral sensitization have been presented. One is the energy transfer mechanism of Franck and Teller [46], the other the electron-transfer mechanism of Gurney and Mott [47]. Both have been proposed for the silver-halide photographic effects.

In the electron transfer model, the electron is transfered to the zinc oxide conduction level from the excited dye molecules when the dye is irradiated by light. The following energy situation is necessary for this model, as shown in Figure 12. The excited level of dyes should lie above the zinc-oxide conduction band or a little below the conduction band in which case thermal assistance is necessary for the injection of the electron from the dye into zinc oxide.

Terenin and Akimov showed that the levels of excited dyes lie below the zinc oxide conduction band by observation of the photo-electron emission from dyed zinc oxide [48]. This is a strong evidence against the electron transfer model and for the energy transfer model.

Mitchell tried to solve this objection in the case of silver halide sensitization [49]. He suggested that the excited dye molecule first captures an electron from a surface halide ion. The level of the excited anion is thus raised to allow transfer of the electron to a surface acceptor level, followed by thermal excitation into the conduction band.

The essential difference of Mitchell's from the Gurney and Mott model is the sequence of events, because the former requires the sensitizer to be restored by the transfer of the trapped electron from the substrate to the oxidized dye molecule. Terenin and Akimov [50] claimed that a fast two-electron interchange cannot in principle be distinguished from excitation energy transfer.

Nelson [51] has supported the electron transfer mechanism, but he mainly observed the photoeffects of dyed cadmium sulfide, not of dyed

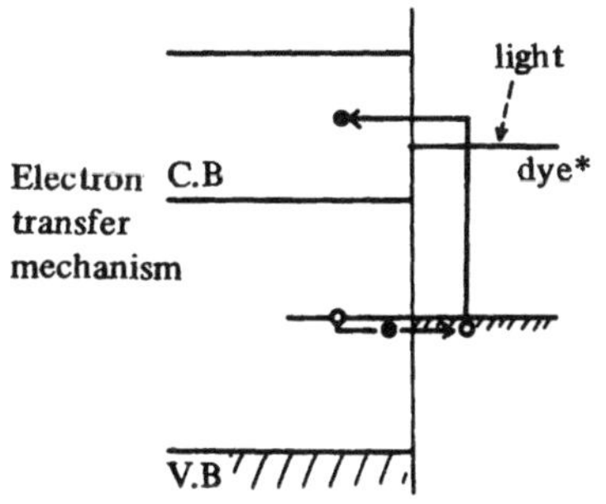

Fig. 12. Scheme of the sensitization mechanism.
After *Meier* [30–c]

zinc oxide. He showed that the best sensitizers are those in which the level of the optically excited electron is above the substrate conduction band. From the works of Nelson [52] and Scheibe [53], the energy situation between zinc oxide and dye is shown in Figure 13. Here, the first excited electronic state of many organic molecules is situated 3.4 eV below the vacuum level, according to Scheibe and co-workers.

Tani and Kikuchi [54] estimated by means of polarographic half wave potentials, excitation energies and HMO calculations that the energy levels of the excited dye molecules are above the conduction band of zinc oxide. No correlation was found between work function and sensitizing ability of dyes, contrary to suggestions by Nelson [55] and Meier [30].

From precise measurements of energy levels in sensitizer dyes adsorbed on cadmium sulfide, Nelson showed that the energy of a conduction electron in the dye phase is 0.1 to 0.4 eV higher than that of the excited "optical electron" of the dye molecule, so that charge-carrier separation in the dye phase is unlikely. The mechanism by which sensitization due to adsorbed molecules and monolayers takes place by tunneling of the excited "optical electron" through a barrier into the substrate is consitent with the measured energies and is analogous to the generation of charge carriers in the solid dye [55].

From this model, p-type dyes can sensitize n-type semiconductors. Also he thought that both single molecule and monolayer aggregates can operate by the same mechanism. Nelson however, was not directly concerned with the spectral sensitization of zinc oxide, in which the role of adsorbed electronegative molecules should be considered. Also we have to note that the replenishment of electrons back to the oxidized dye molecules is an essential aspect of the electron transfer model.

On the other hand, the energy transfer model postulates that the excitation energy of the sensitizer molecules is transfered by a resonance process to an electron at the crystal surface, raising it to the conduction level, as shown in Figure 14.

This model has been supported by Terenin et al., because it can explain a process in which the electron level of excited dyes is below the level of the conduction band in zinc oxide [48]. The energy transfer model also appears to be consistent with many characteristics of spectral sensitization mentioned above, such as items (5), (6), (7), (8), (10), (11) and (12).

However, these characteristics are no direct proof for the energy-transfer model. At present, it is difficult to decide on a model definitely from energy observations because the measurements are strongly affected by the measurements ments are strongly affected by the samples, their treatment, and other experimental conditions.

There have been some informations on the dye sensitization of zinc oxide photoconduction other than those of energetics.

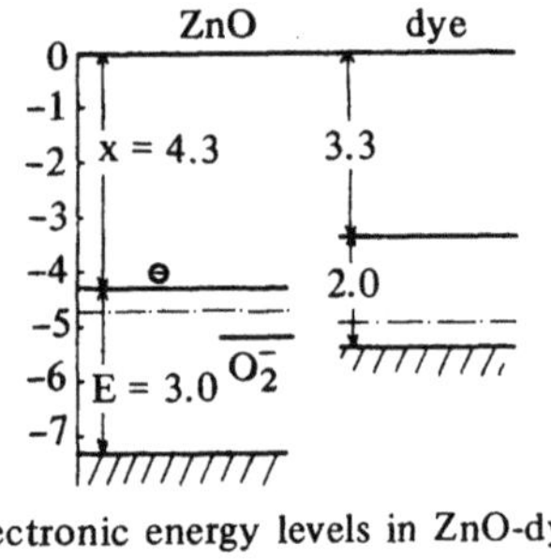

Fig. 13. Electronic energy levels in ZnO-dye. After *Nelson* [52] & *Scheibe* [53]

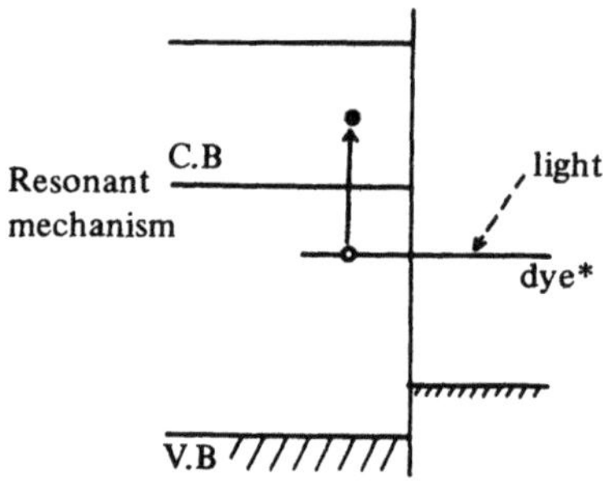

Fig. 14. Scheme of sensitization mechanism. After *Meier* [30–c]

For instance, Meier raised the question of the driving force pushing the electron from dye to zinc oxide or other semiconductors in the energy or charge transfer process. He tried to explain it by the idea of a p–n junction between semiconductor and dye aggregate [30].

As I mentioned before, the electrical behavior of dye on zinc oxide is changed strongly with the density of dye, as shown in Figure 11 [43]. It may be better to discuss separately the sensitization mechanism for dyes in the molecular state and for dye aggregates. I would like to consider here only the molecular state of the dye sensitization of the zinc oxide photoconduction.

Hauffe [56] demonstrated that in the observation of photocurrents of dyed zinc oxide single crystals by an electrolytic method the electron transfer mechanism will occur by injection of the electron from the dye into zinc oxide by illumination, the oxidized dye getting replenished with an electron from Cl ion in solution or oxygen anion on the surface.

Gerischer et al. [57] observed that the sensitization photocurrents can be observed with a zinc oxide crystal placed in contact with an electrolyte to which a suitable dye has been added, when its surface is irradiated with the light of a wave length which is absorbed only by the dye. For rose bengal, the effect of super-sensitization could be observed when reducing agents such as hydroquinone, Na_2SO_3 or H_2O_2 were added.

In the continuation of this work [58, 59] he suggested direct electron transfer to occur between the excited dye molecule and the conduction band of zinc oxide. In the process of super-sensitization, the efficiency of this electron injection is increased either by charge transfer interaction between the dye and the super-sensitizer or by direct reduction of the excited dye molecules, as shown in Figure 15. Both processes retard the deactivation of the electron in the excited state, increasing in this way the probability of electron transfer.

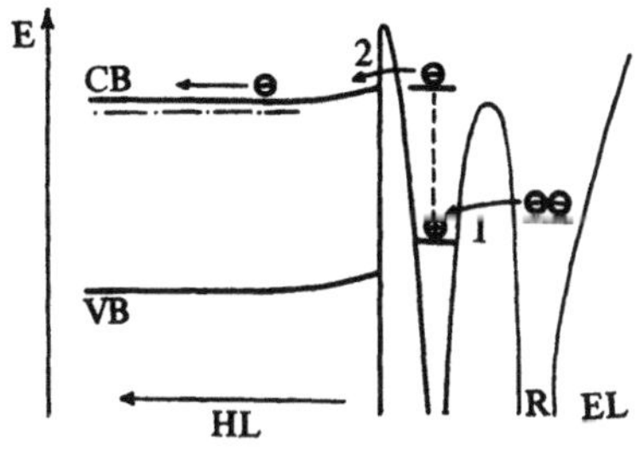

Fig. 15. Model of super-sensitization.

After *Gerischer & Tributsch* [45]

a) Sensitization: $S^* \rightarrow S + e^-$

b) Super-sensitization through molecular-complex
$S^* + R \rightarrow (S^{\delta-} \cdots R^{\delta+}) \rightarrow S + R^+ + e^-$

c) Super-sensitization through reduction of excited sensitizer.
$S^* + R \rightarrow S^- + R^+; S^- \rightarrow S + e^-$

We also confirmed that reducing agents such as n-butylamine can act as the super-sensitizer for spectral sensitization [43] in dry powder zinc oxide.

Gerischer assumed that the excited dye undergoes charge-transfer interaction with the reducing agent and that the electron is transfered into the vacant lowest level of dye from the reducing agent. Consequently, the electron of the excited dye can be easily transfered into the conduction band of zinc oxide. However, in this process, the role of the electron affinitive molecules adsorbed on zinc oxide is not considered.

As we see from the features of spectral sensitization such as [6, 7, 8, 9, 10 and 11], the electron affinitive molecules interact with dyes in the dark. There may be some interaction such as a weak charge-transfer between dye and chemisorbed electron affinitive molecules in the dark because the chemisorbed molecule is expected to be an electron donor, as are the reducing agents. When the reducing agent such as n-butylamine is adsorbed on zinc oxide, the dark resistivity is increased up to one order of magnitude. This fact shows that the replacement of chemisorbed oxygen by a reducing agent is difficult. Consequently, there are on the zinc oxide surface three kinds of species such as chemisorbed oxygen, reducing agents and dye.

Along the above lines, there will be competition for charge-transfer interaction between the dye and the chemisorbed molecule or the reducing agent. The strength of the donors may depend upon the characteristics of the species. We also have to consider the interaction between zinc oxide and chemisorbed molecules. The chemisorbed molecules have interactions both with zinc oxide and the dye. When the dye sensitized zinc oxide is irradiated by visible light, the dye, in the interacted state of dye-chemisorbed molecule(or reducing agent)-zinc oxide will be excited. After excitation, the electron is transferred into the conduction band from the interacted system, as shown in the following scheme.

$$(\text{ZnO-Donor-Dye}) \xrightarrow{h\nu} (\text{ZnO-Donor-Dye}) \rightarrow$$
$$\rightarrow \text{ZnO(electron)} + \text{Acceptor} + \text{Dye}$$

Here, Donor is the chemisorbed molecule or the reducing agent, and Acceptor is the electron affinitive molecule before adsorption or the oxidized reducing agent. In this process, it is difficult to see the difference between energy transfer and electron transfer models. The situation will be similar to the case of Mitchell's two-electron interchange model, in which we find it difficult to see the difference from the energy transfer model [48]. The efficiency of spectral sensitization will depend upon the relative concentration of donor to dye.

5. New Model for Spectral Sensitization

Hereafter, I would like to show you our latest work which is concerned with a new model of spectral sensitization on zinc oxide photoconduction. The work was directed towards elucidating the role played by chemical interaction between zinc oxide and adsorbed dyes in the sensitization process.

5.1. Experimental

Common zinc-oxide powder for Electrofax was used as sample and dyed by an ethanolic solution of a known concentration of dye. The concentration of adsorbed dye was adjusted to 0.004 ± 0.0005 weight percent in all samples. This corresponded to an adsorption density of approximately 1.5×10^{-4} molecules/A^2.

The dyed zinc oxide was coated on glass plates to a thickness of 100 ± 10 microns, after drying. On each of the plates, a pair of an electrodes, 10 mm long, were evaporated to make an electrode gap, 5 mm wide. A d.c. voltage of 10 V was applied. For photoconductivity measurements, monochromatic excitation of 1.5 x x 10^{-4} W/cm^2 was applied, the wave length of which was chosen to coincide with the absorption maximum.

5.2. Results

The dark conductivity of zinc oxide was greatly changed by the adsorption of dyes. When a relatively low density of dye was adsorbed, like in this experiment, the extent of the change appeared to represent the sensitizing power of the dye, as shown in Table III.

Good sensitizers depressed the conductivity to lower levels and poor ones raised it to higher levels, as shown in Figure 16. The temperature dependence of the dark current was measured for each sample, at over 60° C, to establish the equilibrium, as shown in Figure 17.

Plotting i_d vs 1/T from the dark current, the activation energy E_a was obtained, as shown in Table II and Figure 16. Also the photocurrents i_p are presented in Figure 16.

The dark current of zinc oxide, whether it was dyed or undyed, is expressed by the following conductivity equation

$$\sigma_d = \sigma_0 \exp(-E_a/kT) \qquad (1)$$

where σ_d is the dark conductivity and σ_0 is a constant.

The photocurrent i_p and its initial slope $(di_p/dt)_{t=0}$ at 60° C are presented in Table III. The photoresponse curves are shown for each sample in Figure 18. Steep rise response indicates good sensitization. The initial slope of

Table 3. Conductivity data of dyed zinc oxide

	dye (abs. max. : nm)	abbrevia-tion	i_d(A)	E_a(eV)	i_0(A)	$(di_p/dt)_{t=0}$ (A/sec)	i_p(A)
anionic	rose bengal (570)	RB	7.94×10^{-12}	0.88_0	166	8.0×10^{-8}	1.2×10^{-7}
	erythrosine B (566)	ES	2.41×10^{-11}	0.85_2	191	5.8×10^{-8}	6.3×10^{-8}
	fluorescein (512)	FC	3.26×10^{-11}	0.84_3	192	3.2×10^{-8}	4.9×10^{-8}
(undyed zinc oxide)			6.50×10^{-11}	0.84_3	247	–	–
cationic	rhodamine B (567)	RhB	1.12×10^{-10}	0.81_5	250	2.6×10^{-9}	1.6×10^{-8}
	phenosafranine (559)	PS	1.62×10^{-10}	0.81_0	303	1.9×10^{-10}	3.2×10^{-9}
	crystal violet (616)	CV	2.07×10^{-10}	0.80_5	320	8.2×10^{-11}	7.2×10^{-10}
	malachite green (640)	MG	3.09×10^{-10}	0.79_7	364	2.1×10^{-11}	4.3×10^{-10}

photocurrent which corresponds to the sensitizing efficiency is increased proportionally to the excitation intensity, while the stationary current varied approximately with the square root of intensity, as shown in Figure 19. The slope was found to be expressed as a function of the initial dark current i_0, as illustrated in Figure 20.

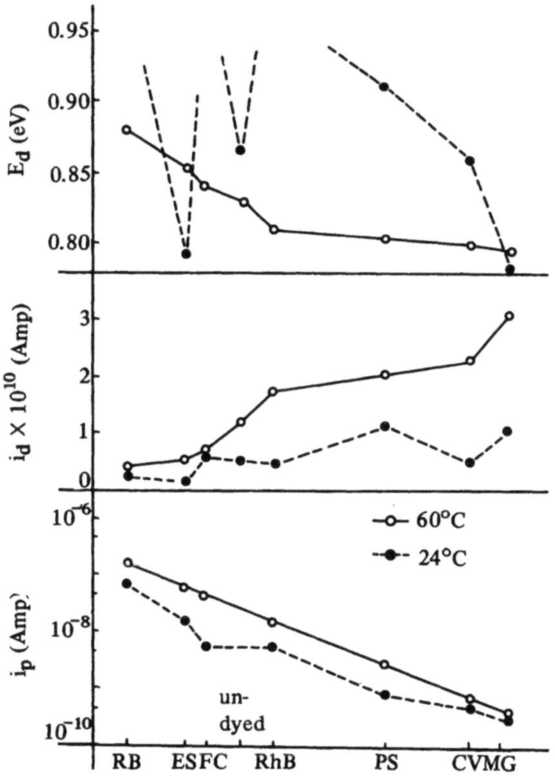

Fig. 16. Illustrative representation of conductivity data of dyed zinc oxides

$$(d\sigma_p/dt)_{t=0} = AI \exp(-\alpha\sigma_0) . \quad (2)$$

Here, I is the light intensity and A and α are constants.

The initial slope of photocurrent was plotted against 1/T, in Figure 21. This Arrhenius plot indicates that sensitization has an activation process. The constant α in Eq. (2), as a consequence, can be replaced by α'/kT, using a new constant:

$$(d\sigma_p/dt)_{t=0} = AI \exp(-\alpha'\sigma_0/kT) . \quad (3)$$

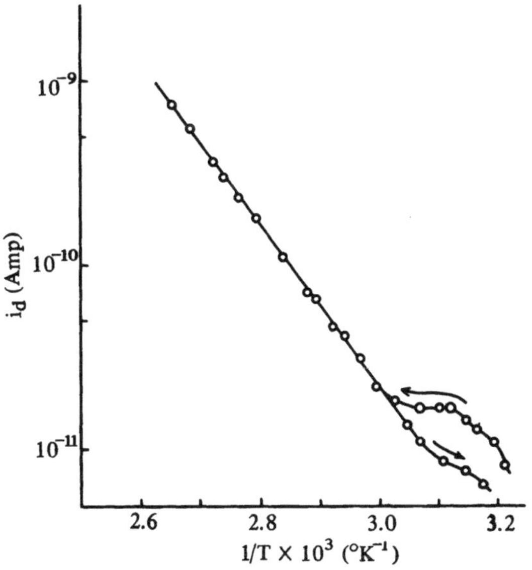

Fig. 17. A typical i_d vs 1/T curve (ES). The arrows indicate the direction of temperature change

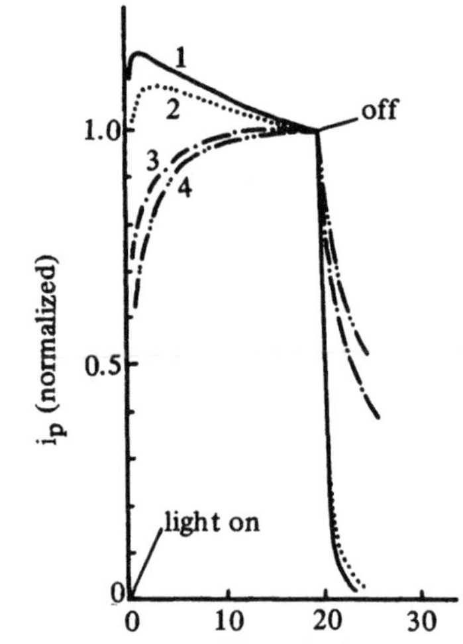

Fig. 18. Curves of photocurrent growth (normalized) 1:RB, 2:RhB, 3:CV, 4:MG

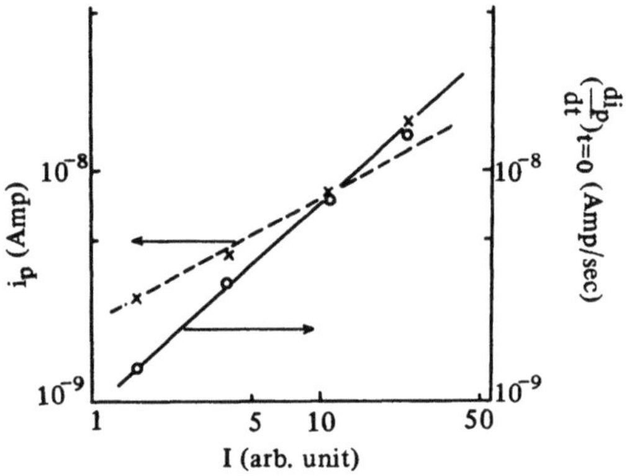

Fig. 19. Light intensity dependences of stationary level and initial slope of photocurrent (RhB)

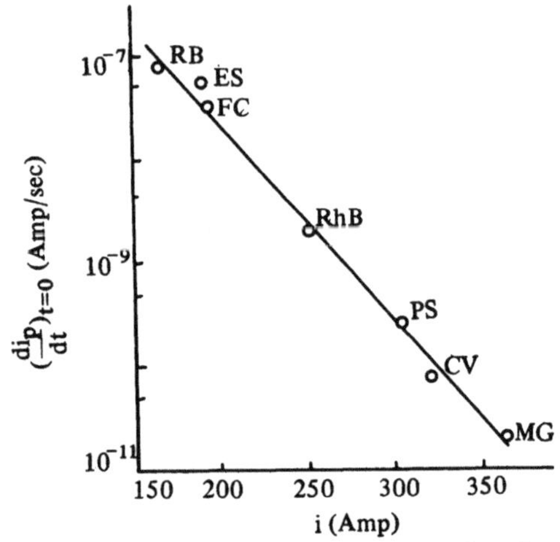

Fig. 20. Initial slope of photocurrent as a function of i_0

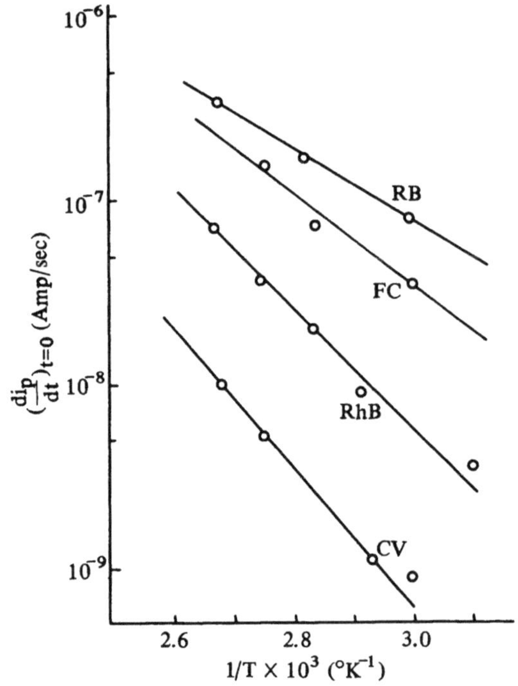

Fig. 21. Temperature effect on spectral sensitization

This equation was checked by examining an anticipated linear relationship between the activation energies calculated from Figure 21 and i_0, as shown in Figure 22.

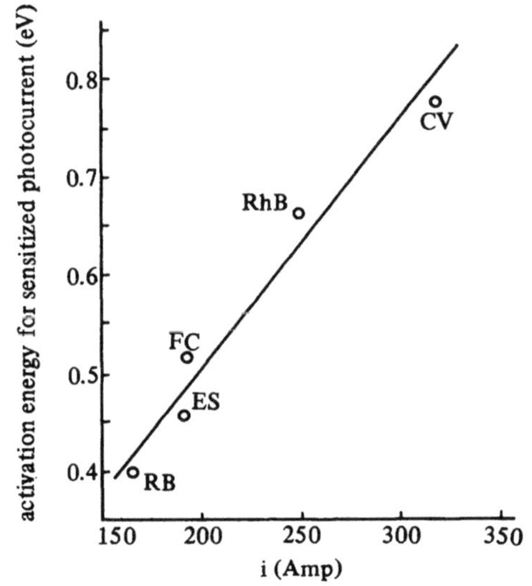

Fig. 22. Relation between activation energy for sensitization process and i_0

5.3. Discussion

We know the sensitizing efficiency to be controlled by equation (3). Weiss [60] and Tischer [61] found that the initial slope of photocurrent in a thin layer of undyed zinc oxide was proportional to the dark current of the layer. For this result, the following explanation was given. The photoconductivity is determined by the photodesorption process of adsorbed oxygen ions. The conductive layers adsorb more oxygen, placing more charges on to the surface. Consequently, the photorelease of surface-trapped electrons is related to dark conductivity. This way of interpretation will be valid in the present case, although the situation is more complicated. Evidence of oxygen photodesorption during the spectral sensitization process has been reported by us [41].

The dark conductivity of the zinc oxide powder system will be described as follows, if the donors are wholly ionized:

$$\sigma_d = (en_cM/NkT)\exp(-E_b/kT)\,. \qquad (4)$$

Here, N is the number of crystallites per unit length, E_b is the barrier height, M is a parameter depending on the type of barrier, e is the electron charge and n_c is the density of conduction electrons [62, 63].

If we can simply assume that the experimentally obtained activation energy E_a represents the barrier height E_b in equation (4), i.e. $E_a = E_b$, the initial dark current i_0 or the initial dark conductivity σ_0 would be proportional to n_c. Here, we also suppose that the adsorbed dye molecules increased the number of surface trapped electrons to $n_s^0 + \Delta n_s$ from n_s^0. The superscript 0 denotes the quantity of undyed zinc oxide. From the neutrality condition, Δn_s must be equal to the change in n_c, which corresponds to $-\Delta n_c$. If the quantities at equiribrium in the dark are distinguished with an asterisk, σ_0 in equation (3) can be rewritten in terms of $-\Delta n_s^*$, and new constants A' and β:

$$(d\sigma_p/dt)_{t=0} = A'I\exp(\beta\Delta n_s^*/kT) \qquad (5)$$

$$A' = A\exp(-\beta n_c^0/kT) \propto \exp(\beta n_s^0/kT)\,. \qquad (6)$$

The situation is shown in Figure 23.

This relationship indicates that the population of surface states essentially determines sensitization efficiency.

Apparently, the quantity $\beta\Delta n_s^*$ in equation (5) has to represent a certain energy. An attempt will be made to give physical significance to this energy. We know that the adsorption of oxygen on zinc oxide surface produces electron traps [3, 4, 5]. Recently, Williams and coworker were able to determine the distribution of electron levels on the surface of zinc oxide single crystals [19]. The levels were distributed over a wide range of energy dephts. Let the energy density of surface levels be $\gamma(E)$.

When the dye molecules will be adsorbed on the zinc-oxide surface, they extract the conduction electrons to the surface, or return the surface electrons to the bulk.

The number of electrons moved is Δn_s^*. This means that the surface states corresponding to Δn_s^* are occupied or emptied. Thus, the quantity Δn_s^* can be regarded as a measure of interaction between zinc oxide and the dyes.

If Fermi-Dirac statistics can be applied to treat the surface levels of zinc oxide, one may define a quasi-Fermi energy for the electrons in the surface states. The symbol E_{sf}^0 will be used to denote such an energy in undyed zinc oxide. Consequently, the dye adsorption changes the quasi-

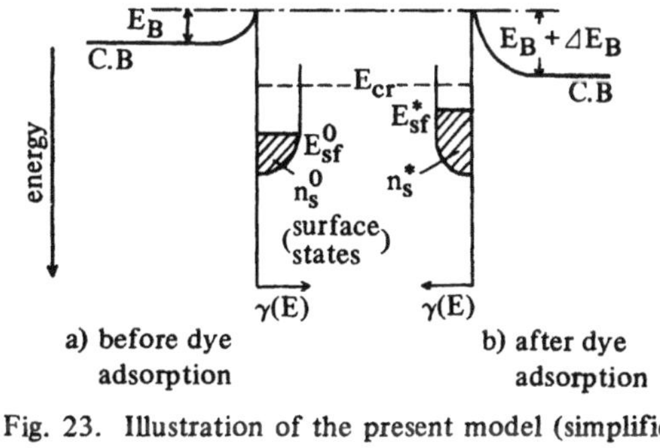

Fig. 23. Illustration of the present model (simplified)

Fermi energy of the surface electron to $E^*_{sf} = E^0_{sf} + \Delta E^*_{sf}$. If it is acceptable to postulate that $\gamma(E)$ does not vary violently with energy in the range of interest, then ΔE^*_{sf} may be treated as a quantity approximately proportional to Δn^*_s:

$$\Delta E^*_{sf} = \Delta n^*_s/\gamma(E) \; . \quad (7)$$

On the other hand, the experimental result in which an activation process is involved in sensitization, seems to imply that all surface electrons do not equally contribute to sensitization. The surface states contributing to sensitization are not exactly defined. However, some energy limitation for sensitization may exist.

Here, we assume a critical surface energy level E_{cr}, above which electrons can be injected into the zinc oxide by excited sensitizers. The sensitizing efficiency can be defined as $(dn_c/dt)_{t=0}$. Then, it is expressed as follows.

$$\begin{aligned}(dn_c/dt)_{t=0} &= -(dn_s/dt)_{t=0} = \\ &= k\,I_{abs}\exp[-(E_{cr} - E^*_{sf})/kT] = \\ &= k\,I_{abs}\exp[-(E_{cr} - E^0_{sf})/kT] \times \\ &\quad \times \exp(\Delta n^*_s/\gamma kT) \; . \quad (8)\end{aligned}$$

Here, I_{abs} is the number of photons absorbed by dye molecules and k is the rate constant of sensitized electron injection. This equation gives a theoretical basis for equation (3).

The quantity E_{cr} does not appear to depend on the type of dye. This is understandable if one considers that the critical level is determined only by properties of zinc oxide and oxygen, but not by the energetic properties of the dye.

Meier [64] classified a number of dyes into p-type and n-type, according to the type of photoconduction being influenced by oxygen and hydrogen. Oxygen sensitizes photocurrents of p-type dyes and desensitizes those of n-type dyes. The sign of Δn^*_s in our observation was always positive in his p-type dyes and negative in his n-type dyes.

All these considerations lead to the suggestion that the sensitizing power of a dye is determined mostly by the magnitude and the type of interaction with zinc oxide. The interactions modify the surface state population. The surface states are probably formed from chemisorbed oxygen ions and molecules.

The model developed here may be explored only at low coverages of dyes (e.g. 1.5×10^{-4} molecules/A^2). If the dye molecules cover too much of the surface, the excess molecules will produce other effects.

In another experiment of us [43] where higher concentrations of dye were adsorbed on zinc oxide thin layers, an increase in dark current was observed, even with cationic dyes. The mechanism is under investigation. Generally, the optimum concentration of dyes for spectral sensitization has been known to be 0.003 to 0.005 mole percent.

The sensitization model discussed here only shows that the electrons occupying the surface levels above E_{cr} can be injected into zinc oxide, when the dyes are irradiated. The density of surface electrons above E_{cr} is changed with temperature affecting the sensitization of photoconduction.

6. Conclusion

In this paper, the current problems of dye sensitization of zinc oxide photoconduction are pointed out and discussed briefly. Also, in our latest work on this problem, a new model is presented.

The efficiency of sensitization of photocurrents differs greatly with the type of dye. The interaction between zinc oxide and dye changes the population density of surface electrons on zinc oxide. This change determines sensitization efficiency. Good sensitizers increase the population and poor ones decrease it. The efficiency depends exponentially upon the number of electrons displaced by the interaction. A model accounting for this postulates a critical surface level, below which the surface states do not contribute to the sensitization process. The temperature dependence of sensitization efficiency is explained by the change with temperature of the density of surface electrons above the critical level.

I would like to thank Prof. H. Kokado for his earnest discussion on this Paper.

References

[1] W. Ruppel, H. J. Gerritsen and A. Rose, Helv. Physica Acta, **30,** 495 (1957).
[2] H. Hasebe and E. Inoue, Nippon Kagaku Zasshi (Japan), **86,** 807 (1965).
[3] K. Hirota and M. Chono, J. Catalysis, **3,** 196 (1964); also (2).
[4] T. J. Barry and F. S. Stone, Proc. Roy. Soc., **A255,** 124 (1960).
[5] W. Doeffler and K. Hauffe, J. Catalysis, **3,** 156 (1964).
[6] S. R. Morrison, Advances in Catalysis, **7,** 259 (1955).
[7] D. J. Bevan and J. S. Anderson, Disc. Faraday Soc., 8, 238 (1950).
[8] D. A. Melnick, J. Chem. Phys., **26** 1136 (1957).
[9] G. Heiland, E. Mollwo and F. Stöckman, Solid State Physics, **8,** 278 (1959) Academic Press, London.
[10] I. Miasnikov and S. Y. Pshezhetski, Dokl. Akad. Nauk USSR, **99,** 125 (1954).
[11] D. B. Medved, J. Chem. Phys., **28,** 870 (1959); J. Phys. Chem. Solids, **20,** 255 (1961).
[12] Y. Fujita and T. Kwan, Bull. Chem. Soc. Japan **31,** 379 (1958).
[13] H. Hasebe and E. Inoue, Nippon Kagaku Zasshi (Japan), **87,** 332 (1966).
[14] E. Molinari, F. Cramarossa and F. Pennicia, J. Catalysis, **4,** 415 (1965).
[15] A. M. Peers, J. Phys. Chem., **67,** 2228 (1963).
[16] T. I. Barry and H. Klier, Disc. Faraday Soc., **31,** 219 (1961).
[17] R. J. Kokes, J. Phys. Chem., **66,** 99 (1962).
[18] H. Hasebe and E. Inoue, Nippon Kagaku Zasshi (Japan), **87,** 332 (1966).
[19] R. Williams and A. Willis, J. Appl. Phys., **39,** 3731 (1968).
[20] V. I. Lyahenko and O. V. Snitko, Dokl. Trudy Inst. Fiz., Akad. Nauk, Ukr. USSR, **91,** 1071 (1953).
[21] E. K. Putzeiko and A. N. Terenin, Doklady Akad. Nauk SSSR, **101,** 645 (1955).
[22] E. K. Putzeiko and A. N. Terenin, Problemy Kinetiki i Kataliza, Akad. Nauk SSSR, **8,** 53 (1955).
[23] H. Kokado, E. Inoue, T. Yamaguchi and K. Takahashi, Bull. Chem. Soc. Japan, **34,** 705 (1961).
[24] E. Inoue, H. Kokado and T. Yamaguchi, J. Phys. Chem., **69,** 767 (1965).
[25] T. Yamaguchi, H. Kokado and E. Inoue, Nippon Kagaku Zasshi (Japan), **88,** 139 (1967).
[26] T. Kurita, Japan J. Appl. Phys., **5,** 977 (1966).
[27] E. Inoue, H. Kakado, T. Yamaguchi and H. Kato, Electrophotography (Japan), **8,** 16 (1958).
[28] E. Putzeiko and A. Terenin, Zhur. Fiz. Khim., **23,** 676 (1949).
[29] C. J. Young and H. G. Greig, RCA Rev., **15,** 469 (1954).
[30] a) J. W. Weigl, "Photographic Science" (Ed. W. F. Berg, Focal Press, London (1963), p. 345.
b) J. Bourdon, J. Phys. Chem., **69,** 705 (1965).
c) H. Meier, J. Phys. Chem., **69,** 719 (1965); Spectral Sensitization, Focal Press, London (1968).
d) H. Frieser and M. Schlesinger, Phot. Korr. „Elektrophotographie", Diskussionstagung Hann., Münden, May 12–13 (1966) Special Edition, No. 8, p. 43.
e) L. I. Grossweiner, Photochem. Photobiol., **8,** 411 (1968).
[31] A. Terenin and I. Akimov, Z. Phys. Chem., **217,** 307 (1961).
[32] A. Terenin and I. Akimov, J. Phys. Chem., **69,** 730 (1965).
[33] S. J. Dudkowski, A. G. Kepta and L. I. Grossweiner, J. Phys. Chem. Solids, **28,** 485 (1967).
[34] R. Matejec, Z. Elektrochem., **65,** 783 (1962).
[35] H. Tsukahara, Electrophotography (Japan), **23,** 17 (1969).
[36] T. Nakayama, H. Kokado and E. Inoue, Kogyo Kagaku Zasshi (Japan), **73,** 2077 (1970).
[37] N. N. Markievich and E. K. Putzeiko, Kinetiki Kataliz, **4,** Np. 2, 307 (1963).
[38] E. Putzeiko, Dokl. Akad. Nauk SSSR, **91** (1953).
[39] E. Inoue and T. Yamaguchi, Bull. Chem. Soc. Japan, **34,** 705 (1963).
[40] I. Akimov, J. Nauch. Prikl. Phot. Kinem., **4.** 64 (1959).
[41] [24].
[42] S. Namba and Y. Hishiki, J. Phys. Chem., **69,** 774 (1965).
[43] T. Nakayama, H. Kokado and E. Inoue, to be published.

[44] [33]; L. I. Grossweiner and W. D. Brennan, J. Photo. Sci., **17,** 189 (1969).
[45] a) H. Gerischer and H. Tributsch, Ber. Bunsenges. Phys. Chem., **72,** 437 (1968).
b) H. Tributsch and H. Gerischer, Ber. Bunsenges. Phys. Chem., **73,** 582 (1969).
c) H. Tributsch, Ber. Bunsenges. Phys. Chem. **73,** 582 (1969).
[46] J. Franck and E. Teller, J. Chem. Phys., **6,** 861 (1938).
[47] R. W. Gurney and N. F. Mott, Proc. Roy. Soc., **A164,** 151 (1938).
[48] A. Terenin and I. Akimov, Z. Phys. Chem., **217,** 307 (1961).
[49] J. W. Mitchell, J. Photo. Sci., **6,** 57 (1958).
[50] [s. 32].
[51] R. C. Nelson, J. Opt. Soc. Am., **46,** 1016 (1956).
[52] R. C. Nelson, J. Chem. Phys., **27,** 864 (1957).
[53] G. Scheibe, D. Bruck and F. Dörr, Chem. Ber., **85,** 867 (1952).
[54] T. Tani and S. Kikuchi, Kogyo Kagaku Zasshi (Japan), **71,** 168 (1968).
[55] R. C. Nelson, J. Phys. Chem., **71,** 2517 (1967).
[56] K. Hauffe, V. Martinez, H. Pusch, R. Schmidt and R. Stechemesser, Appl. Optics, "Electrophotography", Supplement **3,** 34 (1969).
[57] [s. 45] a).
[58] [s. 45] b).
[59] [s. 45] c).
[60] H. Weiss, Z. Physik., **132,** 335 (1952).
[61] G. Tischer, Dissertation, Univ. Erlangen, (1954); Solid State Phys., **8,** 279 (1959), Academic Press, London.
[62] E. Inoue, H. Kokado and T. Yamaguchi, Appl. Optics, Suppl., **3** "Electrophotography", 91 (1969).
[63] R. H. Bube, "Photoconductivity of Solids", J. Wiley & Sons, New York 1960, p. 88.
[64] H. Meier, „Die Photochemie der Organischen Farbstoffe", Springer, Berlin 1963, p. 170.

Discussion

Boroky: In the summary of your paper as appears on pp. 35 of the brochure of this colloquium you referred to a possible chemical interaction which may occur between certain types of dyes and ZnO. You cited a particular merocyanine dye as a good sensitizer and a triphenyl methane derivate as a densitizer. Would you please elaborate on the possible reaction mechanism of the phenomena described?

Would you think that zinc dye intermediate salt formation is responsible for the effect as a cosentiztizing new dye similar to what Prof. Hauffe described in his azo-dye example in the previous paper, or has this compound an EAM effect?

E. Inoue: I think spiropyrans will form new sensitizing dye with zinc. But 2–3·5 triphenyl tetrazolium chloride did not react with zinc. It was oxided photocatalytically on ZnO to formazan dye. In this case intrinsic photoconduction will be desensitized.

On the Mechanism of Dye-Sensitization of Photoconductivity of Poly-N-Vinylcarbazole

H. Meier, W. Albrecht and U. Tschirwitz

Abstract

For any practical application of organic polymeric photoconductors in electrophotography it is necessary to extend their spectral response through the visible spectrum. Spectral sensitization may be realized by the addition of dyes.

The mechanism of dye sensitization of organic photoconductors is still under discussion. Therefore, we investigated dye sensitization of the model system poly-N-vinyl-carbazole with different methods: photo-induced discharge technique, sandwich photocells, vidicon arrangement (which is also used in television pick-up tubes), and photoflash method.

From the results, which are supplemented by measurements of dark conductivity and by experiments on three-component systems (PVCA/dye/electron acceptors), it can be seen that spectral sensitization of the photoconductivity of electrophotographically interesting polymers is the result of an electronic process. The mechanism is discussed.

1. Introduction

Analogous to electrophotography with zinc oxide [1–4], the question of a displacement of the spectral response into the visible region is of great importance also in the use of organic materials. Organic compounds, which can be used in electrography, in particular all heterocyclic polymers, absorb only in the ultraviolet spectral region, so that the applicability of these substances would be strongly restricted without spectral sensitization.

Whereas in the case of the sensitization of inorganic materials we have a lot of data (cf., e.g., [5]), the known results on the spectral sensitization of organic materials are rare and partly contradictory.

On the one hand, this fact can be explained by experimental problems which make research under clear conditions difficult. On the other, the unsolved questions which arise in connection with the mechanism of organic photoconductivity lead to neglect the special problems of spectral sensitization. Moreover, the photoelectric conductivity observed on dyed polymers, as, e.g., poly-N-vinylcarbazole or polyvinylbutyral, was, by many authors, associated with the photoconductivity of the built-in dye particles and not looked upon as the result of a real sensitization of the photoconductivity of the organic substrates [6].

The present knowledge of organic semiconductors leads us to discuss the spectral sensitization of organic electrophotographic layers in a similar way as the discussion of the sensitization of inorganic semiconductors. It is well known that the conductivity of organic semiconductors, such as, e.g., dyes, high polymers, polycyclic aromatic substances or charge-transfer-complexes, can—analogous to inorganic semiconductors—be referred to the formation and migration of electronic charge carriers [7–9]. That is to say

Table 1. Survey on the Dye-Sensitized Organic Semiconductors

Semiconductor	Preparation	Measuring Method	Sensitizer	Reference
anthracene	K	E	rhodamine B	[10–12]
perylene	K	E	rhodamine B pseudoisocyanine	[13]
poly-acetylene	M	K	Mg-phthalocyanine	[14]
Cu-phenylacetylene	M	K, L	methylene blue pinacyanole	[6, 14]
poly-N-vinylcarbazole	S	X	rose bengale, methylene blue	[15–17]
triphenylamine	B	B	malachite green	[18]
protein	S	S	fluorescein	[19]

The second column refers to the form of the material under investigation: (K) crystals, (M) microcrystalline layers, (B) binding layers, (S) other layer materials; for the measurements: (E) electrochemical procedure, (K) condenser photoeffect, (X) electrostatic method, and (L) measurement in sandwich–and surface cells.

the discussion of the spectral sensitization of organic semiconductors can be concentrated on the central question: whether and by which mechanism electronic charge carriers are formed under incident light radiation in spectral regions that lie outside the absorption of the polymer.

2. Previous Results

To illustrate the studies that have been published on the spectral sensitization of organic semiconductors, Table 1 gives a clear survey of the known results.

Table 1 demonstrates that several organic semiconductors (aromatic hydrocarbons, polymers) can be sensitized spectrally by means of cationic, anionic and neutral dyes. The sensitization effect is detectable by several methods: firstly by a method which corresponds to the electrophotographic process in which the photoinduced discharge of electrostatically charged layers is measured; secondly by the measurement of the charge carrier injection from dyes into crystals lying between electrolyte contacts: furthermore, by means of the condenser photoeffect, monitoring the photovoltage induced under alternating illumination in a condenser filled with dyed semiconductor material.

Up to now, there have been only a small number of measurements of sensitized photoconductivity of organic materials by surface and sandwich type cells. Besides, the sensitization experiments were mainly carried out with large crystals or microcrystalline layers, so that the sensitization is suggested to be only a boundary effect in which the charge carriers have been produced in the semiconductor material by an interaction between the excited dye and the surface of the semiconductor [20]. The possibility of sensitizing homogeneous semiconductor films that are only a few microns thick has not yet been systematically examined in undisturbed sandwich type cells. One should not forget—in the case of high ohmic polymers—the impossibility of completely excluding the influence of the conductivity of dye aggregats, as relatively high dye-concentrations have often been used for sensitization. Nor is it possible to make any statements about relations between the sensitization effect on the one hand and the dye-constitution, or dye-concentration, on the other. In addition, the practical applicability of the dye-sensitization of organic materials seems to be limited, since some investigations suggest that certain conditions must be fulfilled to permit the sensitization reaction: For example spectral sensitization of anthracene crystals lying between electrolyte electrodes could only

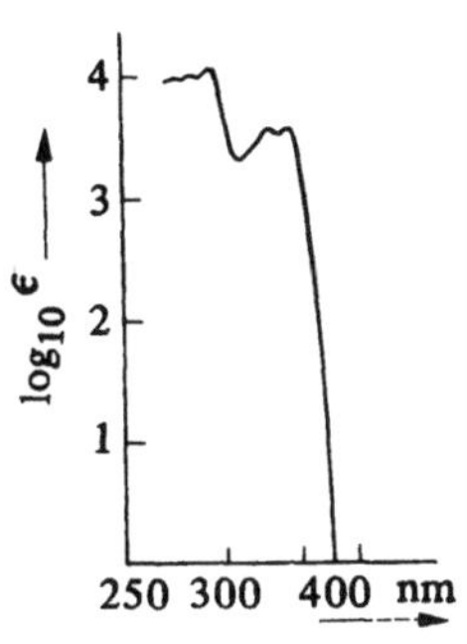

Fig. 1. Absorption of poly-N-vinylcarbazole (PVCA)

be effected by the addition of dyes into the region of the positive electrode [12].

Thus, the previous data do not settle a lot of questions connected with the problem of the sensitization mechanism of organic semiconductors as well as with the practical possibilities of the sensitization of this semiconductor class. In order to elucidate these questions, a systematic research of this effect was started.

3. Methods of Investigation

3.1. Model Substance

As a model substance poly-N-vinylcarbazole (Luvican® of BASF) was used in this work. On the one hand, poly-N-vinylcarbazole belongs to the most sensitive photoconducting polymers the xerographic applicability of which has already been known for a long time [15]. On the other hand, this compound absorbs only in the ultraviolet spectral region (Fig. 1), so that consequences of sensitization effects can well be recognized in the visible region.

For our investigations commercial polymers without any special purifying procedures, were used the first time as the materials in this form are used for xerographic purposes. For the electric measurements, poly-N-vinylcarbazole was prepared either as a thin, uniform film some microns thick or as a foil about 0.1 mm thick. This preparation was carried out by dipping the electrode for a short time into a solution of the polymer with chloroform, for example, as solvent, or by pouring it on glass plates and subsequently evaporating the solvent by heating. For the purpose of sensitization, the dyes were added to the Luvican solutions before the production of the layers, so that a uniform distribution of the sensitizing molecules was possible.

3.2. Measuring Methods

For the measurement of photoconductivity we applied two methods which can be looked upon as parallels to the electrophotographic method.

1. The sandwich-method, in which the layer of the polymer was situated between two aluminium or silver electrodes. Since the electrode that has to perform the function of the charge carrier layer formed by corona charge was transparent, the photoconductivity can be easily measured. The action spectra of photoconductivity were determined by a unit consisting of a monochromator (Jarrell-Ash-0.25 M-Ebert-monochromator) and a xenon lamp (Jarrell-Ash-XLS-J), whose different wavelength–regions were brought to a constant energy.

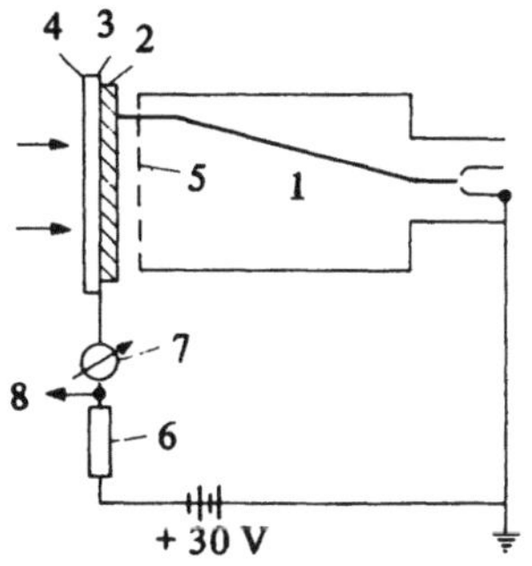

Fig. 2. Scheme of the vidicon-arrangement
1. Scanning electron beam
2. Photoconductor
3. Transparent electrode (Al, Ag et al.)
4. Glass
5. Anode (+300 V)
6. Resistance
7. Ammeter
8. To the Video-amplifier or television-tube

Without any doubt this is a method by which the possibility of the spectral sensitization of the organic semiconductor material can be investigated. The results, however, cannot be applied to the xerographic process without taking into consideration two factors: firstly, the polymer foils that are used in the cells are, for technical reasons, thicker by a factor 10^2 than the electrophotographic layers; secondly, the otherwise free surface which, in the course of the corona discharge is electrostatically charged by negative or positive charges (O_2^- or N_2^+), is covered with a transparent electrode. No indications as to the existence of surface conductivity that might possibly disturb the resolution of the picture can be obtained.

2. For our investigations still another method which has recently been applied with great success in the examination of the action spectra of organic semiconductors, was used: the vidicon method [21, 22]. The principle of this method, which is used in vidicon television tubes, corresponds to the sandwich-method. An essential difference, however lies in the fact that the lightelectric active layers which are fixed on an aluminium, silver or SnO_2-electrode, do not possess a counterelectrode Analogous to xerography, the counterelectrode is replaced by a charge induced from outside. Whereas in xerography this charge originates by corona discharge, it is in the vidicon method induced by a scanning electron beam (scheme in Fig. 2).

For use in electrophotography and for measurements in vidicon cells, the layer must have analogous conditions: an ability to store the deposited charge, a small dark conductivity, a considerable photo-conductivity in the field direction, but a small conductivity perpendicular to the field. These properties must be achieved in both methods by thin layers which are only a few microns thick.

Thus it can be assumed that a layer examined by means of the vidicon method shows a largely analogous behavior to the xerographic process.

Some remarks on the vidicon measuring principle seem to be necessary: for a measurement in vidicon cells it is essential that the potential of the anode accelerating the electron beam does not, with regard to the potential of the cathode, exceed +300 V. The electron, thereby, strikes the photoconductor layer with such a small speed that a secondary electron emission from the photoconductor layer is largely avoided. Thus the surface takes up electrons from the electron beam as long as the potential of the cathode is achieved, i.e., for example, zero volts in the grounded case. If the signal electrode, on which the semiconductor is situated, is charged to about +30 V against this potential, a photocurrent originates on illumination analogous to sandwich cells. This photocurrent can simply be measured by help of a galvanometer or be watched as a signal on a television screen; for more details on this process see [22].

With the help of the vidicon method the action spectra of organic compounds can quickly be measured [22]; this method seemed to be particularly appropriate for the examination of sensitization effects. The characteristic properties that in many points correspond with electrophotography make this method very interesting also in connection with electrophotographic problems.

4. Results of Spectral Sensitization

4.1. On the Problem of Charge-Transfer-Sensitization

When the shift of the photoconduction spectrum of organic semiconductors into the visible region is discussed, it is necessary to consider not only the problem of dye-sensitization, but also the problem of sensitization by the charge-transfer-effect. That is to say the formation of photoconductivity caused by a light-induced electron transfer from an electron donor to an acceptor must be taken into account.

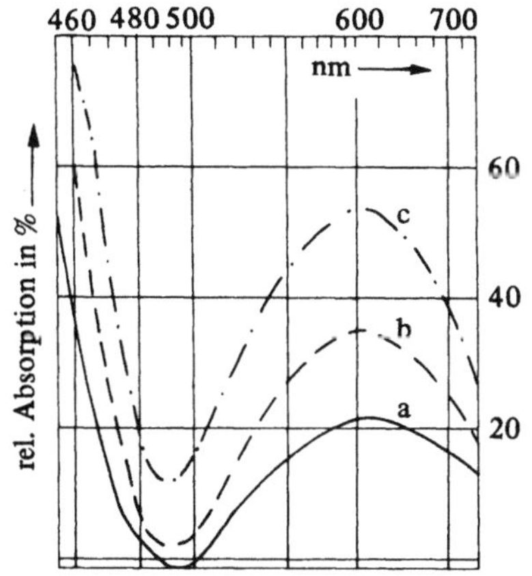

Fig. 3. Absorption spectrum of PVCA doped with 7.7.8.8.-tetracyanochinodimethan (TCNQ)
(a) PVCA + 0.12% TCNQ
(b) PVCA + 0.25% TCNQ
(c) PVCA + 0.5 % TCNQ

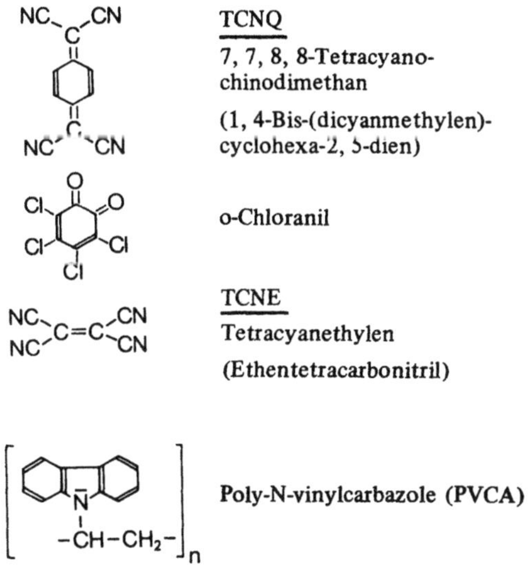

Fig. 4. Formulae of TCNQ, TCNE, o-chloranil and PVCA

A possible increase of the photoconductivity by electron acceptors was observed by Hoegl et al. [15, 23, 24] on electrophotographically interesting polymers, by Kearns, Tollin and Calvin [25] on phthalocyanines, and by ourselves on merocyanines [26]. Several authors have considered this increase of photoconductivity to be a result of an electron transfer reaction that is connected with an excitation of the charge transfer band. (cf. survey in [27]). As the CT-bands often occur in the visible region, one is inclined to apply the electron-acceptor doping as a convenient method for the spectral sensitization of organic semiconductors. And indeed, some studies deal with this possibility [17, 28, 29], some results, however, provide evidence against it [30–32]. To elucidate this problem of CT-sensitization, therefore, further investigations had to be made. Figure 3 demonstrates the structure of a CT-band of poly-N-vinyl-carbazole which was alos observed by other authors [28, 33] either in solution or layers. It can be seen that, for example, on the addition of tetracyanochinodimethan to Luvican a new band is formed in the visible spectral region, which grows in intensity due to an increase of the acceptor concentration. The new band laying at 600 nm can be referred neither to Luvican absorbing in the UV, nor to tetracyanochinodimethan absorbing at 400 nm. Besides, the band width and the missing fine structure indicate that we have it to do with the CT-band of the Luvican–TCNQ-complex. With knowledge of the ionisation energy I_D of the donor, of the electron affinity E_A of the acceptor and of the energy value C corresponding approximately to the Coulomb-energy of the D^+, A^--ionic couple, the CT-band can be calculated [34, 35] by means of the relationship

$$h\nu_{CT} = I_D - E_A + C$$

With I_D = 7.8 eV [33], E_A = 1.7 eV [35, 36], and C = 4.05 eV it follows that $h\nu_{CT}$ = 2.05 eV (600 nm).

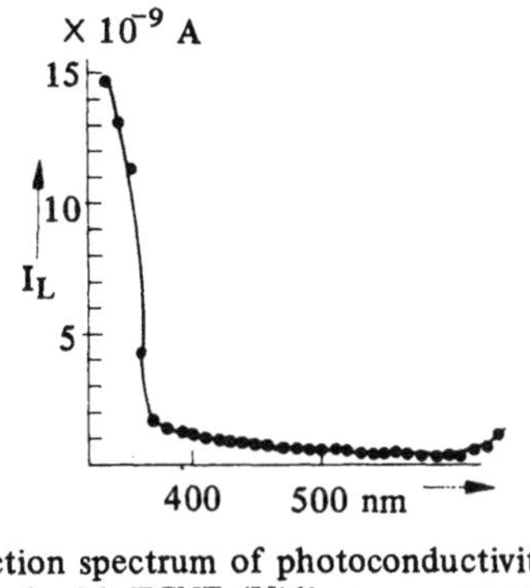

Fig. 5. Action spectrum of photoconductivity of PVCA doped with TCNE (Vidicon-arrangement)

During the change from the solution to the layer the position of the band does not alter, so that a broad maximum should occur in the photoconductivity spectrum. On the addition of doping compounds of o-chloranil (E_A = = 1.37 eV) or tetracyanethylene (E_A = 1.8 eV) the structures of which are represented in Figure 4, a photoconduction band of poly-N-vinylcarbazole, too, is expected in the visible region.

The photoconduction spectrum of Figure 5 that was measured on Luvican doped with tetracyanethylene, however, shows that the CT-band, which is to be expected in the visisble region is missing. And in spite of the use of other acceptors no spectral displacement of the photoconductivity of Luvican could be observed in the visible region of layers that were only a few nm thick. Nor could any unambiguous hint of the formation of charge carriers in the CT-band on illumination be obtained from the nearly 0.1 mm thick layers that were measured in sandwich cells.

An account of these results it seems to be problematic to look upon the addition of electron acceptors as technique for a spectral sensitization of organic semiconductors. Further investigations will aim at finding out the reason for the deviations between these results and the results in other publications.

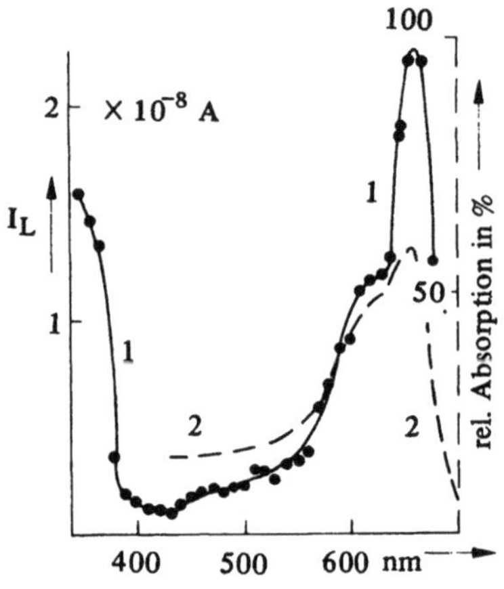

Fig. 6. Action spectrum of photocurrent and absorption spectrum of sensitized PVCA
(1) $I_{phot} = f(\lambda)$: PVCA + methylene blue
(2) absorption spectrum of methylene blue (in chloroform)

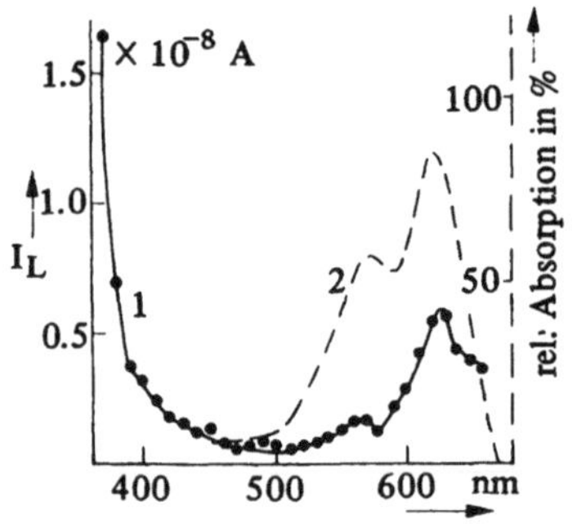

Fig. 7. Action spectrum of photocurrent and absorption spectrum of sensitized PVCA
(1) $I_{phot} = f(\lambda)$: PVCA + pinacyanol
(2) absorption spectrum of pinacyanol

4.2. On the Problem of Dye-Sensitization

Contrary to the last-mentioned results, and, analogously to the sensitization of inorganic semiconductors, we could observe that on the addition of dyes photoconductivity occurred in a region where poly-N-vinylcarbazole itself exhibits no absorption. Figure 6 illustrates this spectral sensitization of photoconductivity by the example of a methylene blue dyed layer.

Comparison with the absorption spectrum of the dye indicates that the situation and form of the action spectrum of the sensitized photoconductivity follows the absorption spectrum of the dye in solution. Luvican becomes photoconducting at 680 nm in the region of the methylene blue absorption maximum which is characteristic for the monomolecular form of the dye. Light absorption in dye-dimers, too, causes a sensitized photocurrent as is indicated by the low maximum at 620 nm.

An analogous agreement between the absorption spectrum of the dissolved dye and the photoconduction spectrum of the sensitized polymer was also found during the use of other dyes as sensitizers. By way of example, Figure 7 shows a sensitization effected with pinacyanole. Here too, two bands (i.e., at 620 nm and 570 nm) occur in solution and in the photoconduction spectrum. They indicate that the dye-molecules, being either molecular or dissolved in the polymer as aggregates, cooperate in the sensitization process. It is noteworthy that the maxima of the photoconduction spectrum largely agree with the maxima of the dissolved dyes i.e., that they are not characterized by those small long-wave displacements (of about 15 nm) that have often been observed on adsorbed layers. This can be taken as a hint at the fact that there is a real distribution of dye-molecules in the Luvican film which does not materially differ from the case with the solution in chloroform or any other solvent.

An analogous spectral sensitization was also measured in sandwich-type cells, e.g., during the sensitization with crystal violet, methylene blue or rhodamine B. Thus, the dye-sensitization of poly-N-vinylcarbazole is not only confined to thin layers or special arrangements. Moreover, it is not important for the sensitization effect of sandwich-type cells whether we illuminate from the negative or the positive electrode.

On long illumination periods, however, the dye-sensitized poly-N-vinylcarbazole layers showed a decrease of photoconductivity which may be associated with the bleaching of the dyes that are not very stable to light, and to a continuous polymerization of the Luvican [37, 38]. In search of methods to minimize this photochemical decomposition depending on the kind of dye an interesting additional spectral sensitization effect that hitherto has been unknown was found on polymer semiconductors.

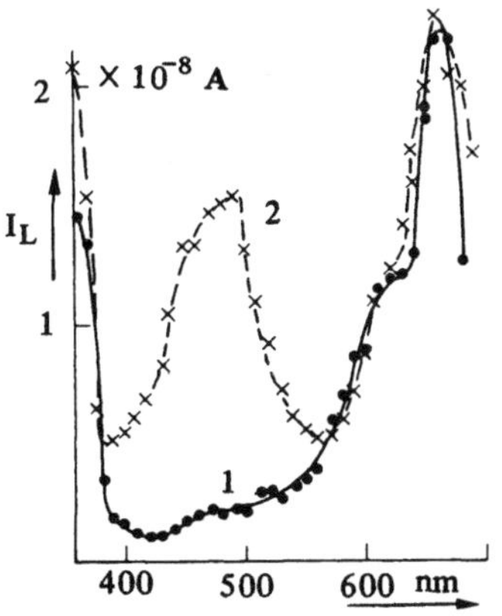

Fig. 8. Action spectrum of photocurrent of sensitized PVCA
(1) PVCA + methylene blue
(2) PVCA + methylene blue + TCNE

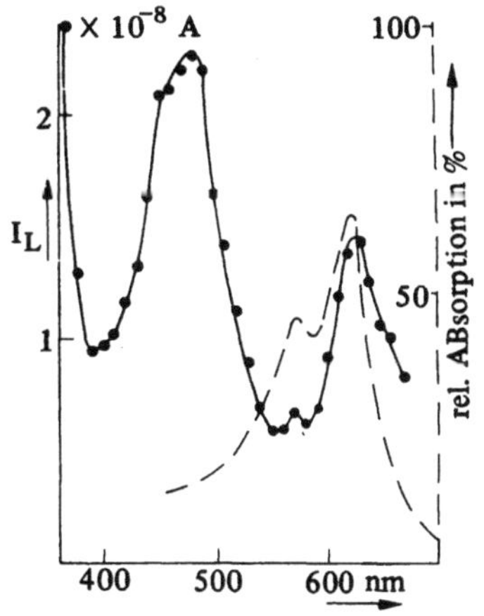

Fig. 9. Action spectrum of photoconductivity of sensitized PVCA
(1) $I_{phot} = f(\lambda)$: PVCA + pinacyanol + TCNE
(2) absorption spectrum of pinacyanol

4.3. On Spectral Sensitization with Dye Electron Acceptor-Combinations

Poly-N-vinylcarbazole layers containing one of the above-mentioned electron acceptors and a dye, was characterized by a spectrum that differs from the normally dye-sensitized photoconduction spectrum and has several characteristic properties.

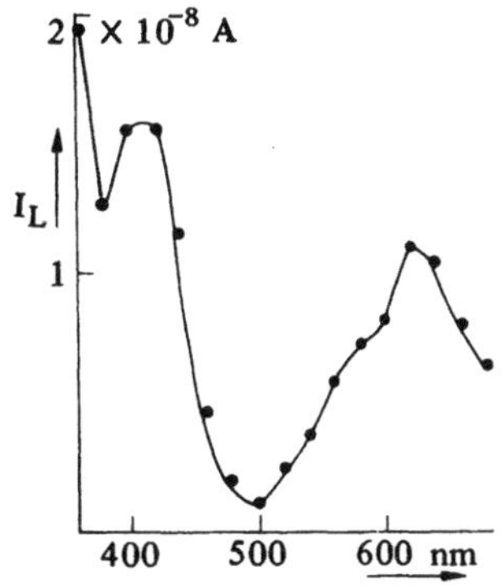

Fig. 10. $I_{phot} = f(\lambda)$: PVCA + TCNQ + pinacyanol

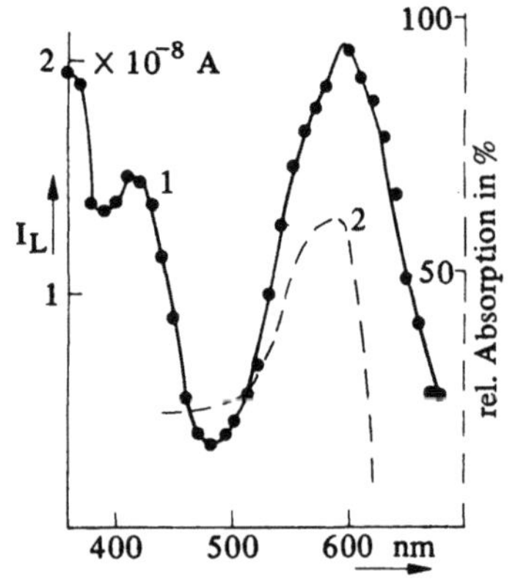

Fig. 11. Action spectrum of photoconductivity of sensitized PVCA
(1) $I_{phot} = f(\lambda)$: PVCA + TCNQ + crystal violet
(2) absorption spectrum of crystal violet

1. The photoconduction spectrum of Figure 8 of a PVCA-layer sensitized with methylene blue + TCNE demonstrates that the band due to the PVCA absorption, and the sensitization band belonging to methylene blue are accompanied by still another band. As the additional photoconductivity maximum lies at 490 nm, i.e. between the PVCA-and the dye-band, it can easily be measured.

2. By a comparison with the absorption spectrum of TCNE dissolved in PVCA we can see that the additonal peak in the photoconduction spectrum cannot be related to the charge transfer-band laying at 600 nm.

3. Nor can the additional band be related to a band that is formed by a charge-transfer between dye and electron acceptor. For the position of the photoconduction peak remains unaltered in spite of a change of the sensitizing dye, as can be seen in Figure 9 from the example of a pinacyanole–TCNE–sensitized layer. A maximum lies at 490 nm also with sensitization with crystal violet and other dyes.

4. The position of the photoconduction maximum depends on the added electron acceptor: with tetracyanchinodimethane the maximum lies at 410 nm as is shown by the photoconduction spectra with pinacyanole and crystal violet of Figure 10 and Figure 11.

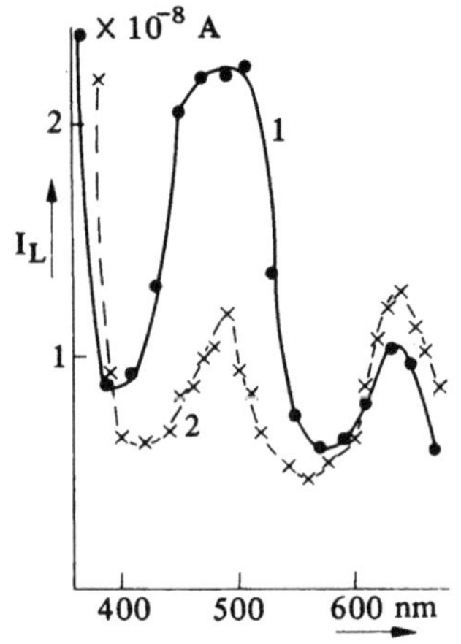

Fig. 12. $I_{phot} = f(\lambda)$ poly-N-vinylcarbazole + TCNE + + pinacyanol
(1) $5 \cdot 10^{-9}$ mol pinacyanol (in relation to the PVCA-layer)
(2) $7 \cdot 10^{-8}$ mol pinacyanol (in relation to the PVCA-layer)

5. The additional photoconduction peak cannot be referred to the absorption of the neutral acceptor. The individual absorption, for example, of tetracyanethylene lies at 280 nm, whereas the additional band appears at 490 nm.

6. According to former investigations, the intensity of tne additional band seems to depend on the dye. For example, the band exhibits greater intensity in the case of pinacyanole than in the case of crystal violet. This dependence, however, must still be confirmed by further studies.

7. The intensity of the additional band depends on the concentration of the dye: with growing dye-concentration the intensity of the additional band increases as the dye-sensitization band decreases. Figure 12 demonstrates this fact for the case of a sensitization with pinacyanole + TCNE.

This last-mentioned effect of a decrease of sensitivity of sensitized photoconductivity in the dye-absorption-region with an increase of the dye-concentration is in agreement with observations of Mylnikov and Terenin [6] on methylene blue sensitized copper-phenylacetylid. The sensitization efficiency of dyes seems to decrease because of a formation of dye-dimers and aggregates.

5. Comparison of the Different Spectral Sensitization Reactions

In summary, it is clear that the photoconductivity of organic semiconductors can spectrally be sensitized by dyes. But the question arises whether the sensitization reaction observed on organic semiconductors can be compared with the spectral sensitization of inorganic photoconductors and with the spectral photographic sensitization, or whether certain distinctions between the sensitization reactions must be assumed.

1. A study of the dyes which sensitize organic semiconductors shows that there is no connection between the photographic applicability of dyes and the sensitization ability discussed in this work. Both photographic sensitizers and desensitizers sensitize organic photoconductivity (Table 2). Therefore, here we have a result similar to that of inorganic photoconductors the photoconductivity of which can also be displaced into the region of dye-absorption with sensitizing–and moreover–desensitizing dyes [1, 5, 39–41].

Table 2. Survey on Dyes Hitherto Used for Spectral Sensitization of Poly-N-Vinylcarbazole

Dye	Class Group	Conduction Type	Photographical Application
pinacyanol	cyanine	n (p)	S
crystal violet	triphenylmethane	n	D
rhodamine B	phencarboxonium	n	S/D
methylene blue	phenothiazein	n	D
rose bengale	fluoresceine	p	S
ceresgelb GGB*)	azo	p	
methyl violet	triphenylmethane	n	D

S = sensitizer, D = desensitizer
* formula

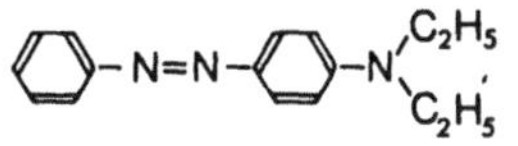

2. From the measurements known hitherto, no desensitization of the conductivity of the organic semiconductors is observed by an addition of a photographic desensitizer. This statement agrees with observations on the sensitized photoconductivity of inorganic semiconductors [42].

3. The relative quantum yield Φ_r, which represents the ratio between the number of quanta absorbed in the organic semiconductor to produce a certain photocurrent, and the number of quanta that are absorbed in the dye for bringing about the same photocurrent, is of the order of 0.5, as is illustrated in Figure 13 with poly-N-vinylcarbazole sensitized with crystal violet.
Similar values are also measured on ZnO-layers sensitized with dyes during the electrophotographic sensitization, which indicates an analogous efficiency of the dyes in the different sensitization reactions [42].

4. The decrease in photoconductivity, observed in the sensitization region with an increase of dye-concentration, is not confined to organic semiconductors. Sensitized ZnO, too, shows this decrease, and the decrease of the relative quantum yield of sensitized photographic layers due to an increase of the dye concentration can be looked upon as a further analogous case [42, 43, 44].

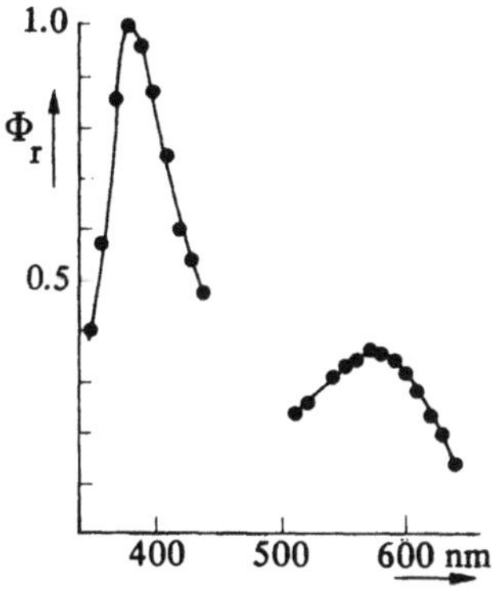

Fig. 13. Relative quantum yield of poly-N-vinylcarbazole photoconductivity sensitized with crystal violet
$[\phi_r = f(\lambda)]$

5. The various sensitized systems differ essentially from one another in the way that the dye is built into the sensitized material: whereas on silver halide grains or zinc oxide microcrystals the dye-molecules are adsorbed as a monomolecular layer or as aggregates, they occur uniformly dissolved in the polymer semiconductors of these investigations.

6. The occurrence of the additional band in the presence of electron acceptors and dyes is characteristic only for organic sensitization. It is an effect by which the exceptional conditions of the sensitization reaction possible for organic semiconductors, are stressed.

From the above-cited examples one can derive similarities, but also differences between the sensitization of organic semiconductors on the one hand and between the sensitization of inorganic photoconductors as well as of the silver halide photolysis on the other. The differences between the processes suggest there should be a mechanism which is specific for the sensitization of organic dyes.

6. On the Mechanism of the Spectral Sensitization of Organic Polymer Semiconductors

6.1. Decisive Experimental Data

In discussing the sensitization mechanism, the following experimental facts must be considered:

1. The spectral sensitization of the polymer semiconductor cannot be caused by the photoconductivity of the built-in dyes, a fact that is –among others–proved by the decrease of the sensitized photoconductivity with the increase of dye-concentration. If the photoconductivity of the dye would be measured, the relationship would be opposite (Fig. 12).

2. As the photoconductivity of the organic semiconductors has its origin in the formation and migration of electronic charge carriers, dye sensitization can be looked upon as the result of the formation of electron charge carriers in the organic photoconductor material following an excitation of the dye.

3. The polymer semiconductor materials containing the sensitizing dye molecules uniformly dissolved, represents analogous systems to those utilized in dye-sensitized photopolymerisation. The discussion of the formation of sensitized charge carriers, therefore, must also consider the possibility of photochemical reactions.

6.2. The Different Sensitization Mechanism

The discussion of the sensitization mechanism can be based on the following general question: "Is sensitization, i.e. the formation of charge carriers, caused by an energy transfer from the excited dye to the organic semiconductor or by an electron or hole transfer? " For more details on these mechanisms one should refer to the surveys in [5, 8, 45, 46].

In the first case, an electron, by means of energy transfer, as simply is illustrated by the scheme of Figure 14, would be shifted from a filled defect of the organic semiconductor into the conduction band. The sensitization would result in an increase of conductivity without influencing the dye which is found to be again in its ground-state after donating the energy. The existence of energy-transfer processes cannot be doubted. The experiments of Kuhn and co-

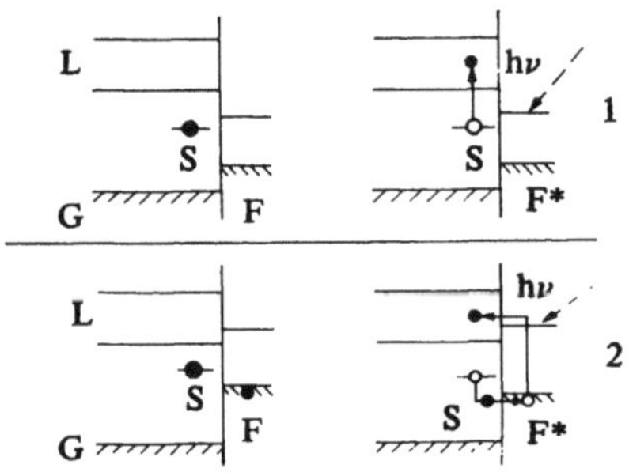

Fig. 14. General schemes of the mechanism of spectral sensitization
(1) Energy transfer
(2) Electron transfer
L = conduction band, G = valence band, F, F* = non-excited and excited dye, S = defects (–•– filled; –o– emptied)

workers [47, 48] clearly prove this fact. But there are many of observations which, for example in the case of poly-N-vinylcarbazole sensitization, point to an electronic mechanism.

6.3. Electronic Sensitization Mechanism

In the electronic mechanism, a photo-injection of electrons or holes under direct influence of the dye must be assumed. But how can we experimentally prove the participation of an electron exchange in the sensitization process?

1. With the help of the Spear-Kepler-Le Blanc-method [49, 50] the formation of electrons induced on the excitation of the dye can be directly proved. The principle of this method which is schematically given in Figure 15, consists in the oscillographic registration of the charge carrier migration caused in the organic material by a flash of about 1 μsec.

The charge carriers, formed by a short duration flash in a thin surface layer of the surface cell, migrate to the dark electrode under the influence of an electric field of suitable sign, thus inducing a short current pulse. This pulse can be caused by holes if the illuminated electrode is positive, or electrons if the front-electrode is negative. In the ideal case of an undisturbed migration, the pulse has a rectangular form. It is true that the pulses are disturbed by space charges and traps, but, nevertheless, they permit valuable conclusions [51]: By way of example, the relative height of the pulses

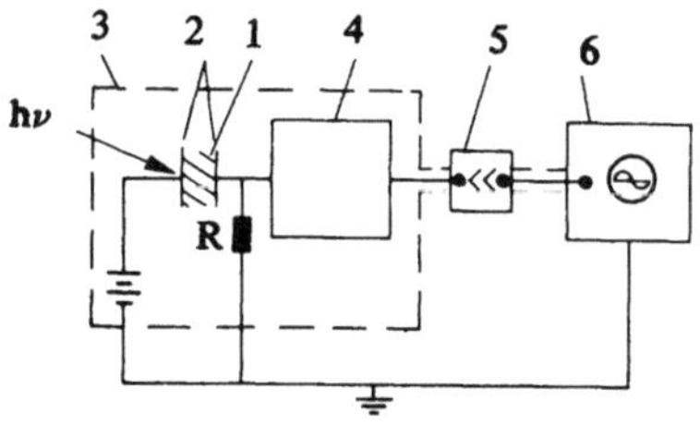

Fig. 15. Photo-flash method (scheme of arrangement)
(1) photoconductor
(2) electrodes
(3) screen
(4) amplifier
(5) oscillograph

can be used for the determination of optically produced charge-carrier ratios because, according to the charge of the exposed electrode, the equation is:

$$(+)\ I_p = n_\oplus(e/L) \cdot E \cdot \mu_\oplus \text{ , and}$$

$$(-)\ I_n = n_\ominus(e/L) \cdot E \cdot \mu_\ominus$$

$n_\oplus$, $n_\ominus$ = hole, or electron concentration; L photoconductor thickness; E field; $\mu_\oplus$, $\mu_\ominus$ drift mobility of the holes or electrons. In the case of $I_p/I_n > 1$ and of nearly equal mobilities and equal conditions, holes must be the majority carriers, whereas in the case of $I_p/I_n < 1$ electrons are the majority carriers. Among other phenomena, this procedure permitted the following conclusions:

A. Poly-N-vinylcarbazole proved to be a p-type conductor, since $I_p/I_n > 1$ (Fig. 16). This statement confirms former electrophotographically obtained findings [17].

B. In layers that are sensitized with dyes, e.g. with crystal violet, the ratio I_p/I_n decreases with an increasing dye concentration (Fig. 17, curve 2).

The analysis of the connection between the I_p- and I_n-maxima with dye-concentration shows that by means of the dye the electron concen-

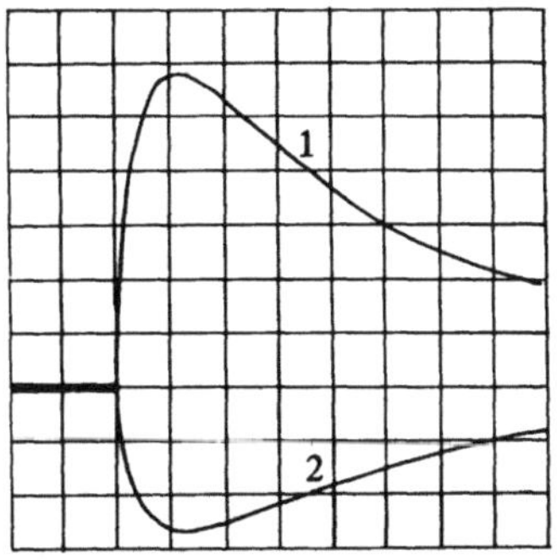

Fig. 16. Short-time photocurrent of poly-N-vinyl-carbazole + o-chloranil
U = 380 V
abscissa: 0.1 msec/scale division (= sd)
(1) hole transport (illuminated electrode: ⊕, ordinate: 200 mV/sd.)
(2) electron transport (illuminated electrode: ⊖, ordinate: 50 mV/sd.)

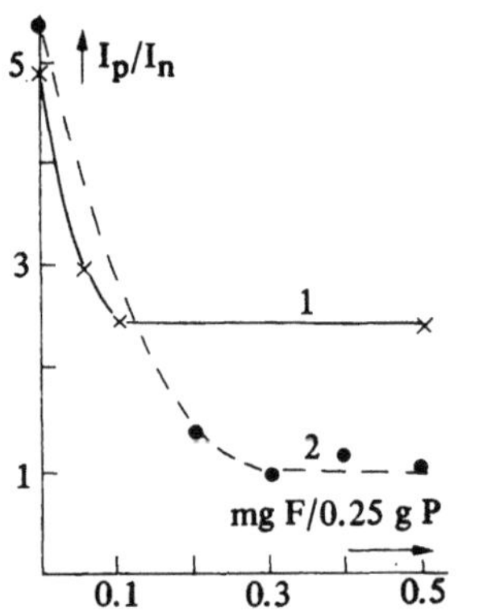

Fig. 17. I_p/I_n = f (concentration of dye)
(1) poly-N-vinylcarbazole + o-chloranil + crystal violet
(2) poly-N-vinylcarbazole + crystal violet

tration, formed on illumination, increases in the polymer semiconductor, whereas the hole concentration decreases.

C. Even in the presence of an electron acceptor, a decrease of the I_p/I_n-ratio is effected by the dye-sensitization, however to a smaller degree (Fig. 17, curve 1).

2. The fact that the photoconductivity of the sensitized polymer decreases on intense illumination, speaks in favour of the electron-transfer mechanism. This decrease can most probably be looked upon as the result of a photochemical reaction between dye and polymer, as is generally known. Oster et al. [52–54] for example with the help of this reaction carried out dye-sensitized polymerisation and there is no doubt that these photoreactions are the consequence of an electron-transfer from the polymer to the excited dye (cf. survey in [27, 55]).

3. It is remarkable that the photochemically conditioned decrease of the photocurrent could be prevented by adding o-chloranil and other compounds. These compounds are partly negatively charged by electron acceptance and therefore have become electron donors. Some indication of this is obtained by measurements of dark conductivity of the doped substrate. From these observations we can infer that electron donation from these reduced molecules to the excited dyes is favoured with the consequence that the polymer has a lesser share in the photochemical reaction leading to the irreversible decomposition.

4. In this context it must still be mentioned that the I_p/I_n ratio is nearly the same on irradiation into the luvican absorption region as that on direct excitation of the dye. The next table shows the corresponding values.

Table 3. I_p/I_n Ratio in Dependence of Wave-Length

Layer	Absorption Region	I_p/I_n
PVCA + crystal violet	dye (600 nm)	1.25
	PVCA	1.34
PVCA + o-chloranil	dye (600 nm)	3.07
+ crystal violet	PVCA	3.17

It is not clear to what extent this observation can be connected with the observation of Mulder and de Jonge [10] concerning the increase of photocurrent in the absorption region of anthracene after dyeing it with rhodamie B. It must, however, be supposed that part of the UV-radiation absorbed by poly-N-vinylcarbazole is transferred to the dissolved dye-molecules which, analogous to direct excitation, form the sensitized photocurrent. That is to say that the reaction chain can be discussed as

$$\text{PVCA} + h\nu = \text{PVCA}^*$$

$$\text{PVCA}^* + \text{F} = \text{PVCA} + \text{F}^*$$

Very recently, an analogous transfer of electronic excitation energy was proved on layers of N-isopropylcarbazole doped with perylene [57]. The additional bands measured in the polymer on the cooperation of dye and doping compounds, as, e.g., TCNQ or TCNE (Fig. 8–12), can possibly be explained by a similar mechanism. It is possible that primarily a reduced form of the acceptor is excited which, afterwards, passes the energy with a high yield onto the photoelectrically active dye. This hypothesis will be examined in further studies.

But how can excited dye molecules lead to the formation of charge-carriers in the organic semiconductor?

It is possible that, in the sensitization reaction, the redox behavior of the excited dye which is different from that of the unexcited dye, is of great importance. For the photocatalytic behavior it is characteristic (cf. survey in [55, 58]).

1. that the redox-potential of dyes is displaced by an excitation (particularly in the presence of polymer substances [59]) to more positive values, which can be of the order of 1 volt [60]. Thus, excited dyes can–compared to unexcited dye-molecules–develop a high tendency towards electron acceptance (e.g., from cysteine, compounds with secondary and tertiary nitrogen atoms);

2. that the dye in its reduced form after electron acceptance represents a strong reducing agent, i.e., an electron donor with a relatively negative redox-potential (for example −1,3 V for proflavine [61]), trying to transfer electrons to other compounds and thus returning to the state it had before illumination.

The photocatalytic reactions we know, in which dissolved dyes, on illumination, transfer electrons from electron donors to electron acceptors [55], suggest the formation of charge carriers according to an analogous redox mechanism. The donors operating in the sensitization process are formed by monomer units of the poly-N-vinylcarbazole and by negative doping molecules (e.g., $TCNQ^-$). These are situated between the energy bands of the polymer that are responsible for the migration of the charge carriers and are affected by an overlapping of the π-electron-orbitals. Whereas the photoconducting material decomposes to some extent irreversibly after the electron donation (from the non-binding electron couple of the carbazol-nitrogen) to the dye, the doping molecules will, after their oxidation, again assume the original, reduced state of equilibrium by hole-formation. The illumination of the dye/PVCA-system can therefore be followed by a photochemically conditioned decrease of photocurrent and, in the presence of the substance TCNQ, TCNE and others, by a hole-formation. It is particularly important that the molecular orbitals of the dye being reduced after electron acceptance, are displaced to higher energy values according to the photographic sensitization mechanism [62, 63] of Mitchell by a decrease of the ionic potential. From there the electron of the half-filled dye-molecular orbital can travel to an empty MO, or to the conduction band of the polymer, if the relative position of the co-operating energy levels permits. Figure 18 illustrates the corresponding conditions.

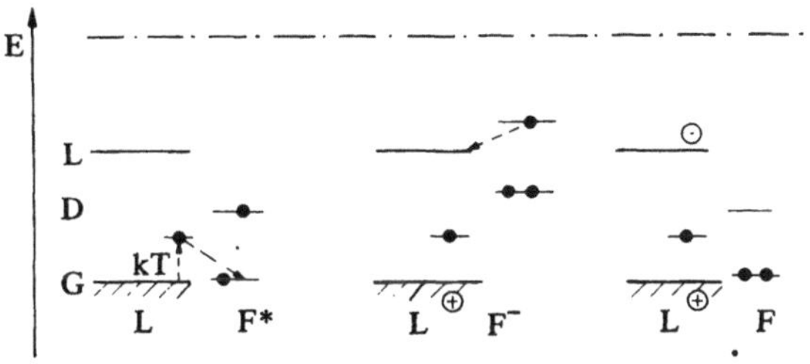

Fig. 18. Scheme of redox-mechanism of sensitization L = conduction band; G = valence band; F, F*, F⁻ = = unexcited, excited and reduced dye; D = donator; ⊕, ⊖ holes and electrons.

For the various reaction steps, therefore, the following scheme can be discussed:

I Excitation

direct $\quad F + h\nu \rightarrow F^*$ (1)

indirect $\quad PVCA + h\nu \rightarrow PVCA^*$ (2a)

$PVCA^* + F \rightleftharpoons PVCA + F^*$ (2b)

$X + h\nu \rightarrow X^*$ (3a)

$X^* + F = X + F^*$ (3b)

II Photoreduction of the dye

$F^* + PVCA = F^- + PVCA^{\dagger}$ (4)

$F^* + D^- = F^- + D$ (5a)

$D + PVCA \rightleftharpoons D^- + \oplus$ (5b)

III Electron transfer of the dye

$F^- + PVCA = F + \ominus$ (6)

(Note: X = still unknown species (e.g., $TCNQ^{\dot{-}}$), which is responsible for the additional band; $PVCA^{\dagger}$ = primary photochemical oxidation product of PVCA, which [eventually also under

H^+-donation] decomposes irreversibly; D, D^- = added disturbed molecules, or disturbed molecules present as impurities; ⊕, ⊖ = holes, or electrons, resp.; F*, PVCA*, X* indicates the excited state, the possibility of a triplet-formation by "inter-system-crossing" [especially in connection with an energy transfer and a photoreduction of the dye which often takes place via the triplet state] being included).

The mechanism we propose explains among other phenomena the above-cited observations such as additional bands, decrease of the photoconductivity without doping substances, different I_p/I_n ratios in the presence or absence of the electron acceptors. It must be noted that the investigations of Gerischer and co-workers [64–66] also point out the importance of reducing agents (e.g. hydroquinone) for spectral sensitization processes. In our studies, however, these donors are substances being dissolved in the PVCA and which partly exist in a reduced form, as dark conductivity measurements indicate. In this context, it should be mentioned that disturbed bromide ions are significant for photographic sensitization; see Eggert, Berg [67, 68].

It is up to further studies to confirm the above-mentioned mechanism. Measurement of the ϕ_r-dependency on dye-constitution, concentration, and so on, as well as the research of a possible co-operation of triplet states are envisaged.

7. Summary

Finally it must be stated that the investigations prove that the electrophotographically interesting poly-N-vinylcarbazole can be sensitized by dyes in every spectral region. Further investigations must try to extend the dye-sensitization-effect, as well as the additional sensitization that was observed in the presence of doping substances, to other polymers.

The authors are indebted to the Deutsche Forschungsgemeinschaft and to the Fonds der Chemischen Industrie for financial support.

Reference

[1] H. Frieser and M. Schlesinger, Photogr. Korresp. 8. Sonderheft (Elektrophotographie), 1966, S. 43–57.
[2] K. Hauffe, V. Martinez, J. Range and R. Schmidt, Photogr. Korresp. **104,** 113 (1968).
[3] D. D. Taft and St. C. Heidecker, Tappi **50,** A 36 (1967).
[4] E. G. Giaimo: R. C. A. Rev. **23,** 96 (1962).
[5] H. Meier, Spectral Sensitization, Focal Press, London-New York, 1968.
[6] V. Mylnikov and A. Terenin, Molecular Physics, **8,** 387 (1964).
[7] H. Meier, Dissertation, Mainz 1954, S. 107ff.
[8] H. Meier, Die Photochemie der organischen Farbstoffe, Springer-Verlag, Berlin-Göttingen-Heidelberg, 1963, S. 127ff.
[9] F. Gutmann and L. E. Lyons, Organic Semiconductors, John Wiley + Sons, Inc., New York-London-Sidney, 1968.
[10] B. J. Mulder and J. de Jonge, Proc. Kon. nederl. Akad. Wetensch., Ser. B, **66,** 320 (1963).
[11] B. J. Mulder, Philips techn. Rdsch. **25,** 195 (1963/64).
[12] J. W. Steketee and J. de Jonge, Proc. Kon. nederl. Akad. Wetensch., Ser. B, **66,** 76 (1963).
[13] H. Gerischer, M. E. Michel-Beyerle and F. Rebentrost in, Symposium on Spectral Sensitization, Bressanone 1967 (ed. W. F. Berg, U. Mazzucato, H. Meier, G. Semerano), The Focal Press, London-New York, 1969, p. 140.
[14] A. Terenin, I. Akimov, J. phys. Chem. **69,** 730 (1965).
[15] H. Hoegl, O. Süs and W. Neugebauer, DP 1068115, 1957/1967 (Kalle AG); DP 1111935, 1961 (Kalle AG).
[16] H. Hoegl, E. Lind and H. Schlesinger, USP 323775, 1966 (Kalle AG).
[17] H. Hoegl, J. Phys. Chem. **69,** 755 (1965).
[18] W. C. Needler, J. chem. Phys. **42,** 2972 (1965).
[19] A. Szent-Györgyi, Science **93,** 609 (1941); Nature (London) **157,** 9 (1941).
[20] H. Meier, s. [5], Seite 110, 179ff.
[21] W. Albrecht and H. Meier, Ber. Bunsenges. phys. Chem. **72,** 1056 (1968).
[22] H. Meier and W. Albrecht, Ber. Bunsenges. phys. Chem. **73,** 86 (1969).
[23] H. Hoegl, Belg. Pat. 591164, 1960 (Kalle AG).
[24] P. M. Cassiers, Belg. Pat. 589994-6 (Gevaert).
[25] D. R. Kearns, G. Tollin and M. Calvin, J. chem. Phys. **32,** 1020 (1960).
[26] H. Meier and W. Albrecht, Z. phys. Chem. [N.F.], **39,** 249 (1963).

[27] J. Bourdon and B. Schnuriger in, Physics and Chemistry of the Organic Solid State, Vol. III (ed. D. Fox, M. M. Labes, A. Weissberger), Interscience Publ., 1967, S. 69 ff.
[28] M. Lardon, E. Lell-Döller and J. W. Weigl, Mol. Crystals, **2**, 241 (1967).
[29] S. Oka, T. Mori, S. Kusabayashi, A. Taniguchi, Y. Yamamoto, M. Ishiguro and H. Mikawa, Mem. Inst. Sci. Ind. Res. Osaka Univ. **21**, 169 (1964).
[30] H. Akamoto and H. Kuroda, J. Chem. Phys., **39**, 3364 (1963).
[31] H. Kokado, K. Hasegawa and W. G. Schneider, Can. J. Chem. **42**, 1084 (1964).
[32] J. H. Sharp, J. phys. Chem. **71**, 2587 (1967).
[33] W. Klöpffer and W. Willicks, Makromol. Chem. 115, 156 (1968).
[34] G. Briegleb, Elektronen-Donator-Acceptor-Komplexe, Springer-Verlag, Berlin-Göttingen-Heidelberg (1961).
[35] G. Briegleb, Angew. Chem. **76**, 326 (1964).
[36] W. Klöpffer and H. Rabenhorst, J. chem. Phys. **46**, 1362 (1967).
[37] M. Hatano, H. Nomuri and N. Tamura, High Polymers, **18**, 823, 905 (1967).
[38] S. Tazuke and S. Okamura, J. Polym. Sci., A-1, **6**, 2907 (1968).
[39] H. G. Greig, USP 3052540, 1959 (RCA).
[40] J. E. Jones and P. H. Stewart, USP 3047384, 1959 (Eastman Kodak).
[41] C. J. Claus, Image Techn. **11**, 12 (1969).
[42] H. Frieser and M. Schlesinger, Photogr. Sci. Engng. **12**, 17 (1968).
[43] E. C. Giaimo, RCA Rev. **23**, 96 (1962).
[44] H. Pietsch, D. Jäkel. W. Schneider and D. Treue in, Symposium on Spectral Sensitization, Bressano 1967 (ed. W. F. Berg, U. Mazzucato, H. Meier, G. Semerano), Focal Press, London-New York 1969.
[45] F. Dörr, Spektrale Sensibilisierung in, Die photographische Empfindlichkeit, Band 3, Akademische Verlagsges., Frankfurt/M., 1968, S. 1021–1084.
[46] W. Noddack, G. Eckert and H. Meier, Z. Elektrochem., Ber. Bunsenges. phys. Chem. **56**, 735 (1952).
[47] K. H. Drexhage, M. M. Zwick and H. Kuhn, Ber. Bunsenges. phys. Chem. **67**, 62 (1963).
[48] H. Kuhn, Naturwissenschaften **54**, 429 (1967).
[49] O. H. Le Blanc, J. chem. Phys. **33**, 626 (1960).
[50] R. G. Kepler, Phys. Rev. **119**, 1226 (1960).
[51] H. Meier and W. Albrecht, Z. Naturforschung **24a**, 257 (1969).
[52] G. Oster, Nature **173**, 300 (1954).
[53] G. Oster and Y. Mizutani, J. Polymer Sci., **22**, 173 (1956).
[54] S. Chaberek, R. J. Allen and A. Shep, J. phys. Chem. **69**, 2842 (1965).
[55] H. Meier, Photochemistry of Dyes in "The Chemistry of Synthetic Dyes" (ed. K. Venkataraman), Academic Press, New York, 1971, 398–515.
[56] H. Meier, W. Albrecht and U. Tschirwitz (in preparation).
[57] W. Klöpffer, Battelle Information **4**, 26 (1969).
[58] G. Oster, Sci. Amer. **219**, 158 (1968).
[59] G. Oster and J. S. Bellin, J. Amer. chem. Soc., **79**, 294 (1957).
[60] G. Oster, Photogr. Sci. Engng. **4**, 237 (1960).
[61] G. Oster and N. Wotherspoon, J. Amer. chem. Soc. **79**, 4836 (1957).
[62] J. W. Mitchell and N. F. Mott, Philos. Mag. **2**, 1149 (1957).
[63] F. Dörr and G. Scheibe, Z. wiss. Photogr., Photophysik, Photochem. **55**, 133 (1961).
[64] H. Gerischer, Vortrag auf dem II. Elektrophotographischen Kolloquium, München, 10. 4. 1969.
[65] H. Tributsch, Dissertation, München, 1968.
[66] H. Gerischer and H. Tributsch, Ber. Bunsenges. phys. Chem. **72**, 437 (1968).
[67] J. Eggert, M. Biltz and F. G. Kleinschrod, Z. wiss. Photogr., Photophysik, Photochem. **39**, 140, 155 (1941).
[68] W. F. Berg, Phys. Soc. Rep. Progr. Physics, **11**, 264 (1946/1947).

Discussion

M. Schlesinger: Dr. Meier, you proposed in your paper for the sensitizing action of crystal-violet on Polyvinyl-N-carbazole an electron-transfer mechanism. Your arguments had been, that by increasing exposure the photoconductivity does decrease.

Question: Could it be possible that by the longer exposure time a dye-bleaching is happening and an energy transfer is still possible?

H. Meier: Because of an analogy with dye-sensitized polymerization reported in literature, which are photochemically induced redox reactions, we conclude that an electron exchange occurs in dye sensitization of PVCA-photoconductivity.

Dye Sensitization of Zinc Oxide by Means of the Electrochemical Cell Technique

K. Hauffe, H. Pusch, J. Range and D. Rein

Abstract

The lack of understanding of the mechanism of sensitized electron transfer through the interface photoconductor/adsorbed charged species and photoconductor/paper represents at present an unpleasant impediment to the development of high-speed electrophotographic layers. If one considers that the interface photoconductor/paper is equivalent to the interface photoconductor/electrolyte, it is obvious to employ the electrochemical cell technique for the study of electron transfer and its sensitization by organic dyes dissolved in the electrolyte. Therefore, the following two electrochemical cells were employed in this study:

+ In | ZnO Single Crystal | Electrolyte + Dye | Platinum –

+ Ag | NESA Glass | ZnO–Resin+Dye | Electrolyte |
| Platinum –

The dark and photocurrent were measured with a potential difference of between 1 and 2 volts, the zinc oxide being the anode. The electrolyte was composed of 90 vol% methanol and 10 vol vol% water in which 10^{-2} mole KC1 per liter was dissolved.

This paper deals with the sensitizing ability of morin, rhodamine B and an azo dye amployed as single components and as binary and ternary mixtures. The most effective sensitizer is morin. The sensitizing efficiency, however, is strongly dependent on the pH of the electrolyte. Thus for instance, it has been found that the sensitizing ability of morin and of the azo dye is in a neutral electrolyte and is significantly decreased in an acidic electrolyte. The pH dependence of sensitizing efficiency of the binary dye mixture azo dye-rhodamine is remarkable. Optimum sensitization of this dye mixture was found unexpectedly in a basic electrolyte, indicating strikingly the mutual interchange of the dyes adsorbed on zinc oxide. The strong decrease in sensitizing ability of morin in the simultaneous presence of rhodamine B is a further example indicating strikingly the mutual interchange of the dyes adsorbed on zinc oxide.

The justification for the application of the electrochemical cell technique as a useful tool for the investigation of the electron transfer sensitized by light-excited dyes was confirmed by discharging experiments with zinc oxide-resin layers doped with the same dye mixtures employed in the electrochemical cell. Here was found roughly the same wave-length dependence of the speed of discharge under illumination. A direct comparison, however, becomes difficult if the resin exhibits an influence on the sensitizing ability of the dyes. This fact was detected by means of a dyed zinc oxide-resin layer operating as anode in the electrochemical cell.

1. The Problem

The quality of an electrofax paper® is mainly characterized by the speed of the charge-transfer from the charged surface to the grounded support under light exposure. Since zinc oxide acting as photoconductor has a rather large forbidden band gap of 3.2 eV, illumination with light of <380 nm is required for the production of electron-hole pairs responsible for the discharge. In order to shift the discharge and the production of charge carriers into the visible light region—because light sources in the visible region are cheaper and more reliable—the application of sensitizers is necessary. Organic dyes, as for instance morin, rhodamine B, bromphenol blue, eosin and acridine orange, adsorbed on zinc oxide exhibit a significant sensitization of the charge transfer and the production of charge carriers when illuminated in the optical absorption maximum of the corresponding dye. In order to avoid the complications appearing with polycrystalline zinc oxide embedded in a resin binder as in electrofax layers®, most of the experiments dealt with in this paper have been carried out with ZnO single crystals. A few experiments, however, have been performed with zinc oxide resin layers. These were dyed with a three-component system (morin – rhodamine B – azo dye), employed in the electrochemical cell experiments with ZnO single crystals, in order to test the usefulness of the electrochemical technique for the development of efficient sensitizer systems. As was reported earlier [1, 2], the electrochemical cell technique is appropriate for the study of the sensitization of the electron transfer through the interface zinc oxide/adsorbed dye. A ZnO single crystal was used as anode in the following electrochemical cell:

+ Indium | ZnO Single Crystal | Electrolyte +
+ Dye | Platinum –

An adsorption equilibrium between the ZnO surface and the dye mixture in the electrolyte may be assumed. The illumination took place from the rear of the crystal. Therefore, mainly dye molecules adsorbed on zinc oxide will be effective for the sensitization.

As demonstrated by several authors [3–9], certain organic dyes with conjugated double bonds and charge resonance are especially effective for the sensitization of the discharge of negatively charged zinc oxide resin layers. Additionally, we have been able to demonstrate that the formation of a complex between a zinc ion at the surface and the adsorbed dye enhances its sensitizing ability which shall be dealt with elsewhere. Furthermore, it may be expected that in an electrolyte the sensitizing efficiency of a dye mixture is decisively determined by the mutual electronic interchange between the single dyes and between the dyes and the ZnO surface.

It seemed desirable to use ZnO single crystals at the beginning of this study because of the complications introduced by the application of polycrystalline zinc oxide and the resin. For example, in the polycrystalline systems the terminal groups of the resin may react chemically. Also the radiation intensity is not well defined because of the optical shielding of an unknown amount of ZnO crystallites. With single crystals it appeared possible to achieve defined experimental conditions . In earlier measurements only single dye systems (e.g. rhodamine B) were used [10]. In the present paper the mutual influence of binary and ternary dye mixtures on their sensitizing ability has been studied. As will be seen below, the mutual influence of the dyes could not be predicted in most cases. Sometimes a strong decrease and somtimes an increase has been found in the sensitization of the anodic photocurrent, the parameter used to monitor the electrophotographic sensitivity of zinc oxide.

We believe that in our present study the sensitization is mainly caused by charge transfer and not by energy transfer.

® Registered trademark of RCA for a paper coated with a zinc oxide resin layer.

2. Experimental

As the anode in the electrochemical cell, single crystals of pure zinc oxide in the form of platelets with a thickness of about 1 mm were used. Only the anodic current was measured. The (0001) face, area 0.2 cm^2, was contacted with the electrolyte and the (000$\bar{1}$) face was evaporated with a thin and semi-transparent indium film in order to get a barrier-free contact. Both faces of the crystal were polished with diamond paste of grain sizes down to 0.25 μ. This was very important for penetration of the light. Then the prepared crystal was cemented with Apiezon W wax in a small window of the electrochemical cell (Fig. 1) so that for all measurements the (0001) face was in contact with the electrolyte. It was composed of 90 vol% CH_3OH and 10 vol% water containing 10^{-2} mole KCl per liter. For long-duration experiments the wax was substituted by a teflon packing. The experimental set-up can be seen from the schematic sketch in Figure 1. A potentiostat served as the 1.5 volt supply and a picoammeter A and an electrometer V were used for current and voltage measurements, respectively. The light from a 75 watt xenon high-pressure lamp was passed via a monochromator through the rear of the crystal. The light intensity of the light source was determined by means of a calibrated thermopile. Since the transparency of the indium film and the ZnO crystal as function of wave length was known, the light intensity and the incident quanta/cm^2-sec at the zinc oxide/electrolyte interface were calculated.

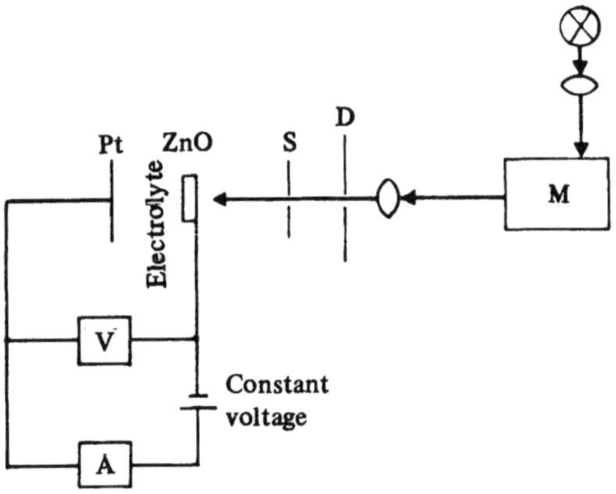

Fig. 1. Schematic representation of the electrochemical cell technique. A = picoammeter and V = = electrometer, M = monochromator, S = shutter

The pH dependence of the sensitized anodic photocurrent was caused by addition of diluted HCl and NaOH. Buffered electrolytes were not used because of their complex influence on the sensitization. In other experiments, the adsorption of the dyes has been measured using ZnO powder columns. The experimental results some of which will be used in this paper, shall be presented elsewhere.

For comparison, and in order to demonstrate the usefulness of these measurements for the elucidation of the mechanism of sensitization, charging and discharging experiments have been carried out with electrophotographic layers in the dark and under illumination. These electrophotographic layers consisted of ZnO powder embedded in resin and dyed with the same sensitizers used for single crystals. The charging and discharging experiments were made using an experimental set-up with a rotating turntabe (132 revolutions/minute) on which a ZnO resin layer repeatedly passes a corona discharge unit and a transparent probe. After charging up only the probe was used to measure the dark and light-decay.

For this study, the following dyes singly and in a mixture have been employed:

1. Morin = 2′, 3, 4′, 5, 7 penta-hydroxyflavone
2. Azo dye. = 1(2-hydroxyphenyl azo)2-naphtol
3. Rhodamine B = 9-(o-carboxyphenyl)-6-diethyl amino-3-diethyl ammonium chloride-xanthene.

Since the solubility of the azo dye in pure water is too small the above mentioned methanol-water electrolyte was employed. Generally, the concentration of the dye amounted to 5 x 10^{-4} mole/liter each.

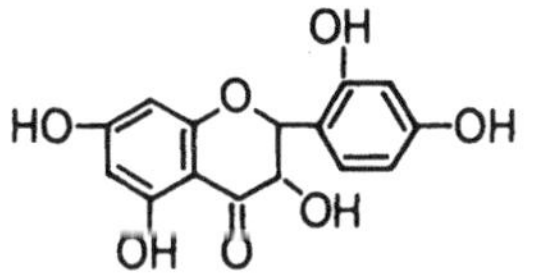

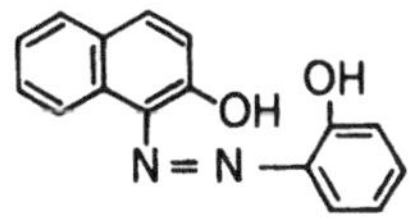

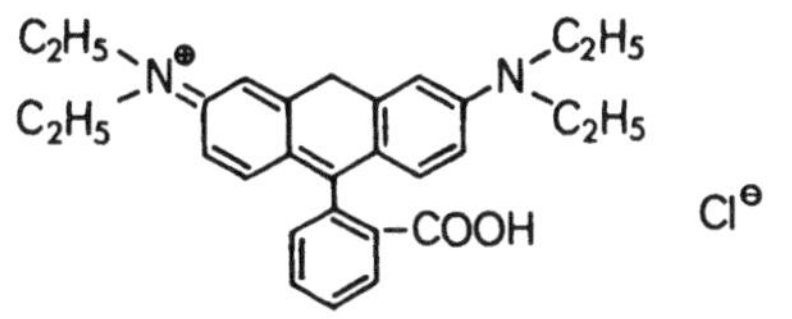

The photocurrent is plotted in the Figures in relative units versus the wave length of the irradiated light. All data have been standardized on a constant number of photons calculated as follows:

$$\text{Reduced photocurrent (plotted)} = \frac{\text{photocurrent (measured)}}{\text{wavelength x incident light intensity x transmission of ZnO}}$$

Therefore, a direct comparison of the sensitizing ability of the single dyes and their mutual influence in dye mixtures is possible. With respect to the absolute values of the photocurrents, the current density corresponding to the value of 100 on the relative scale amounted to 3×10^{-8} amperes/cm^2 of ZnO surface when illuminated with 5×10^{13} photons/sec-cm^2. Usually, the anodic dark-current was between 2×10^{-10} and 5×10^{-10} amperes/cm^2.

Besides the three dyes mentioned above, all of which exhibited a significant sensitizing ability, both nile blue and auramine were examined but did not exhibit any sensitization in the pH region between 3 and 10. The sensitizing ability of indigo carmine was poor. Worth mentioning, however, is the increase in the anodic dark-current in a quasi-neutral electrolyte in the presence of this dye from 2×10^{-10} to 6×10^{-9} amperes/cm^2. Obviously, indigo carmine is able to inject electrons into zinc oxide already in the dark indicating that the electronic exchange level (= ground state) of the adsorbed dye should be located near the edge of the conduction band of zinc oxide so that an appreciable electron transfer is caused by thermic energy already at room temperature. Because of the low sensitization these dyes have not been further investigated.

3. Experimental Results

In order to study the sensitizing ability of morin, rhodamine B and of the azo dye on the electron transfer through the interface zinc oxide/electrolyte, both ZnO single crystals with the dyes in the electrolyte and dyed zinc oxide silicone resin layers were used as anode of an electrochemical cell. At first we shall describe the experimental results obtained with single crystals.

3.1. Results with ZnO Single Crystals

As can be seen from Figure 2, morin is a significant sensitizer, particularly in a neutral and in a weakly basic electrolyte. The sensitizing ability of morin for the electron transfer from the electrolyte into zinc oxide was almost destroyed in an electrolyte with a pH between 3 and 4. The anodic photocurrent–indicating the sensitizing ability–in a basic electrolyte (pH = 9–10) was rather large but only half of the value obtained in a neutral solution. On the basis of the dependence of the anodic photocurrent of the three dyes on wave length, the superior sensitizing ability of morin in a neutral solution is apparent. Under the optimum experimental conditions, the sensitizing ability of the single other dyes–azo dye and rhodamine B–is only about 30% of that of morin. However, these dyes belong also to the group of the very good sensitizers. The different behavior of the dyes in their sensitizing ability when the pH region was changed is worth noting. For instance, rhodamine B exhibited an excellent sensitization both in neutral and acidic electrolytes while at a pH about 9 the sensitization of

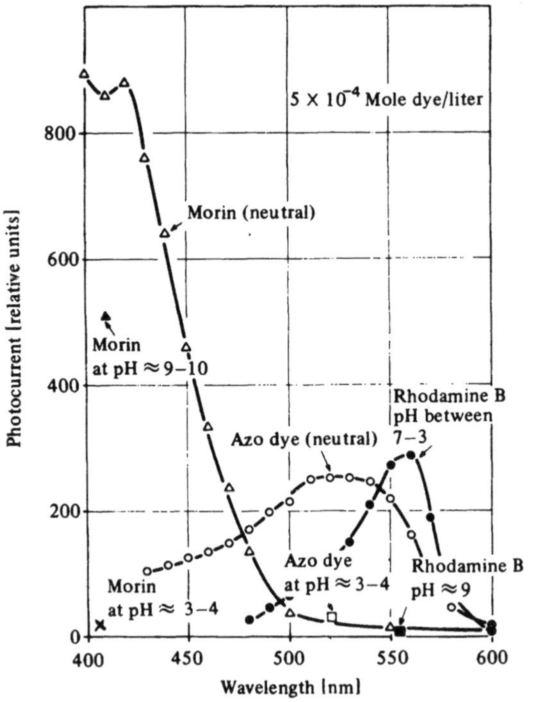

Fig. 2. Wave-length dependence of the photocurrent of various dyes in a 10^{-2} molar KCl solution, consisting of 90 vol% methanol and 10 vol% water, at various pH. Concentration of the single dyes = 5 x 10^{-4} mole/liter each

the anodic photocurrent was strongly decreased. In contrast to this behavior, the azo dye exhibited its best sensitizing ability in a neutral and basic electrolyte while the sensitization in an acidic electrolyte was only poor. The influence of the pH of the electrolyte on the sensitizing efficiency of the single dyes is of great importance to the corresponding quantity for dye mixtures as shall be demonstrated below.

In the following Figures 3a, 3b, 4 and 5, the mutual influence on the sensitizing ability of various binary dye mixtures is represented as a function of the wave length. As can be seen from Figure 3a, the sensitizing ability of morin on the anodic photocurrent is significantly decreased by the simultaneous presence of rhodamine B in an electrolyte of pH = 4. The sensitizing ability of rhodamine B, however, is slightly increased by the presence of morin.

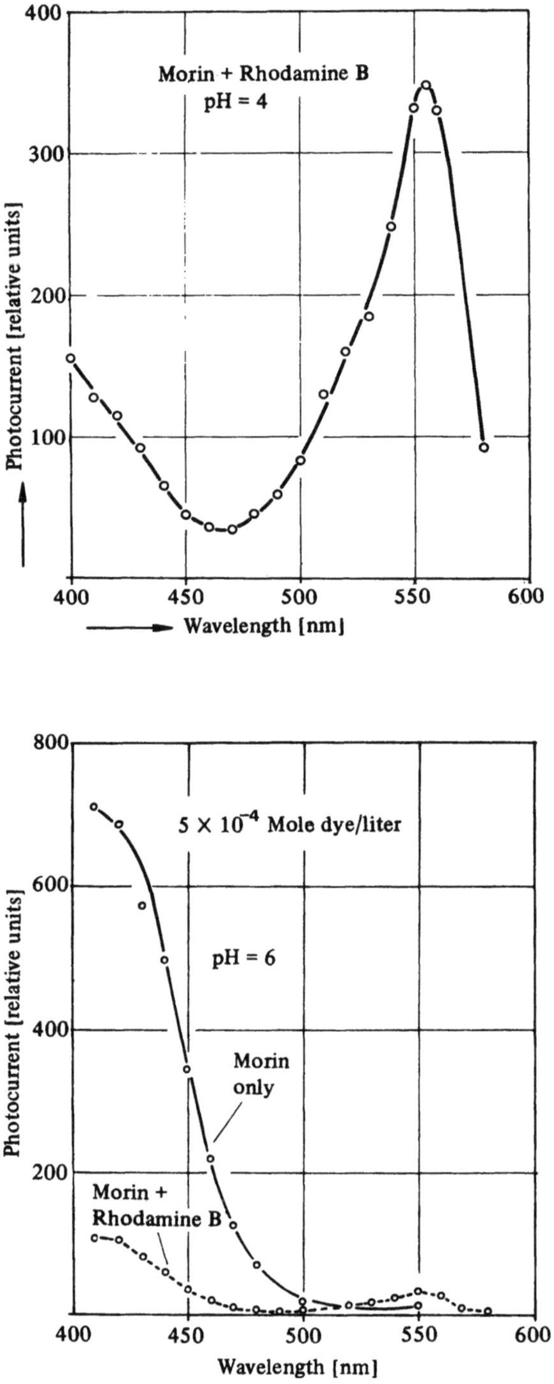

Fig. 3. Wave-length dependence of the photocurrent in an equimolar mixture (5 x 10^{-4} mole dye/liter each) of morin and rhodamine B in the same electrolyte as in Fig. 2 (Fig. 3a: pH = 4 and Fig. 3b: pH = 6)

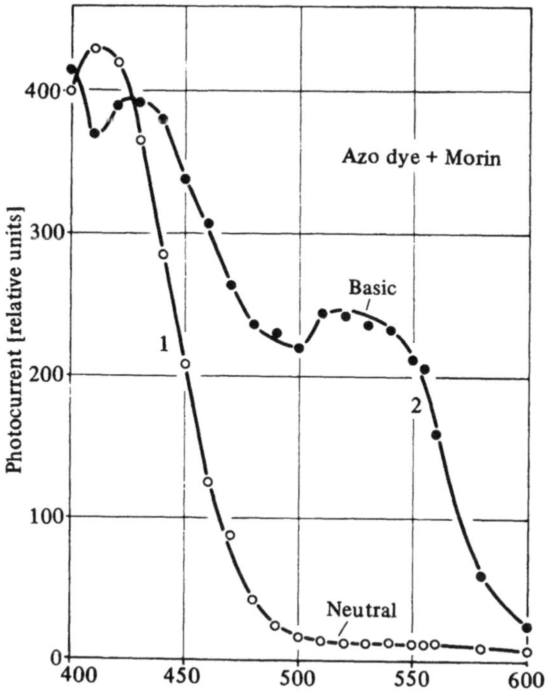

Fig. 4. Wave-length dependence of the photocurrent in an equimolar mixture (5 x 10^{-4} mole dye/liter each) of morin and the azo dye in a basic and in a neutral electrolyte of the same composition as in Fig. 2

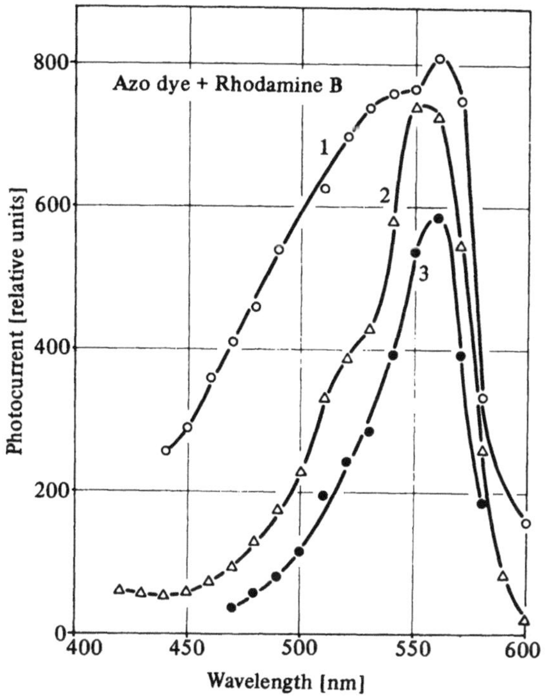

Fig. 5. Wave-length dependence of the photocurrent in an equimolar (5 x 10^{-4} mole dye/liter each) of rhodamine B and the azo dye in the same electrolyte as in Fig. 2.
1. basic, 2. neutral and 3. acidic

Noticeable is, however, the strong decrease of the sensitizing ability of rhodamine B when the electrolyte has a pH = 6. As shown in Figure 4, the sensitizing ability of morin is not decreased so strongly by the presence of the azo dye either in a basic or in a neutral electrolyte. While in a basic electrolyte the sensitizing properties of the azo dye are maintained, they are no longer detectable in a neutral electrolyte with morin present. This result is surprising insofar as in the absence of morin a good sensitization by the azo dye both in a neutral and in a basic electrolyte has been found.

Another puzzling result was obtained with an equimolar dye mixture of the azo dye and rhodamine B as can be seen from Figure 5. Here, it was found that the joint sensitizing efficiency was superior in a basic, a neutral and in an acidic electrolyte. While the sensitizing ability of rhodamine B remains significantly independent of the change of the pH, the sensitizing efficiency of the azo dye, however, was decreased in neutral and annihilated completely in an acidic electrolyte. As represented by curve 3 in Figure 5, the sensitization caused by rhodamine B was twice as high as in the absence of the azo dye. Particularly worth mentioning is the unexpected large increase of the sensitized anodic photocurrent in a basic electrolyte by rhodamine B in the presence of the azo dye because in the absence of the azo dye rhodamine B is a poor sensitizer in this pH region. According to our hypothesis [10], the azo dye might act as a cosensitizer.

Further information on the mutual influence of the single dyes on their sensitizing ability has been obtained by photocurrent measure-

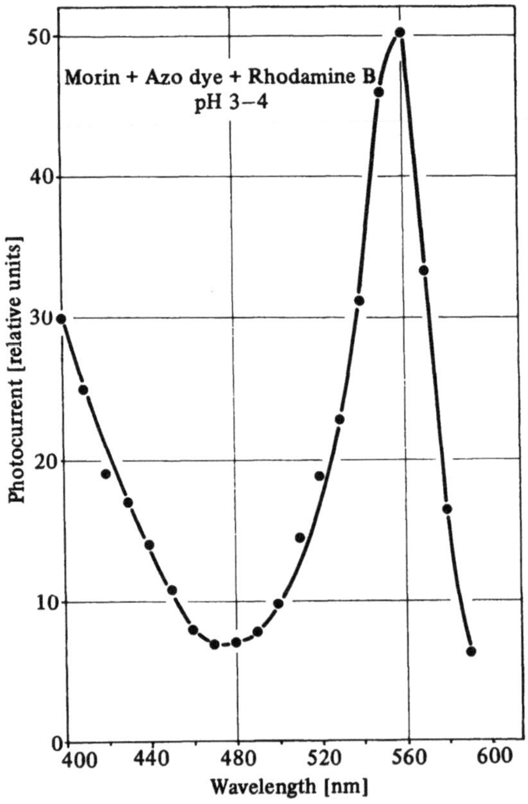

Fig. 6. Wave-length dependence of the photocurrent in an equimolar (5 x 10^{-4} mole dye/liter each) of morin, rhodamine B and the azo dye in an acidic electrolyte of the same composition as in Fig. 2

ments in the presence of three dyes in the electrolyte. Also here, an equimolar concentration of the dyes (5 x 10^{-4} mole/liter each) was used. The wave-length dependence of the photocurrent in various pH regions is represented in the Figures 6–9. The weakest sensitization was obtained in an acidic electrolyte with a pH between 3 and 4 (Fig. 6). Also in a neutral electrolyte, the sensitizing efficiency of morin was only 10% of that obtained in the absence of rhodamine B and the azo dye. Furthermore, the excellent sensitization behavior of the binary dye mixture of rhodamine B and the azo dye was drastically decreased by the simultaneous presence of morin (compare Figs. 5 and 7). As the electrolyte becomes more basic going from a pH between 8 and 9 (curve 1 in Fig. 8) to between 9 and 10 (curve 2 in

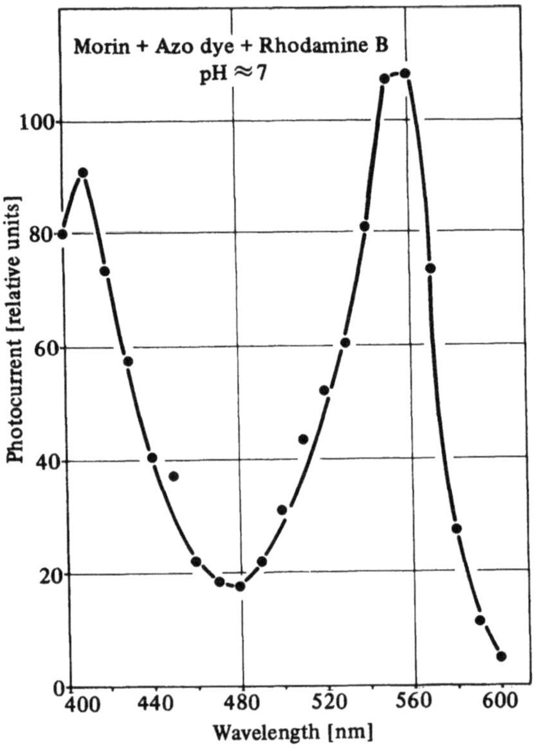

Fig. 7. Data obtained as in Fig. 6 but in a neutral electrolyte

Fig. 8), the sensitizing ability of morin is increased to a value of about 2/3 of the value obtained with morin only. However, the sensitizing ability of the other two dyes remains unsatisfactory. Therefore, we may conclude that the ternary dye mixture employed in our study exhibited only a poor sensitizing ability in a neutral electrolyte and cannot be recommended as a ternary dye mixture for electrophotographic purposes.

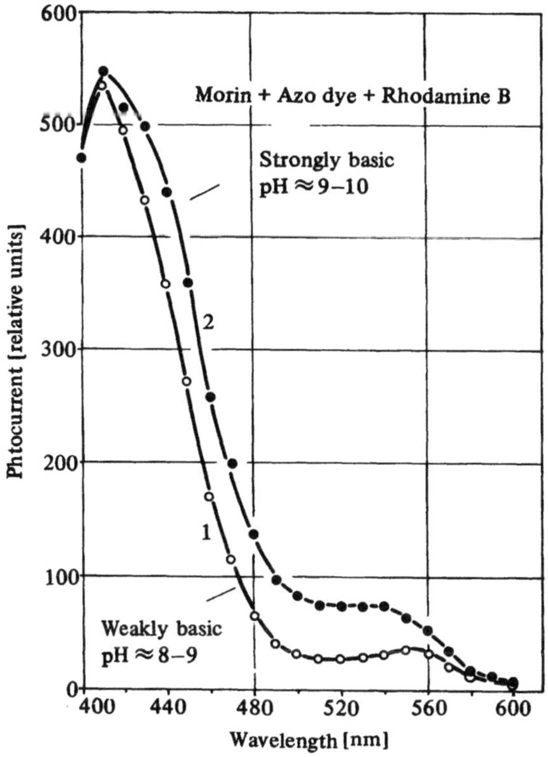

Fig. 8. Data obtained as in Fig. 6 but in a weakly basic and in a stronger basic electrolyte

3.2. Experiments with Dyed Zinc Oxide Resin Layers

To continue the study on the sensitizing ability of various dyes on the electron transfer through the interface zinc oxide/adsorbed dye, the zinc oxide single crystal was now replaced by a thin zinc oxide silicone resin layer. The following cell was used:

+ Ag | NESA Glass | ZnO Resin Layer +
+ Dye | Electrolyte | Pt –

Anodic current experiments with this system have been carried out in order to get direct information on the influence of the resin on the sensitizing ability of the dyes. The substitution of the ZnO single crystal by an electrophotographic layer coated on an indium conductive glass surface introduces, however, some difficulties not understood at present in spite of the large experimental effort. One of the puzzling phenomena, for instance, with a freshly prepared layer is the very slow adjustment of the anodic dark-current, characterized by a slow decay from the initial to a steady-state current. Generally, the decay achieved its final value after 3 to 4 hours. The composition of the zinc oxide resin layer is very decisive in determining the ratio of the dark and photocurrent. Furthermore, the initial leakage current is often so large that the current-voltage measurements could not be made. Leakage currents can sometimes be observed which are several orders of magnitude larger than the normal anodic current. Small pores and large grain boundaries in the layer caused by a too high drying temperature seem to be responsible for this. Therefore, nearly all layers were dried at

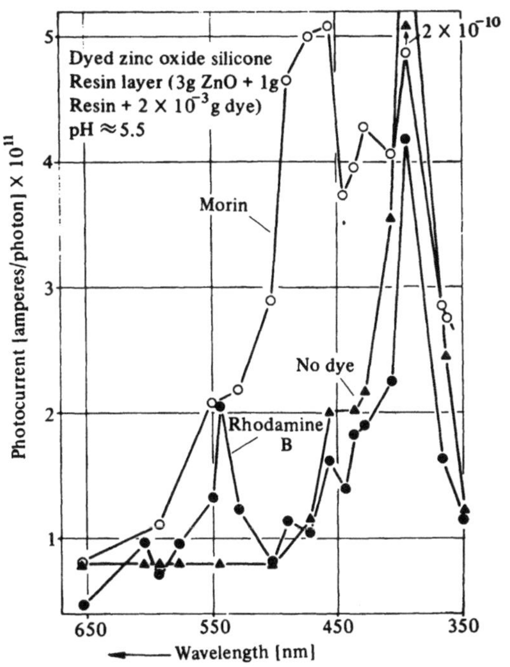

Fig. 9. Wave-length dependence of the photocurrent through a zinc oxide silicone resin layer undyed and dyed with morin or rhodamine B in an aqueous electrolyte containing 0.1 mole KCl/liter and an acetate buffer.

room temperature in air. A rather high solvent content must therefore be expected in the layer.

Measurements were made of the wave-length dependence of the photocurrent at 2 volts with a zinc oxide resin layer under three conditions: without any dye, with rhodamine B and with morin. The results are shown in Figure 9. The layer was composed of 3 parts by weight of ZnO (Fotosiegel A)*) and of 1 part silicone resin and of 2 x 10^{-3} parts of the dye. The aqueous electrolyte consisted of 0.1 m KCl and of a small amount of an acetate buffer (pH ≈ ≈ 5.5). As can be seen from Figure 9, the maximum of the curve for the layer dyed with rhodamine B was shifted from 560 to 540 nm while the maximum of an identical layer dyed with morin was shifted from 420 to 470 nm. The latter represents a significant shift of the maximum of the action spectrum. Also by these measurements, the sensitization of the photocurrent by morin is about a factor of 2 larger than the corresponding value obtained by rhodamine B. The photocurrents in Fig. 9 were normalized to the same incident quanta/cm^2-sec.

These results demonstrate very drastically the large influence of the resin on the sensitizing ability of the dyes adsorbed on zinc oxide embedded in a resin. Encouraged by these results, further experiments will be performed in the near future.

3.3. Charging and Discharging Experiments with Zinc Oxide Resin Layers

In order to find out whether the experimental results obtained by means of the electrochemical cell technique give useful information on the sensitizing ability of the dyes in an electrophotographic layer, charging and discharging experiments have been performed with zinc oxide silicone resin layers dyed with the same dyes in an equimolar concentration. By means of a corona discharge the layers were charged negatively. Subsequently, both the dark-decay and the speed of the discharge under illumination was measured by a transparent probe as described elsewhere [11]. Due to the method with the revolving turntable which carries the electrophotographic layer, no continuous illumination is possible but only a short time illumination followed by a relatively large dark period when the turntable has performed one revolution. Therefore, the light-decay curves represented as 50 seconds are identical to those of about 2 seconds under a continuous illumination.

The curves of the light-decay represented in the Figures 10 and 11, were measured with inter-

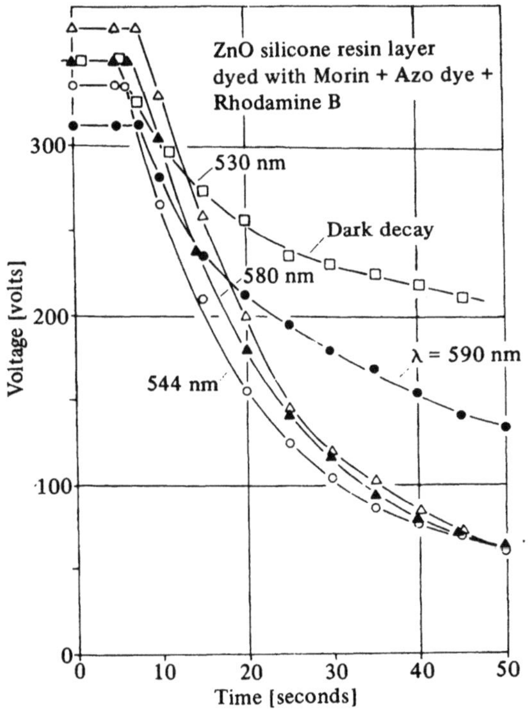

Fig. 10. The light-decay under various wave-lengths of a corona charged zinc oxide resin layer dyed with an equimolar mixture of morin, rhodamine B and azo dye (1 g ZnO with about 3 m^2/g + 10^{-6} mole of each dye + g silicone resin)

*) We appreciate very much the supply of this zinc oxide by the Zinkweiss-Forschungsgesellschaft at Oberhausen, West Germany.

ference filters at various wave lengths under a constant number of quanta/cm^2-sec. In Figure

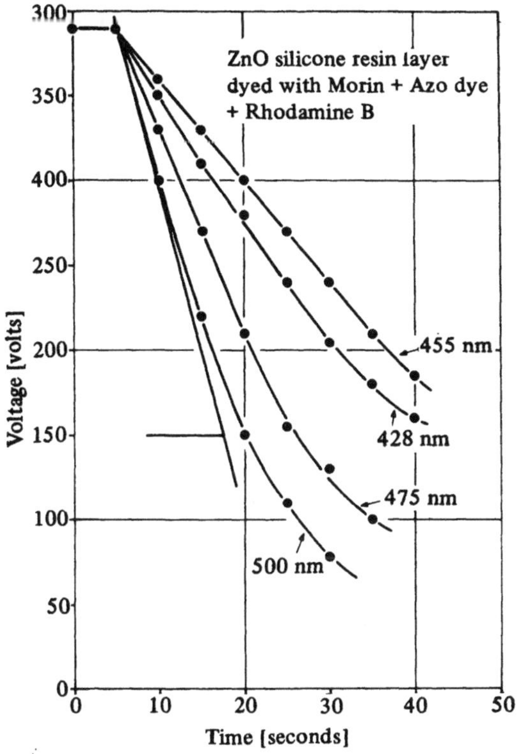

Fig. 11. Similar light-decay measurements as in Fig. 10 but under illumination with other wave-lengths

12, we plot the rate of discharge at the beginning of discharge as indicated by the line in Figure 11. This rate as function of the wavelength of the incident light, represented in Figure 12, is in a remarkably good agreement with the wave-length dependence of the anodic photocurrent, as can be seen in Figure 7. Because of the measuring technique employed, with the revolving turntable and the intermittent illumination, the speed of discharge was plotted in relative units.

On the basis of these preliminary results, it seems permissible to employ the results of the sensitization experiments obtained with ZnO single crystals in an electrochemical cell for the development of a sensitizing dye system applicable to electrophotographic zinc oxide resin layers. Further experiments to demonstrate this point of view are in progress.

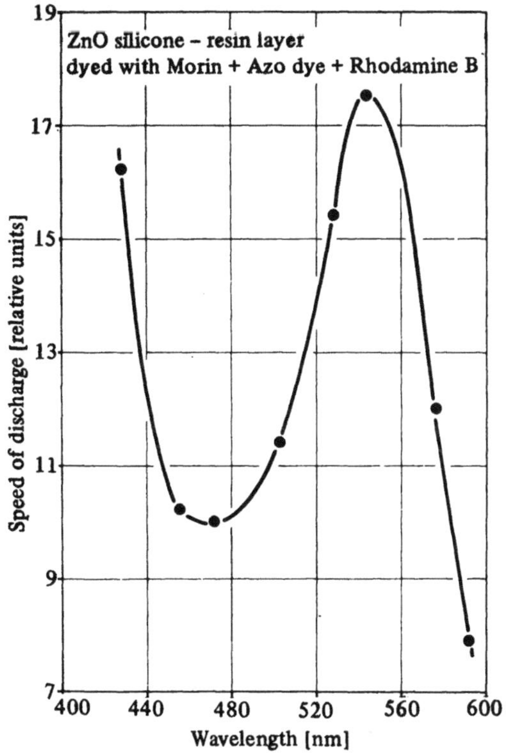

Fig. 12. Wave-length dependence of the speed of discharge under illumination at various wave-lengths with a constant number of photons/cm^2-sec

4. Discussion

Because of the fact that the anodic dark-current in an electrochemical cell with a zinc oxide single crystal as anode is several orders of magnitude smaller than the cathodic dark-current, it may be assumed that an electron transfer from a negatively charged adsorbed species, as for instance O_2^-, OH^-, Cl^- or generally M^-, into the conduction band of zinc oxide is a rare statistical event. This means the reaction

$$O_2^-(ads) \rightarrow O_2(aq) + e' \quad (1)$$

or

$$OH^-(ads) \rightarrow OH^x(ads) + e' \quad (1a)$$

or

$$Cl^-(ads) \rightarrow Cl^x(ads) + e' \quad (2)$$

is very slow in contrast to the inverse reaction

$$O_2(aq) + e' \rightarrow O_2^-(ads) \ . \quad (3)$$

This result is understandable if we consider the barrier layer in zinc oxide near the interface ZnO/electrolyte which is characterized by an appreciable exhaustion of free electrons causing a positive space-charge accompanied by a bending up of the conduction band edge [12, 14]. This is schematically represented in Figure 13. At anodic polarization, the energy difference $\Delta E = E_c + eV_D - E_{M^-}$ is too large for an electron transfer into zinc oxide. Therefore, the observed anodic current amounts to about 10^{-10} amperes/cm^2 while the cathodic current attains values of about 10^{-5} to 10^{-4} amperes/cm^2. The unavoidable dissolution of the ZnO anode occuring independently during the current measurements, has no influence on the current-voltage measurements. It has been found [13] that the so-called electronic "leakage current" amounts only to 0.2% of the total exchange calculated from the neutrally dissolved equivalents of zinc oxide. On the basis of this study [13], we may assume that the main portion of the anodic dark-current is caused by the donors (= zinc ions in interstitial position, Zn)–present in the dissolved portion of the crystal–which enter the electrolyte during the dissolution of zinc oxide according to:

$$Zn^{\cdot} \rightarrow Zn^{++}(aq) + e' \quad (4)$$

The freed electron e′ migrates through the zinc oxide crystal towards the anode of the applied battery. If we may assume that the negative surface charge remains constant in spite of the continuous destruction of the surface of the crystal, also the space-charge layer has to remain

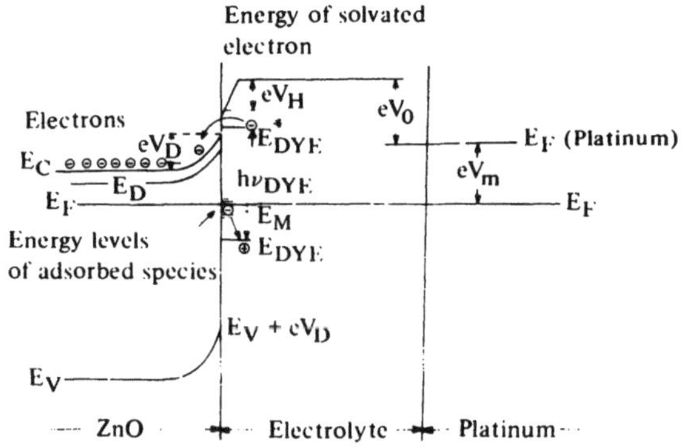

Fig. 13. Schematic representation of the dye-sensitized electron injection from a negatively charged species M^- adsorbed on ZnO into zinc oxide by means of the band model and energy levels notation in the system: ZnO | Electrolyte + Dye | Platinum in accordance with the representation of Morrison [14]. The used symbols as follows:

E_F = Fermi-potential of zinc oxide
E_F(Platinum) = Fermi-energy in the platinum electrode
E_C = energy of the conduction band edge in the bulk
E_V = energy of the valence band edge in the bulk
E_D = energy of the donor level
V_D = diffusion potential between the surface and the bulk
e = electronic charge
V_H = Helmholtz double layer potential difference
V_m = potential difference measured on the voltmeter
eV_0 = energy difference between an electron at the Fermi-energy of platinum and a solvated electron in the electrolyte
E_{M^-} = energy level of a negatively charged species adsorbed on ZnO (O_2^-, OH^-, Dye^- etc)
E_{DYE} = energy level of the adsorbed dye in the ground state
E^*_{DYE} = energy level of the adsorbed dye in the excited singlet or triplet state

constant. If this is correct, no donor diffusion is necessary. Only an equivalent number of electrons which belong to the donors entering the space-charge region, are shifted into the interior of the crystal and disappear via the anode contributing to the measured dark-current. Hence for every dissolved donor two electrons are flowing to the anode. The corresponding steps at the platinized platinum electrode operating as cathode may be assumed as follows:

$$2\,H_3O^+ + 2\,e^-(Pt) \rightarrow H_2(aq) + 2\,H_2O \quad (5)$$

or

$$2\,H_2O + 2\,e^-(Pt) \rightarrow 2\,OH^-(aq) + H_2(aq) \quad (6)$$

From the results of the quoted paper [13], the unsensitized photocurrent in the electrochemical cell under ultra-violet light ($\lambda < 380$ nm) of the zinc oxide crystal is mainly caused by the following reaction of the zinc ions $Zn^+(ads)$ available on the surface, with the light-generated electron-hole pairs according to

$$Zn^+(ads) + e' \sim |e|^{\cdot} \rightarrow Zn^{++}(aq) + e' \qquad (7)$$

accompanied by the subsequent step

$$O^-(ads) + e' + H_3O^+(aq) \rightarrow OH^-(aq) + H_2O \qquad (8)$$

It might be assumed that the supply of $Zn^+(ads)$ and $O^-(ads)$ necessary for the reaction steps (7) and (8), is guaranteed by the fast surface reaction

$$ZnO \rightarrow Zn^+(ads) + O^-(ads)$$

At present, the details of the mechanism are not quite understood.

However, the continuous disintegration of the crystal has no significant influence on the dye-sensitized anodic photocurrent. In the subsequent discussion, we employ the energy level representation of the electrochemical cell according to Morrison [14] and complete this model to describe the dye sensitization of the electron transfer through the interface ZnO/electrolyte. On the basis of this model, the sensitized electron transfer may occur during irradiation with photons, the energy of which corresponds to the maximum of the optical absorption of the adsorbed dye. By means of illumination, electrons will be brought up to the first excited singlet state $E_{DYE} = E(^+DYE^-)$:

$$h\nu_{DYE} \rightarrow {}^+DYE^-(ads) \qquad (9)$$

and from there into the conduction band of zinc oxide if the excited singlet or triplet state is located very close to or above the conduction band edge, as indicated in Figure 13. The subsequent step may be assumed to be either by an electron injection into ZnO

$$^+DYE^-(ads) \rightarrow {}^+DYE(ads) + e'(ZnO) \qquad (10a)$$

or by a recombination step as follows:

$$^+DYE^-(ads) + O_2^-(ads) \rightarrow DYE^-(ads) + O_2(aq) \qquad (10b)$$

or better by

$$^+DYE^-(ads) + Cl^-(ads) \rightarrow DYE^-(ads) + {}^1/_2 Cl_2(aq) \qquad (10c)$$

If the electron transfer (10a) occurs more rapidly than (10b), then for electroneutrality an electron trapped by an adsorbed species fills up the electronically unoccupied ground state of the dye $^+DYE(ads)$ by recombination, as for instance

$$^+DYE(ads) + Cl^-(ads) \rightarrow DYE(ads) + {}^1/_2 Cl_2(aq) \qquad (11a)$$

or

$$^+DYE(ads) + OH^-(ads) \rightarrow DYE(ads) + {}^1/_2 H_2O_2 \qquad (11b)$$

If reaction step (10b) is faster than the step (10a), then the following electron transfer may occur

$$DYE^-(ads) \rightarrow DYE(ads) + e' \,. \qquad (11c)$$

The reaction of Cl^- as a cosensitizer and the resulting increase in sensitization was investigated by Pusch [19].

For the occurence of the mechanism described in the Eqs. (9)–(11), the following requirements have to be fulfilled:

1. As mentioned above, the excited singlet or triplet state $^+DYE^-$, $E(^+DYE^-)$, must be located near or above the conduction band edge at the surface, given by $E_c + eV_D$.
2. The energy levels of the adsorbed species, E_{M^-}, which are generally below the Fermi-potential E_F, must be located above the ground state of the adsorbed dye, E_{DYE}.

Deviations from these necessary requirements will decrease significantly the sensitizing ability of the dye. Such a simple model, however, must be extended to explain the details of the different sensitizing ability of the single dyes.

Unfortunately, at present no such extending model is available by which the speed of the sensitized electron transfer could be predicted because parameters, for instance the activation energy and the effective cross section for the electronic reactions, are at present largely unknown. The sensitizing efficiency (= the speed of the electron transfer) seems to be very dependent on the electronic interaction between the adsorbed dye and zinc oxide in the dark and also on the tendency for complex formation between the dye and zinc ions available on the surface. As was demonstrated by Danzmann [15], in a more extensive study, particularly morin and the azo dye exhibit a complex formation with zinc ions. This was demonstrated by a study of the fluorescence and of the electron spectra of the dyes in solution which were significantly changed when zinc ions were added to the electrolyte. This work will not be discussed here in detail but it should be mentioned that the excellent sensitizing ability of the azo dye in a neutral or weakly basic electrolyte is probably with the tendency to form a complex with zinc ions as follows.

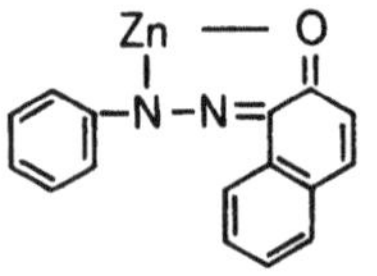

This monomer complex was discussed by Schetty [16]. A polymer complex structure has been assumed by Brändli [17]. Also with morin, a complex formation with zinc ions may be supposed with an optimum extinction at a pH ≈ 6.5. These results are consistent with the well-known complex formation with aluminum ions in the analytical chemistry.

A further important observation is the mutual influence of the single dyes in binary or ternary dye mixtures adsorbed on zinc oxide from electrolyte. Adsorption experiments with single dyes and binary dye mixtures using ZnO powder as adsorbens give evidence that a significant portion of the decrease of the sensitizing ability of morin is mainly caused by a significant decrease of the adsorption of morin in the presence of rhodamine B acting as acid. Detailed results shall be published elsewhere [18]. A similar behavior might be expected by the simultaneous presence of rhodamine B and the azo dye in an acidic or neutral electrolyte. A further decrease could be caused by a mutual electronic interaction of the adsorbed dyes with participation of zinc oxide. We might suppose that rhodamine B acts as a quencher for the excited singlet state of adsorbed morin according to

$$\mathrm{MORIN(ads)} + h\nu_{\mathrm{MORIN}} \xrightarrow{1} {}^{+}\mathrm{MORIN^{-}(ads)} \qquad (12)$$

$$\begin{aligned} {}^{+}\mathrm{MORIN^{-}(ads)} + \mathrm{RH\text{-}B(ads)} \xrightarrow{2} \mathrm{MORIN^{-}(ads)} &+ \\ + \mathrm{RH\text{-}B^{+}(ads)} & \qquad (13a) \end{aligned}$$

$$\mathrm{MORIN^{-}(ads)} \xrightarrow{4} \mathrm{MORIN(ads)} + e'(\mathrm{ZnO})$$

Sensitization (14)

and

$$\mathrm{RH\text{-}B^{+}(ads)} + \mathrm{M^{-}(ads)} \rightarrow \mathrm{RH\text{-}B(ads)} + \mathrm{M(ads)} \qquad (13b)$$

or the step for the quenching

$$\begin{aligned} \mathrm{MORIN^{-}(ads)} + \mathrm{RH\text{-}B^{+}(ads)} \xrightarrow{3} \mathrm{MORIN(ads)} &+ \\ + \mathrm{RH\text{-}B(ads)} & \qquad (15a) \end{aligned}$$

Also the following sequence is possible

$$\mathrm{MORIN^{-}(ads)} + \mathrm{M^{-}(ads)} \rightarrow \mathrm{MORIN^{-}(ads)} + \mathrm{M(ads)} \qquad (13c)$$

followed by step (14) in the case of sensitization or if a quenching occurs:

$$\begin{aligned} \mathrm{MORIN^{-}(ads)} + \mathrm{RH\text{-}B(ads)} \rightarrow \mathrm{MORIN(ads)} &+ \\ + \mathrm{RH\text{-}B^{-}(ads)} & \qquad (15b) \end{aligned}$$

$$\mathrm{RH\text{-}B^{-}(ads)} + \mathrm{M(ads)} \rightarrow \mathrm{RH\text{-}B(ads)} + \mathrm{M^{-}(ads)} \qquad (15c)$$

Here M^- denotes a negatively charged molecule such as O_2^- or Cl^- etc. and RH-B the abbreviation for rhodamine B. On the basis of our results, the quenching step is large in an acidic ambient and weak in a basic one. The quenching mechanism is schematically represented in Figure 14. On the basis of our experi-

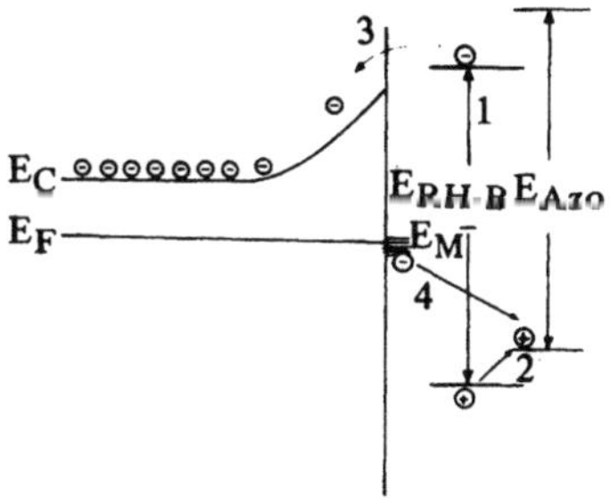

Fig. 14. Schematic representation of the quenching of the morin sensitized electron transfer by rhodamine B. E_{MORIN} and $E_{RH\text{-}B}$ are the singlet excitation energies of morin and rhodamine B, respectively. 1, 2, 3 and 4 indicate the sequence of the single steps. Besides a direct recombination step 3, it may be occur also via the excited singlet level of rhodamine B

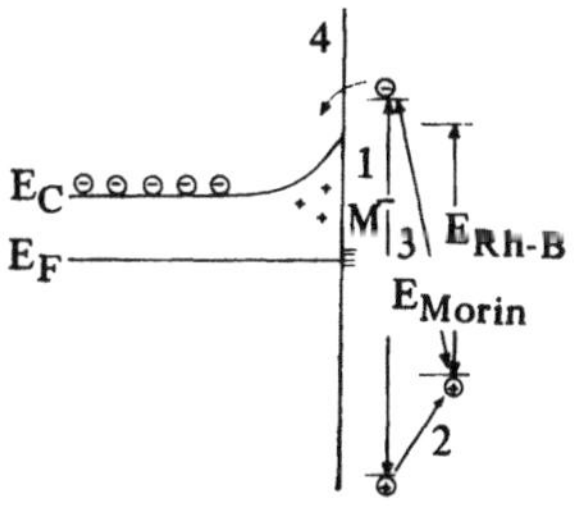

Fig. 15. Schematic representation of the increase of the sensitizing ability of rhodamin B on the injection of electrons into zinc oxide by the simultaneous presence of the azo dye. The used symbols have the same meaning as in Fig. 13. E_{AZO} and $E_{RH\text{-}B}$ are the singlet excitation energies of the azo dye and of rhodamine B, respectively. The figures 1 to 4 indicate the chronological sequence of the single steps

mental results, the recombination step 3 seems to be faster than the injection step 4.

Particularly puzzling was the large enhancement of the poor sensitizing ability of rhodamine B in a basic electrolyte due to the simultaneous presence of the azo dye. According to the cosensitizer mechanism recently discussed by us [10], we might propose tentatively the following sequence of reaction steps:

$$\text{RH-B(ads)} + h\nu_{\text{RH-B}} \xrightarrow{1} {}^{+}\text{RH-B}^{-}\text{(ads)} \quad (16)$$

A recombination of the excited singlet in the rhodamine B molecule seems to be prevented by the hole emission step 2, schematically signified in Figure 15 where an adsorbed azo dye molecule is positively charged according to:

$$\begin{aligned} {}^{+}\text{RH-B}^{-}\text{(ads)} + \text{AZO(ads)} &\xrightarrow{2} \text{RH-B}^{-}\text{(ads)} + \\ &+ {}^{+}\text{AZO(ads)} \end{aligned} \quad (17)$$

which is assumed exothermic with a large effective cross section so that the injection of the electron into zinc oxide may occur preferably:

$$\text{RH-B}^{-}\text{(ads)} \xrightarrow{3} \text{RH-B(ads)} + e'\text{(ZnO)} \quad (18)$$

followed simultaneously by step 4:

$$\text{M}^{-}\text{(ads)} + {}^{+}\text{AZO(ads)} \xrightarrow{4} \text{M(aq)} + \text{AZO(ads)}\,. \quad (19)$$

This tentative mechanism for the interesting and stimulating experimental results is only a first attempt to find a possible understanding of the rather complex phenomena occuring during the sensitization. The adsorption of the single dyes on zinc oxide from binary and ternary dye solutions and their mutual influence are under investigation. Also rather large efforts have been undertaken in the study of the electronic structure of the dyes adsorbed on zinc oxide as a single compound or in binary and ternary mixtures. By these results, expected in the next future, we hope to improve our present qualitative mechanism of sensitization.

A further complication with respect to the understanding of the sensitization mechanism is introduced when resin is employed as binder in the electrophotographic zinc oxide layer. As can be seen from Figure 9, the maximum of the sensitization of morin in a silicone resin ambient is shifted from 420 towards 460–470 nm. Because of the low drying temperature (25° C), there is a residual content of solvent in the layer which may interact with the dye. For the elucidation of the details, however, further experiments are necessary.

Acknowledgement

We are grateful to Dr. S. R. Morrison for stimulating discussions.. We are also indebted to Professor Dr. H. Zollinger for the supply of the azo dye. Furthermore, we appreciate the sponsorship of this research by the Deutsche Forschungsgemeinschaft concerning the grant for the experimental equipment and for the financial support to H. Pusch and D. Rein.

Reference

[1] K. Hauffe and J. Range, Z. Naturforsch. **23b**, 736 (1968). – K. Hauffe, V. Martinez, J. Range and R. Schmidt, Phot. Korr. **104**, 113 (1968). – K. Hauffe, H. Pusch and J. Range, Z. phys. Chem. (NF) **64**, 122 (1969).
[2] H. Gerischer and H. Tributsch, Ber. Bunsenges. phys. Chem. **72**, 251 (1968). – E. Michel-Beyerle, H. Gerischer, F. Rebentrost and H. Tributsch, Electrochim. Acta **13**, 1509 (1968). – H. Tributsch and H. Gerischer, Ber. Bunsenges. phys. Chem. **73**, 251 (1969).
[3] S. J. Dudkowski and L. I. Grossweiner, J. opt. Soc. **54**, 486 (1964). A. N. Terenin and I. A. Akimov, J. Phys. Chem. **69**, 730 (1965), J. Phys. Chem USSR **217**, 307 (1961). – I. A. Akimov, J. Nautsch, Prinkl. Phot. Kine **4**, 64 (1959). – I. A. Akimov, V. M. Bentsa, F. I. Vilesov and A. N. Terenin, Dokl. Akad. Nauk, USSR **172**, 371 (1967). – See also the comprehensive book of H. Meier, „Die Photochemie der organischen Farbstoffe", Springer-Verlag Berlin 1963.
[4] D. D. Taft and S. C. Heidecker, Tappi **50**, 36A (1967). – R. Memming and H. Tributsch, J. Phys. Chem. **75**, 562 (1971).
[5] R. C. Nelson, J. Phys. Chem. **71**, 2517 (1967).
[6] E. Inoue, H. Kokado and T. Yamaguchi, J. Phys. Chem. **69**, 767 (1965).
[7] N. N. Markevich and E. K. Putzeiko, Zh. Fiz. Khim. USSR **36**, 2393 (1962), Soviet Phys. Solid State **5**, 868 (1963).
[8] H. Frieser and M. Schlesinger, Phot. Korr. **101**, 69, 133 (1965).
[9] T. Sakata, S. Kikuchi and Y. Takahashi, Electrophotography, Japan **5**, 38 (1963).
[10] K. Hauffe, H. J. Danzmann, H. Pusch, J. Range and H. Volz, J. electrochem. Soc. **117**, 993 (1970).
[11] K. Hauffe and R. Stechemesser, J. Phot. Sci. Eng. **11**, 145 (1967). Phot. Korr., 8 Sonderheft, 7 (1967).
[12] For a detailed discussion see K. Hauffe and R. Stechemesser, Zur Randschicht-Theorie der Adsorption und Katalyse an Halbleiter-Katalysatoren, in electronic Phenomena in Chemisorption and Catalysis on Semiconductors, ed. by K. Hauffe and Th. Wolkenstein, Walter de Gruyter-Verlag Berlin 1969, pg. 1 ff.
[13] H. Erbse, K. Hauffe and J. Range, Z. phys. Chem. (NF) 74, 248 (1971).
[14] S. R. Morrison, Surface Phenomena Associated With The Semiconductor/Electrolyte Interface, Progr. Surface Sci. **1**, 105 (1971)
[15] H. J. Danzmann, Unpublished results shall be published elsewhere.
[16] G. Schetty, Helv. Chim. Acta **53**, 1437 (1970).
[17] R. Brändly, Doctor Thesis, ETH Zürich 1969.
[18] K. Hauffe and S. Ishikawa, unpublished results.
[19] H. Pusch, doctor thesis Göttingen 1972.

Discussion

Goodman: In your last slide you have presented an energy level diagram with which the observed effects are explained. Is there any independent evidence for the energy levels shown for the adsorbed Azo and Rhodamine B dyes?

K. Hauffe: As far as I know, there is presently no independent evidence for the energy levels of adsorbed dyes. Therefore, the positions of the levels are tentatively assumed.

J. W. Weigl: Is there any evidence that the same kinds of co-sensitization and mutual desensitization, which you have described for the electrolyte system, occur in electrophotographic ZnO layers?

K. Hauffe: At present no experiments with dyed electrophotographic ZnO resin layers on the co-sensitizing influence of a second dye are available. However, we suppose that such effects should also occur in electrophotographic layers. Experiments in this direction are in progress.

K. Verhille: In the experiments you described only a narrow selection of wavelengths was transmitted to the ZnO. Was in your viewpoint the fact that Morin absorbed at an entirely different part of the spectrum not important with respect to the different experimental evidence shown in the two systems Morin + Rhodamin B and Azo dye + Rhodamin B?

K. Hauffe: We believe that the different positions of the level of the excited singlet and of the ground state of the two dyes in their adsorbed state should be responsible whether the second dye will react preferably either as cosensitizer or as quencher. This is schematically represented in the Figures 14 and 15 of my paper.

However, stimulated by this question we shall perform sensitization experiments in the case with the dye mixture morin – rhodamine B both with light of 420 and of 560 nm.

W. F. Berg: One is much impressed with the elegance of the electrochemical cell technique, but, following Carroll, one would ask for more experimental evidence before speculating on the detailed mechanisms of the super- and anti-sensitization phenomena described, such as:

Is there cooperative adsorption?
Is there competitive adsorption?
Do components act as anti-sensitizers or as impeders of anti-sensitization?
Do they break up dye-aggregates?
Are the absorption spectra of the dyes changed by the presence of the others or by changed conditions?
Do any of the dyes act as desensitizers?

K. Hauffe: These questions are indeed very important and could be recommended as items on the next meeting. As fas as we know at present, the adsorption may have also a large influence on the sensitizing ability of a dye. In the case of rhodamine B, the adsorption of dye-aggregates seems to be preferred. Furthermore, it should be principally expected that dyes may act as consensitizers and also as desensitizers. Experiments are in progress.

H. Meier: What can be said about the influence of adsorbed dyes on the barrier layer formed between oxygen and ZnO?

J. Range: The space-charge and the charge on the surface cannot be influenced by absorption of dye molecules in ionic form. The diffusion voltage is given by other species on pH-value in the electrolyte. In other case with respect to monolayer-coverage, the space charge would be too high.

P. Rys: 1. What indications do you have that the mutual dye interchange you found is not simply explained by competitive adsorption effects?

2. How did you—if you did—bufferd you alkaline electrolyte solutions? It is known that acid is produced by adsorbing azo dyes on ZnO. This would make your system acidic and therefore, rhodamine B would sensitize again.

K. Hauffe: 1. On the basis of our preliminary adsorption experiments both with single dyes (rhodamine B, azo dye and morin) and with corresponding binary dye mixtures, we may assume that both adsorption competition (rhodamine B – morin) and electronic interchange (azo dye – rhodamine B) can be responsible for the significant difference of the sensitization.

2. Since the area of the surface of the ZnO single crystal is very small (0.2 cm^2) in comparison to the amount of the electrolyte, no noticeable change in the pH was detected. Furthermore, we did not use buffered electrolytes because of the experimental difficulties (no reproducible photocurrent etc.). Only small additions of HCl and NaOH were employed.

The Present Status of Organic Photoconductors in Electrophotography

David L. Stockman

Abstract

Significant recent advances have been made in understanding the physical phenomena involved in organic photoconductors for use in electrophotography. In particular, the generation and transport properties of these materials are substantially different from conventional inorganic photoconductors. Generation is strongly controlled by the molecular nature of the photoconductor and is limited, particularly at low fields, by the many competitive photophysical processes which can occur in the organic solid state. The efficient generation of free charge carriers in these materials is controlled by the field as it is for selenium. However, there is a wide variation in the field dependence for different materials, and some recent experimental results and theoretical progress in this area are reported.

Due to the weak intermolecular interaction existing in these materials, carrier mobilities are generally low (10^0 to 10^{-8} cm^2/volt-sec for practical materials) and field dependent, but are found to vary dramatically with the physical state of the system. Of special significance is the large difference in transport characteristics exhibited by single crystals and their vinyl polymer counterparts using the same aromatic chromophore. Unlike single crystals, the polymers show no well defined drift mobility. Alternatively, the weak coupling is a major deterrent to impurity trapping, and extremely long life-times against deep trapping are observed; trap-free space charge limited currents are often encountered, Recent theoretical progress is outlined which relates transport properties to the electronic states of the disordered polymer.

These generation and transport properties can limit or enhance the electrical characteristics of the photoreceptor and have a major influence on practical properties such as charging, dark decay and potential buildup during cycling. Using these properties and their relationship to electrophotography, a new assessment of the performance requirements for an organic photoconductor is presented.

1. Introduction

The purpose of this review is to examine in a qualitative, but critical way, the recent experimental and theoretical progress in the understanding of the properties of organic materials as they relate to their photoconductivity for use in electrophotography. The intent is to show where the subject stands and to identify areas where experiment and theory agree and conversely where more work needs to be done. In order to do this, a point of view has been adopted in each section of the paper that is believed to represent the best present ideas of how energy moves, photogeneration occurs and transport proceeds in crystalline and amorphous organic materials. To do this, a great deal of reliance is placed (hopefully correctly) on single crystal studies where the quantitative data are on the soundest footing and an attempt is made to transfer the ideas of the single crystal to the amorphous state where data are extremely limited. The data are inadequate to definitely prove any of these points of view, but it is hoped that this will serve as a focal point for further research to resolve the important issues.

The paper consists of four technical sections: Section 2 considers the spectroscopy and photophysics of these materials; Section 3 considers photogeneration and Section 4 is a discussion of the charge transport properties. Section 5 assesses the impact of these properties on the behavior of these materials as photoreceptors in electrophotography.

2. Absorption, Energy Migration and Photophysical Processes in Organic Materials

The past 20 years have produced an extensive body of spectroscopic details on a variety of polyatomic molecules in the solid state and the essential facts are now well established. The materials of concern here include molecular crystals, charge transfer complexes and polymers in which the predominate binding energy of the solid arises from Van der Waals forces which are due to the correlated motion of electrons on different molecules. The Van der Waals energy stabilizing two interacting molecules, A and B, is given approximately by

$$E = -\frac{3}{2} \cdot \frac{{}^{I}A\,{}^{I}B}{{}^{I}A + {}^{I}B} \cdot \frac{{}^{\alpha}A\,{}^{\alpha}B}{R^6}, \qquad (1)$$

where ${}^{I}A$ and ${}^{\alpha}A$ are the ionization potential and polarizability of molecule A, and R is the separation between A and B. The weakness of this force and the characteristic separation of the molecules (~3.5 Å) is perhaps the dominant feature of organic solids and helps to control the spectroscopy, the motion of the excited states, influences the subsequent photophysics observed in these systems and is of prime importance in generation and transport of free charge.

In covalent or ionic solids, the original atomic wave functions are merged into wave functions (bands) associated with the entire crystal. Absorption of light is thus a transition from one band to another band normally producing electron/hole pairs free to move through the crystal. Electron exchange dominates over electron phonon interactions, which are treated as a small perturbation producing occasional scattering of the carrier. In molecular crystals, electron exchange is much weaker due to the Van der Waal's bonding; the electron phonon interaction is now significant and cannot be treated as a perturbation.

The observed spectra of organic solids are, to first order, the absorption spectrum of the isolated molecule. Four significant perturbations on this isolated are observed in the solid and have been discussed by McClure [1] and by Craig and Walmsley [2]. These include a general displacement of the spectrum, a removal (usually) of any degenerate states, violations of the selection rules for the isolated molecule and splitting of the transitions due to intermolecular resonances. The magnitude of these influences is normally rather small and to an excellent first approximation, the features observed for the lowest electronic transitions are those of the transitions of the free molecule. The materials of interest to this paper are characterized in simplest terms by extensive delocalization of some of the electrons. These levels lie highest in energy and the absorption of a photon of visible or ultraviolet light by a molecule promotes an electron from such an orbital to an unfilled one. Such a transition may involve more than electronic excitation, since vibrations and rotations of the molecule can be simultaneously excited.

The high energy transitions of molecules result in ionization, and there are as many such transitions as there are filled orbitals in the ground state of the molecules. Typical values of the lowest ionization potentials of polyatomic molecules are in the range of 7–12 eV. At energies close to ionization, Rydberg transitions are clearly observed in the vapor phase. These transitions correspond to excitation into large hydrogen-like orbitals. Since these orbitals are spatially large, they are heavily perturbed in solids and the Rydberg transitions are broadened into a continuum.

The lower energy transitions, which are of primary interest here, are specific to particular molecules, since they depend cricitally upon the molecular structure. As this type of excitation is into orbitals which are not large spatially, they are perturbed in solids to a much smaller extent than the Rydberg transitions.

The absorption corresponding to purely electronic transitions of a polyatomic molecule in a vapor at low pressure should be sharp and intense as in atomic transitions, and the absorption spectrum should consist of a series of sharp lines corresponding to allowed transitions between the stationary states. Of course, the molecule possesses internal motion in the form of 3n–6 vibrations (n = No. of atoms) and also rotations. This internal motion, whose characteristic energy is small compared to the electronic transition energy, will modify the sharp absorption since rotational, vibrational and electronic levels can be simultaneously excited. Thus the transition appears at many discrete frequencies. This set of discrete frequencies (or levels) is often termed the manifold. Figure 1 illustrates qualitatively these levels. Within a given electronic transition, the density of vibrational states increases rapidly as the vibrational energy is increased. For a molecule with more than 10 to 12 atoms, the density of states of the vibrational motion built onto an electronic transition can be so large as to render the spectrum apparently continuous.

The intensity distribution of the vibrational manifold of the transition is explained by the Franck-Condon principle, which states that since the nuclear motion is very slow compared to electronic motion, the geometry of the molecule immediately after absorption of the photon must be almost identical to that prior to absorption. The Franck-Condon principle does not influence the total transition probability, but only the relative importance of the sub-bands. At the moment of the absorption, the electronically excited molecule's geometry is almost identical to the ground state. This initial excited state is termed the "Franck-Condon State". Most molecules in their ground state have filled orbitals with all electrons paired and are thus singlet states. Allowed transitions preserve the spin (and also require appropriate orbital symmetry between initial and final states) so that the intense absorptions are singlet-singlet transitions. For each singlet state, there is a corresponding state with triplet spin character lying lower in energy.

Following excitation into the Franck-Condon state, a variety of radiative and radiationless transitions can occur as illustrated in Figure 1.

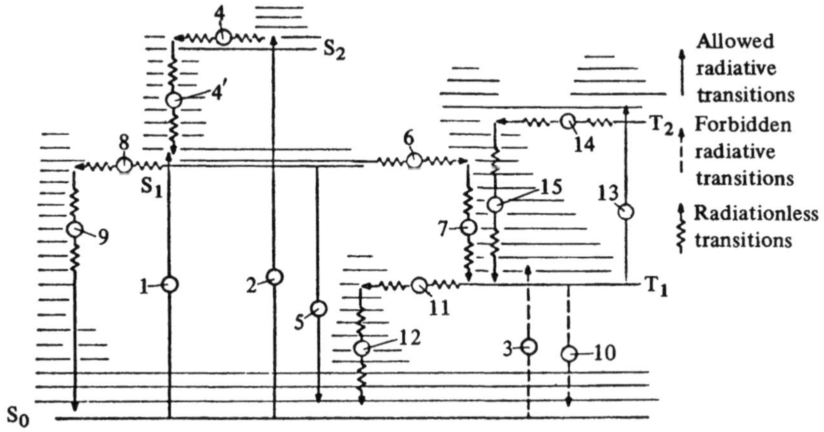

Fig. 1. Lowest electronic states of a polyatomic molecule and its transitions

The vertical lines represent radiative transitions; allowed transitions by solid lines, forbidden transitions by dashed lines. The wavy lines represent nonradiative transitions. In this discussion, the molecules are considered to be large with at least 10–12 atoms. This ensures that the density of states associated with the internal degrees of freedom of the molecule is large as discussed by Robinson [3]. The recent book by Birks [4] on photophysics of aromatic molecules gives a more complete discussion than is possible here. Steps 1, 2, and 3 represent the absorption transitions. Steps 4 and 4′ represent internal conversion between the higher excited singlet states. Process 5 corresponds to the emission of fluorescence, which normally occurs after the excited molecule has come into thermal equilibrium with its surroundings. The radiative lifetime for fluorescence is 10^{-9} – 10^{-7} sec. The emission of fluorescence competes with the intersystem crossing steps, 6 and 7, in some cases with internal conversion (steps 8 and 9) to a high vibrational level of the ground electronic state, and with bimolecular annihilation in the bulk or unimolecular annihilation at a surface. Normally, the triplet state, T_I, is populated by allowed singlet transitions, 1 or 2, followed by intersystem crossing. Phosphorescence (radiative lifetime 10^{-3} – 10^{+1} sec.) which corresponds to step 10 competes with nonradiative degradation, 11 and 12, and with bimolecular annihilation in the decay of the lowest excited triplet state. The absorptive transitions and relaxation of the triplet state are illustrated by steps 13 to 15.

Radiationless transitions have received close attention in the last years. The basic experimental facts concerning radiationless transitions of molecules can be summarized as follows:

1. Radiationless transitions occur in a molecule in all phases, especially if the molecule is large,
2. The rates of the radiationless transition are first order, even between states of different multiplicity,
3. The rate constants depend upon the energy difference between the two states involved, but are almost independent of temperature,
4. Deuteration of the compound grossly retards the radiationless rate while having an almost negligible effect upon the radiative rate,
5. Heavy atom and paramagnetic atom effects are about as efficient whether produced intra- or intermolecularly.

Robinson and Frosch [5] first proposed that radiationless transitions were largely intramolecular in nature. They started by assuming that the small interaction terms, such as spin orbit coupling, produce non-stationary states of the molecule. The initial state (the excited state undergoing the transition) is weakly coupled to a very large number of final states (the vibrationally excited lower electronic state of the transition isoenergetic with the initial state). This large number of final states of the system then undergoes vibrational relaxation and the two steps combine to make the process irreversible and thus capable of being described by a rate constant.

Siebrand and coworkers [6] have advanced these ideas in developing a theory which predicts the lifetime of the lowest triplet and singlet state of a large number of normal and perdeuterated hydrocarbons. For the triplet state, Siebrand's work is based on an empirical relationship between the triplet lifetime and the energy of the transition. The triplet radiative lifetimes all seem to be about 30 sec. From the fact that the perdeuterated hydrocarbons all seem to have approximately a triplet lifetime which is purely radiative, the C-H vibrations would seem to be the vibration which is predominant in converting the electronic energy into vibrational energy. An excellent correlation between the triplet lifetime and a function which includes the energy of the triplet, first published by Siebrand, is shown in Figure 2. The denominator of the abscissa contains a number which is related to the number of C-H vibrations in the molecule. The 4000 cm^{-1} comes from the fact that the phosphorescence spectrum of these molecules

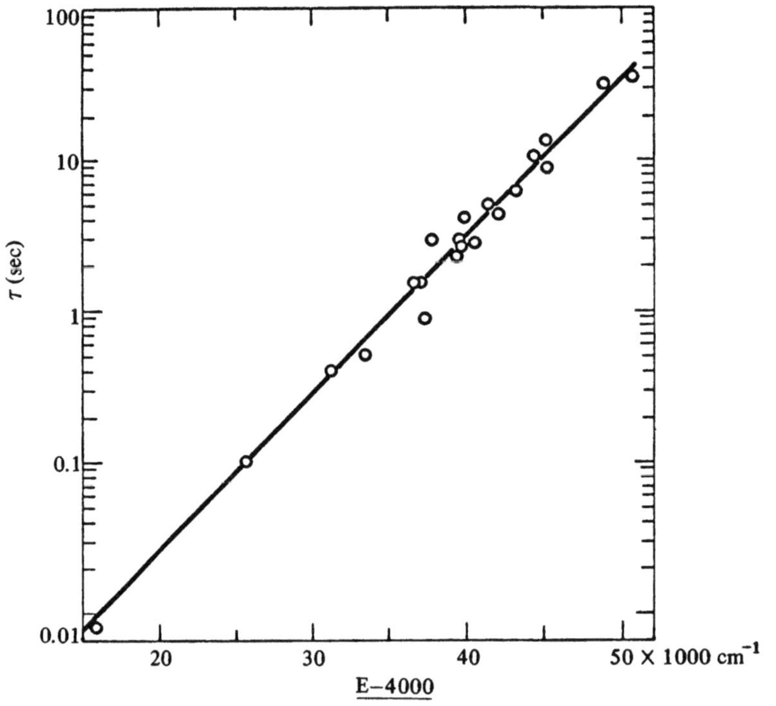

Fig. 2. Nonradiative triplet lifetime for aromatic hydrocarbons after Siebrand [6]. η is a factor that is determined by the relative number of hydrogen and carbon atoms

is typically 4000 cm^{-1} wide and the vibronic pattern observed is the C-C stretching frequency. The Franck-Condon factors for this transition fall off quite rapidly after the third or fourth overtone. Only the electronic energy difference greater than 4000 cm^{-1} will be dissipated by C-H vibrations since the Franck-Condon factors become larger than for C-C vibrations only above 4000 cm^{-1}. Siebrand used these facts to calculate the radiationless rate constants for the aromatic hydrocarbons. The additional assumption he used, which is implied by the plot in Figure 2, is that the spin orbit coupling factors must be the same for all of the aromatic hydrocarbons. When this assumption is included, the calculation reduces to a computation of the relative Franck-Condon factors for this class of molecules. The results of his calculations predict the radiationless rate constant, and thus the mean lifetime, to within a factor of two for over 30 molecules. In addition, he has calculated the rate constants for internal conversion $S_1 \rightsquigarrow S_0$, and finds that with a few exceptions, the rate constants are such that the internal conversion cannot compete with fluorescence and intersystem crossing for decay of the lowest excited singlet state as known from experiment.

In addition to these intramolecular processes, intermolecular processes can occur. As stated earlier, the treatment of the excited state is that of a tightly bound hole and electron on the same molecule, i.e., a Frenkel exciton. This does not mean that there is no significant interaction of the exciton with its neighbors. The oscillator strength for allowed transitions in these molecules is typically ~1 or more with characteristic excited state radiative lifetimes of $\sim 10^{-8}-10^{-9}$ sec. The average time that the excitation can be associated with a single molecule is $\tau \approx h/\Delta E$ when ΔE is the energy of interaction between molecules. For $\Delta E \sim 10^2 cm^{-1}$, a relatively weak interaction compared to the 10–30,000 cm^{-1} energy for the electronic transition, the time is $\tau \approx 10^{-13}$ sec. Even a 1 cm^{-1} interaction, insignificant spectroscopically, would imply $\tau \approx 10^{-11}$ sec. Thus, in spite of these energetically weak interactions, the exciton will move through the crystal a significant amount during its lifetime.

Three general regimes of energy coupling can be considered. In the first regime, the molecules are separated by distances comparable to or greater than the wavelength of light. Ordinary fluorescence emission by one molecule and reabsorption by the second molecule occurs. This process is of no importance in solids except for doped materials where the transferring molecules are in very low concentrations. In the second regime, the separation of the molecules are small compared to the wavelength of light but large compared to the molecular dimensions. For these allowed transitions, the usual description of the process is a transition dipole-induced dipole coupling with a typical $1/R^6$ dependence. For allowed transitions (large dipoles), transfer distances of ~50–100 Å are possible with excited states that live $\sim 10^{-8}$ sec. The transfer is radiationless and is termed resonance transfer [7]. At very small distances (< 10 Å), characteristic of molecular solids, the third regime applies and the dipole approximation breaks down badly in calculating the rate of transfer of the exciton. Knox and Chang [8] have treated this problem. They note that the resonance transfer rate constant is determined by the square of the electric field associated with the transition dipole. If the magnitude of the transition dipole is comparable to the separation between the interacting molecules, the dipole approximation must be replaced by the discrete components of the charge distribution associated with the transition. Depending on the relative orientation of the molecules and their spacing, rather large correction factors (20–100 fold) to the isotropic transfer rate can be obtained at distances comparable to the molecular crystal spacing. This complicates the analysis of diffusion of the singlet excitons.

Mulder [9] has carefully measured the singlet exciton diffusion lengths for anthracene and his data are summarized in Table 1, along with the calculated diffusion coefficients. As Jortner has discussed [10], singlet excition motion is best characterized by a strongly scattered random-walk model. In a random-walk model, the diffusion coefficient is related to the mean displacement expected for each jump, $\langle r_{ave} \rangle$, and the number of displacements per unit time, k_{ave}, by [11]

$$D = \frac{k_{ave} \langle r_{ave}^2 \rangle}{6} . \qquad (2)$$

Using $\langle r_{ave} \rangle \approx 3.5$ Å and the data from Table 1, $k_{ave} \approx 1 \times 10^{13}$ sec^{-1}. Since the lifetime of the excited singlet of anthracene is $\sim 2 \times 10^{-8}$ sec ($k_s = 5 \times 10^7$ sec^{-1}), the average number of sites, n, visited during an exciton lifetime by random-walk is

$$n = \left[\frac{k_{ave}}{k_s}\right]^{1/2} \approx 450 . \qquad (3)$$

Table 1. Diffusion Coefficients and Lengths in Anthracene Single Crystals [(a)]

Crystallographic Axis	Electronic State	D[(b)] cm^2 sec^{-1}	L Microns
a	Singlet	1.6×10^{-3}	0.08
b	Singlet	4.9×10^{-3}	0.14
c*	Singlet	4.0×10^{-4}	0.04
a	Triplet	1.5×10^{-4}	27.0
b	Triplet	1.8×10^{-4}	30.0
c*	Triplet	1.2×10^{-5}	7.7

a Data from references 9 and 13.
b For singlets, L is reported, D is calculated from

$D = kL^2/2$ with $k_s = 5 \times 10^7$ sec^{-1}.

For triplets, D is reported, L calculated using $k_T = 4.0 \times 10^1$ sec^{-1}.

The diffusion parameters determined this way are appropriate only in the absence of bimolecular annihilation processes or in the absence of trapping of the exciton at an impurity or defect site in the crystal. As Mulder [9] has shown, the diffusion lengths measured experimentally can vary quite widely due to trapping. The results reported in Table 1 are probably limited by crystal purity and defects and may not represent intrinsic values.

For triplet states, the transition dipole is zero to a first approximation, and rather weak coupling is expected. Swenberg calculated rate constants for triplet-triplet annihilation and showed that large anisotropies in diffusion are expected [12]. This was verified experimentally by Ern [13] whose values for the diffusion coefficients are also shown in Table 1. The same analysis given above leads to a number of displacements per unit time $\sim 10^{12}$ sec^{-1} and using the triplet decay rate constant $k_T = 4 \times 10^1$ sec^{-1}, an average number of sites visited is $\sim 1.5 \times 10^5$. It is interesting that the rate of hopping of the triplet exciton is only a factor of ten smaller than the rate for the singlet even in the absence of dipole-dipole coupling. The triplet exciton problem has been recently discussed theoretically by Munn and Siebrand [14]. They showed that the electron-phonon coupling that produces exciton motion is controlled by the out of plane vibrations of the molecules.

Virtually no literature exists on the corresponding properties of singlet and triplet excitons in polymers or charge transfer complexes although studies of energy transfer of polymers in dilute solutions have been reported [15] and a study of the excimer emission of poly-n-vinyl-carbazole (PVK) has been published [16]. Carnes [17] has studied decaying photocurrents in PVK using bulk excitation and his analysis is consistent with a long range for the exciton in the material although it was not established what spin state is involved. It would seem appropriate for extensive measurements on these materials to be made in order to obtain a better picture of the dynamics of excitons in disordered systems.

Other exciton properties include bimolecular annihilations of the excited state. Singlet-singlet and triplet-triplet reactions have been reviewed recently [18].

Another type of interaction of excited states is excimer (excited dimer) formation. Excimers result from the interaction between an excited singlet state and an unexcited molecule to produce an excited dimer. It forms only when one of the molecules is excited and usually dissociates upon emission as shown by the broad structureless emission spectrum characteristic of a repulsive ground state. In crystals, excimers can be observed when the crystal structure is such that adjacent molecules are parallel and have significant overlap of the delocalized electrons. Anthracene, which does not meet this structure requirement shows no excimer emission while pyrene, with a favorable structure, shows exclusively excimer emission. The best correlation of structure to excimer emission is perylene which exists in two crystalline forms, the α form having excimer fluorescence and the β form having only normal fluorescence [19]. The polymer poly-n-vinylcarbazole (PVK) exhibits excimer emission [16], indicating the presence of significant inter- or intramolecular coupling of the carbazole pendents. Excimer emission has also been observed in polystyrene [20], and polyvinylnapthalene [21]. Since the excimer state has slightly less energy, it can act as a trap for singlet exciton motion and then should influence the diffusion parameters, although this has not been studied in any detail. An extensive review article by Birks [22] summarizes the energetics and kinetics of excimers.

Are these considerations of any value in understanding the properties of these materials as photoconductors? Since these photophysical processes compete with and are precursory to photogeneration, it seems clear that they warrant major attention. As we shall see, the dependence of the quantum efficiency for photogeneration on field must be reflected in corresponding changes in the energetics and/or kinetics of the photophysical steps. This subject has not been investigated in any detail. The presence of excimers in polymers has provided a clue (Section 4) to the transport properties of polymers and this area deserves detailed study as well. A complete understanding of the photophysics of these materials is a prerequisite to development of a complete model of organic photoconductors.

3. Photogeneration Properties

3.1. Introduction

Early photogeneration studies on molecular systems were hampered by the lack of recognition of the dependence of the generation process upon the applied electric field, and by the failure to clearly separate electrode and transport effects. Additionally, there are a variety of competing generation steps. some intrinsic and others extrinsic with excitation intensities, wavelengths, crystal purity, surfaces and electrode materials all interacting to control which step(s) dominate in any given experiment. It has only been very recently that this situation has begun to be sorted out and for organic materials, quantitative data exists only for single crystals and until very recently on one material, anthracene. The best data on generation in amorphous molecular solids exists for selenium and its alloys with arsenic, although some photogeneration studies have been reported on poly-n-vinylcarbazole (PVK) and very recently on a complex of PVK with trinitrofluorene (TNF).

3.2. Energetic Considerations

In an isolated molecule, absorption of light of sufficient energy can lead to the ejection of the electron from the molecule. The lowest energy at which this can occur, termed the ionization potential, I_G (Fig. 3), represents a transition of the electron from the highest occupied orbital of the ground state to an unbound state with the electron removed to a distance such that the Coulomb energy of the two charges approaches zero. This energy level is called the vacuum level. In typical organic molecules, I_G = 7–12 eV, with the lowest energies normally associated with large aromatic molecules. In the solid state, a crystal vacuum level can be defined

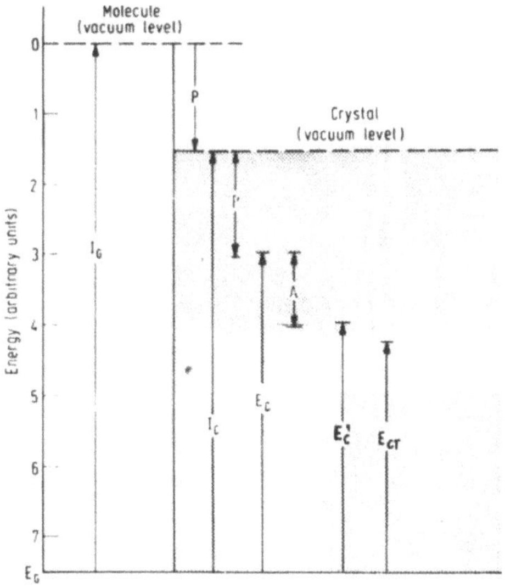

Fig. 3. Electronic "Conduction Levels" after Lyons [23]

which is smaller than I_G by an amount P, the energy of polarization of the crystal by a single point charge. In most organic crystals, $P \approx 1.5–2.0$ eV.

Lyon's [23] first discussed conduction band levels in organic crystals. Lyon's defined a "proper conduction band" as the state in which the electron is far away from the hole *and* in which it is not attached at any time to another molecule in the crystal.* This "proper conduction band", E_c, was calculated by Lyons to be

$$E_c = I_G - 2P = I_e - P \,. \tag{4}$$

There is also the possibility of conduction levels at lower energies when the Coulomb attraction of the electron and hole is negligible ($\leqslant kT$), but where the electron is bound to a distant molecule by the electron affinity A. This level, E_c', also shown in Figure 3 is

$$E_c' = E_c - A \,. \tag{5}$$

There will be other lower lying states in a regime of energies where the electron is removed from the molecule, but is bound by Coulomb

* In the absence of knowledge of transport parameters of the crystal (in particular the mobilities and the lifetime of the carriers against deep trapping), the definition is somewhat philosophical.

attraction. It is of interest to calculate the magnitude of the separation at which the Coulomb attraction is just kT. For solids, this separation, r_s, is (24)

$$r_s = \frac{e^2}{2\pi \epsilon\epsilon_0 kT}, \qquad (6)$$

where e is the electronic charge, ϵ_0 is the permitttivity of free space, and ϵ is the dielectric constant. Using $\epsilon = 3.4$ (for anthracene) $r_s \approx 300$ Å at 25° C. If the solid were *isotropic*, this would represent a sphere of interaction radius r_s within which the states are electrostatically bound, but they are not bound in the spectroscopic sense discussed earlier. The highest energy state of this grouping of energies is E_c' and the lowest energy member represents the situation where the electron and hole are on adjacent molecules and, if thermalized are rather tightly bound. This state is frequently denoted by E_{ct} and termed the charge transfer exciton state. Its energy was first calculated by Lyons. This group of states extending in energy from E_{ct} to E_c' are all charge transfer exciton states and are the organic crystal equivalent of Wannier excitons. As will be shown, the experimental evidence on photogeneration suggests that these are the autoionization states that are the precursor to complete separation of electron and hole.

If the density of such final states is large enough, then efficient bimolecular annihilation of singlet and triplet excitons or monomolecular processes such as photoionization of singlets and triplets and direct autoionization of a Frenkel exciton could represent essentially irreversible reactions with the irreversibility due to the large density of final states, relative to the density of the initial states. In general, the density of final states available will depend on the bound spectroscopic state's spin, symmetry and the crystal structure. In a highly anisotropic crystal, the specific states available for autoionization can be anisotropic by a factor of 100 or more as shown by Swenberg [12] for triplet-triplet annihilation (into the lowest charge transfer state). The particular charge transfer exciton state(s) finally resulting and the "average" separation of the bound hole/electron pair should be dependent on the initial spectroscopic states. Strong evidence for this point of view is shown by the experimental results of Batt, Braun and Hornig [25] which shows that the average initial separation of hole/electron pairs in anthracene is dependent on the photon energy.

3.3. Generation Mechanisms

Since this paper is mainly concerned with these materials for electrophotography, where sensitivity and thus the free carrier yield is important, most of the emphasis of this section will be on generation mechanisms in the regime of high generation efficiencies.

As noted earlier, both extrinsic and intrinsic photogeneration have been observed. The extrinsic process involving surface dissociation of a Frenkel singlet exciton was first proposed by Carswell and Lyons [26]. Kepler's now famous pulsed-light experiments demonstrated [27] that in the case of anthracene,

1. The quantum yield for both electron and holes followed the absorption spectrum in the region 3400–4100 Å.
2. A linear intensity dependence was found.
3. The quantum yield and the ratio of holes to electrons varied widely with the state of the crystal surface.

All of these observations are consistent with diffusion of the exciton to the surface followed by dissociation. This technique was extended further into the UV by Castro and Hornig [28] who observed that the electron quantum yield showed a peak at 4.4 eV which did not correspond to the absorption spectrum and who proposed a direct band-to-band transition. Subsequently, Geacintov and Pope [29] extended these measurements into the range of 6 eV and found another peak at 5.4 eV. This result and subsequent experiments by the same authors

[30] showed that both the 4.4 and 5.4 eV peaks were due to intrinsic generation via autoionization of a singlet Frenkel exciton. However, in the low photon energy range ($\lesssim 3.7$ eV), it is quite clear that the process is extrinsic and at the surface. Recent experiments by Johnston and Lyons [31] have firmly established this point.

A variety of intrinsic generation processes have been proposed and confirmed. These include one exciton ionization following one [28] or two [32] photon excitation, double exciton interaction of both the singlet-singlet [33] and singlet-triplet [34] kind and photoionization of the lowest singlet [35] and triplet [36] excitons. A natural question that occurs is whether the autoionization states reached by the various kinetic routes available to the excitons are the same or whether they involve different final states. At least for the case of anthracene, Pope and Burgos [37] argue convincingly that the result of annihilation of two low energy singlet excitons produces the same autoionized state as a single higher energy (6.3 eV) singlet exciton. It is by no means certain that this will be true for all cases.

Photogeneration studies have recently been extended to materials other than anthracene, notably benzene, naphthalene [38], tetracene, pyrene and biphenyl [39] as well as in metal and metal-free phthalocyanines [40] where a variety of kinetic processes precursory to charge carrier separation have been observed. The case of pyrene [39] is particularly interesting since the intrinsic process is a one photon mechanism and the onset of photoconduction with wavelength does not coincide with absorption edge. A similar situation was found by Sharp [41] for n-isopropyl carbazole. In this case, the absorption edge ist at 3.57 eV while the onset of photoconduction occurred at 4.7 eV. Sharp also studied the crystals of the charge transfer complex of n-isopropyl carbazole with picryl chloride and showed that carrier generation still required 4.7 eV, and was of the same efficiency and no carriers were generated by excitation into the charge-transfer absorption band. This latter result is important since it shows that this type of charge transfer state does not assist in carrier generation.

The importance of the electric field in photogeneration in organic materials was not recognized for a long time. The demonstration of the importance of this effect will lead us quite naturally into a viewpoint of the generation processes in the amorphous state, where the chief present constraint limiting the use of these materials is the efficiency of generation. The first reported field dependent generation in anthracene was by Castro and Hornig [28] who found that the photocurrent for holes was linearly dependent on field. This has subsequently been confirmed by Batt et al. [25], and Johnston and Lyons [31] for both electrons and holes for an intrinsic generation step. Similar results have been found for metal-free phthalocyanine by Hackett [42].

In amorphous molecular materials, the effect is well-established for selenium [43] and As_2Se_3 [44]. Additionally, Regensburger [45] and Pai [46] have shown that photogeneration in polyvinylcarbazole is field dependent. For selenium, at least, the field dependence of the photogeneration persists down to very low fields. This is significant in that a low field saturation for the generation step should occur when the field adds only kT in energy to an electron transversing one Coulomb capture diameter. This field is given by [24]

$$E_c = \frac{\pi\epsilon\epsilon_0}{e}\left(\frac{kT}{e}\right)^2. \qquad (7)$$

For selenium, $E_c \sim 7 \times 10^3$ V/cm ($\epsilon = 6$) and no saturation is observed even at fields down to $\sim 1 \times 10^3$ V/cm. The magnitude of the changes induced by the fields is very large and the intensity, temperature and wavelength dependence have all been studied in this system. Using this data, Warter [24] showed that conventional recombination (either at the surface or in the generation region) in the sense of free electrons, holes and recombination centers could not explain the results. He demonstrated that the field controlling step must occur during the initial creation of a free pair. The problem of geminate recombination of this type was first calculated by Onsager [47] whose model is appropriate to a continuous isotropic medium. Warter's model takes account of the anisotropy of the medium and shows that photogeneration can be significantly enhanced when the carrier motion during the generation step is highly anisotropic on a microscopic scale. This is true even when the fields are low enough so that less than kT of thermal energy would be given to a carrier in an isotropic medium.

An alternative description of the anisotropy of the carrier motion during the separation would be to envision the autoionized intermediate states of the generation step as being highly anisotropic. There is experimental evidence which lends support to this idea. A variety of papers have shown that delayed fluorescence and photocurrents in anthracene and tetracene crystals [18] are influenced by the orientation of a magnetic field which influences the triplet excitons. Rusin et al. [48] shows that the results may be accounted for on the basis of an anisotropic model for the charge transfer exciton produced by triplet-triplet annihilation. Their results seem to resolve an earlier controversy [49] centered around the nature of the magnetic field anisotropy effect which had been postulated to resolve the interaction of a triplet exciton and a trapped hole.

The next question in the generation step relates to the amorphous materials. For both PVK [45] and the charge transfer complex PVK-TNF [50], the generation step is field dependent, but it is not obvious that the process is intrinsic. For PVK, the intensity dependence is linear and the quantum yield is proportional to the absorption coefficient, suggesting an extrinsic surface ionization mechanism. This apparently holds true for low fields [17] as well as high fields. Hughes [51] has very recently found that the generation step in PVK excited by x-rays is also field dependent and ascribes his results to geminate recombination using the Onsager model. The published data for PVK-TNF complex shows [50] field dependent generation, linear intensity dependence at low intensities (sublinear at high intensities) and a crude correlation of photoresponse with absorption coefficient. Photogeneration in this material may also be an extrinsic process but direct participation of the charge transfer complex in carrier generation or other intrinsic mechanisms cannot be ruled out.

Thus, the picture of the intrinsic (and probably extrinsic) photogeneration process in organic materials that emerges is that of a single or double exciton autoionization to a possibly anisotropic intermediate charge transfer state in which the electron and the hole are separated and thermalized, but in which significant Coulomb attraction still exists (i.e., in the energy range from E_{ct} to E'_c). In the absence of a field, geminate recombination is very likely and the quantum yield very low. With applied external fields, there is a significant field assisted separation step manifested as field dependent generation. This mechanism will certainly require much further research and analysis before it can be accepted on a broad scale. Since the photogeneration efficiency is the chief present constraint on the use of these materials for electrophotographic layers, this is probably the most important area for future research.

4. Transport Properties

4.1. Introduction

The transport properties of this class of materials show very significant differences between the single crystal and the amorphous state. In the single crystals there is normally a well-defined mobility, usually for both carriers, and a variety of shallow and deep traps associated with gross impurities and defects have been studied. The few amorphous materials studied to date exhibit no well-defined drift mobility, only one carrier may be mobile, and as will be shown, trapping in the conventional sense is absent unless gross impurity levels are present. The measurement techniques employed to study these materials are those used in insulator physics and have been discussed extensively elsewhere [52]. This section explores the reasons for the differences between single crystals and amorphous materials, outlines the arguments against significant trapping in polymers and explores the utility of a new phenomenological theory of transport in polymers recently proposed which may be capable of explaining the observed effects.

4.2. The Band Model

As shown in the previous section, Lyons [23] first calculated a conduction level, E_c, for an organic crystal based on a model in which the electron was completely free of the hole and in which it is not attached at any time to another molecule. There is a variety of strong experimental evidence to suggest that although such bands exist, significant transport does not occur in them. The most significant results (all for anthracene) which argue against a band conduction model are:

1. The mobilities for electron and holes in single crystals are very nearly equal [27]. The values of the mobilities (< 1 cm^{2}volt^{-1}sec^{-1}) suggest narrow conduction and valence bands. The anisotropies of the electron and hole mobilities with crystal direction and the temperature and pressure dependence [27] are consistent with this view. Also, consistent with this viewpoint is the fact that the experimentally determined density of states for the valence band requires a band width of $<$kT [27].
2. "Hot" hole injection from metals [53] shows a very narrow valence band ($\leqslant 0.1$ eV) with vibrational broadening.
3. The rates of radiative recombination of electrons and holes [54] are consistent with a model involving very short mean free paths for the carriers.
4. For a "pure" crystal, the number of trapping centers seem to be the same for electrons and holes and can be directly related to the number density of dislocations present adventitiously or introduced deliberately [55].
5. There is evidence that subtle structure changes in the crystal change the temperature dependence of the mobility [56].

The conclusion appears inescapable that a band conduction model is not appropriate for these materials even in the crystalline state.

4.3. The Small Polaron

The alternative model is the small polaron model originally developed by Holstein [57]. In this transport model, the carrier interacts quite strongly with the vibrational modes of the molecule and this produces distortions which induce a "self" field that traps the carrier, in a manner analogous to the strong self trapping observed in some inorganic oxides. The transfer of the carrier from molecule to molecule is caused by non-zero electronic interactions between adjacent molecules.

Munn and Siebrand [14] have recently extended the small polaron model to include anharmonicities in the vibrational modes which yields improved results for anthracene. They discuss two new limits; the "slow phonon" limit and the "slow exciton" limit.* It is beyond the scope of this paper to discuss their results in detail, except to say that the anisotropies and temperature dependence for the mobility of electrons and holes in anthracene can be explained by their model. This lends considerable support to the idea that the small polaron model is the appropriate description of motion of carriers in a crystalline molecular solid.

4.4. Trapping Effects

In order to complete the discussion of single crystals and provide the framework for a discussion of transport in the amorphous state, it is

* Their results are also applicable to motion of triplet excitons, discussed in Section 2.

appropriate to mention trapping effects. Trapping of both carriers has been observed and can be divided into two categories. There is trapping due to chemical impurities for which tetracene in anthracene is the best studied case. Here the trap depths, 0.42 eV for holes [58], 0.17 eV for electrons [59] and the measured cross sections (~3 x 10^{-14} cm^2) are consistent with the trapping occurring because of electronic energy differences (higher A or lower I_c) of the trapping molecule. The absolute energy level data of Pope for tetracene and anthracene [60] are combined with the trapping levels in Figure 4, to show the internal consistency of this argument.

In single crystals of reasonable purity, say $\leqslant 10^{14}$ impurities/cm^3, a different type of trapping is evident. Here the relevant experimental facts [61] are:

1. An exponential distribution of trapping energies is observed.
2. The number of traps is usually the same for both carriers.
3. The number of traps of the same type can be varied by introducing known defects into the crystals.
4. The estimated size of the crystal distortions (~400 Å) introduced by the dislocation is consistent with the trapping cross sections observed.

All of these facts are consistent with the viewpoint that the binding energy associated with these traps is due to disortion polarization energy and not from enhanced electronic binding energy of impurity molecules.*

4.5 Experimental Transport in Amorphous Materials

The most significant differences between single crystals and the amorphous state of organic materials is found in their transport properties. Almost all of the data on polymers is for PVK. A suitable model crystal for the monomer is N-isopropylcarbazole, mentioned earlier. Sharp [41] found well-defined drift mobilities for both holes (μ_D = .33 cm^2 $volt^{-1}$ sec^{-1}) and electrons (1.0 cm^2 $volt^{-1}$ sec^{-1}) in this study. The data suggest that both carriers move $\gg 50\,\mu$ without

* This viewpoint cannot explain all observed results. Helfrich [62] has summarized the trapping situation much more completely.

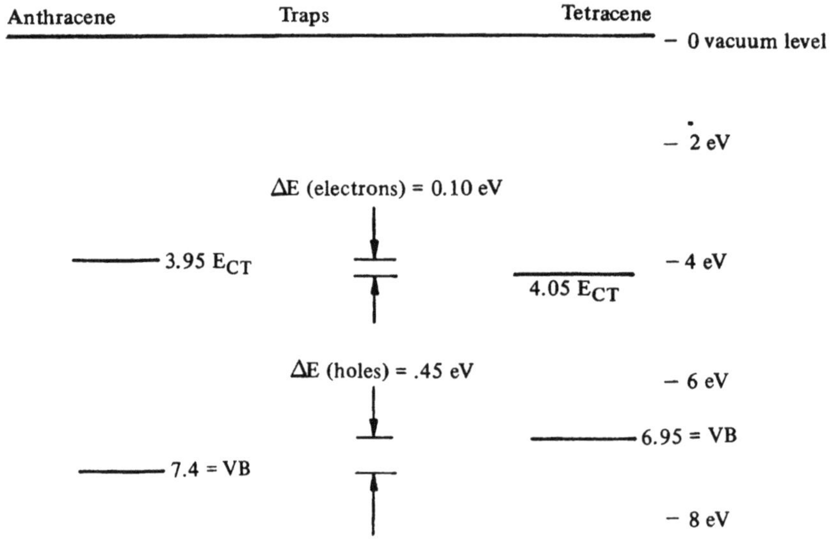

Fig. 4. Electronic and trapping levels in anthracene and tetracene. The levels chosen were the valence band and E_{ct}. However, the data of Pope [60] show that the trapping levels are consistent for all choices of electron conduction levels

significant deep trapping at the fields employed, i.e., a range of $\geqslant 5 \times 10^{-5}$ cm^2 volt^{-1}. By contrast, several studies on the polymer PVK, show significant range only for holes. Carnes [17] established a carrier range for holes $\geqslant 4 \times 10^{-6}$ cm^2 volt^{-1}. The mobility of the carriers is not well defined, i.e., it is a function of field. Regensburger [45] first reported this field dependent mobility and found

$$\mu_D = 5.0 \times 10^{-18} E^2 \ (\text{cm}^2 \ \text{volt}^{-1} \ \text{sec}^{-1}), \qquad (8)$$

with E in units of volts/cm. Pai [46] characterized the transport by an effective transit time. Lakatos and Mort [63] studied transient photoemitted carriers from gold into PVK and Mort observed trap-free space charge limited currents injected from a thin selenium generating layer [64]. The last two studies both characteristize the drift mobility as

$$\mu_D = \mu_D^0 \left(\frac{E}{E_0}\right)^2 , \qquad (9)$$

with $\mu_D^0 \cong 3.7 \times 10^{-7}$ cm^2 volt^{-1} sec^{-1} and $E_0 = 10^5$ V/cm. Similar field dependent mobilities have been observed for As_2Se_3 [44], selenium at low temperatures [65] and very recently [66] for the PVK-TNF complex. In all of these materials, the mobility is quite low, but low electron mobilities (10^{-4} cm^2 volt^{-1} sec^{-1}) are observed in liquid and orthorhombic sulphur (via hopping between adjacent S_8 rings) and no field dependence is observed [67]. Some authors have explained their results on the basis of field dependent thermally assisted hopping [45, 46], while others [66] have characterized the transport in terms of multiple trapping.

The concept of trapping to characterize such dispersive transport is, in the present situation, somewhat philosophical. When trapping occurs and the field serves to help empty the traps, then a well defined carrier transit time (as measured in the current mode [52]) can only be observed over a fairly narrow range of fields.* Additionally, at low fields, one expects some charge to be lost via trapping and further, the distribution of the carrier velocities should depend on the field. The data on PVK show that:

1. At all fields studied (a range of 2 decades), a well-defined "current mode" cusp is observed [64] which characterizes the velocity of the fastest carriers,
2. All of the charge injected is transported (within experimental error) over the same range of fields [64], and
3. Most important, the *distribution* of carrier velocities does not change with field.

The structure of a thin layer of selenium on a film of PVK first studied by Regensburger [45], illustrated in Figure 5, is an ideal model structure for studying charge transport in a polymer. Since both selenium and the polymer have high resistivity, the field division is controlled only by the dielectric thicknesses. Further, the field dependent generation of carriers known for selenium ensures that only one CV's ($\equiv \epsilon\epsilon_0 E_0$) worth of charge carriers can be produced and the holes can be injected easily into the PVK. Using this structure, Mort [64] observed trap-free space charge limited currents (space charge sheets in Helfrich's [62] terminology). Figure 6 shows "current mode" results obtained by Sliva [68] for the transport of the charges in this structure. The data show a well-defined transit time at all fields and the transit time has been normalized based on the value of this well-defined time at each field. The absolute values of transit times vary by 100 fold in this field range. In all cases, a CV's worth of charge was

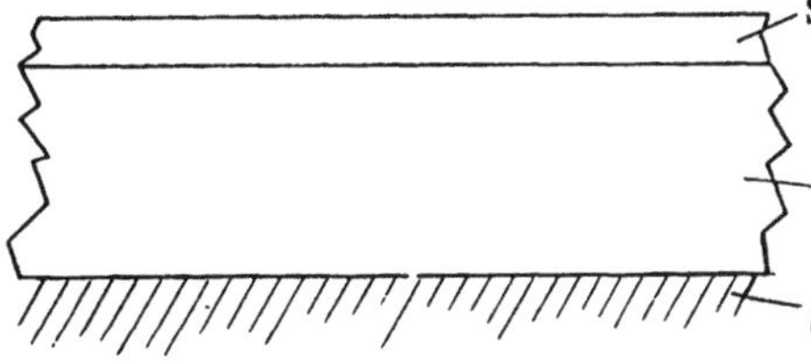

Fig. 5. A model structure for use in charge transport studies

* I am indebted to M. Tabak for this particular important point.

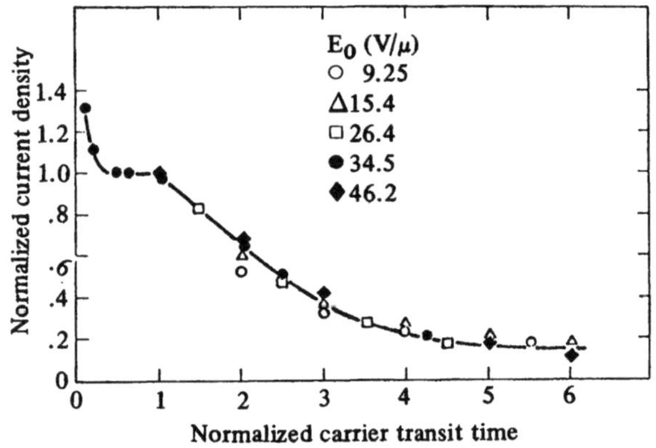

Fig. 6. Dispersion of carrier transit times as a function of field for the structure of Fig. 5

injected* so that the absolute currents vary by a factor of about five. Over this range of fields, the distribution of velocities is the same within experimental error, Recent results obtained by Seki and Gill [66] on electron transport in PVK-TNF do not show such well-resolved transits although they imply that the transit mark is observed over a wide range of fields. This author does not see how multiple trapping level models can explain such results. The data on polymers is so limited that it is difficult to assess whether the experimental results on PVK will be broadly applicable to this class of materials. The influence of charge transfer complexes on polymer transport is a very important area for future research since the PVK-TNF material reported by Schaffert [50] shows significant carrier ranges for both holes and electrons suggesting that, as in the case of selenium [69], electron and hole transport may proceed by significantly different mechanisms.

4.6. Theoretical Aspects of Transport in Amorphous Materials

The magnitude of the difference in transport parameters between single crystals and polymers appears at first glance to require a new mechanism for charge transport. Recently, however, Chen and Mort [70] have proposed a model for transport in PVK which qualitatively explains the observed field dependent mobility. This model is an extension of the Holstein small polaron model. In their paper, Chen and Mort note that previous calculations on the electronic states of polymers have treated the ideal case of transport down an infinite chain with perfect periodicity, and that this is not appropriate to an amorphous solid. The interaction between chains is the dominant characteristic and as they note, there are significant chances that two carbazole pendant units from different segments form a sandwich dimer with large overlap and essentially parallel molecular planes. The evidence for such chances is demonstrated by the excimer fluorescence observed in PVK by Klopffer [16]. Because of the large overlap between the orbitals of the two pendants, an electron (or a hole) in such a dimeric configuration has much lower energy. For example, if two naphthalene molecules form a sandwich dimer with an intermolecular plane separation of 3 Å, the lower energy orbitals of an excess electron and an excess hole are stabilized by 0.5 eV and 0.3 eV respectively. These values are much larger than the corresponding values for a perfect chain, where the overlap between adjacent molecules is small. Thus, the "dimeric states" arising from the entangled chains lie much lower in energy than the "perfect chain states", and therefore charge carriers are postulated to move (or hop) through the dimeric state. Klopffer estimated the density of excimer states in PVK to be 10^{-3} mol/mol of pendant units. That means, on the average, these states are separated by a distance of 10 pendant units which is of the order of 30–50 Å. The key implicit assumption used by Chen and Mort is that it is the formation of these adventitious dimers that is chiefly responsible for the excimer emission observed by Klopffer.

Chen and Mort then proceed to extend Holstein's small polaron model. For small polarons, the mobility of the hopping carriers is proportional to the square of the transfer integral, J,

* The transit time marking is not a consequence of the space charge limitation as the same shape is observed for smaller current densities.

between initial and final states which is

$$J = \langle \Phi_i \, | V_j | \, \Phi_j \rangle, \tag{10}$$

where V_j is the molecular potential due to the jth molecule at i. The presence of an external field, E, adds a potential eEx where x is the intermolecular separation. Thus J can be written in two parts

$$J = J_0 + J_I E \,. \tag{11}$$

For single crystals or where the distance between hopping states is small (< 10 Å), the second term is negligible, and well-defined mobilities are expected. Figure 7 shows the results of their calculation on the relative importance of these two terms as a function of the separation between the hopping states (the adventitious dimer). Qualitatively, this type of mechanism could explain the observed field dependence. Such a model is capable of being critically tested by the use of model polymers and should prove to be a source of fruitful research in the next few years. Certainly the key assumption of a 1 : 1 correspondence of adventitious dimer states and excimer emission is worthy of study. The model itself needs to be modified to include the recent extension of Holstein's model by Munn and Siebrand [14] discussed earlier. Such a model is certainly appealing since it implies only an extension of the small polaron model appropriate to single crystals.

5. Applications to Electrophotography

This section explores the application of organic materials to electrophotography. This discussion is restricted to single layer devices and does not consider multiple-layer devices where the properties of the photoreceptor are often controlled by the nature of the charging step(s) as discussed by Batra and Schechtman [71] and Schaffert [72]. While many insulating photoconductive materials can be used to produce some sort of electrostatic latent image, very few materials come close to satisfying the rigid requirements necessary for a modern, large area photoreceptor. These requirements certainly include sensitivity, speed of response, uniformity, freedom from defects over large (~5000 cm^2) areas, and the intrinsic capability to cycle rapidly. Ideal photoreceptor performance requirements have been discussed by Warter [73] and more recently by Tabak et al. [74].

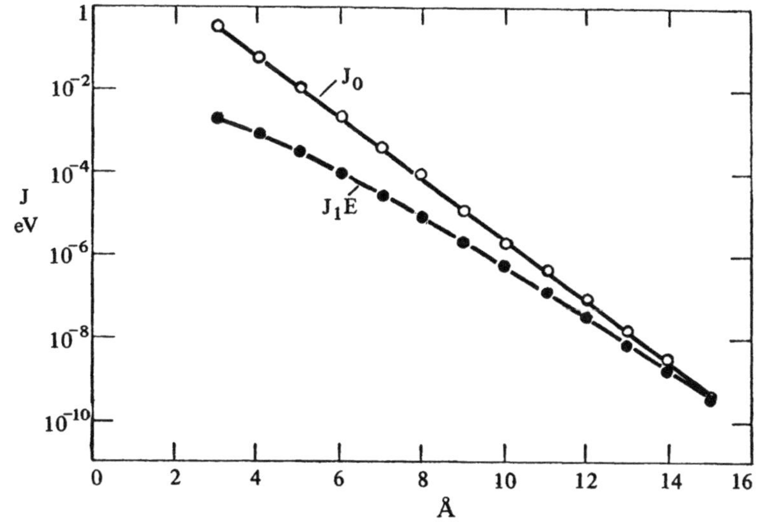

Fig. 7. Relative importance of the two terms in the small polaron hopping model (Chen and Mort [70]). E = 10^6 volts/cm.

Certainly, an important feature in device performance is dark decay, which can occur from thermal generation in the bulk or from injection at either surface. An excellent discussion of bulk thermal carrier generation in organic materials has been given by Sharp and Smith [75]. They show that, except for strongly interacting charge transfer complexes, the low field resistivities measured are not due to intrinsic thermal carrier generation. As Warter has shown [73], resistivity per se has little significance for estimating dark decay and assessing the chargeability of the photoreceptor. The actual resistivities measured for this class of organic material is very high and experimentally [46, 50], the charging proceeds according to the geometrical capacitance of the layer. The dark decays observed after corona charging are typically in the range of $\sim 10^{-2}-10^{0}$ volts μ^{-1} $\sec^{-1}$ at fields of $\sim 10^5$ v/cm [50], and thus do not represent an important problem if sufficiently blocking interfaces are used.

Photogeneration of an "ideal" photoreceptor can be described quite simply; each quantum absorbed results in the transport of one carrier completely through the photoreceptor layer. This "ideal" case produces a potential drop in direct proportion to the exposure. However, this type of discharge will produce a brightness acceptance range (BAR) [76] which may not be optimum for all development systems, particularly in continuous tone applications. The field dependent generation observed in organic materials can often be used to extend the BAR. Consider a system with no field dependent generation step. The rate of discharge of the plate will be given by

$$\frac{dE}{dt} = K \phi I, \tag{12}$$

where ϕ is the quantum efficiency, I, the intensity and E is the field. Introducing a field dependent generation step of the form

$$\phi = \phi_0 E^{\alpha}, \tag{13}$$

where α is a constant, we have

$$\frac{dE}{dt} = K \phi_0 I E^{\alpha}. \tag{14}$$

The potential drop across such a photoreceptor* is shown in Figure 8, for this equation with

* This model assumes generation in a thin layer compared to the thickness of the photoreceptor and no transport limitations (either in carrier range or mobility).

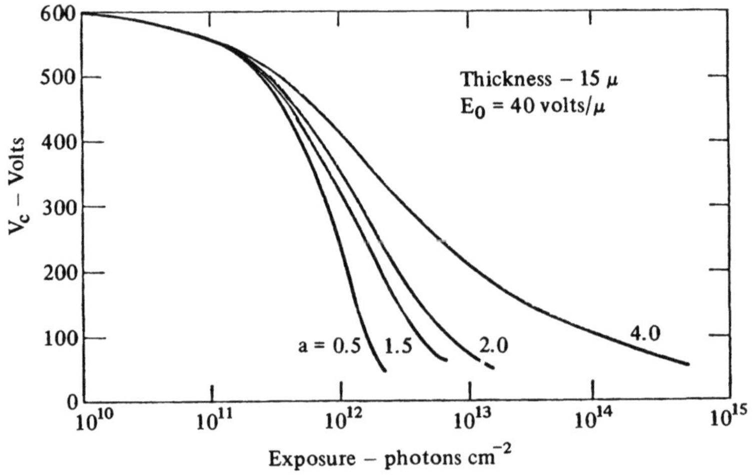

Fig. 8. Contrast potential as a function of exposure for various parameter values of field dependent generation step

varying α. Note that the BAR can be significantly affected by variations in α. This should prove to be very useful in producing continuous tone photoreceptors.

A possible limitation that is evident is a mobility limitation. The very low mobilities associated with these materials can produce a situation in which the carrier velocity is low enough to require a significant amount of time for development of the contrast potential. For a photoreceptor with a well-defined mobility μ at field E and of thickness L, the transit time, t, of the carriers produced near the top surface is

$$t = L/\mu E . \qquad (15)$$

For a field of 2×10^5 v/cm, L = 20μ and if we require onset of development of the contrast potential within 0.1 sec, $\mu \geqslant 10^{-7}$ cm^2 volt^{-1} sec^{-1}. A simple extension of this calculation can incorporate the effective mobilities and distribution of carrier velocities discussed in Section 4.

An important attribute of this class of materials is that the large carrier ranges measured imply no significant bulk trapping of charge. It is too early to tell whether this will hold true for these materials as a group, but it should be relatively easy to obtain good cycling characteristics from these photoreceptor materials. A principal reason for failure of many materials to cycle well is cumulative residual buildup due to bulk trapped charge which cannot be released during the time required for one cycle in a machine.

The greatest promise of these devices for electrophotography would seem to be in the wide variety of generation and transport effects observed which should permit the design of photoreceptors with properties tailored to meet the demands of the development system and the photographic latitude desired. The present limited data are sufficient to be optimistic, but not nearly complete enough to assess all of the possibilities for these materials.

6. Acknowledgement

The author wishes to acknowledge helpful discussions with many of his colleagues at the Xerox Laboratories, in particular, Mark Tabak, Harvey Scher, Philip Sliva and Algird Leiga, and to acknowledge the great assistance of Frank Belli who performed the extensive literature search used in preparing this paper.

Reference

[1] McClure, D. S., Solid State Physics, Vol. 8, p. 1, Academic Press, New York, 1959.
[2] Craig, D. P., and S. H. Walmsley, Excitons in Molecular Crystals, W. A. Benjamin, Inc., New York, 1968.
[3] Robinson, G. W., J. Chem. Phys. **47**, 1967 (1967).
[4] Birks, J. B., Photophysics of Aromatic Molecules, Interscience Monographs on Chemistry, Phys. Chem. Section, John Wiley & Sons, New York, 1970.
[5] Robinson, G. W., and R. P. Frosch, J. Chem. Phys. **37**, 1962 (1962); **38**, 1187 (1963).
[6] Siebrand, W., J. Chem. Phys. **47**, 2411 (1967) and preceding papers.
[7] Förster, Th., Ann. Physik **2**, 55 (1948).
[8] Knox, R. S., and J. C. Chang, Proc. Rochester Symposium on Atomic and Molecular Optics, January, 1971.
[9] Mulder, B. J., Philips Res. Depts **21**, 283 (1966).
[10] Jortner, J., Phys. Rev. Lett. **20**, 244 (1968).
[11] Chandrasehkar, S., Rev. Mod. Phys. **15**, 1 (1943).
[12] Swenberg, C. E., J. Chem. Phys. **51**, 1753 (1969).
[13] Ern, V., Phys. Rev. Lett. **22**, 343 (1969).
[14] Munn, R. W., and W. Siebrand, J. Chem. Phys. **52**, 47, 6391 (1970).
[15] Lashkov, G. I., and V. L. Ermolaev, Opt. and Spect. **22**, 462 (1967).
[16] Klopffer, W., J. Chem. Phys. **50**, 2337 (1969).
[17] Carnes, J., PhD Thesis, Princeton Univ., 1970.
[18] Avakian, P., and R. E. Merrifeld, Molecular Crystals **5**, 37 (1968).
[19] Tanaka, J., Bull. Chem. Soc. Japan **36**, 1237 (1963).
[20] Hirayama, F., PhD Thesis, Univ. of Michigan, 1963.
[21] Vala, M. T., J. Haebig and S. Rice, J. Chem. Phys. **43**, 886 (1965).
[22] Birks, J. B., Progress in Reaction Kinetics, Vol. 5, p. 181, G. Porter, Ed., Pergamon Press, Oxford, 1970.

[23] Lyons, L. E., J. Chem. Soc. **1957,** 5001 (1957).
[24] Warter, P. J., Proc. Third Int. Conf. Photoconductivity, p. 311, Ed. Pell, E., Pergamon Press, Oxford, 1971.
[25] Bott, R. H., C. L. Braun and J. F. Hornig, J. Chem. Phys. **49,** 1967 (1968).
[26] Carswell, D. J., and L. E. Lyons, J. Chem. Soc. **1955,** 1734 (1955).
[27] Kepler, R. G., Organic Semiconductors, p. 1, Ed. by J. Brophy and J. Buttrey, MacMillan, New York, 1962.
[28] Castro, G., and J. F. Hornig, J. Chem. Phys. **42,** 1459 (1965).
[29] Geacintov, N., and M. Pope, J. Chem. Phys. **45,** 3884 (1966).
[30] Geacintov, N., and M. Pope, J. Chem. Phys. **47,** 1194 (1967).
[31] Johnston, R. G., and L. E. Lyons, Aust. J. Chem. **23,** 1571 (1970).
[32] Strome, F. C., Phys. Rev. Lett. **20,** 3 (1968).
[33] Braun, C. L., Phys. Rev. Lett. **21,** 215 (1968).
[34] Fourny, J., G. Delacote and M. Schott, Phys. Rev. Lett. **21,** 1085 (1968).
[35] Kepler, R. G., Phys. Rev. Lett. **18,** 951 (1967).
[36] Holzman, P., R. Morris, R. C. Jarnagin and M. Silver, Phys. Rev. Lett. **19,** 506 (1967).
[37] Pope, M., and J. Burgos, Mol. Cryst. **3,** 215 (1967).
[38] Braun, C. L., and G. M. Dobbs, J. Chem. Phys. **53,** 2718 (1970).
[39] Castro, G., IBM J. Res. Dev. **15,** 27 (1971).
[40] Heilmeier, G. H., G. Warfield and S. E. Harrison, J. Appl. Phys. **34,** 2278 (1963).
[41] Sharp, J. H., J. Phys. Chem. **71,** 2587 (1967).
[42] Hackett, C., J. Chem. Phys. **55,** 3178 (1971).
[43] Tabak, M. D., and P. J. Warter, Phys. Rev. **173,** 899 (1968).
[44] Scharfe, M. E., Phys. Rev. B **2,** 5025 (1970).
[45] Regensburger, P. J., Photochem. Photobio. J. **8,** 429 (1968).
[46] Pai, D. M., J. Chem. Phys. **52,** 2285 (1970).
[47] Onsager, L., Phys. Rev. **54,** 554 (1938).
[48] Rusin, B. A., B. M. Rumyantsev, I. V. Alexandrov and E. L. Frankevich, Phys. Stat. Sol. **34,** K103 (1969).
[49] Geacintov, N., M. Pope and S. Fox, J. Phys. Chem. Solids **31,** 1375 (1970).
[50] Schaffert, R. M., IBM J. Res. Dev. **15,** 75 (1971).
[51] Hughes, R. C., Chem. Phys. Lett. **8,** 403 (1971).
[52] Spear, W. E., J. Non-Crystalline Solids **1,** 197 (1969).
[53] Williams, R., and J. Dresner, J. Chem. Phys. **46,** 2133 (1967).
[54] Helfrich, W., and W. G. Schneider, Phys. Rev. Lett. **14,** 229 (1965).
[55] Baessler, H., G. Herrmann, N. Riehl and G. Vaubel, J. Phys. Chem. Solids **30,** 1579 (1969).
[56] Fourny, J., and G. Delacôte, J. Chem. Phys. **50,** 1028 (1969).
[57] Holstein, T., Ann. Phys. (N. Y.) **8,** 325, 343 (1959).
[58] Hoesterey, D. C., and G. M. Letson, J. Phys. Chem. solids **24,** 1609 (1963).
[59] Oyama, K., and I. Nakada, J. Phys. Soc. Japan **24,** 792 (1968).
[60] Pope, M., J. Polymer Sci. Part C, No. 17, 233 (1967).
[61] Thomas, J. M., and J. O. Williams, Mol. Cryst. Liquid Cryst. **9,** 59 (1969), and preceding papers.
[62] Helfrich, W., Physics and Chemistry of the Organic Solid State, Vol. 3, p. 2, Ed. by D. Fox, et al., Interscience, John Wiley and Sons, New York, 1967.
[63] Mort, J., and A. Lakatos, J. Non-Crystalline Solids **4,** 117 (1970).
[64] Mort, J., Phys. Rev., to be published.
[65] Tabak, M., Phys. Rev. B **2,** 5025 (1970).
[66] Seki, H., and W. D. Gill, Conf. on Low Mobility Materials, Eilath, Israel, April, 1971.
[67] Gibbons, D. J., and W. E. Spear, J. Phys. Chem. Solids **27,** 1917 (1966).
[68] Sliva, P., unpublished results.
[69] Tabak, M, Proc. Third Int. Conf. Photoconductivity, p. 87, Ed. E. Pell, Pergamon Press, Oxford, 1971.
[70] Chen, I., and J. Mort, submitted to Phys. Rev.
[71] Batra, I. P., and B. H. Schechtman, J. Phys. Chem. solids **32,** 769 (1971).
[72] Schaffert, R. M., Phot. Sci. Eng. **15,** 148 (1971).
[73] Warter, P. J., Appl. Opt. Suppl. **3,** 65 (1969).
[74] Tabak, M. D., S. W. Ing and M. E. Scharfe, Special Issue IEEE Trans. Electron Dev. (to be published).
[75] Sharp, J. H., and M. Smith, Physical Chemistry, Vol. 10, p. 435, Academic Press, Inc., New York, 1970.
[76] Bixby, W. E., P. F. Andrus and L. E. Walkup, Phot. Sci. Eng. **5,** 195 (1954).

Discussion

A. Matulionis: Field-non-controlled photogeneration has been observed in amorphous selenium that is made in a high-polymeric form [1]. It can be explained by longer free path of a freshly generated ("hot") carriers in a high-polymeric substances. The same is, possibly, true for single crystals, for example Si.

[1] Juska, G., A. Matulionis and J. Viscakas, Phys. stat. solidi (a) 1971.

D. L. Stockman: As far as I know, this has not been observed in our laboratories. Could your results be due to small crystalline phases in the selenium?

A. Matulionis: Field-non-controlled photogeneration cannot be ascribed to the effect of microcrystals in amorphous phase as Dr. Stockman suggests. Internal photoemission of carriers into amorphous selenium is a field-controlled process. (The same is true for the internal photoemission into all materials investigated). Therefore, if carriers were photogenerated in microcrystals without field control, their emission would have been field-controlled and the resultant yield would have been field-controlled. As this is not true the field-non-controlled photogeneration is a property of a high-polymeric amorphous Se.

D. L. Stockman: Field controlled photogeneration is also observed in single crystals of anthracene, metal-free phthalocyanine and naphthalene which are the only materials I am aware of where it was looked for and found. I expect it to be a rather general phenomena in these materials. It is, of course, also observed in both inorganic and organic amorphous systems as I noted.

R. Schaffert: I would like to point out that the use of amorphous Se as an argument against a trapping model may not be justified since we can have many different kinds of amorphous Se layers, depending on purity and preparation conditions.

D. L. Stockman: I agree that there is a variety of types of Selenium. The selenium used in xerography normally shows good range for holes and a variable range for electrons. Tabak [1] has recently discussed this problem in some detail. In selenium, there is a well defined drift mobility for the holes and a remarkable absence of long term trapping is observed. In my view, trapping occurs only if the carrier is trapped for a time long enough to remove it from the observation being made. For the materials I have discussed, this does not seem to occur in spite of the fact that the motion of the charge is dispersive.

[1] Tabak, M., Conf. on Low Mobility Materials Israel, April 1971.

H. Moroson: Are there conditions of structure, doping or field wherein present day organic photoconductors may have a photographic efficiency approaching that of Selenium or Zinc oxide?

D. L. Stockman: Yes. The PVK-TNF photoreceptor recently described by Schaffert (ref. 50) is the same "speed" as Selenium within a factor of about 2. It also can be used in the cycling mode.

H. Meier: What is your opinion on the influence of trapping on mobility? Do you agree with the Shockley-Read model in organic photoconductors?

D. L. Stockman: I have outlined in my talk three facts about the transport properties of PVK which, taken together, seem to argue strongly against multiple trapping models. There is not sufficient data yet available to assess the generality of the PVK observations. Clearly, in materials like polystyrene, rapid trapping occur, but is is well known that these materials are not of practical use.

H. Meier: What is your opinion on the generation-step in dark conductivity? We could show that in doped PVK the rules of solid state physics can be used for explaining e.g. the relation between thermoelectric power and conductivity. What do you think about the hypothesis of doping?

D. L. Stockman: The best discussion that I have seen of dark decay is due to Sharp and Smith (ref. 75). They demonstrate quite conclusively that, excepting strong charge transfer complexes, dark conductivity in these materials is not intrinsic. Our charging measurements on relatively pure PVK layers suggest capacitive charging, with proper thickness scaling laws obeyed, at fields of up to at least 5×10^5 V/cm. With respect to doping, I believe that certain types of doping could very likely produce bulk generated carriers in the dark, but we have not observed it in undoped layers.

H. Meier: Some experiments of ourselves (on PVK) demonstrate that with the aid of recombination models many facts can be explained without taking into account excitons, for instance, the disagreement and agreement, resp., between action spectrum of photoconductivity and absorption spectrum observed on PVK layers be mentioned. What is your opinion?

D. L. Stockman: The reason why conventional recombination cannot account for the photogeneration have been discussed in some detail by Warter [24]. The reasons given by him seem appropriate to the materials I have discussed. The response curves for discharge of PVK layers reported by Regensburger [45] and the agreement with the absorption spectrum suggest as I noted, an extrinsic mechanism.

Dependence of Photoconductivity on Molecular Weight of Poly-N-Vinyl-Carbazole

R. H. Epping

Abstract

Conductivity of Poly(N-vinyl carbazole) – LUVICAN M 170 – is measured with coatings of different molecular weight ($3 \cdot 10^5$–$7 \cdot 10^6$). To ensure good adhesion of coatings to metal a method is applied that causes the thinner to evaporate preferentially in a tangential direction.

If there are local impurities of higher conductivity, metal contacts will show strongly varying conductivity values due to the free movement of charge carriers within the electrode area (Sandwich type).

Coronoid contacts will avoid this effect. But a relationship between dark–and photo-conductivity and the amount of charge having been injected during electrostatic charging can be found. Positive and negative charging will result in different characteristics which seem to be due to hole conduction and negatively charged traps. In this case with longer electrostatic charging of positive polarity an ohmic contact is formed, whereas with negative polarity the originally present blocking contact will be maintained.

Boeck's breakdown mechanism of electronic insulators by AC-corona stress is applied to the DC-corona. In this way one arrives at an equation that allows to determine the ratio between trap concentration and trap occupation probability. Dark conductivity during electrostatic charging and photoconductivity after charging as well were found to be independent of molecular weight. Therefore traps and centres of recombination will not preferably accumulate at the ends of the molecular chains.

Experimental Results and Discussion

An investigation of the problem whether the photoconductivity of PVCA is influenced by molecular weight is of interest, because the ends of the molecular chains can act as active centres in connection with the carrier transport mechanism. Also in the technology of preparing films for electrophotographic layers the dependence of photoconductivity on molecular weight is important, because the speed of solution, the bonding strength and the adhesion decreases with increasing molecular weight.

Figure 1 shows a device for evaporation of the PVCA solution in vertical direction. With this process we found only a poor adhesion of the film to the base. It was possible to avoid a strong shrinking of the material during the drying by raising the temperature and varying the solvent, but the adhesion was not improved.

We obtained very good adhesion, when we prepared layers with the dipping pulling process (Figure 2). This method for obtaining good adhesion for PVCA-layers was first reported in a contribution from the Battelle-Institute. To reach a thickness in the order of 5–10 μm, we need solutions with high viscosity.

It was necessary to vary this method, because only small quantities of fractionated PVCA were available. In Figure 3 is shown a closed container in which the base plate of the PVCA layer has an angle of about 30° to the horizontal.

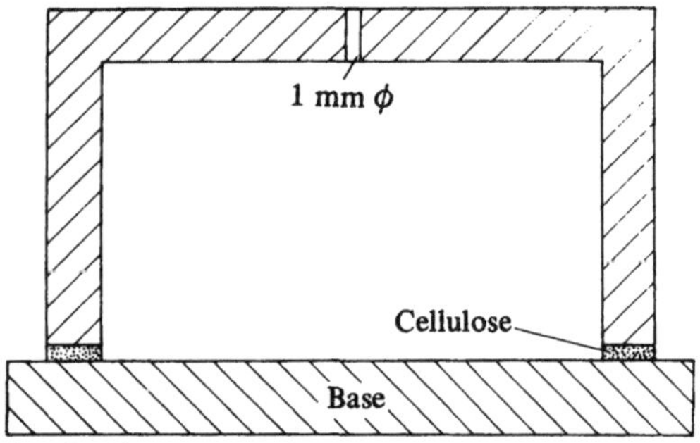

Fig. 1. Evaporation in vertical direction

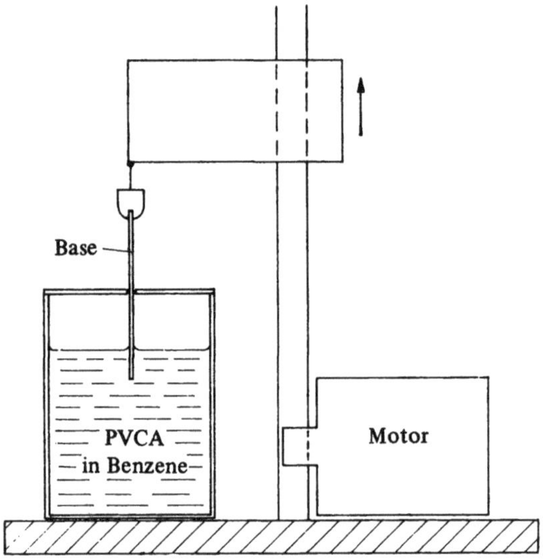

Fig. 2. Pulling dipping process

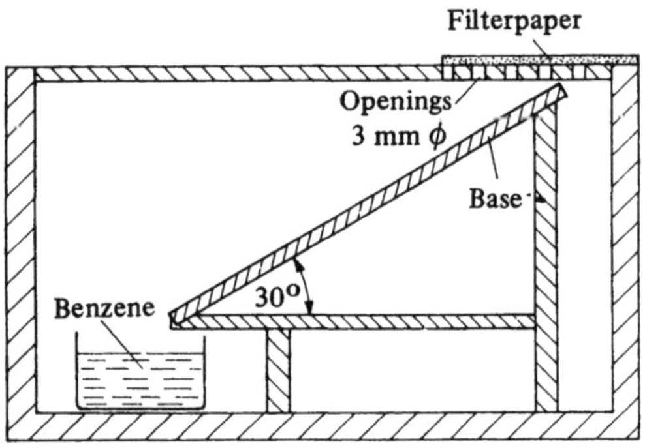

Fig. 3. Variation of pulling dipping process

A filled trough with solvent at the lower level of the layer generates a higher vapor pressure of the solvent than at the upper side. Here some openings are provided partially closed by filter paper. We used only about 1 cm^3 solution for a layer of about 5 cm in diameter and a thickness of 5 μm. The preparation time was between 15 and 30 minutes. Drying of the layer occurs continuously from the upper to the lower level. The adhesion was nearly the same as with the dipping process.

For base plates we used quartz with evaporated aluminum electrodes.

First we measured dark- and photocurrents in "sandwich"-cells, using a second electrode prepared by evaporating aluminum in vacuum. These metal contacts in "sandwich"-cells led to non-reproducible dark currents, however, produced by local impurities of higher conductivity. Because the ratio of the molecular weight of the fractions was measured over the range 10 : 1, we needed good reproducible

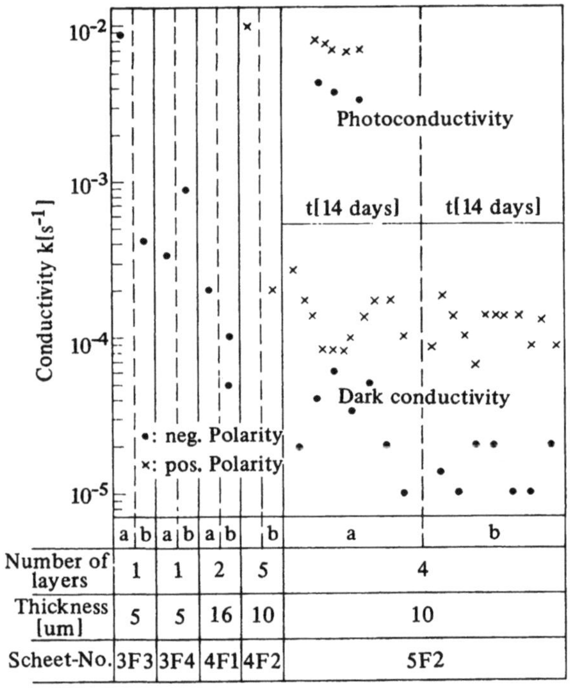

Fig. 4. Dark- and photoconductivity with metal contacts

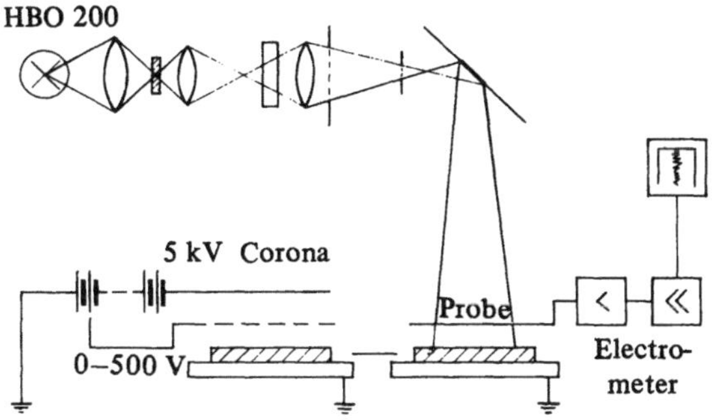

Fig. 5. Measurement of corona current after charging

measurements from layer to layer to determine the dependence of photoconductivity on molecular weight. As it is shown in Figure 4, the currents also changed when multi-layers were measured.

Metal contacts show strongly varying conductivity values due to the free movement of charge carriers within the electrode area. Coronoid contacts avoid this effect.

In Figure 5 is shown the apparatus for this measurement. There is a corona electrode in a fixed position with three wires at a distance of 10 mm. A metallic screen grid with 6.400 mesh/cm^2 serves to produce a uniform electrostatic charge and to limit the maximum charge.

We exposed with a mercury high pressure lamp combined with double band interference filters. The surface potential was measured by an induction electrometer with semitransparent electrode.

Only positive charging gives an exponential light decay curve. The example in Figure 6 shows this between 100 and 1 Volt surface potential. Negative charging led to space charge effects and the light decay curves are not exponential.

The determination of molecular weight was made with an automatically operating membrane osmometer. The results are shown in Figure 7. Six fractions between 300.000 and 4 Million were prepared; the last value has an uncertainty of a factor of 2.

The results are shown in Figure 8. The curves 1, 4 and 5 represent the photoconductivity

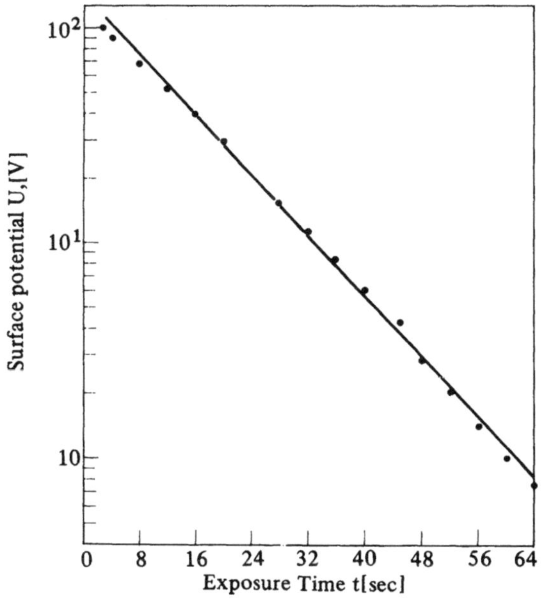

Fig. 6. Light induced charge decay with positive charging

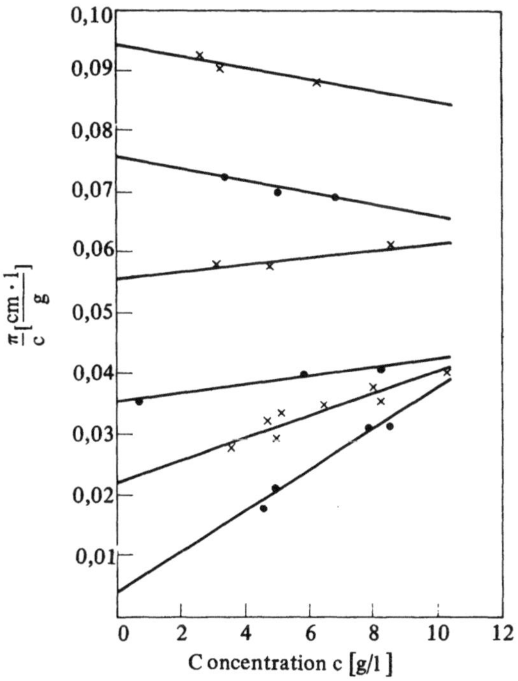

Fig. 7. Determination of molecular weight by osmose

after positive charging, using freshly prepared sheets.

Curve 4 shows results when layers were subjected to repeated positive and negative charging cycles including exposures.

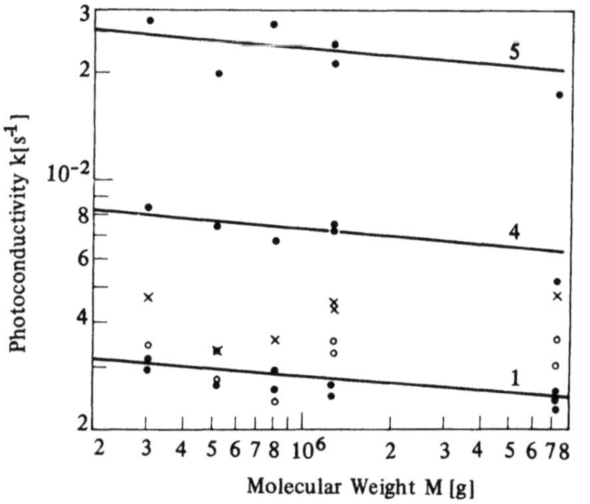

Fig. 8. Dependence of photoconductivity on molecular weight

Measurements on layers after very long charging and high light exposure are shown in curve 5.

The conclusion is that photoconductivity of PVCA seems to be independent of molecular weight. Therefore traps and recombination centres for this process are not preferably located at the ends of the molecular chains.

Discussion

J. Rochlitz: I would like to know, how you have prepared the different PVCas with different molecular weights and how you have measured the molecular weights.

R. H. Epping: Fractions of PVCA was made by solvent previpitation. We took Benzene as solvent and Methanol as previpitation liquid. Molecular weight was measured by an automatically operating osmometer.

R. Moroson: Was the increase in field caused by strong light exposure of the polyvinylcarbazole ever reduced following storage in the dark?

R. H. Epping: Curves 4 and 5 of Figure 8 were measured after dark adapting of the pre-exposed layers.

This paper is an abstract of the thesis: „Zur Frage der Abhängigkeit der Photoleitung vom Molekulargewicht beim Poly-N-Vinylcarbazol.“ Universität München 1970.

Electrophotographic Behavior of Bromopyrene Resin Acceptor Complexes

Bernd Löhr, Reinhold Arneth and Detlef Winkelmann

Abstract

Formaldehyde condensation of mono-bromopyrene yields a photoconductive polymer. The spectral sensitivity of electrophotographic layers comprising this photoconductor depends from the polarity of charge and is limited to the blue respectively to the blue and UV-region.

By activation with suitable acceptors, layers of high electrophotographic sensitivity are obtained which can be cycled several thousand times.

The electrophotographic behaviour of complexes with nitrated 9-fluorenones and nitrated 9-dicyanomethylene-fluorenes is described in more detail. With increasing electron affinity of the acceptors, the maximum of light sensitivity is shifted to longer wavelengths and broad plateaus are formed.

The values of quantum yield increase with increasing electric field. Maximum quantum yields of 0.6 have been measured.

From light decay curves obtained by strongly absorbed light, it is concluded that bromopyrene resin and its charge transfer complexes with the acceptors investigated are p-type material.

The discharge process of the layers is explained on the basis of a range limited model.

The range of holes during the discharge process amounts to 3.5–4 μ, that of electrons to about 1–1.5 μ.

1. Introduction

The photoconductivity of some polycyclic aromatic systems like anthracene, tetracene, pyrene and perylene [1] has been studied extensively and has been described in the literature in detail. It has been shown, that the photoconductivity could be enhanced by the addition of so-called electron acceptors [2] like chloranil, tetracyanoehtylene or trinitrobenzene, and moreover the light sensitivity could be shifted, into the visible region of the spectrum. This enhanced light sensitivity in the visible region is very important for electrophotography. The sensitivity can be explained by the formation of so-called charge transfer-complexes, which are formed from the interaction of the aromatic hydrocarbons and the electron acceptors.

The charge transfer is responsible for the generation of the charge carries. An electron from an occupied orbital in the donor, like the polycyclic aromatics, is transferred to an empty orbital in the acceptor releasing the charge transfer energy E_{CT}. The charge carrier generation can be described by the following simplified equation [3].

$$(D \ldots A \leftrightarrow D^{+} \ldots A^{-}) \overset{h\nu_{CT}}{\rightleftharpoons} (D^{+} \ldots A^{-} \leftrightarrow D \ldots A) \qquad (1)$$

D = donor
A = acceptor

Whereas in the ground state of CT-complexes the nonionic form is usually more abundant, in the excited state the equilibrium lies on the side of the ionic form. Generally, the energy needed for the intermolecular charge transfer transition can be expressed by the following equation [3]

$$E_{CT} = I - E_A - E_C + R_E \qquad (2)$$

where I is the ionisation potential of the donor, E_A the electron affinity of the acceptor, E_c the coulomb attractive energy between A^- and D^+ ions and R_E represents the resonance energy of interaction between the state (D^+ ... A^-)-and (D ... A). In the first approximation one can say that the excitation or charge transfer energy depends on the two values I and E_A. At a constant value of I, the charge transfer energy can be reduced by increasing the electron affinity. This is equivalent to shift of the absorption spectrum to longer wavelengths. From equation (2) it follows that by means of electron acceptors with increasing electron affinity a spectral sensitization of electrophotographic layers should be possible.

2. Survey of the Photoconductive Systems Investigated

The possibility of extending the light sensitivity of electrophotographic layers into the visible spectrum by using electron acceptors with increasing electron affinity, can be tested using 3-bromopyrene-fomaldehyde condensation polymers. The condensation of bromopyrene with formaldehyde opens a fascinating possi-

Nr.	Structure of the Acceptor	Color of the Ct-Complex	E_A [eV]
I	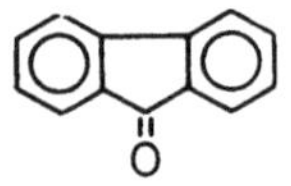	yellow	–
II	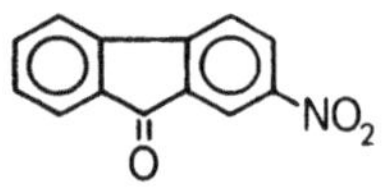	yellow	–
III	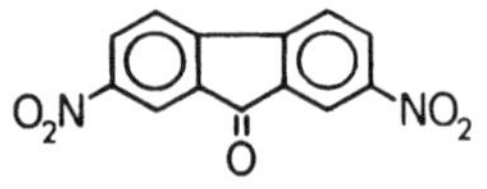	orange	0.765
IV	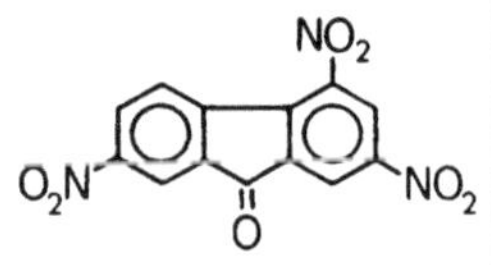	reddish brown	0.945
V	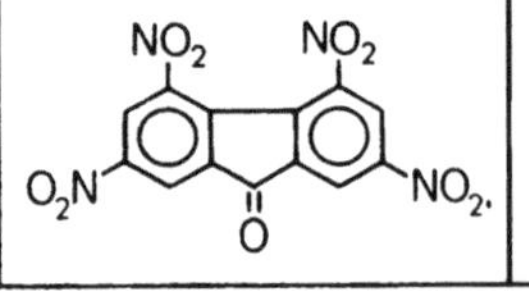	green	1.16

Table 1. Compounds derived from Fluorenone used as Electron Acceptors

bility to get very soluble CT-complexes. Pyrene or bromopyrene forms insoluble or only weakly soluble CT-complexes with most electron acceptors which are described in this report. This fact makes the application and investigation very difficult if not impossible. Naturally the use of the corresponding vinylpolymers is possible, but it has been pointed out that these polymers, which are synthesized with a considerably larger effort, did not provide any advantage over the condensation polymers.
The condensation of 3-bromopyrene with formaldehyde leads to very soluble polymers with the following structure, which could be confirmed by NMR-investigation.

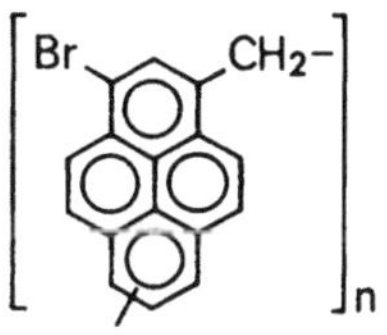

The condensation degree of n = 6–10 is sufficiently high to guarantee a very good solubility of the CT-complexes.

The electron acceptors used have been derived from fluorenone (Tab. 1) and from dicyanomethylene fluorene (Tab. 2). Their electron affinity was increased by successive nitration up to the tetranitro compounds.

Nr.	Structure of the Acceptor	Color of the CT-Complex	E_A [eV]
VI	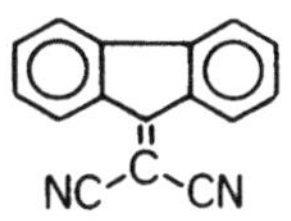	white red	0.70
VII	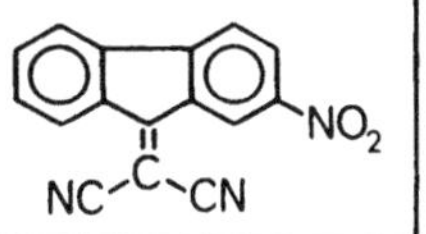	reddish brown	0.94
VIII	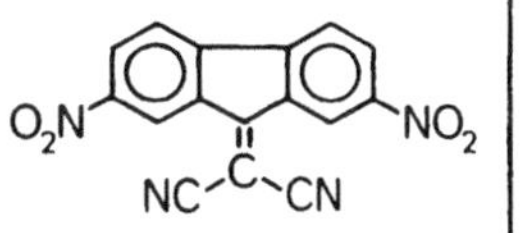	yellow green	1.03
IX	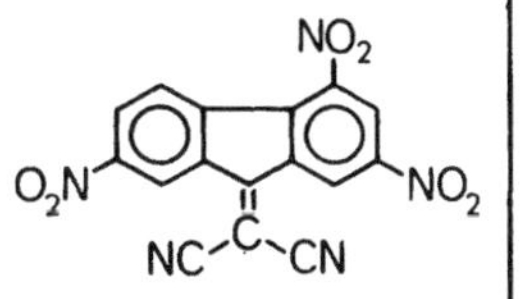	green	1.31
X	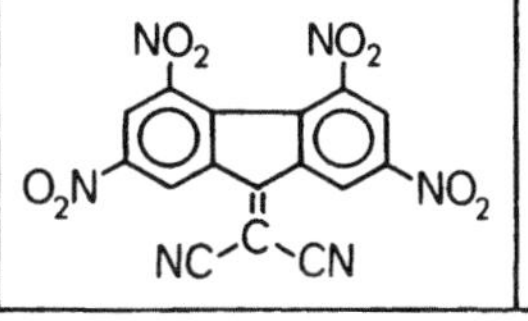	green	1.52

Table 2. Compounds derived from Dicyanomethylene Fluorene used as Electron Acceptors

Tables 1 and 2 illustrate the structure of the electron acceptors, the color of the CT-complexes in the photoconductive layer and the electron affinity as far as they are cited in the literature. The values of the electron affinity of fluorenone and 2-nitrofluorenone could not be found, but there is no doubt that the succession indicated is correct.

As is seen from the tables, the color of the CT-complexes changes from yellow to green with increasing electron affinity. At the same E_A the same color of the CT-complexes is to be expected. Thus the relationship of equation (1) can be verified visually by two independent acceptor families. The proof is given by the following optical and electrical investigation.

3. Absorption Spectra

The absorption spectra of the photoconductive layers investigated are shown in Figures 1 and 2.

Light-absorption measurements have been carried out using layers coated on polyester foils Hostaphan®. The thicknesses of the electrophotographic layers ranged between 3 μ and 15 μ. The absorption coefficient $K = \frac{1}{d} \ln \frac{I_0}{I}$ (d = = layer thickness, I_0 = incident, I = transmitted light intensity) has been plotted against wavelength.

Curve a represents the absorption of the pure bromopyrene resin layer. As could be expected from the yellow color this layer absorbs in the blue region of the visible spectrum. Activation with the acceptor fluorenone causes a weak charge transfer band which is continuously broadened due to increasing nitration of the fluorenone molecule.

Similar but broader absorption bands are obtained using dicyanomethylene fluorene and its nitrated derivatives (Fig. 2).

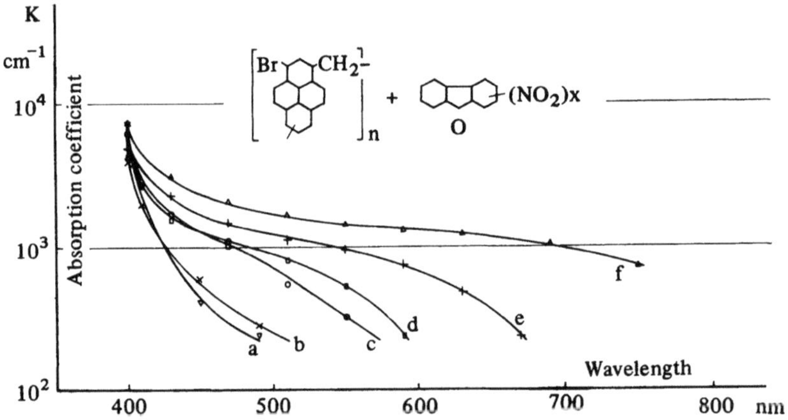

Fig. 1. Absorption in the longer wave-length region of the absorption edge of bromopyrene resin. Broad chargetransfer bands are formed

a) bromopyrene resin
b) bromopyrene resin + fluorenone
c) bromopyrene resin + mononitrofluorenone
d) bromopyrene resin + dinitrofluorenone
e) bromopyrene resin + trinitrofluorenone
f) bromopyrene resin + tetranitrofluorenone

In the case of b) to f) 0,1 mole acceptor has been added to 1 mole of the monomeric unit of bromopyrene resin

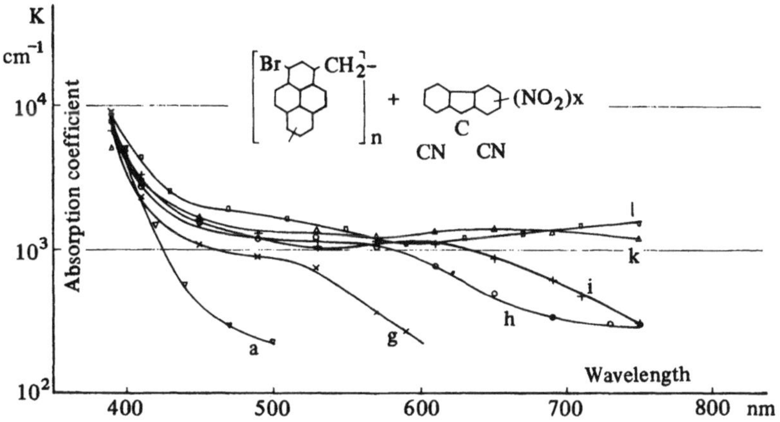

Fig. 2. Absorption in the longer wave length region of the absorption edge of bromopyrene resin. By adding higher nitrated dicyanomethylenfluorenes charge transfer bands originate in the near UV and extend into the infrared
a) bromopyrene resin
g) bromopyrene resin + dicyanomethylenefluorene
h) bromopyrene resin + dicyanomethylenenitrofluorene
i) bromopyrene resin + dicyanomethylenedinitrofluorene
k) bromopyrene resin + dicyanomethylenetrinitrofluorene
l) bromopyrene resin + dicyanomethylenetetranitrofluorene
In the case of g) to l) 0,1 mole acceptor has been added to 1 mole of the monomeric unit of bromopyrene resin

4. Spectral Sensitivity

In spectral sensitivity large differences arise in the short wave-length range with respect to the sign of charge of the layer. This fact is illustrated in the Figures 3 to 6. The reason for these discrepancies will be discussed later in section 6. The spectral sensitivity is expressed as the reciprocal of the light energy per unit area ($\mu Ws/cm^2$) to reduce the surface potential by 50% at a specified wavelength. This light energy equals $T_{1/2} \cdot I_0$ ($T_{1/2}$ = halftime of light decay curve). The light source was a xenon lamp, used with interference filters. The energy of radiation impinging on the surface of the layers ranged between 1 and 4 $\mu Ws/cm^2$. Measurements of spectral response have been carried with 6–9 μ thick layers coated on aluminized Hostaphan®.

4.1. Negatively Charged Layers

As can be seen from Figure 3 bromopyrene exhibits very small light sensitivity when negatively charged. Spectral response is limited within the region between 400 nm and 500 nm with a maximum at about 410 nm.

Addition of 0.1 mole fluorenone per mole bromopyrene unit in the polymer causes a considerable enhancement and broadening of the spectral sensitivity. This tendency continues up to the acceptor trinitrofluorenone, by application of which maximum photosensitivity is reached in this series. In the presence of tetranitrofluorenone the over-all sensitivity decreases again but the extension of the spectral response still persists into the infrared. From these results the following conclusions can be drawn: With increasing electron affinity of the acceptor, the spectral range of sensitivity is monotonically expanded to longer wave-lengths. The short wavelenght cut off remains unchanged. The maximum of spectral response also shifts to longer wave-lengths or a broad plateau appears. The maxima of spectral response increase up to the tinitrofluorenone and then decrease due to further enhancement of electron affinity of the acceptor.

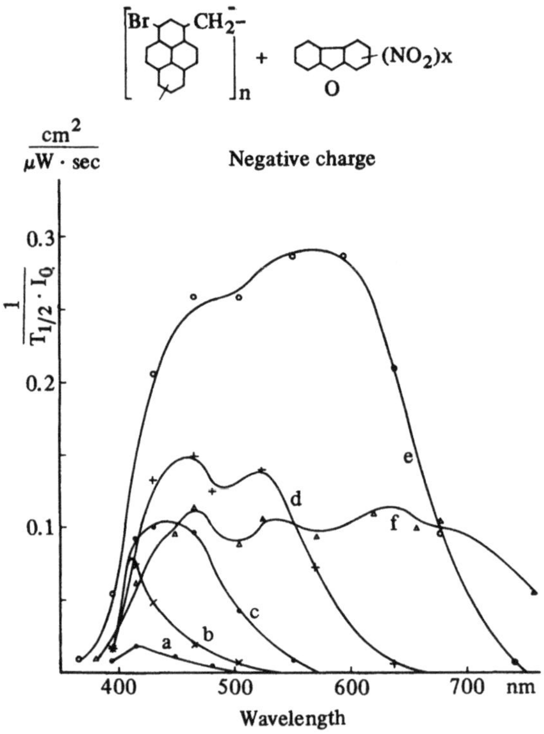

Fig. 3. Spectral sensitivity of bromopyrene resin + + fluorenone or the nitroderivatives of fluorenone. Negatively charged layers. Characterization of the curves by letters is the same as that of Fig. 1.

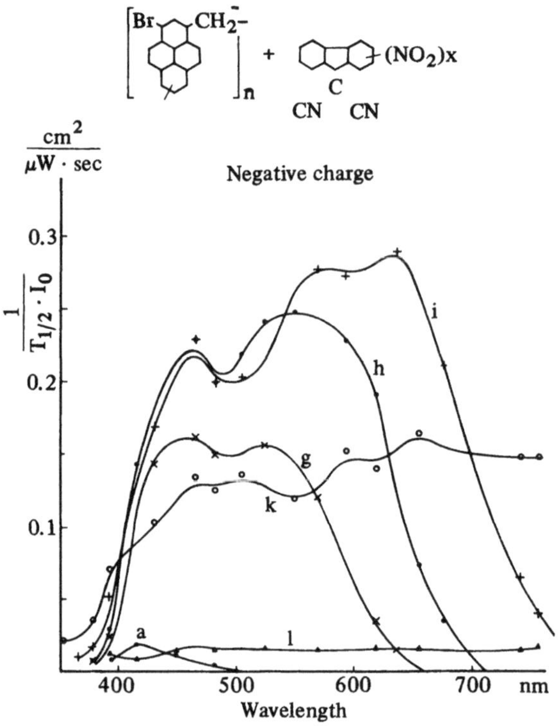

Fig. 4. Spectral sensitivity of bromopyrene resin + + dicyanomethylenefluorene or the nitroderivatives of dicyanomethylenefluorenes. Characterization of the curves by letters is the same as that in Fig. 2.

The same trend is observed with respect to the over-all sensitivity against unfiltered light of a xenon lamp.

With dicyano-methylene fluorene or the nitroderivatives of this similar results are obtained (Fig. 4). From the curves of spectral response we conclude that a CN-group as substituent has the same efficacy as a NO_2-group. Thus the curve of spectral response is almost the same using dicyanomethylenefluorene or dinitrofluorene. Furthermore the spectral sensitivity of dicyanomethylene-mononitrofluorene corresponds so that of trinitrofluorenenone. But the maximum of light sensitivity in the acceptor family of fluorenes is attained with dicyanomethylenedinitrofluorene.

At this point we must emphasizes that for comparison all investigations cited have been carried out in the 0.1 molar ratio of acceptor to donor. For each acceptor, however, the optimum molar ratio of sensitivity is quite different and ranges between 0.1 and 0.4. Nevertheless if for each acceptor we use its own optimum molar ratio, the maximum overall sensitivity to unfiltered light of a xenon lamp is obtained with trinitrofluorenone.

4.2. Positively Charged Layers

Contrary to negatively charged layers where the spectral response is vanishingly small in the UV i.e. in the region of strong absorption, we find almost the same or even higher sensitivity in this region using positively charged layers. In the range of weak absorption ($\lambda > 450$ nm) no large differences between the negative and positive mode can be detected.

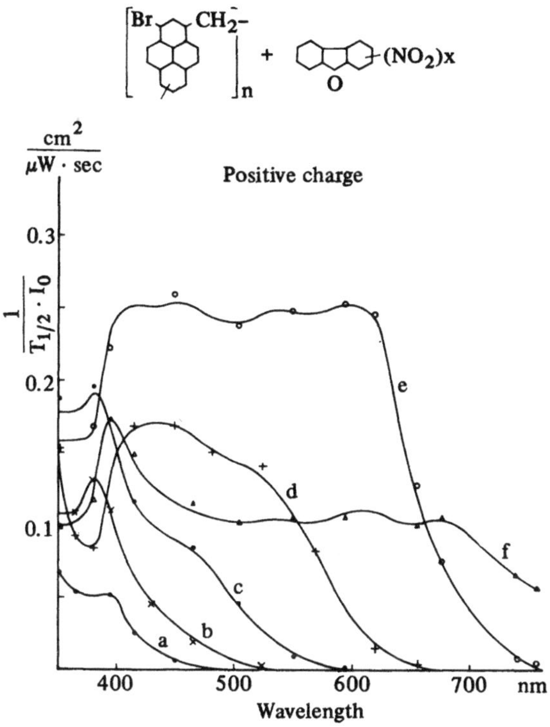

Fig. 5. Spectral sensitivity of bromopyrene resin + + fluorenone or the nitroderivatives of fluorenone. Positively charged layers. Characterization of the curves by letters is the same as that of Fig. 1.

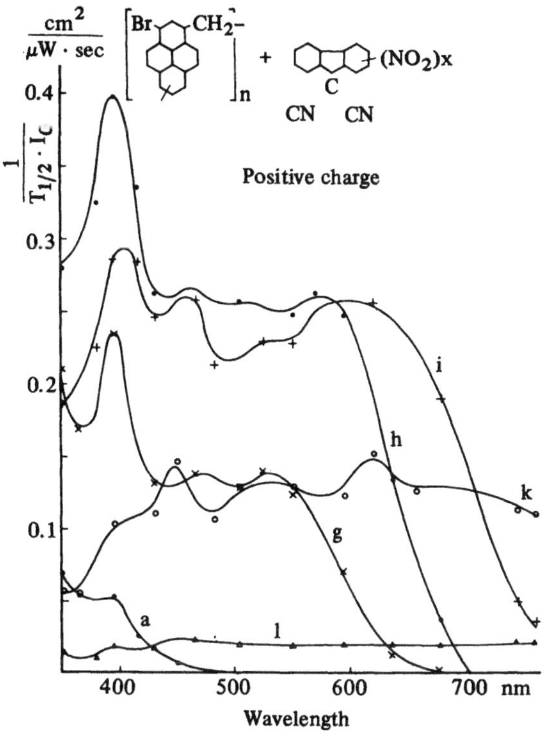

Fig. 6. Spectral sensitivity of bromopyrene resin + + dicyanomethylenefluorene or the nitroderivatives of dicyanomethylenefluorene. Positively charged layers. Characterization of the curves by letters is the same as that in Fig. 2.

5. Electrophotographic Quantum Yield

The electrophotographic quantum yield η is defined as:

$$\eta = \frac{\text{number of displaced elemental charges } N_e}{\text{number of absorbed photons } N_{ph}}$$

We use for determination of η the initial slope of the voltage discharge curve under illumination. Then the following relation holds:

$$\eta = \frac{\frac{C}{e} \cdot \left(\frac{dU}{dt}\right)_0}{\frac{I_{abs}}{h\nu}} \qquad (4)$$

where

C = capacity of the photoconductive layer per unit area (cm^2)

$\left(\frac{dU}{dt}\right)_0$ = initial voltage discharge rate

$\frac{I_{abs}}{e}$ = absorbed photon flux $\left[\frac{\text{photons}}{cm^2 \cdot sec}\right]$

e = elemental charge

The measured values of quantum yield increase with increasing electric field strength. In Figures 7–10 the values measured at maximum charge acceptance of the layers have been plotted as a function of wave length. Commonly the charge acceptance of organic photoconductive layers increases more slowly than linearly with increasing thickness. Therefore larger maximum η-values are obtained using thinner layers of the same photoconductor. As already mentioned, all layers investigated on their electro-

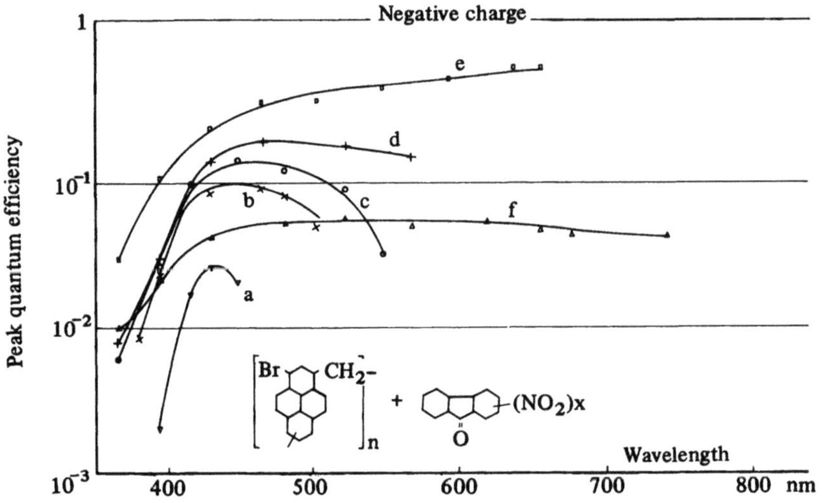

Fig. 7. Electrographic quantum efficiency of bromopyrene resin + fluorenone or the nitroderivatives of fluorenone. Maximum values of negatively charged layers. Characterization of the curves by letters is the same as that of Fig. 1.

photographic behavior were 6–9 μ thick. In this range deviations from proportionality of charge acceptance with layer thickness are very small. For this reason the values of the peak of the quantum efficiencies measured for the different photoconductors are directly comarable.

The shape of the curves for quantum efficiency is quite different from that of the curves for spectral sensitivity because quantum efficiency is referred to absorbed photons, whereas spectral sensitivity, to allow practical application, is referred to the energy of incident radiation.

The maximum quantum efficiency of negatively charged bromopyrene resin-layers amounts to 0.03 at 430 nm. Similar to spectral sensitivity the quantum efficiency increases with increasing

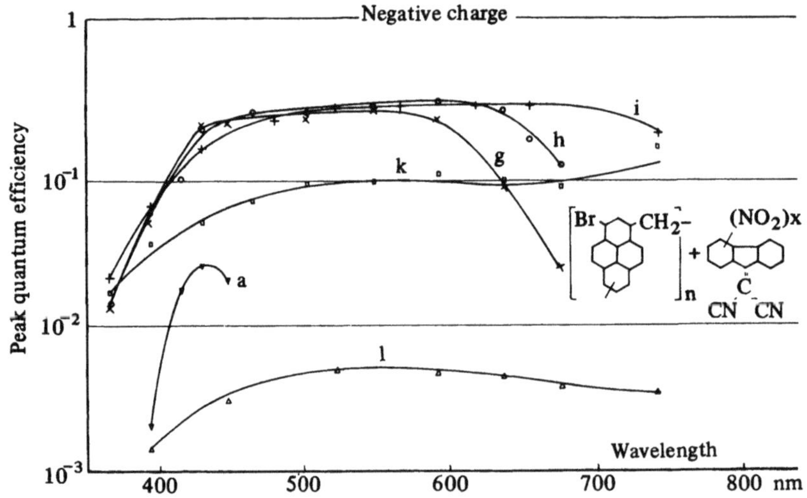

Fig. 8. Electrophotographic quantum efficiency of bromopyrene resin + dicyanomethylenefluorene or the nitroderivatives of dicyanomethylenefluorene. Maximum values of negatively charged layers. Characterization of the curves by letters is the same as that of Fig. 2.

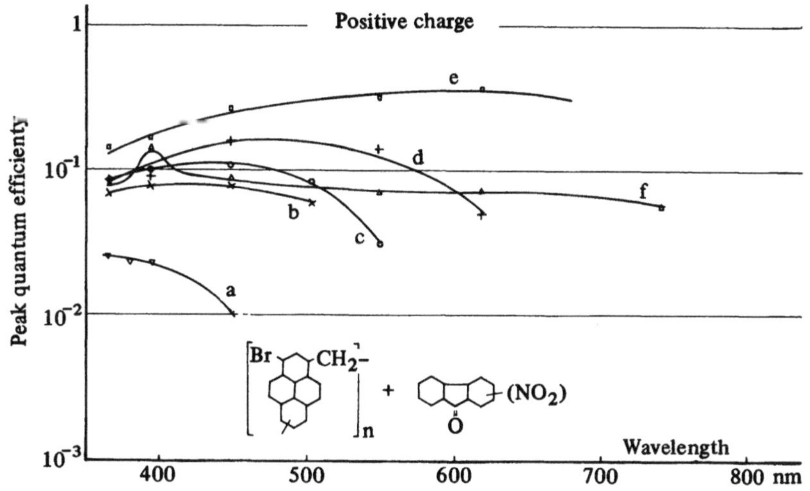

Fig. 9. Electrophotographic quantum efficiency of bromopyrene resin + fluorenone or the nitroderivatives of fluorenone. Maximum values of positively charged layers. Characterization of the curves by letters is the same as that of Fig. 1.

electroaffinity of the acceptor up to trinitrofluorenone (Figs. 7 and 9) respectively dicyanomethylenemono- or dinitrofluorene and decreases due to further nitration of the corresponding acceptors. The quantum efficiencies of layers containing dicyanomethylenetetranitrofluorene as acceptor are even smaller than those of bromopyrene resin. Remarkable is the abrupt drop in the UV-region at negative charge. Here we are faced with quite another situation when using positively charged layers. The quantum efficiencies of bromopyrene resin increase continuously to shorter wave lengths within the measured range. Moreover with nitrofluorenones as acceptors almost the same quantum efficiencies and with dicyanomethylenefluorenes in part even higher quantum efficiencies are obtained than in the visisble region.

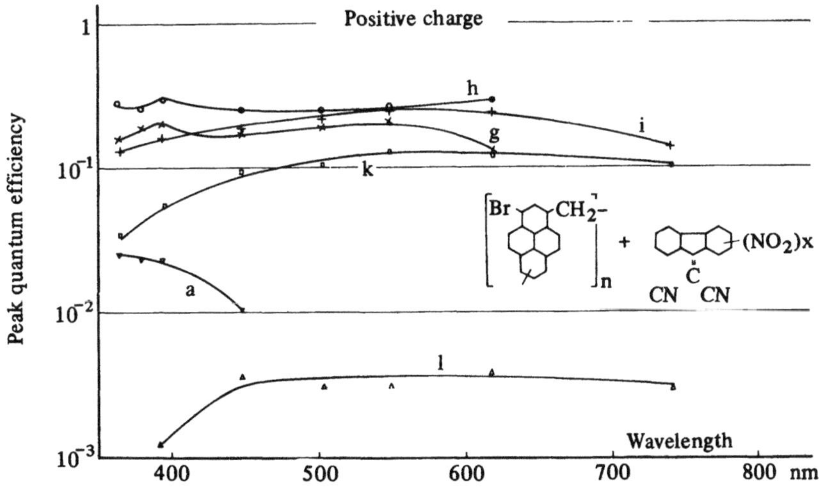

Fig. 10. Electrophotographic quantum efficiency of bromopyrene resin + dicyanomethylenefluorene or the nitroderivatives of dicyanomethylenefluorene. Maximum values of positively charged layers. Characterization of the curves by letters is the same as that of Fig. 2.

6. Conductivity Type and Transport Properties of the Charge Carriers

The type of the conductivity of electrophotographic layers can be determined from the light decay curves using strongly absorbed light (penetration depth ≪ thickness of the layer) for exposure. This fact is explained in Figure 11.

The photons absorbed in the photoconductor surface create electron-hole pairs. In the case of negatively charged layers the holes neutralize the negative surface charge while the accompanying electrons are forced by the influence of the electric field to move towards the backside of the layer. The light decay curve is due only to the transport of electrons. Analogously one observes the transport properties of holes in the case of positively charged layers.

In a special experiment, coatings have been made using the photoconductor bromopyrene resin/dicyanomethylenefluorene (in the 0.1 molar ratio of acceptor to donor) with layer thicknesses of 7 μ, 3.5 μ and 1.5 μ. These layers have been provided with the desired charge in the dark by a corona device. After charging each layer has been exposed to monochromatic radiation first of wave length 365 nm and subsequently of wave-length 577 nm. The decay curves thus obtained are shown in Figure 12.

The photon flux impinging on the surface amounted to $1.28 \cdot 10^{13}$ photons/cm² · sec at 365 nm and to $7.1 \cdot 10^{12}$ photons/cm² · sec at 577 nm. The penetration depth of light with 365 nm is less than 1 μ. If the negatively charged 7 μ thick layer is exposed to this radiation, (Fig. 12 left row above) we observe a relative fast initial discharge rate, which is immediately slowed down. Even after 20 sec exposure the surface potential has been reduced to only 2/3 of its initial value. On subsequent illumination with monochromatic light of 577 nm and with essentially lower number of photons there is a rapid discharge within few seconds.

The 3.5 μ thick negatively charged layer is already discharged to 25% of its initial potential when illuminated with the UV-radiation of 365 nm and finally the 1.5 μ thick layer is completely discharged (Fig. 12 left row beneath).

At positive charge acceptance, the 7 μ thick layer is already discharged to 25% of the initial potential (Fig. 12 right column above) and the 3.5 μ thick layer is completely discharged. The same results as described above are obtained with all other layers. Therefore the following conclusions can be drawn:

Bromopyrene resin and its CT-complexes with the aforementioned acceptor families are p-type material. The range of electrons is about 1–1.5 μ. In negatively charged layers thicker than 1.5 μ, electrons created by UV-excitation are trapped in the bulk. These trapped electrons can be freed by weakly absorbed radiation (volume excitation!). The range of the holes

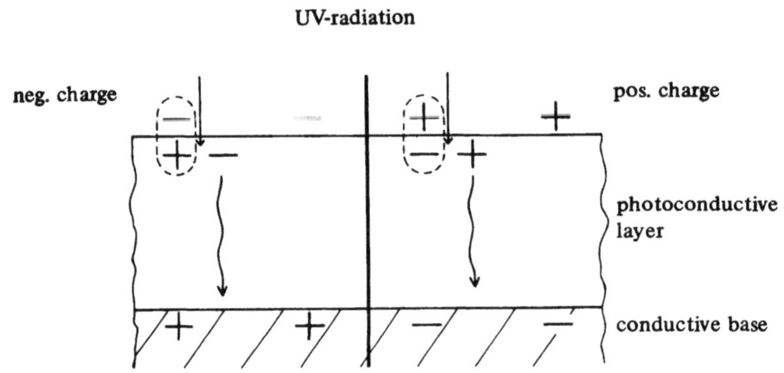

Fig. 11. Schematic illustration for the method of determining conductivity type.

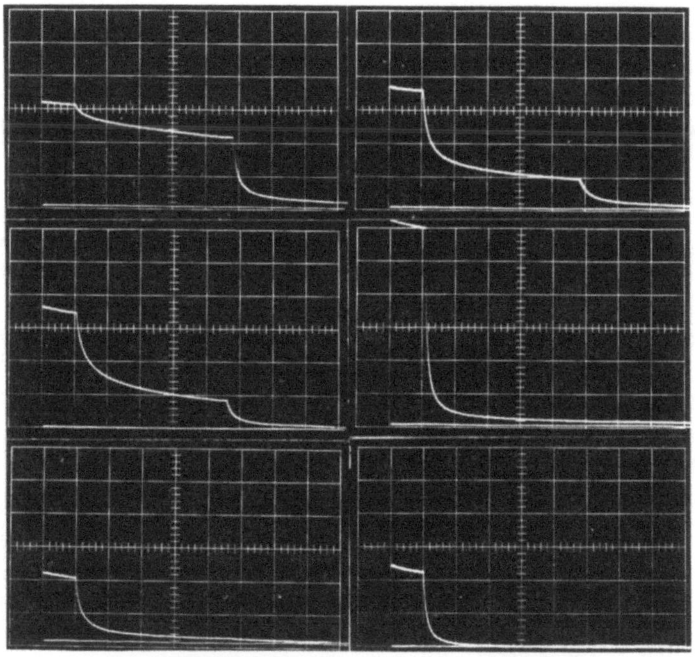

Fig. 12. Light decay curves of the layer bromopyrene resin/dicyanomethylenedinitrofluorene. Oscilloscope pictures. Layer thicknesses:

above: 7 μ; middle: 3.5 μ; beneath: 1.5 μ.
Left row: negative charge; Right row: positive charge.
Vertical deflection above: 200 V/division
middle: 100 V/division
beneath: 50 V/division
Horizontal deflection: 5 sec/division in all cases.

At the points marked with arrows (↓) light of wavelength 365 nm $\left(1.28 \cdot 10^{13} \frac{\text{photons}}{\text{cm}^2 \text{ sec}}\right)$ or 577 nm $\left(7{,}1 \cdot 10^{12} \frac{\text{photons}}{\text{cm}^2 \text{ sec}}\right)$ has been applied

is about 3.5–4 μ. This is 3 times larger than that of the electrons.

With the knowledge of these facts the very small light sensitivity of negatively charged bromopyrene resin layers in the UV can be explained. It is due to the short range of electrons because of bulk trapping.

Furthermore the nearly unchanged spectral response of positively charged layers in the UV and in the visible region is understandable, because the range of holes is comparable with the layer thickness used. Therefore we get the same discharge curve using strongly or weakly absorbed light.

There is still another effect that can be realized from Figure 12. Using the 7 μ thick negatively charged layer there is still some noticeable discharge process even after long exposure times. This discharge process is considerable faster than dark decay, as could be verified by switching off the UV-radiation. It can not be caused by longer wave-length radiation because the interference filter for 365 nm has been applied in combination with a 2 mm thick UG 2-filter and a 3 mm thick KG 3-filter from Schott & Gen. We also exclude the possibility that UV-radiation was able to penetrate into the bulk material through cracks and fissures because the coatings looked very homogeneous. For these reasons we incline toward the following explanation:

After all surface charge carriers have penetrated into the bulk, the electron-hole pairs that are still generated in the surface by the incident UV-radiation are in a field-free space. They can only migrate into the volume by diffusion. This diffusion may occur in the form of excitons. In the bulk the excitons get into the influence of the electric field by which they are separated (Fig. 13). After dissociation the hole recombines with a trapped electron (if the volume charge is negative) while the corresponding electron of the separated exciton moves to the backside of the layer and vice versa. By this process the observed slow decrease of voltage on further UV-illumination can be explained.

Upon subsequent irradiation of light with wavelength of 577 nm, the penetration depth of which is larger than the thickness of the layer, photons are absorbed also in the range of trapped carriers. Consequently elimination of the volume charge can occur rapidly and a fast and complete discharge of the layer is possible. In the same manner, the discharge process of UV-irradiated positively charged layers can be explained, the thickness of which is larger than the range of holes.

The experiments demonstrated in Figure 12 show that in bromopyreneresin-layers excited with strongly absorbed radiation charge carriers are trapped in the bulk if layer thickness is larger than 5 μ. Therefore, it is reasonable to explain the field-dependence of quantum efficiency as being caused by the field dependence of mobility and recombination with traps for the charge carriers.

Tabak and Warter [4] concluded from their measurements with amorphous selenium layers that in this material charge carrier generation by absorbed light is field dependent and thus causes the field dependence of quantum efficiency. This effect may exist in bromopyrene resin layers too but it can not be predominant. If one assume that the discharge process of a negatively charge 7 μ layer due to irradiation of strongly absorbing light is drastically slowed down—as demonstrated in Figure 12—because of lack in field—aided photogeneration, the same argument should have been valid for subsequent irradiation of weakly absorbed light. But this is not true! Although only 55% of the

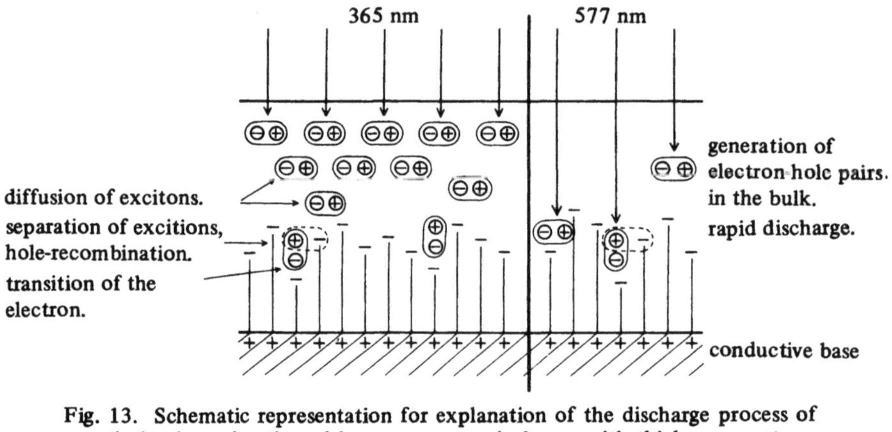

Fig. 13. Schematic representation for explanation of the discharge process of negatively charged activated bromopyrene-resin layers with thickness greater than the range of electrons when illuminated with radiation of 365 nm

UV-photon flux was applied with light of 577 nm, a fast voltage decrease was observed at reduced electric field strength, if no bulk trapping is assumed.

7. Synthesis of the Materials

3-bromopyrene was synthesized by a procedure of Org. Synth. [5]. With this compound the 3-bromopyrene resin layer was prepared as follows:

40 g 3-bromopyrene and 4.5 g paraformaldehyde were dissolved in 400 ml glacial acetic acid at 95–100° C. To the stirred solution 2 g $ZnCl_2$ (anhydrous) were added. After a short time the polymer began to precipitate out and the condensation was complete in 3–4 hours. The polymer was successively washed with water and methyl alcohol. After drying the polymer was dissolved in THF, filtered and reprecipitated from methyl alcohol. This purification was repeated twice.

Nitrofluorenones were synthesized and purified as described in the literature [6].

Dicyanomethylfluorene compounds were prepared by a procedure published by Mukherjee [7].

The authors would like to thank Miss B. Sindermann for her valuable efforts in the preparation of samples and J. Emig and H. Trittler for their skilled technical assistance and performance of measurements.

References

[1] Kommandeur, J., J. Phys. Chem. Solids **22**, 339 (1961).
Inokuchi, H., and K. Ohki, Bull. Chem Soc. Japan **36**, 105–7 (1963).
Iguchi, K., J. Phys. (Paris) **24**, 367–73 (1963).
Mulder, B. J., Rec. Trav. Chim. 713 (1965).

[2] Kuroda, H., K. Yoshihava and H. Akamatsu, Bull. Chem. Soc. Japan **36**, 1365–6 (1963).
Akamatsu, H., and H. Kuroda, J. Chem. Phys. **39**, 3364–8 (1963).
Tobin, M. C., and D. P. Spitzer, J. Chem. Phys. **42**, 3652–8 (1965).

[3] Briegleb, G., „Elektronen-Donator-Akzeptor-Komplexe" Springer-Verlag Berlin 1961.

[4] Tabak, M. D., and P. J. Warter, Phys. Rev. **173**, 899 (1968).
Tabak, M. D., Applied Optics 1969, Supplement on Electrophotography, 4–7.

[5] Org. Synth. V **48**, page 30 (1968).

[6] Schultz, G., A **203**, 103.
Ray, Francis, J. Org. Chem. **8**, 52 (1943).

[7] Mukherjee, T. K., and L. A. Levasseur, J. Org. Chem. **30**, 644 (1965).
Mukherjee, T. K., J. Phys. Chem. **70**, 3848 (1966).

Aging of Organic Photoconductors

W. Lanker

Abstract

An introductory literature survey deals with various aging mechanisms occurring in organic materials. In reusable organic electrophotographic PVCa layers corona aging (by corona charging-light discharging cycles) leads to a decrease of charge acceptance V_0 and dark resistivity ρ_0. The influence of humidity and corona aging on V_0 and on recovery effects, i.e. variations of V_0 after interruption of aging, was investigated. A pronounced decay in the humidity dependence V_0(R. H.) appears with aged layers (humidity decay). From these results and further experiments including surface treatments and peeling off a top layer after aging, it was concluded, that a chemical aging mechanism, due to the gas discharge of the corona, leads to decomposition of surface material and, upon adsorption of water, to the formation of a thin surface layer with altered electrical properties, which is responsible for the decrease in V_0 and ρ_0. The electrical properties of the bulk, e.g. the photoconductivity, on the other hand remain comparably unaffected. This aging mechanism appears not to be restricted to the model substance PVCa only.

1. Introduction: Survey on Aging Mechanisms in Organic Material

If we understand by aging in a very general sense, all influences and actions which lead to changes in organic material—especially in polymers—and its electrical properties, an immense amount of information can be found in literature on such fields as insulation technique, electronics and macromolecular chemistry. Some typical titles are given in references 1–15. Based on literature studies it is tried to give a short survey and to sketch a general aging scheme, which might be relevant with respect to possible aging mechanisms in organic electrophotographic layers, in order to allow rough localization and discussion of electrophotographic aging, which was investigated on the example of polyvinylcarbazole (PVCa).

1.1. General Aging Scheme

As shown in Table I, a general aging scheme is outlined here by ordering its elements into the following four main categories: aging mechanisms; parameters of influence on these mechanisms; material and its parameters: and finally resulting changes in material and properties due to aging. A more detailed description of the aging mechanisms and their effects is given in Table II where the following groups are distinguished: Origins of aging A producing various aging elements AE which, by the aging mechanisms AM, give rise to the actions W in the organic material with the effect of corresponding changes in its properties. Connections between the groups A, AE, AM and W are indicated by arrows. It has to be mentioned that this classification is not strict, since e.g. the material parameters can also be parameters of influence. Some connections in Table II are further described in the next section.

Table 1 General Aging Scheme

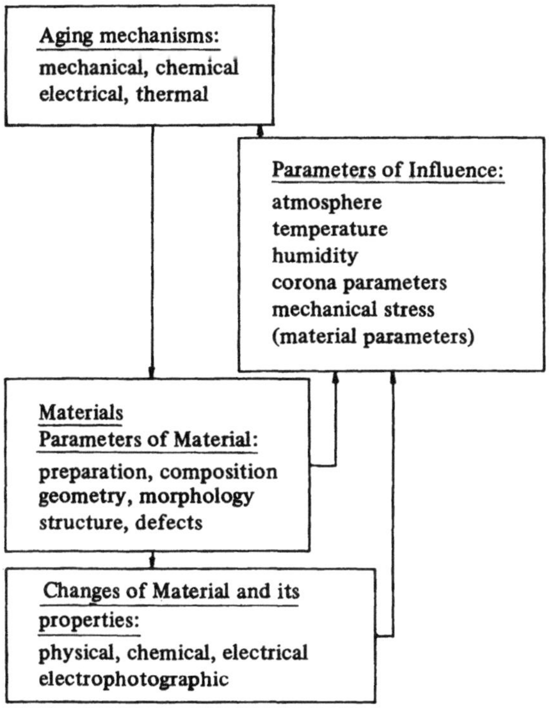

1.2. Aging Mechanisms

1.2.1. Aging by Externally Applied Energy

a) *Mechanical aging:* mechanical energy of particles (e.g. charge carriers, toner) and mechanical stress can lead to erosion of the surface, induce cracks and defects, where secondary effect (e.g. gas discharges) may occur. Chemical and structural changes can result, too.

b) *Chemical aging:* chemically active agents (e.g. reactive gases and UV light from gas discharges) cause reactions, decomposition and structural changes of the material.

c) *Electrical aging:* application of electrical field and charge–as well as photoeffects–can result in the formation of space charge and boundary layers and give rise to trapping effects and changes in contact properties. Reversible memory effects and fatigue appear. In combination with thermal and chemical mechanisms also irreversible changes (e.g. breakdown) can occur.

Table 2 Aging – Mechanisms

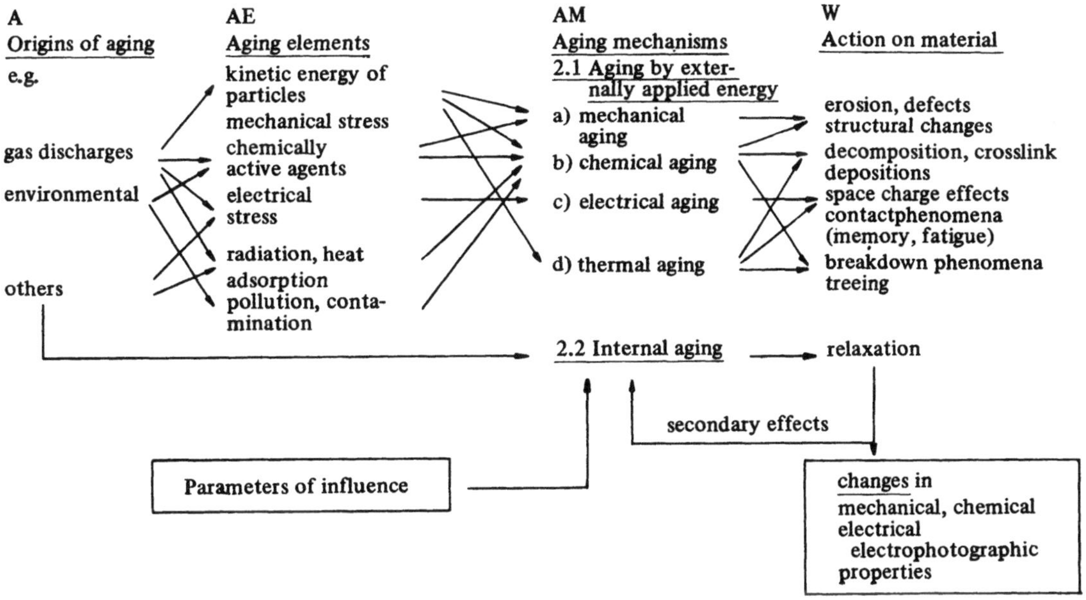

d) *Thermal aging:* application of thermal energy can lead to chemical reactions, structural changes, relaxation of stress and rearrangement of molecules.

1.2.2. Internal Aging

Changes in organic layers also occur without externally applied (additional) energy, simply by the effect of time: e.g. sorption phenomena, sedimentation, creep recovery and similar effects as with thermal aging.

The aging mechanisms and their effects on the material depend upon many parameters as illustrated in the Tables I, II. It is very importtant to note that the elements of this whole aging system (consisting of aging elements, aging mechanisms, effects, parameters of influence and material) are correlated in many ways, so that changing of one element of the system may lead to direct or secondary changes of any other element.

2. Electrophotographic Aging

In electrophotography aging effects due to processing of electrophotographic layers are well known under such terms as fatigue (charge-, plate- and light-fatigue), memory and ghost images, which are described by Schaffert [16], Dessauer [17] and many others [18–20]. Fatigue of selenium layers is mainly related to electrical mechanisms such as trapping and barrier effects, whereas oxygen sorption e.g. is supposed to be responsible for light fatigue of ZnO-binder layers. However, according to the literature mentioned in Part 1, it must be expected that, with organic materials, mechanisms which are distinctly different from the above mentioned memory and fatigue effects may play an important role. This was investigated with the model substance Polyvinylcarbazole (PVCa), which is well known in literature [21–24], serving as reusable organic layer. In the following the term aging will be used only for electrophotographic aging by corona charging–light discharging cycles.

3. Experimental (Standard Conditions)

Standard layers consisting of PVCa (Luvican M170 from BASF) sensitized with 3% of tetranitrofluorenone were coated from benzene solution on aluminum foils in a thickness of 5–6 μ. After a storage time of 3 to 8 weeks, the layers in a width of 90 mm were placed on drum 1 in the aging machine shown in Figure 1, where they were aged at a rate of 20 cycles/min. The layers were charged with negative polarity by the two wire corona 2. The resulting surface potential V was measured about 2 seconds later by means of the capacitive probe 4. Light discharge with an exposure of 2000 luxsec was carried out by incandescent light at position 3. The corona voltage was adjusted so that a constant charging current of 75 μA (through the layer) was maintained.

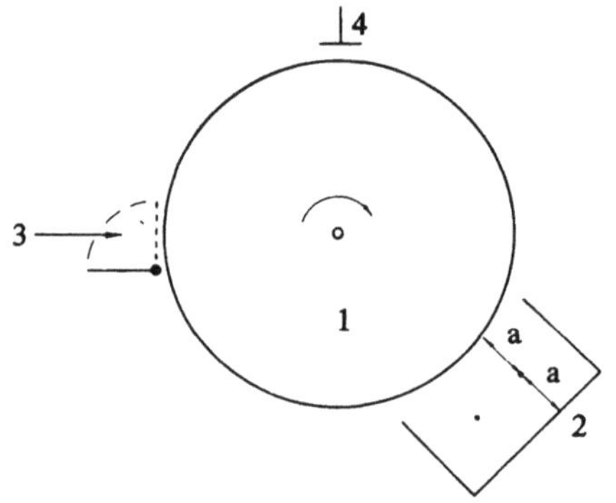

Fig. 1. Aging machine: (1) drum, diameter 150 mm; (2) corona, wire diameter 0.1 mm, a = 20 mm; (3) light exposure; (4) capacitive probe, diameter 15 mm

Hereby the resulting surface potential V after one corona passage almost reached the saturation value V_0 (charge acceptance). Standard temperature was 25° C.

4. Aging Effects

The main effect of aging appeared to be the following changes in the charge acceptance V_0 and in dark resistivity ρ_0–being characterized by the dark decay (T_0) to $V_0/2$:

a) Decrease of V_0 and ρ_0 with increasing number of aging cycles.
b) Recovery and instability of V_0 (and ρ_0): V_0 changes–in positive or negative sense (i.e. increase or decrease of V_0)–after interruption of aging for a certain recovery time Rt and again after restart of aging.
c) Inhomogeneity of V_0, i.e. local variations of V_0 along the layer appear. Due to the dimension of the probe, these variations could only be measured with a resolution of about 10 mm. Smaller inhomogeneities could be observed by toning the layers.

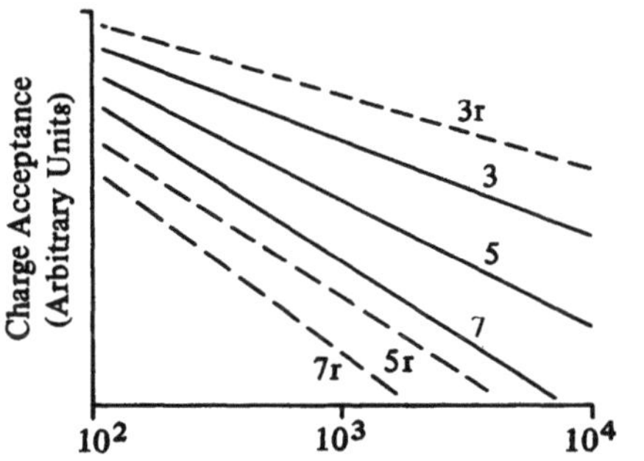

Fig. 2. Dependence of V_0 on the number of aging cycles: without interruption (solid lines), value after recovery for 1–3 days (dashed lines r) at the relative humidities 30% (3), 50% (5) and 70%(7) respectively

All these effects a)–c) show the general tendency to increase with the number of aging cycles (Fig. 2).

d) Photoconductivity and spectral response on the other hand remain comparably unaffected. This can be seen in Figure 3, where new and aged layers are compared with respect to their spectral response, which was determined by the half decay exposures (being about 30 luxsec for incandescent light). The measurements were carried out at a low relative humidity where V_0 was almost the same for both cases, so that the field dependence of the conductivity caused no additional errors.

e) The following influences on the copy quality result by this electrophotographic aging:
- reduction of contrast and resolving power
- instability and inhomogeneity of contrast (varying density, defects, points).

The aging behaviour can therefore be characterized well by the dependence of V_0 on the aging parameters, which is done in the following sections.

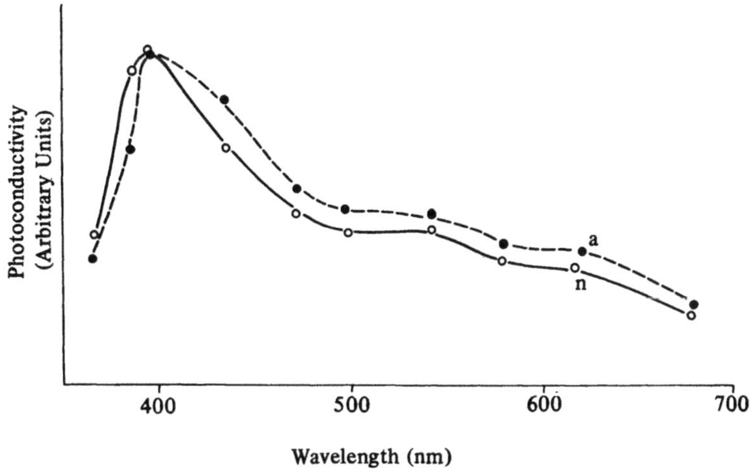

Fig. 3. Spectral dependence of the reciprocal half decay exposure, constant intensity 3 $\mu W/cm^2$, for new layers (n) and aged layers (a); m = 6000 cycles, Rt = 2 months

5. Influence of Humidity

Relative humidity (R. H.) was found to have a drastic influence on the aging effects with PVCa layers. It was tried to analyze this influence, by keeping all other parameters as constant and controlled as possible, while only the humidity was varied.

5.1. Dependence of V_0 on Aging at Various Relative Humidities

Variable: aging cycles
Parameter: R. H. = 30, 50, 70%

As can be seen from the schematic representation in Figure 2 and the examples in Figure 6, the decrease of V_0 with the number of aging cycles, as well as the recovery effect is much more pronounced at high humidities. The recovery effect is normally negative at R. H. higher than about 40% and Rt below about 5 days, but positive at low R. H.

An example for recovery and instability of V_0, after 5000 aging cycles at 50% R. H., is shown in Figure 4. There the following 3 time regions can be distinguished: a negative short time recovery (T_1 about 0–5 days) followed by a positive long time recovery (T_2 about 5–100 days) and an increase of V_0 upon restart of aging, after short time recovery, in about 10 minutes (T_3).

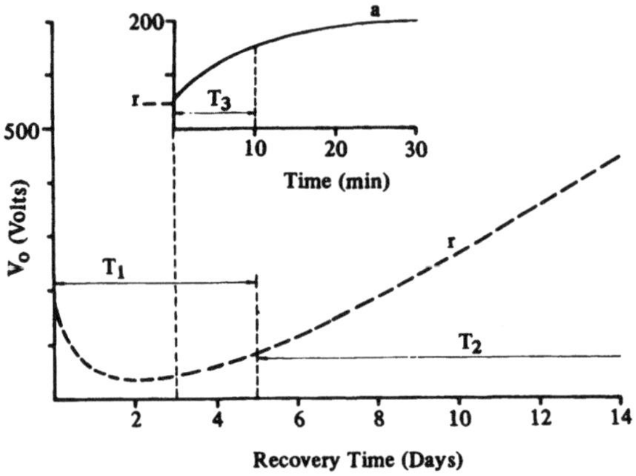

Fig. 4. Recovery and instability of V_0, i.e. V_0 dependence on time after interruption (r) and upon restart of aging (a) at Rt = 3d and 50% R.H.; previous aging was 5000 cycles at 50% R.H.

5.2. Dependence of V_0 on R. H. After Defined Aging

Variable: R. H.
Parameters: number of aging cycles and R. H. at which this aging was carried out.

After aging and recovery the layers were conditioned for at least 15 minutes at each R. H. before V_0 (R. H.) was measured.

V_0 of new layers (V_{00}) is relatively independent of humidity (only near 100% R. H. V_0 decreases sharply), whereas V_0 of aged layers decreases strongly with increasing humidity (Fig. 5). In order to characterize this behaviour, the following definitions will be used in this paper: The value of R. H. where V_0 has decreased to 1/2 of its max. value after aging (or to 1/2 of V_{00max}) is called humidity decay value: Hd (or Hd′ respectively).

A reasonable value for a lifetime L can only be given for defined humidity, since L depends strongly on R. H. (L indicating the number of cycles that can be carried out, without the electrophotographic properties, e.g. V_0, decreasing below a certain limit). This can be seen in Figure 6, where L is defined by: $V_0 \geqslant 0.7\ V_{00}$, after max. negative recovery; then the lifetime at 30% R. H. is about one order of magnitude larger than L at 70% R. H. for identical layers. In Figure 7 the same layers are compared with respect to their humidity dependence after aging. Here, almost no difference is observed in the Hd values. But the influence of the number of aging cycles m performed prior to the determination of the humidity dependence V_0 (R. H.) is clearly shown in Figure 8 (m = 500, 1000, 5000, 10,000 cycles): Hd decreases markedly with increasing m.

Kinetics: The layers seem to reach equilibrium with the humidity atmosphere relatively fast, since V_0 changes fast upon change in R. H. (both during aging, as well as in recovered state) the time constant or delay time T_{ads} is of the order of a few seconds or below.

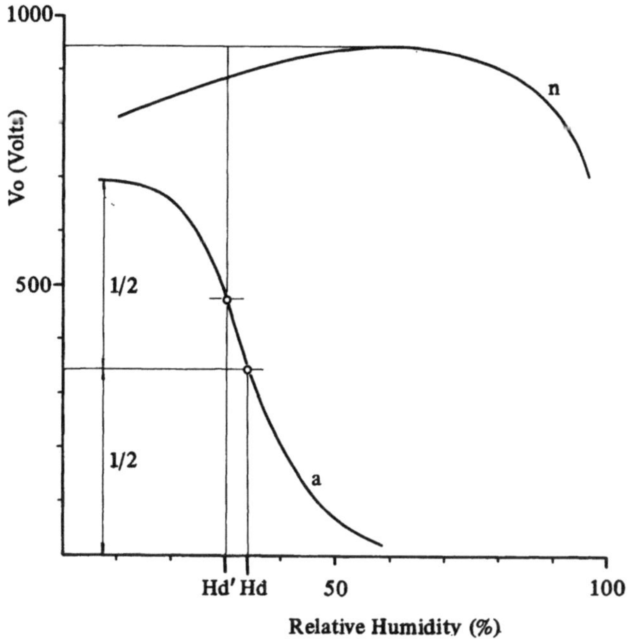

Fig. 5. Dependence of V_0 on relative humidity for new (n) and aged (a) layers. The humidity decay value Hd is defined by the decay of V_0 to 1/2 of its max. value

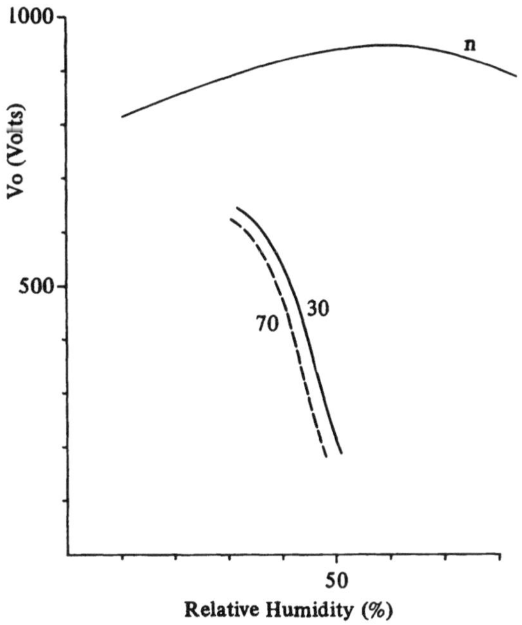

Fig. 7. Humidity dependence of V_0 for the layers of Fig. 6, 8 days after aging for 6300 cycles at 30% and 70% R.H. respectively; n = new layers

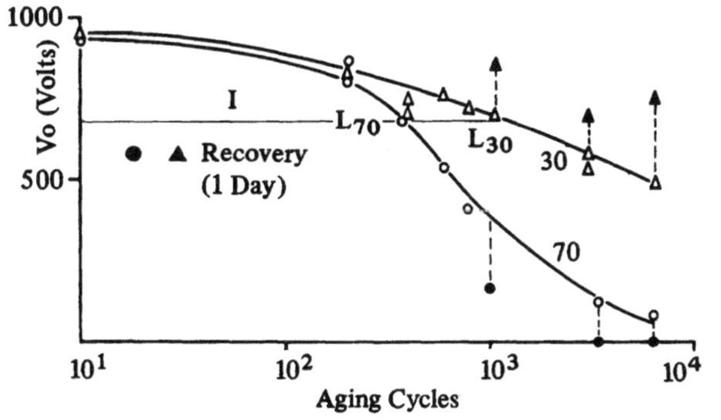

Fig. 6. Dependence of V_0 on aging cycles at 30% and 70% R.H. respectively. Line 1 defines the lifetimes L 30 and L 70 by $V_0(L) = 0.7 \cdot V_{00}$

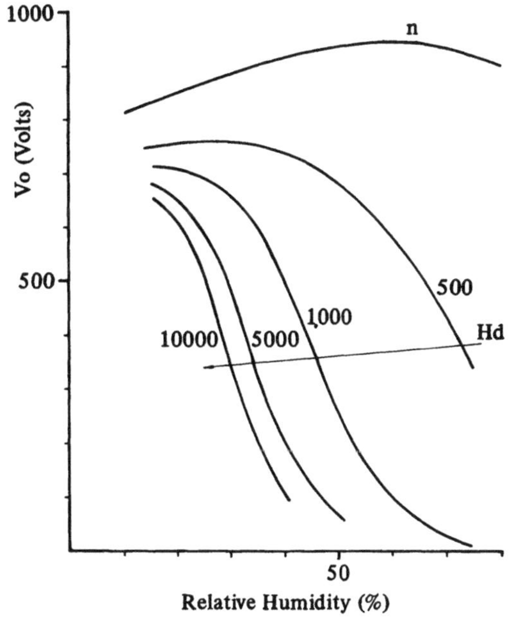

Fig. 8. Humidity dependence of V_0, 3 days after aging for O(n), 500, 1000, 5000, 10,000 cycles. Aging was carried out at 50% R.H.

5.3. Influence of Treatment After Aging

After aging, the layers have been treated in the following way in order to find changes in the humidity behaviour, which could give indications to the electrophotographic aging mechanism:

Treatment:

O: no treatment, i.e. recovery only

D: drying by evacuation or with drying agents (about 0% R. H.) for 3 days.

H, H′: heating to 110° or 180° C, respectively, for 15 minutes.

W: washing with various liquids, not solving PVCa, e.g. cleaning with an ethanol soaked Kleenex tissue.

P: polishing with a fairly hard rotating brush.

All these treatments have practically no influence on V_0 of new layers, whereas after aging distinct effects are observed.

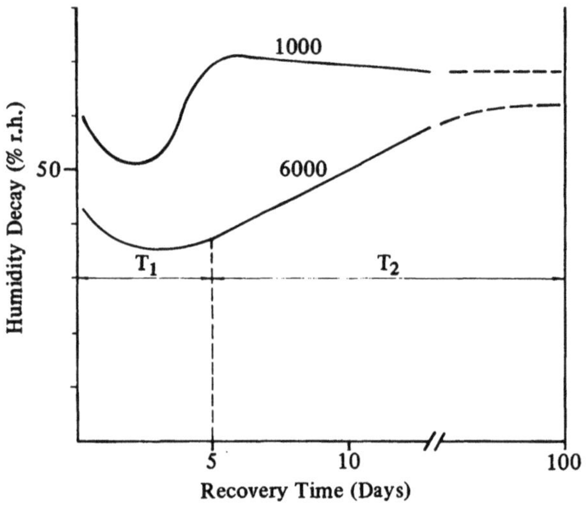

Fig. 9. Dependence of Hd (humidity decay value) on recovery time Rt. Short time recovery in T_1 (Rt = = 0–5 d), long time recovery in T_2 (Rt = 5–100 d) for layers previously aged for 1000 and 6000 cycles carried out at 50% R.H.

Results:

O: The recovery behaviour is shown in Figure 9 by the dependence of Hd on recovery time Rt. Again the same time interval T_1 and T_2 for short and long-time recovery (as described for the example of Fig. 4) can be observed. In addition to the negative short time recovery of strongly aged layers (m ⩾ 5000 cycles), now for light aging (1000 cycles) positive and negative changes in the interval T_1 were found. The long time recovery (in T_2) seems to be positive in any case.

In order to show only additional effects due to the treatments, the curves O, D, H, H′, W, P, are compared after identical aging and Rt = 5d: Figure 10.

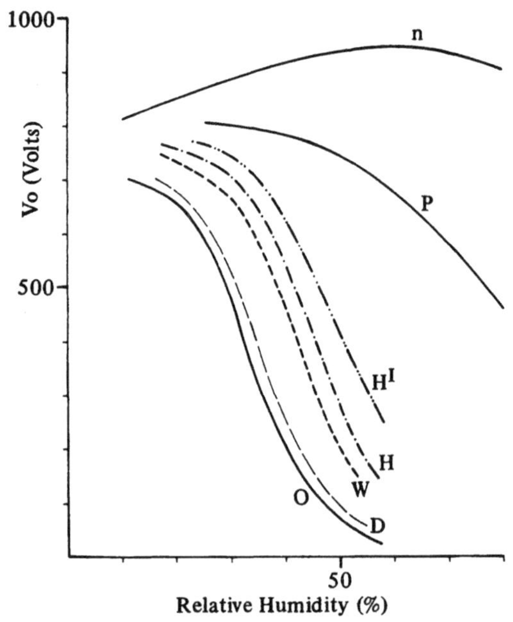

Fig. 10. Humidity dependence of V_0 for aged layers (m = 6000 cycles, Rt = 5 d), subjected to the following treatments: drying (D), heating (H, H′), washing (W), polishing (P), no treatment (O)

D: Drying had no significant or only a very small effect on the humidity dependence.

H, H′: Heating causes a distinct increase of the Hd values, being larger for H′ than for H, but the shape of the curves is hardly altered. Kinetics: An example for the time dependence of V_0 after heating is shown in Figure 11 where a decrease of V_0 in a time interval T_H of the order of 10 minutes is found. ($T_H \gg T_{ads}$).

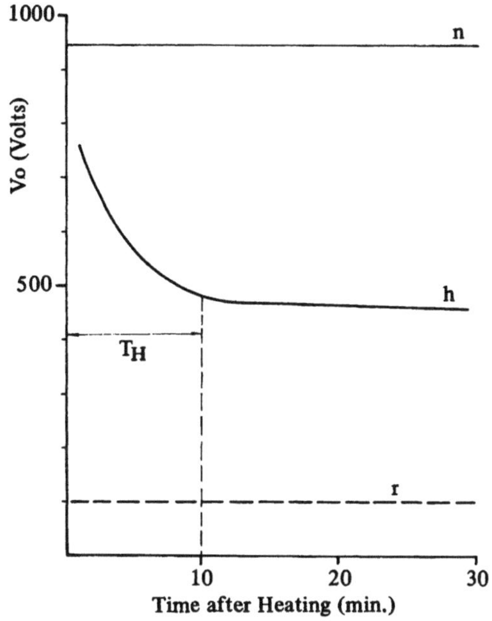

Fig. 11. Dependence of V_0 upon time after termination of heating H: curve (h); (n) and (r) are the values for the new layer and for the aged layer (5000 cycles, Rt = 6 days) prior to heating H; all measurements at 60% R.H. and 25° C

W: Washing has about the same effect as H. Traces of brownish, removed material could directly be observed on the tissue (with aged layers only).

P: Polishing causes a distinctly different and improved humidity dependence–the shape of which approaches the one of new layers–since no marked decay appears anymore. The same difference is found for the dark decay at 80% R. H., which is slow for P and new layer, but fast for O, W, H and H′.

In some cases the charge acceptance for positive charging V_0^+ of such (negatively aged) layers was determined and compared to the charge acceptance for negative charging V_0^- (Fig. 12). V_0^+ always appeared to be somewhat smaller than V_0^-, but the shape of V_0 (R. H.) as well as the differences in Hd upon treatment were qualitatively the same for both polarties, which indicates, that the effect of treatment is not restricted to one polarity.

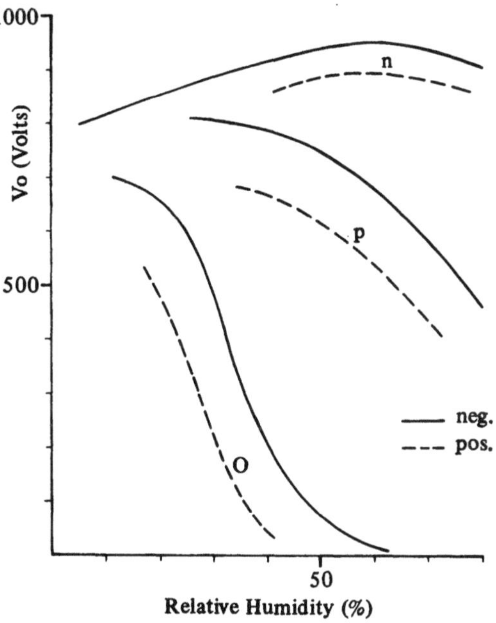

Fig. 12. Humidity dependence of the charge acceptance for positive and negative polarity V_0^+, V_0^- for the same layers as shown in Fig. 10 (n, O, P)

5.4. Removal of a Top Layer

An additional experiment was carried out in order to find the effect that the removal of the top part of an aged layer has on the humidity dependence. For that purpose top layers of identical material and a thickness of 1 μ (in case a) and 3 μ (in case b, c) were placed on standard layers, where they were kept in intimate contact by electrostatic forces (a, b). In case c the standard layer was wetted with a very small amount of an inert oil to improve the contact. The resulting double layers, as well as a standard layer s were aged for 10,000 cycles at 50% R. H. After Rt = 1 day the humidity dependence was measured: 1a, 1b, 1c, 1s in Figure 13.

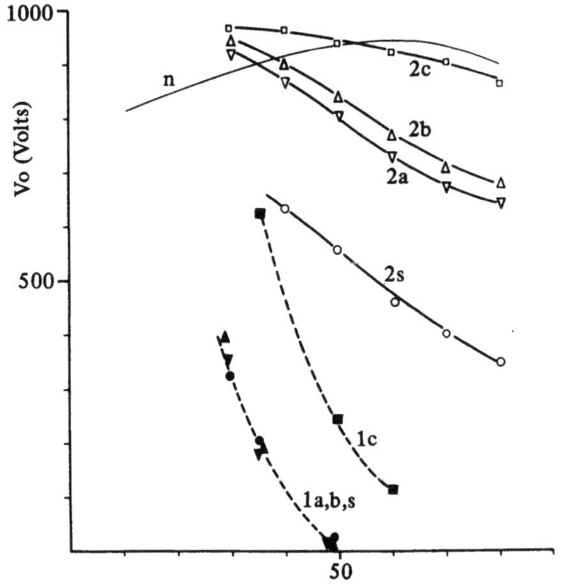

Fig. 13. Humidity dependence of V_0 of aged layers before (1) and after removal of their top layers (2a, 2b, 2c), or after polishing (2s); s = standard layer covered with a top layer of 1 μ (a) or 3 μ thickness (b, c)

Then the top layers were peeled off, s was polished and with the remaining layers, V_0 (R.H.) was measured again. The resulting curves 2a, 2b, 2c showed a completely changed shape, i.e. these layers reached practically the charge acceptance V_{00} of new standard layers–especially 2c–and the slow dark decay of the new state was found to be completely restored for a, b, c. That means, with respect to V_0 and ρ_0, the remaining layers appeared to be practically new.

5.5. Influence of Absolute and Relative Humidity

The dependence of V_0 on R. H. was measured at 16, 26 and 36° C. The Hd values were found to decrease only slightly with temperature, therefore it could be concluded, that the humidity behaviour V_0(hum.) depends mainly upon the relative humidity and hardly upon the absolute humidity.

6. Complementary Experiments

a) Aging with positive polarity charging showed a behaviour qualitatively similar to that with negative polarity: decrease of V_0 on the aging cycles, recovery and instability effects, as well as an analogous humidity dependence V_0 (R.H.) was found.

b) Scanning Electron Microscope pictures of the surface of aged layers (resolving power about 200 Å) showed no signs of erosion, pinholes or any other morphological differences as compared to new layers. Images of polished layers showed removal traces with a depth of the order of 0.1 μ.

c) Aging was found to increase strongly at inhomogeneities and defects of the layers (e.g. cracks), where secondary corona discharges may be enhanced.

d) *Various materials:* The following non-standard layers were investigated with respect to their aging behaviour: PVCa layers of various thicknesses, of purified or unsensitized PVCa, or containing other sensitizers and plasticizers; other polymeric layers: e.g. of acrylate-, epoxy- or silicone resins and mylar. Always the same effects were found: decrease of V_0 with aging, V_0(m), and relative humidity, V_0 (R.H.), and also the positive regeneration effect of treatment P.

7. Relevant Parameters of Electrophotographic Aging

In the course of these investigations of the electrophotographic aging of PVCa layers, the following parameters were found to be of distinct influence on the aging mechanism or its effects: the parameters determining the

- layer (composition, preparation, thickness, morphology, substrate)
- mechanical properties and stress of the layer (adhesion to substrate)
- corona charging conditions (voltage, layer current)

besides the influence of the operating conditions and atmosphere (humidity), which is described in previous sections.

There are certainly more parameters of influence than those mentioned here. A main problem is posed by the fact, that it is very difficult or even impossible to separate and analyze the action and effect of only one parameter, to vary one parameter without changing the others and the mechanism by–at least partly unknown–interrelations and secondary effects. To define an aging experiment completely is therefore very difficult and absolute values (e.g. for a lifetime) might include accordingly large errors. It was rather tried here to find relative values or differences and qualitative dependences, in order to obtain some information on the kind of mechanisms and correlations determining the electrophotographic aging–as regarded in connection with the general aging scheme of **Part 1.**

8. Discussion

From the experiments and in accordance with literature (where the chemical influence of reactive agents from gas discharges is well known) it is concluded, that the dominant mechanism in aging of PVCa layers is the chemical aging mechanism b), due to the gas discharge of the corona. Exposure to daylight and aging without light discharge were found to cause no significant differences.

The following hypothesis was assumed to explain the results: The gas discharge leads to reactions with the PVCa material, mainly at the surface, where decomposed material is produced. Adsorption of water at this decomposed surface results in the formation of a surface layer, in thickness of the order of $0.1\,\mu$ or less, with altered electrical properties, "injecting" charge carriers into the bulk, so that the overall resistivity and the charge acceptance V_0 of the layer decrease strongly, whereas the electrical properties of the bulk remain relatively unaffected. Chemical reactions not leading to observable changes in V_0 and ρ_0 may extend further into the bulk.

No signs of thermal and mechanical aging mechanisms were found, since no morphological changes–such as erosion, pinholes or conducting filaments, larger than about 200 Å **(6b)**–could be observed. As to the electrical mechanism only small reversible memory effects may be superimposed on the dominant (chemical) mechanism in PVCa, which therefore seems to be completely different from the fatigue effects encountered e.g. with selenium.

The experimental results are in good agreement with the supposed mechanism (chemical decomposition and surface effect): the unchanged bulk photoconductivity (4d) and the dark resistivity, i.e. ρ_0 and V_0, hardly depend on aging near 0% R. H. (Fig. 8); as well as the marked decay in V_0 (R. H.) after aging, and the very fast reversible adsorption of water (5.2). A loss in resolution of copies observed upon aging may point to increased surface conductivity. Comparison of Figures 6 and 7 indicates, that it is not primarily the chemical decomposition which depends on humidity, but the subsequent adsorption. The regeneration effect upon treatment (5.3) is thought to be due to removal of deteriorated surface material (up to about $0.1\,\mu$ by polishing). By the treatments H, H′ and W, causing only a limited enhancement of Hd, probably only some of various decomposition products are removed, whereas by polishing P all of them may largely be separated from the bulk–partly being removed from the layer, partly just being smeared along its surface. By peeling off the top layer (5.4), very probably, the whole deteriorated surface layer was removed. The somewhat incomplete regeneration effect in case a and b may be due to internal discharges in small spacings (about $0.1-0.3\,\mu$) between top and underlying standard layer. The decrease of V_0 with the number of cycles m might be related to the amount of decomposed material that is formed. Instability and recovery behaviour (4c, 5.3) also agree well with the assumed chemical mechanism. The changes of V_0 observed in the time intervals T_1 (0–5 days), T_2 (5–100 days), T_3 and T_H (10 minutes)–being large as compared to T_{ads} and dark decay time T_0 (seconds)–may point to the formation of unstable intermediates during the aging process (T_3) and reactions including several reaction steps to end products of decomposed PVCa material after aging (T_1) or after heating (T_H). Some of these end products may slowly disappear by evaporation (T_2)–which would be in agreement with the removal effect of heat treatment (H).

9. Conclusions

Chemical aging mechanisms (caused by gas discharges) can generally be expected to occur with organic photoconductors, and the influence of humidity (adsorption) seems to lead to corresponding changes in the electrical properties of many sorts of organic layers (6d). Measurement of Hd and of the humidity dependence of V_0 and ρ_0 might be useful for the investigation of aging mechanisms and for characterization and comparison of organic layers.

The long-time recovery behavior could have important consequences in application, since the practical lifetime of a layer may be considerably enhanced in commercial use (with interruptions), as compared to the lifetime L for continuous aging. Since the lifetime is strongly limited by the humidity, drying would allow to enhance L considerably.

Acknowledgement

The author would like to thank G. Roos for valuable literature studies and F. Albertin for skilled experimental assistance. Helpful discussions with many colleagues, especially with R. Kern, G. Forgo and L. Svestka are also gratefully acknowledged.

References

[1] Abasow, S. A. et al., Plaste und Kautschuk **15**, 3 (1968).

[2] Bartnikas, R., IEEE Trans. Electr. Insul. **6**, 2 (1971).

[3] Boeck, W., ETZ-A **88**, 26 (1967).

[4] Bruins, P. F. (ed.), Plastics for Electrical Insulation. New York, Interscience, 1968.

[5] Chemical Reactions in Electrical Discharges. Symposium ACS, Div. Fuel Chem. and Div. Phys. Chem., 153rd Mtg., Miami Beach, Fla., April 11–13, 1967. Washington, ACS, 1969.

[6] Conley, R. T. (ed.), Thermal Stability of Polymers, v. 1. New York, Dekker, 1970.

[7] Dakin, T. W., Electro-Technol. **66**, 6 (1960).

[8] Densley, R. J., and B. Salvage, IEEE Trans. Electr. Insul. **6**, 2 (1971).

[9] Hewitt, G. W., and T. W. Dakin, IEEE Trans. Power App. Syst. **82**, Dec. (1963).

[10] Loeb, L. B., Electrical Coronas. Their Basis Physical Mechanisms. Berkely, Univ. California Press, 1965.

[11] Mayoux, Ch., and Y. Teisseyre, C. R. Acad. Sci. Paris, sér. B. **271**, 14 (1970).

[12] Meats, R. J., and A. W. Stannett, IEEE Trans. Power App. Syst. **83**, 1 (1964).

[13] Shahin, M. M., Conference on Electrical Insulation and Dielectric Phenomena, Buck Hill Falls, Pa., October 21–23, 1968. Washington, D.C., Nat. Acad. Sci, 1969, p. 85–89.

[14] Shahin, M. M., Preprints, 3rd Internat. Congress on Reprography, London, March 1970, paper No. 1, 24 p.

[15] Thinius, K., Stabilisierung und Alterung von Plastwerkstoffen. Bd. 1–2. Berlin, Akademie-Verlag, 1969–71.

[16] Schaffert, R. M., Electrophotography. London, Focal Press, 1965.

[17] Dessauer, H. H., and H. E. Clark (eds.), Xerography and Related Processes. London, Focal Press, 1965.

[18] Cassiers, P. M., J. Phot. Sci. **10**, 2 (1962).

[19] Chapman, D. W., F. J. Stryker, Phot. Sci. & Eng. **11**, 1 (1967).

[20] Njunko, L. I., Elektrofotografiya i Magnitografiya. Vilnius, 1965, p. 192–201.

[21] Meier, H., W. Albrecht, U. Tschirwitz, Preprints, Internat. Congress on Photographic Sciences, Moscow 1970, sect. E: Non-silver and Unconventional Photographic Processes, paper E 46.

[22] Regensburger, P. J., Photochem. Photobiol. **8**, 5 (1968).

[23] Schaffert, R. M., IBM J. Res. Develop. **15,** 1 (1971).
[24] Wagner, W. J., and E. L. Gasner, Phot. Sci. & Eng. **14,** 3 (1970).

Discussion

D. L. Stockman: What solvent did you use for coating the PVCa?

W. Lanker: Benzene for the Standard layers. In some experiments other solvents were used also.

D. L. Stockman: Our experience would suggest that the magnitude of the humidity effects observed on PVCa layers can be modified substantively, depending on the solvent used for coating.

W. Lanker: In a few experiments with different solvents no qualitative differences in the aging effects were observed. It is felt that the supposed simple aging mechanism is not principally depending on the solvent used (if this did not cause mechanical defects in course of preparation)—although quantitative differences in the effects may be possible.

W. F. Berg: Did your humidity values relate to the aging treatment or the time of measurement?

W. Lanker: The treatments were normally carried out at about 50% R.H., since it was not this treatment-humidity which was found to be relevant but only, to what extent the decomposed surface material was removed.

EPR Investigations of Photoconducting Charge-Transfer Complexes

Jürgen Rochlitz

Abstract

Charge transfer complexes of polyvinyl-carbazole (Luvican) show EPR-signals if illuminated at low temperatures. Thin layers of Luvican-trinitrofluorenone complexes show a decrease of EPR-signal intensity with decreasing acceptor concentration. Comparing the EPR intensities and the photocurrents of similar electrophotographic layers, a linear relation between EPR intensity and the root of the photocurrent can be found. An activation energy of the transition from the spin uncoupled excited state into the groundstate was determined by the temperature dependence of the EPR-signal intensity. This activation energy is dependant on the acceptor concentration: Neighbouring acceptor molecules stabilize the transferred electron.

1. Introduction

In semiconducting materials conduction is observed only if by energetic activation, quasi-free charge carriers are produced, which are able to move in a certain direction under the influence of an electric field.

The energy states of those charge carriers can be described either by a band model with narrow bandwidth or by a row of localized energy levels. The transport of charge carriers is dependant on the nature of the energy states, whereas the primary processes of photoconduction leading to production of charge carriers are determined by the ionization potentials and electron affinities of the molecules involved as well as by the energies of stabilization of the resulting ions. These ions are radical ions; therefore it should be possible to investigate the primary processes of photoconduction by EPR, especially in the case of CT-complexes.

Indeed Kommandeur et al. [1] have found a correlation between EPR-signal intensity and conductivity together with identical activation energies for spin-concentrations and for the specific conductivity. Contrary results are presented by Labes [2]; he questioned a general connection between EPR and conductivity. In many other investigations the numbers of spins/ cm^3, that is the numbers of radical ions are increased by thermal activation. Hoegl et al. [3] reported an exponential decrease of the spin concentration with lowering the temperature in the case of the p-Chloranil-Carbazole complex. Even during an illumination of CT-complex solutions an increasing signal-intensity with higher temperature was observed by Briegleb [4]. Only Nelson [5] has found an inverse temperature-dependence of the EPR-signal caused by illumination of the photoconductive Rhodamin B. He suggested excited electrons not involved in the conduction process as the cause of the signals.

2. Experimental Results

This investigation is to show the ability of charge-transfer complexes of Luvican (Polyvinylcarbazole) to generate radical ions if illuminated*; moreover it is an attempt to find a correlation between this ability and the photoconductivity. At 20° C, only in the case of the 1 : 1 complex of 2,3-dicyano-naphthoquinone-1,4 (DCN) could an EPR-signal be detected; surprisingly this was possible even without illumination. The other complexes showed a signal at temperatures below –40° C, if illuminated (Fig. 1).

The signals are without any structure and represent singulets with half-widths of 4–9 G. The pure Luvican and the pure 2, 4, 7-trinitrofluorenone (TNF) show no signals under the same conditions.

In view of the correlation between EPR-signals and photoconductivity it is surprising to find a very low voltage when a thin layer of the 1:1-Luvican-DCN-complex is charged negatively by a corona as is used in electrophotography. Contrary to the other complexes of Figure 1, the dark conductivity is large because of the generation of radical ions without any optical excitations. DCN belongs to electron-acceptor molecules with high electron affinity: the electron transfer can take place in the ground state. The TNF complex possesses the most favorable photoconductive properties of the investigated complexes, in accordance with the relatively high signal intensity.

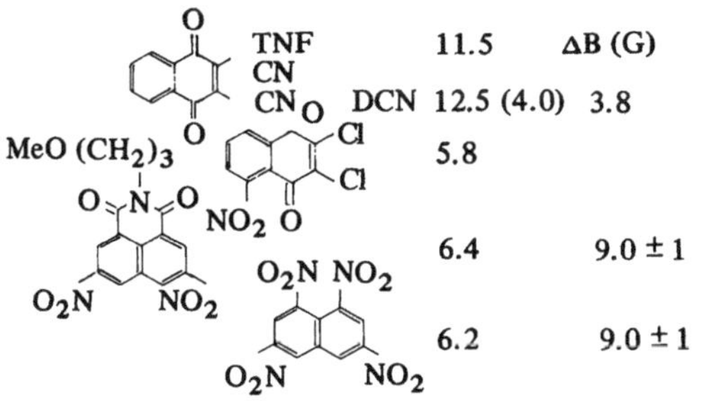

Fig. 1. EPR-signal intensities of 1:1 CT complexes of Luvican (arbitrary units) at 90° K; illumination by Hg-lamp; half-widths of the signals in Gauß. Value in parantheses: signal without illumination.

---

* Illumination: Hg-lamp Philips HPK 125 W without UV-part

EPR-apparatus: Varian E-9

2.1. The Dependance of the EPR-Signal Intensity on the Concentrations of TNF

In order to investigate the influence of the TNF concentration on the EPR signals thin layers of the TNF complex were illuminated at 90° K.

The ERP intensity increases by a one-fourth power relation with increasing TNF/carbazole unit-ratio (Fig. 2).

Analogous layers with an TNF/carbazole unit-ratio of more than 1 show a dependence of the EPR-intensity (normalized to equal extinction at 500 nm) different from that with lower ratios (Fig. 3).

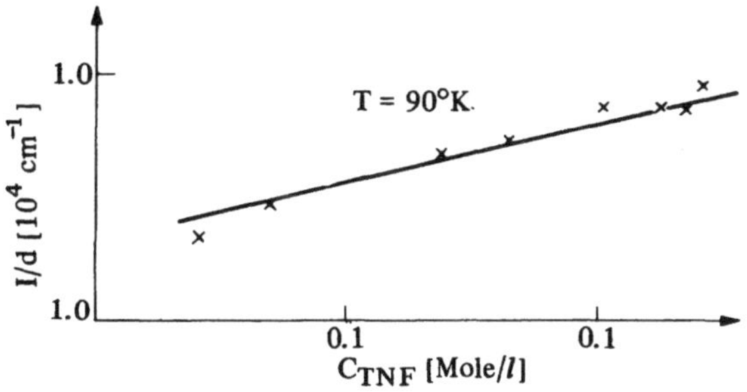

Fig. 2. EPR signal intensities of TNF Luvican complexes in relation to the TNF content (arbitrary units)

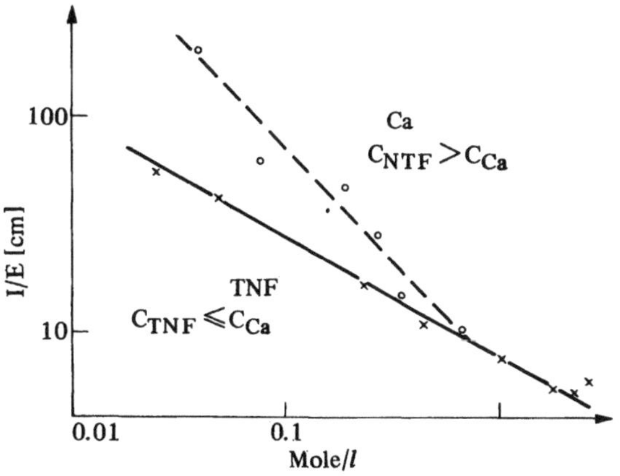

Fig. 3. EPR intensities, normalized on equal extinction at 500 nm (arbitrary units) in relation to TNF concentration ($C_{TNF} \leqslant C_{Ca}$) and carbazole unit concentration ($C_{TNF} > C_{Ca}$); in the latter case the binder material is a polyester

The photocurrents of the layers mentioned above (measured as $\left(\frac{1}{d}\frac{dU}{dt}\right)_{t=0}$ of the voltage decay [6]; the layers have been deposited on aluminized Melinex® *) and show the same tendency in the two systems: with rising complex concentrations the photocurrents (the illuminated side negatively charged) increase (Fig. 4).

The steeper slope at $C_{TNF} \leqslant C_{Ca}$ results from the enhancement of the electron mobilities with decreasing distance between the TNF molecules. If the number of charge carriers were enlarged alone, the slopes of the two functions (with $C_{TNF} \leqslant C_{Ca}$ and $C_{Ca} < C_{TNF}$) would have to be equal.

In comparing the signal intensities at 90° K with the photocurrents of the various TNF complexes the following relation results (Fig. 5):

$$\frac{I}{d} \sim \sqrt{i_{phot}}$$

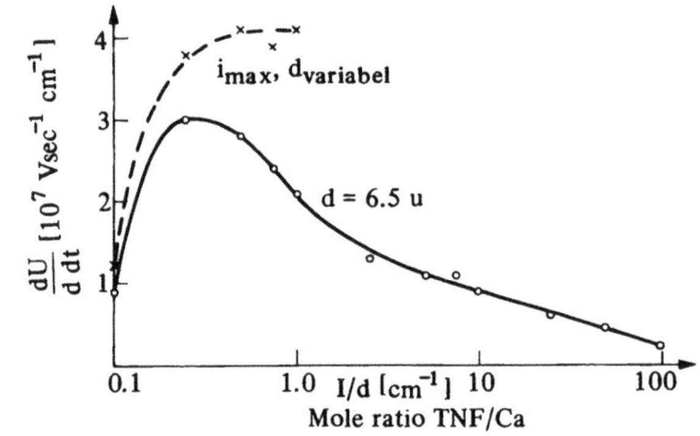

Fig. 4. Photocurrents (measured as function $\left(\frac{1}{d}\frac{dU}{dt}\right)_{t=0}$ of the voltage decay on illumination) of TNF-Luvican complexes with different mole ratios TNF/carbazole unit. The values are given for equal layer thickness and in some cases for variable thicknesses d but maximum photocurrent. Layer voltage: negative

* d: layer thickness; $\frac{dU}{dt}$ at illumination with a Xe-lamp (XBO 150 W) which produces 300 Lux on the layers.

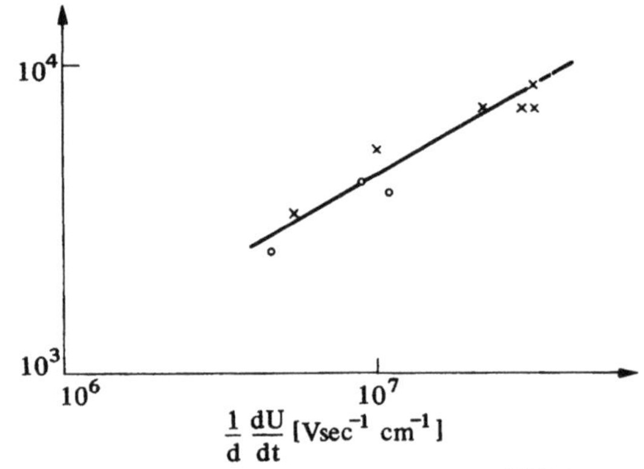

Fig. 5. Relation between photocurrent and EPR intensities normalised on equal thickness.

$C_{TNF} \leqslant C_{Ca}$ ○ $C_{Ca} < C_{TNF}$

Together with the relation of Figure 2 a square root connection between photocurrent and TNF concentration can be obtained:

$$i_{phot} \sim \sqrt{C_{TNF}}$$

which is given by the direct measurements. Between the extinctions of the complexes at 500 nm and the EPR intensities the cube root relation of Figure 6 could be found:

$$\frac{I}{d} \sim \sqrt[3]{\frac{E}{d}}$$

This relation shows that only a part of the CT complexes remains in the excited state with decoupled spins. The probability of the transi-

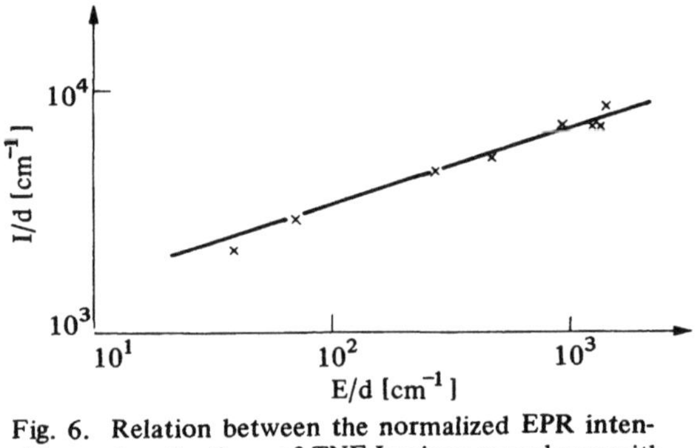

Fig. 6. Relation between the normalized EPR intensities and extinctions of TNF-Luvican complexes with $C_{Ca} \geqslant C_{TNF}$ (λ = 500 nm).

tion into this state diminishes with rising complex concentration. The cause for this phenomenon may be explained by the measurements of the temperature-dependence.

A first hint is given by the decay characteristics after switching off the illumination: Figure 7 shows the complicated logarithmic function. Whereas the EPR signals of triplet states of CT complexes possess halftimes of seconds [7], these signals with CT complexes decay in hours. The decay law of formal high order seems to be caused by a deactivation involving severed molecules.

2.2. Temperature Dependence of the Stationary EPR Signals

The stationary EPR signals (under illumination) increase in intensity if the temperature is lowered. The 1:1-complex doesn't show a significant intensity at room temperature. The exponential functions of Figure 8 permit the estimation of activation energies for the transition back to the ground state. This activation energy is increasing with the enhancement of the TNF concentration.

For the cases with a TNF content of more than one mole per mole carbazole unit no further increase of the activation energy can be observed (Fig. 9).

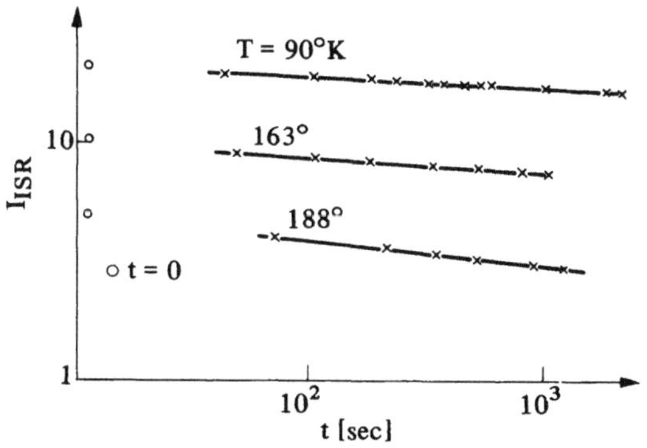

Fig. 7. Decay characteristic of the EPR signal of the 1:1-TNF-Luvican complex after swtiching off the illumination at various temperatures.

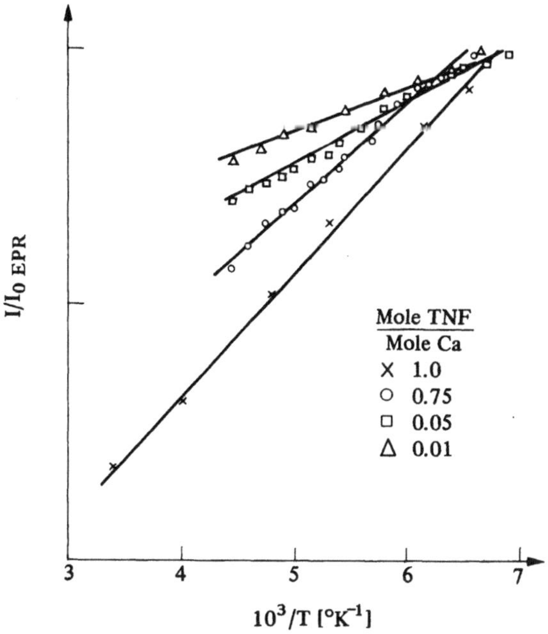

Fig. 8. Temperature dependence of the EPR signals (related to the signal at the lowest temperature) of TNF Luvican complexes with various TNF contents. $C_{TNF} \leqslant C_{Ca}$

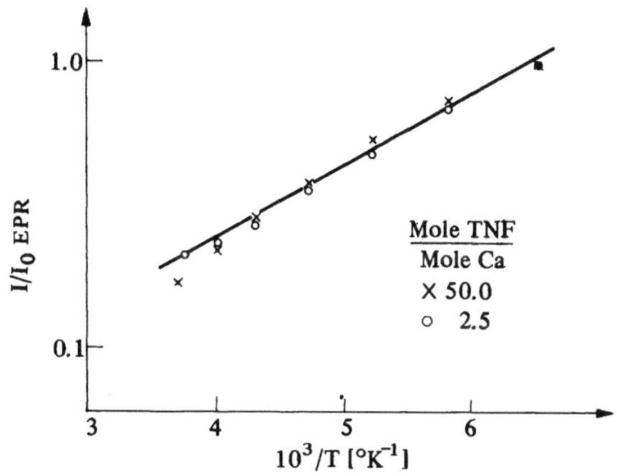

Fig. 9. Temperature dependence of the EPR signals (related to the signal at the lowest temperature). $C_{TNF} > C_{Ca}$

3. Discussion

The variation of the concentrations of TNF have shown the invariance of the structure and position at g = 2.0044 of the EPR signals; therefore one can assume TNF-radical anions as the signal generating species. The enhancement of the TNF concentrations causes a stabilization of the spin-decoupled excited state. Such a stabilization is only possible by electron exchange between neighbouring TNF molecules, as has been described by several authors [8].

This electron exchange may be responsible for the diminishing probability of transition into the excited state (Figs. 2 and 3). It is the cause for the enlargement of the activation energies. Therefore at high TNF concentrations (Fig. 9) a variation of the mole ratio cannot alter the activation energies significantly.

As Kepler [9] has shown in the case of radical ion salts of tetracyanoquinodimethane, the possibility of an electron exchange between neutral molecules and radical ions improve the conductivity by several orders of magnitude.

In analogy an increased stabilization of the excited spin-decoupled state of the complexes should enhance photoconductivity if this state belongs to the first stages of quasi-free charge carriers in the sense of the schematic model of Figure 10.

By this model an excited state by electron transfer is reached, followed by spin decoupling and stabilization of the TNF-radical anion by neutral molecules in the neighbourhood.

Influenced by the electrical field the electrons are released out of this state, whereupon they have to surmount the potential walls between the regions where electron exchange is possible.

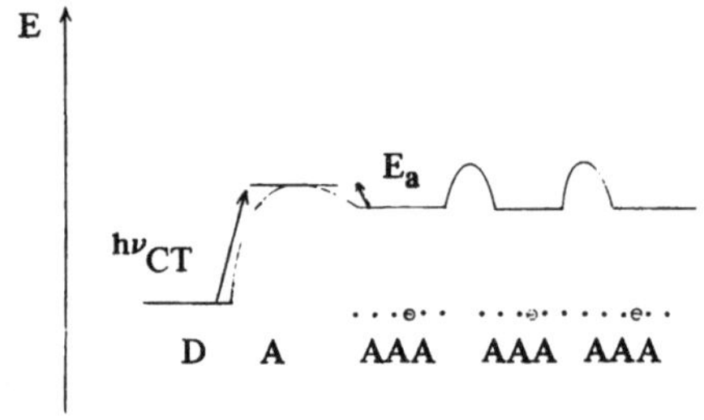

Fig. 10. Schematic model of the energy-levels in TNF-Luvican complexes for charge transport without consideration of field strength.

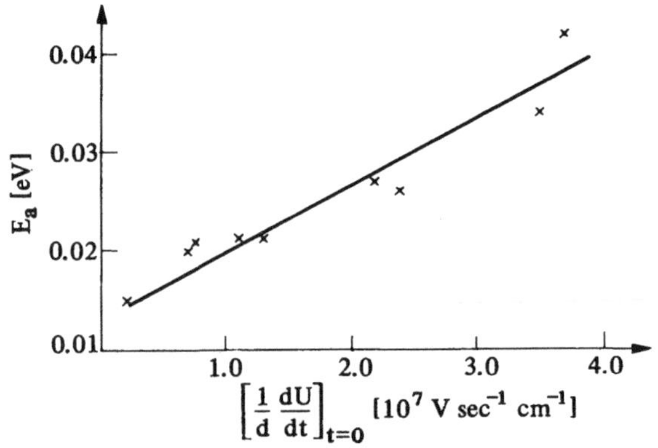

Fig. 11. Relation between the activation-energies of the stationary EPR signals (Activation energy of the reaction back to the ground state) of Luvican-CT complexes and their photocurrents.

If this model is applicable, a relation between the activation energies for the reaction back to the ground state and the photocurrents must exist. In Figure 11 the activation energies are plotted against the values of $\frac{1}{d}\frac{dU}{dt}$ [6] of the same complexes used in electrophotographic layers. The systems investigated show a proportionality between the photocurrents and the activation energies. The agreement would be better if the influence of mobility could be eliminated: the rate of charge carrier production should be directly dependent on the activation energy.

Acknowledgement

The author is very much indebted to Prof. Dr. Hanns Fischer, University of Zürich, and his co-workers for the EPR measurements and for discussion related to them.

Reference

[1] Kommandeur, J., and F. R. Hall, J. Chem. Phys. **34**, 129 (1961). Singer, L. S., and J. Kommandeur, ibid. **34**, 133 (1961). Blomgren, G. E., and J. Kommandeur, ibid. **35**, 1636 (1961).

[2] Labes, M. M., and M. Bose, J. Phys. Chem. 83, 4505 (1961).

[3] Kohin, R. P., K. A. Müller and H. Hoegl, Helv. Phys. Acta **35**, 255 (1962).
[4] Briegleb, G., and G. Lind, Z. Naturforsch. **23a**, 1747 (1968).
[5] Nelson, R. C., B. J. Cho and L. C. Brown, J. Chem. Phys. **39**, 499 (1963).
[6] Regensburger, P. J., Photochem. + Photobiol. **8**, 429 (1968). Arneth, R., B. Lorenz, Reprographie **3**, 199 (1963).
[7] Hayashi, H., S. Iwata, S. Nagakura, J. Chem. Phys. **50**, 993 (1969).
[8] Lewis, I. C., L. S. Singer, J. Chem. Phys. **43**, 2712 (1965). Ishitani, A., S. Nagakura, Bull. Chem. Soc. Japan **38**, 367 (1965). Howarth, O.-W., G. K. Fraenkel, J. Chem. Phys. **52**, 6258 (1970).
[9] Kepler, R. G., P. E. Bierstedt, R. E. Merrifield, Phys. Rev. Letters **5**, 503 (1960).

Discussion

Weigl, J. W.: How is the width of the ESR signal influenced by the concentration of TNF in the complex?

Rochlitz, J.: No measurable change could be detected.

Meier, H.: Can you tell something about the order of enhancement of mobility you had observed.

Rochlitz, J.: The enhancement of mobility was obtained in a qualitative manner by comparison of the slopes of the two photocurrents.

Charge Carrier Formation in Oxadiazole

R. Moraw

Abstract

To study the primary process of charge carrier formation in organic photoconductors, the quantum yields of photoconduction in layers containing the photoconductor 2,5-bis-(p-diethylaminophenyl)-1, 3, 4-oxadiazole, dispersed in a styrene/butadiene-copolymer resin, were measured. The wave-length of the incident light, the strength of the electric field and the thicknesses of the layers were varied. The results agree with the assumption that charge carrier formation is due to tunneling of excited electrons.

1. Introduction

To study the photoconduction of organic compounds in binders, we measured the quantum yields of layers of the photoconductor 2,5-bis-(p-diethylaminophenyl)-1, 3, 4-oxadiazole

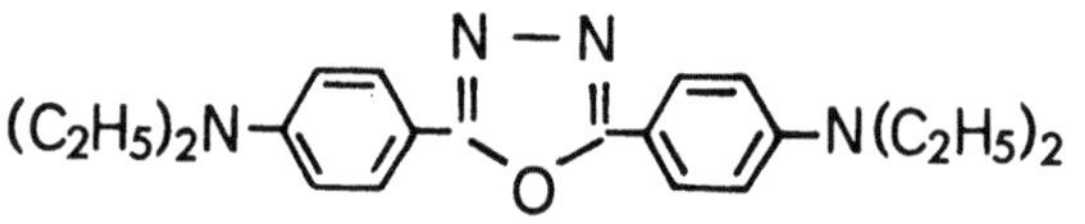

in a resin of a styrene/butadiene-copolymer (Pliolite 5SB). The photoconductor absorbs at 360 nm. Layers of a mixture of the photoconductor and the resin showed no spectral evidence for complex-formation. The layers were coated on a polyester film provided with a vacuum-deposited aluminum layer.

The photoconductive sample was charged by an adjustable corona and was positioned behind a transparent probe. There, the sample was uniformly exposed with filtered light of known intensity from a mercury-vapor lamp. The signal from the probe was applied to a cathode ray tube.

The quantum yield is expressed as the ratio of elemental charges transported through the layer to absorbed light quanta. We calculated the number of transported elemental charges from the voltage drop at the beginning of the exposure.

2. Results

To get consistent results, we varied a great number of parameters:

1. electric field strength (V/cm)
2. polarity of charge (±)
3. penetration depth of light (μ)
4. wave length of light (nm)
5. intensity of light (quanta/cm^2 sec)
6. thickness of the layers (μ)
7. ratio of resin and photoconductor

An example of the experimental curves obtained is given in Figure 1. The quantum yield is plotted on a logarithmic scale as a function of the electric field for several wave lengths of incident light. The curves bend sharply down toward small values of the electric field.

Such curves were measured with layers of different thicknesses between 1 and 17 μ. Therefore, we could combine the quantum yields at given electric fields, for instance at 4.5 x 10^5 V/cm, derived from layers of different thicknesses. Such plots of the quantum yield as a function

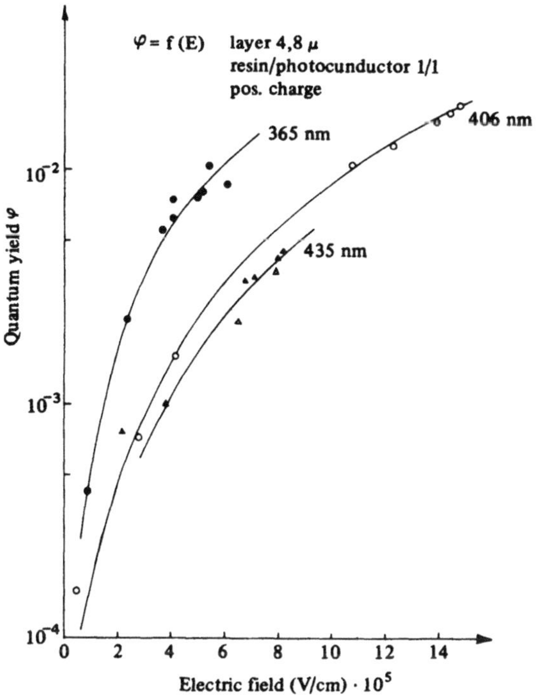

Fig. 1. Quantum yield as a function of the electric field

of the thicknesses of the layers at constant electric field, as given for example in Figure 2, enabled us to extrapolate toward quantum yields which represent values for vanishing thicknesses of the layers.

By the extrapolation toward quantum yields at vanishing thicknesses of the layers, we eliminated to a high degree all the effects involved with the transport of charge carriers through the layer. The extrapolated quantum yields are therefore mainly the quantum yields of charge carrier formation by light. There is evidence that the transport of charge carriers might be without loss at all, because it is possible to remove all charges by light without residual charges left in the layer.

The curves of Figure 2 show nearly constant quantum yields when we used thick layers of several microns, especially at 365 nm. At this wave length the penetration depth of light is of the order of only 0.28 μ. This is very small compared with the thicknesses of the layers of

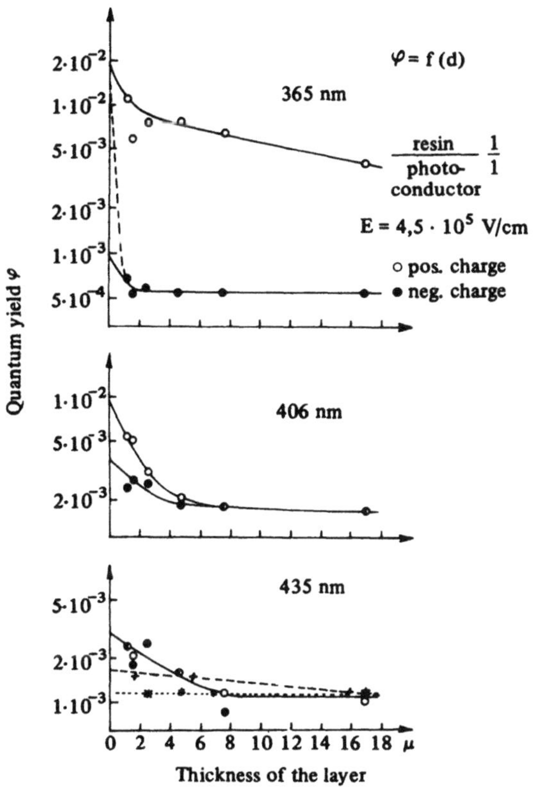

Fig. 2. Quantum yield as a function of the thickness of the layer

several microns. When we used thin layers, the quantum yields showed increased values. This might indicate that besides charge carrier formation inside the layer, charge carriers are partly formed by the formation of excitons in the layer, which, after migration to the surface, decompose and generate additional charge carriers.

There is a difference between the results for layers with positive and with negative charges. Generally, layers with positive charges showed higher quantum yields than layers with negative charges. This effect is very pronounced at 365 nm, where the penetration depth of light is very small, as already mentioned. A reasonable explanation might be different mobilities. Generally, the mobility of holes is larger than the mobility of electrons.

According to our results, there is no dependence of the quantum yield on the incident light intensity between 3.5 x 10^{14} quanta/cm^2 sec and 57 x 10^{14} quanta/cm^2 sec. This might be an indication that the formation of charge carriers is a one photon-process.

We also changed the ratio of resin and photoconductor. Normally, we used equal parts by weight. In one series, we increased the content of resin reaching ratios of 3/1 and 10/1.

resin / photoconductor	calculated mean separation (Å)	quantum yield
1/1	10.7	0.0021
3/1	15.5	0.00086
10/1	23.0	0.00069
neg. charge 435 nm 4.5 x 10^5 V/cm		

With increasing mean separation, the quantum yield decreases. The decrease of the quantum yield is of the same order as one would expect if the charge carrier formation is by tunneling.

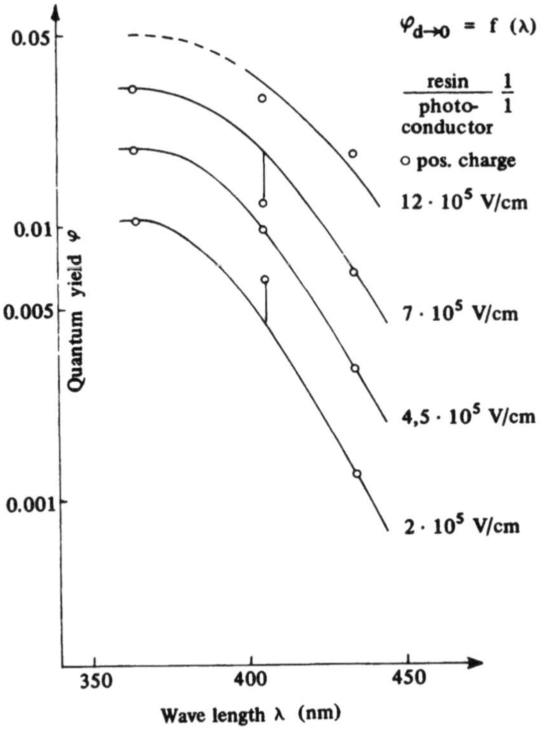

Fig. 3. Quantum yield as a function of the wave length

3. Discussion

In the discussion of charge carrier formation we will only consider those results from positive charged layers, to avoid complications due to the assumed low mobility of electrons and due to the difficult extrapolation toward vanishing thicknesses for negative charged layers. For the discussion, in Figure 3 the extrapolated quantum yields for layers of vanishing thicknesses are plotted as a function of the wave length of incident light. The values of the quantum yield increase when light of a short wave length is used. The values increase, too, when the electric field is increased.

With an appropriate theory, it should be possible to describe quantitatively the increase of quantum yields with decreasing wave length and increasing electric field.

X-ray measurements showed no evidence for crystallization centers in the layers of resin and photoconductor. Therefore, the band model, well suited to describe inorganic semiconductors, seems to be a poor model to describe charge carrier formation in amorphous organic layers. We have attempted to describe the charge carrier formation by the well known tunnel effect. Tunneling means that there is a certain probability that an electron will be found behind a potential barrier, as shown in Figure 4.

This probability is proportional to

$$e^{-\frac{2S}{\hbar}\sqrt{2mU}}$$

U is the height of the potential barrier,
S is the width of the potential barrier,
m is the mass of an electron.

The calculations must be made using the residual height of the potential barrier U_0 after excitation of the electron by absorption of light. It is

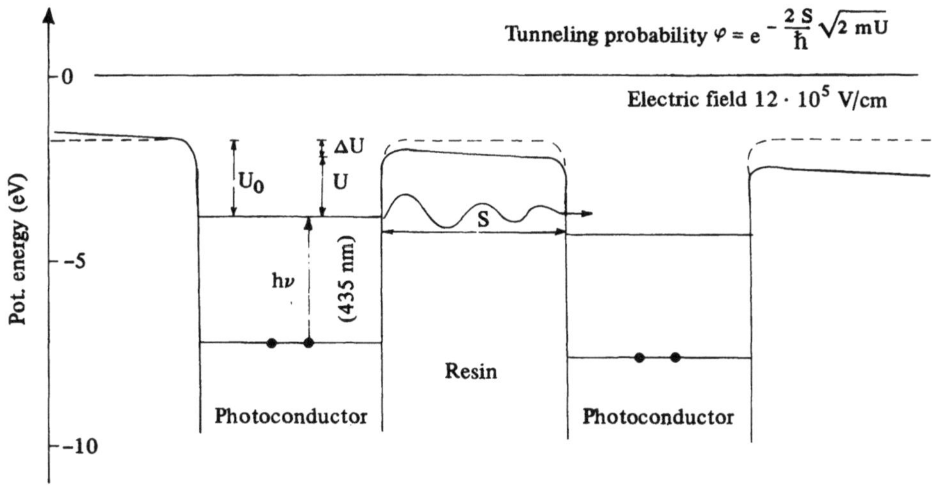

Fig. 4. Potential cross section of a photoconductor layer

necessary also to correct the potential U_0 for the energy gain ΔU by the external electric field. To calculate this gain, we estimated the photoconductor molecules to be 20 Å long.

The thermal relaxation time for loss of vibrational energy of an excited electronic level is 10^{-12} sec or 10^{-13} sec. Were the charge carrier formation to happen after the thermal relaxation, then the quantum yield would be independent of the wave length of the exciting light, similar to the case of fluorescence. According to Figure 3 there is a pronounced dependence of the quantum yield on the wave length of the incident light. Charge carrier formation must therefore be a very fast reaction. It can be assumed that the charge carrier formation takes place at the first approach or after very few approaches of the excited electron to the potential barrier, because otherwise the thermal relaxation would have decreased the probability for tunneling significantly. Therefore, to a good approximation, the measured quantum yields are identical with the probability for tunneling.

For a given set of values of the height and the width of the potential barrier, we calculated these probabilities. When we plotted the potential versus the distance, the probability or the quantum yield was a parameter. The next problem was to fit the experimental quantum yields, obtained at constant wave lengths at different electric fields, to this nomogram and to determine, whether the values for the potentials and for the widths are of resonable magnitude. The procedure is shown on Figure 5. As an example, a set of quantum yields 0,008/0.020/0.032/0.052 at 365 nm is given. These quantum yields were obtained at different electric fields $2 \times 10^5/4.5 \times 10^5/7 \times 10^5/12 \times 10^5$ V/cm, corresponding to different energy gains ΔU. These lines of different energy gains ΔU cross the curves of the quantum yields. Crossing points which belong together are combined by the auxiliary curves A, B, C, D, E and F. The auxiliary curves A, B, C, D, E and F are some of the general solutions of the problem. From these general solutions, we must pick those solutions which can be interpreted physically. A physical requirement is that without external electric field the quantum yields should quickly approach zero. An other physical demand is that with increasing electric field (regard crossing points 1, 2, 3, 4) the width of the potential barrier should not increase. When there is any significant change of the width at increasing electric fields, the width should slowly decrease. The auxiliary curve C fits these demands satisfactorily, with some approximation.
From the crossing points 1, 2, 3 and 4 on curve C, we derive widths for the potential barrier be-

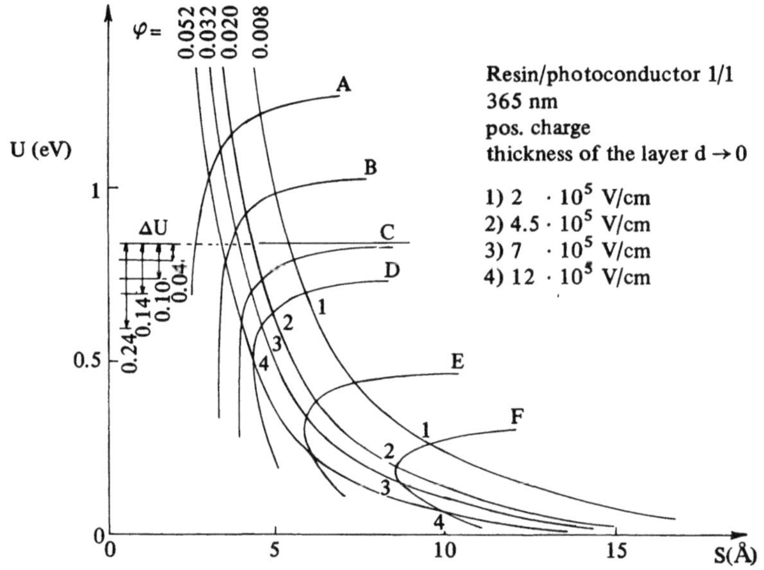

Fig. 5. Tunneling probability φ as function of the height U and the width S of the potential barrier

tween 4 Å and 5.5 Å. These values are in accordance with estimation by means of the calotte model and with the mean distance of the photoconductor molecules in the layer.

From the crossing points 1, 2, 3 and 4 on curve C, we also derive values for the height of the potential barrier of 0.61 eV to 0.80 eV and for the height of the potential barrier without external field of 0.83 eV. This last value for light of 365 nm, together with the corresponding values at 406 nm and 435 nm, is shown in the next table.

λ (nm)	U_0(eV)	ΔU_0(eV)	hv (eV)	Δhv (eV)
365	0.83		3.40	
		0.16		0.34
406	0.99		3.06	
		0.16		0.21
435	1.15		2.85	

The difference of the potentials, obtained from the measured quantum yields by application of the tunnel effect, agrees well with the energy difference of the incident light. It would not be justified to expect very close agreement in these energy values.

It seems therefore possible to describe the charge carrier formation in layers of 2,5-bis-(p-diethylaminophenyl)-1, 3, 4-oxadiazole by tunneling.

Discussion

Stockman, D. L.: How are you certain that the low apparent quantum efficiencies at low fields are not due to a transport limitation? As shown by Mort and Regensburger's structure referred to in my paper, the very high field dependence of the mobility can produce a fall off in apparent quantum efficiency at low fields.

Moraw, R.: Transport effects were neglected for two reasons:

1. The layers were discharged by light without residual charge.
2. The quantum yields as a function of the thickness of the layers between 2 and 17 μ are constant.

Weigl, J. W.: Your model seems to account for the increase in efficiency of photogeneration with decreasing wave-length by actual utilization of excess photon energy to reduce the barrier height ΔU. The very rapid thermalization of excess vibronic energy, which is expected in all complex molecules, would require tunneling to occur with extreme rapidity and efficiency, e.g. within 10^{-13} sec. Does this seem reasonable and is there any independent independence for such fast carrier separation?

Moraw, R.: The charge carrier formation must happen before the thermal relaxation, because the quantum yields are a function of the wave length of the incident light. Therefore, charge carrier formation must be a very fast reaction of less than 10^{-12} seconds.

Organic-Inorganic Photoconductor Systems

H. Meier, W. Albrecht and U. Tschirwitz

Abstract

Besides the inorganic photoconductive insulators in many cases electrophotographic layers contain organic compounds which can lead to different effects:

a) to spectral sensitization of the inorganic photoconductor by an injection of electrons (as in the case of ZnO) or holes (as in the case of selenium);

b) to an increase in photoelectric sensitivity in the manner of electrophotographic pn-systems by profiting from a space charge field formed between the inorganic photoconductor and the organic components,

c) to the formation of bipolar surface charges.

It is obvious to postulate the build-up of an electrostatic double-layer in the boundary phase between dye and semiconductor originating from electron transfer and leading to a compensation of different chemical potentials, thus causing these different effects.

The present paper deals with various experimental results which agree with this hypothesis. The following items will be discussed in detail:

a) the question of an n- or p-type conductivity of organic layers,

b) the positions of the energy levels and, especially, the problem of Fermi potentials or, in other words, chemical potentials of electronic charge carriers in organic systems, and

c) the electric properties of model systems built up from organic and inorganic photoconductors.

Finally, the different effects of inorganic-organic electrophotographic layers will be discussed by using pn- and nn -boundary layer models.

1. Introduction

Electrophotographic layers are often sensitized by various organic compounds. For instance, by addition of dyestuffs the spectral response of the inorganic photoconductor is extended through the visible spectrum, an effect which is the prerequisite for the practical application of the zinc oxide Electrofax system [1, 2, 3, 4].

Yet, it should not be overlooked that, besides this phenomenon, further effects can be attained by additives: Among the various techniques–beginning with the variation of the type of binder and reaching to the addition of organic materials, which are not dyestuffs–there are such which may significantly improve the photosensitivity of electrophotographic layers [5].

In the following sections, however, we shall restrict our discussion to effects which seem to be the result of an electronic interaction between an inorganic photoconductor and an organic layer. With this rectriction it is possible to discuss phenomena which probably come about by the participation of the electrical properties of organic systems. Therefore, the following problems shall be discussed:

1. Different effects in which inorganic/organic junctions may be involved.
2. Experimental results obtained in inorganic/organic model systems.
3. The mechanism of conductivity of organic materials and the question of energy levels in organic insulating materials.
4. The mechanism of photovoltaic effects of inorganic-organic junctions and the possible transfer of the results and mechanisms obtained in model systems to inorganic/organic systems present in electrophotographic layers.

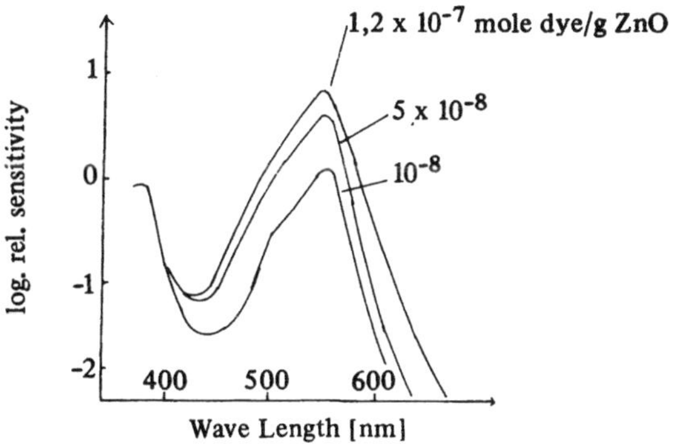

Fig. 1. Spectral sensitivity of ZnO-resin layers sensitized with 3,3′, 9-triethylthiacarbocyanine according to [6]

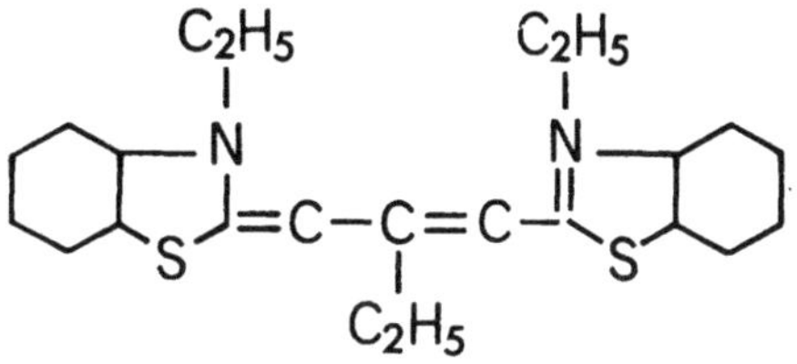

2. Effects Which May be Connected With Inorganic-Organic Junctions

Inorganic-organic photoconductive systems are involved in the following effects:

2.1. Dye Sensitization

In this well known process the energy absorbed by a dye in contact with a solid is transferred to the solid making it to show photoelectric effects. As mentioned above this effect is very important in the Electrofax process in which the electrosensitive layer is made of zinc oxide dispersed in a binder resin [5–10] because the intrinsic absorption of zinc oxide ends at 380 nm [11].

It is remarkable that there is generally an increase of photosensitivity with an increase of dye concentration as shown, for instance, in Figure 1 for zinc oxide-resin layers sensitized with thiacarbocyanine [6].

As a consequence of this rule various photoconductors could also be sensitized with thick dye layers: Sulfur sensitized with pinacyanol layers of different thickness, e.g. 2200 Å and 5000 Å, showed almost the same photosensitivity [12]. In amorphous selenium, sensitization increased with a dye-coating thickness up to 6000 Å [13]; Figure 2. With cadmium sulfide, which can utilized in binder layers [14] and in photoplastic recording films [15], spectral sensitization could be observed with films of merocyanine dyes or pinacyanol having a thickness up to one micron [16, 17]; Figure 3.

In summary, we see that the highest sensitivity is obtained in electrophotographic layers which consist of photoconductor particles coated with monomolecular and also with thick dye films. This fact clearly supports the hypothesis of a charge-transfer mechanism of spectral sensitization [18]. In this context also should be mentioned the adsorption condition and, moreover,

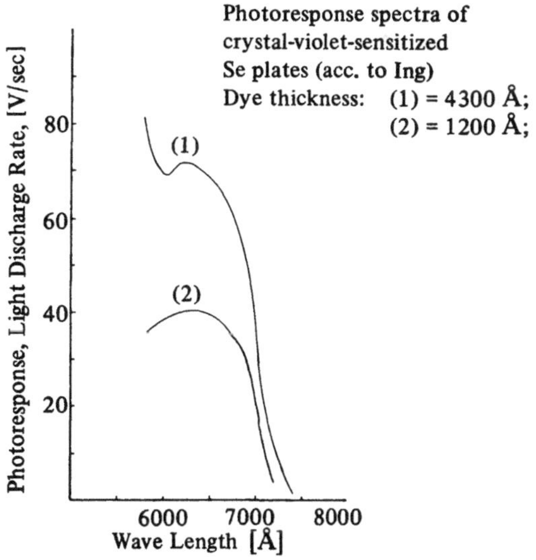

Fig. 2. Photoresponse spectra of selenium sensitized with crystal violet; according to [13]

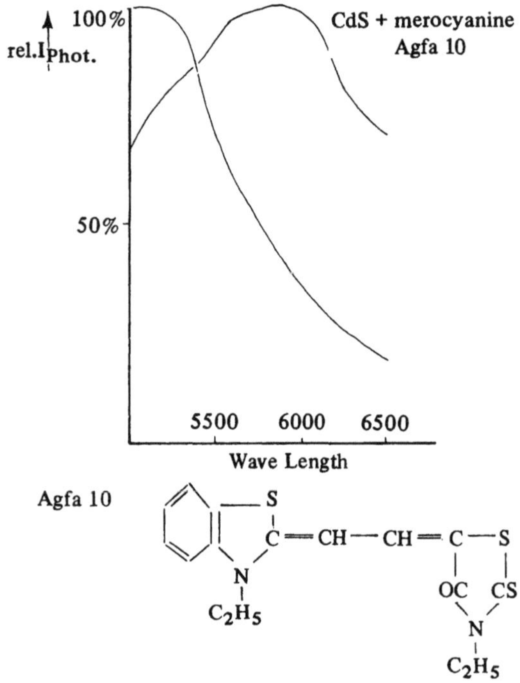

Fig. 3. Photoresponse spectra of sensitized CdS (merocyanine layer ≈ 1 micron)

the observation made during photodischarge experiments in amorphous selenium that by the excitation of the sensitizer methylene blue and phenosafranine holes, and by the excitation of crystal violet and pinacyanol electrons are injected into the photoconductor [13].

2.2. Increase of Photoelectric Sensitivity in the Manner of Electrophotographic pn-Systems

In a process developed by Shepard and Shely [19] a junction between an n-type layer and a p-type layer is used for electrophotographic reproduction using an electrolytic development process. These combinations, for instance of a p-type indium antimonide and an n-type zinc oxide, result in an increased photoconductivity as compared to the use of a single photoconductive layer. The authors did not describe the application of organic semiconductive layers in these photoresponsive elements. However, a xerographic deformation imaging process is reported employing the use of a four-layer plate structure comprising the following layers [20]: a conductive support substrate, a layer of vitreous selenium having a thickness of about 50 micron, an interlayer in the range of 0.1 to 5.0 microns consisting of methylene blue, Indigo blue or nigrosine in a resin binder, and an electrostatically deformable plastic imaging layer (Fig. 4). Experiments showed an enhancement of resolution by the organic layer.

Surely, systems with mixed inorganic and organic photoconductors of the p- and n-types are different to the last-mentioned sandwich structures. However, it should be remarked that according to Hauffe [21] systems consisting of n-type zinc oxide and p-type phthalocyanine are characterized by an enhanced photosensitivity as compared to zinc oxide or phthalocyanine alone.

2.3. Formation of Bipolar Surface Charges

According to experiments of Hauffe and coworkers [21] it is possible to overcome the restrictions on charge retention which are characteristic of the surface of n-type and p-type photoconductors [22]. It is observed that n-type

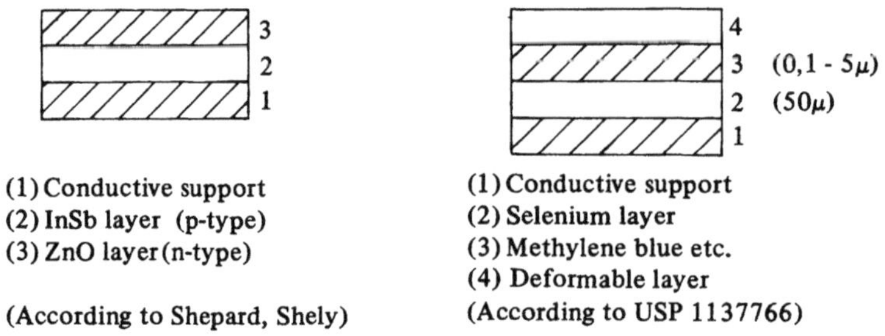

Fig. 4. Diagram showing the composition of electrophotographic copy sheets with pn- and inorganic/organic junctions

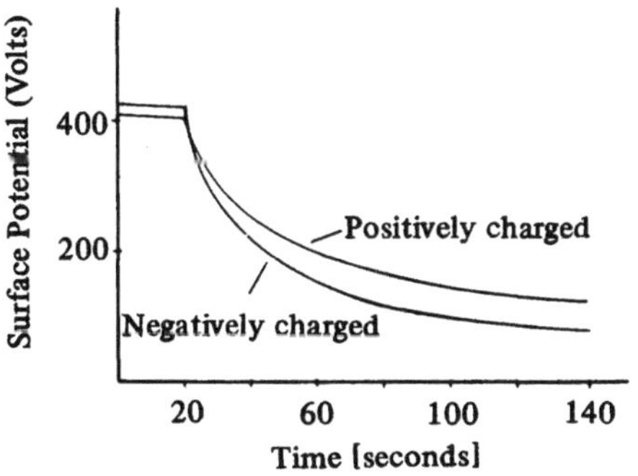

Fig. 5. Discharge characteristic of ZnO/stearic acid according to [23]

zinc oxide layers which, commonly, can be charged only negatively will be capable of accepting and retaining also a positive charge after the addition of p-type copper phthalocyanine. Obviously, this effect seems to be the result of forming boundary phases between the dye and the inorganic semiconductors. An analogous effect could also be observed with zinc oxide after addition of stearic acid [23]; Figure 5.

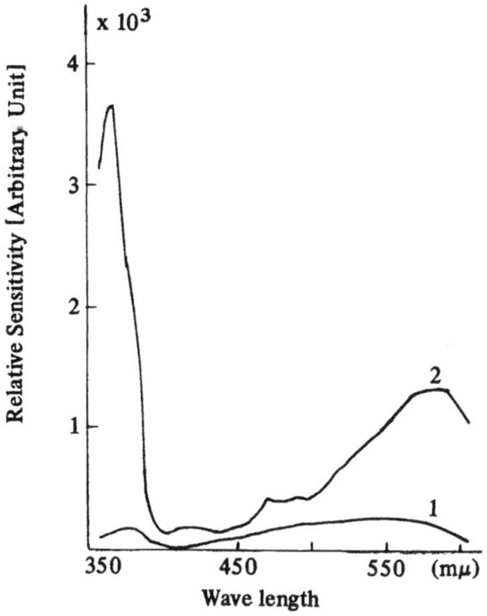

Fig. 6. Photoresponse spectra of rose bengal sensitized ZnO: Hypersensitization according to [24, 25].
1. ZnO
2. ZnO + rose bengal + iodine

2.4. Additional effects

The following effects should be mentioned because an influence of electronic parameters of the dye component in inorganic-organic systems seems probable:

a. The hypersensitization observed by Inoue and coworkers [24, 25] which is characterized by an enhancement of photoresponse of dye-sensitized zinc oxide by the addition of electron acceptors (such as o-chloranil, phthalic anhydride) which are able to increase the dark- and photoconductivity of the used sensitizers [26]; Figure 6.

b. Pre-Exposure effects which describe the changes in electrophotographic properties—above all an enhancement of the background—of dye-sensitized zinc oxide papers [27], and which seem to be the result of an influence of the dye component during the exposure immediately before use.

3. Photovoltaic Effects of Organic-Inorganic Models

The examples given above show that the junction of a photoconductor film and a dye layer may be involved in different electrophotographic processes. In discussing this fact it must be noticed that, upon illumination, models of organic-inorganic junctions produce photovoltaic effects. Nelson and coworkers [28, 29] have studied the photovoltage of a number of ionic dyes on cadmium sulfide, Grossweiner and coworkers [30, 31, 32] investigated zinc oxide-dye bilayers and we ourselves [33, 34, 35] made experiments on simple model photocells consisting of cadmium sulfide-dye and silver halide-dye. The scheme of our arrangements is shown in Figure 7.

The first arrangement (a) consists of a semiconductor (I) and a dye layer (II) placed in series between two electrodes which form a gap between 0.3–0.5 mm. In the second arrangement (b) an inorganic photoconductor deposit and a

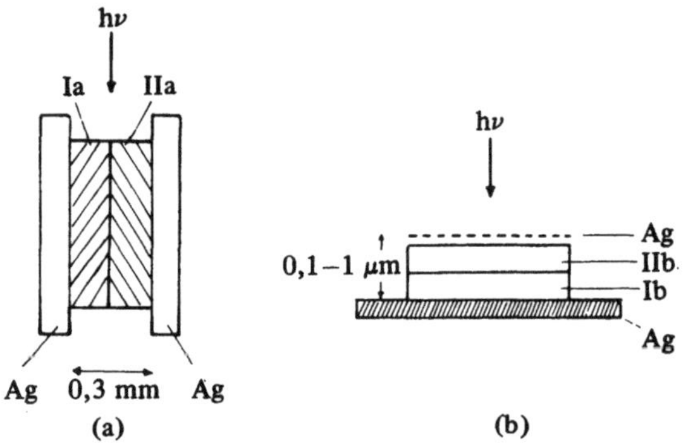

Fig. 7. Arrangements of inorganic-organic model photocells (see text)

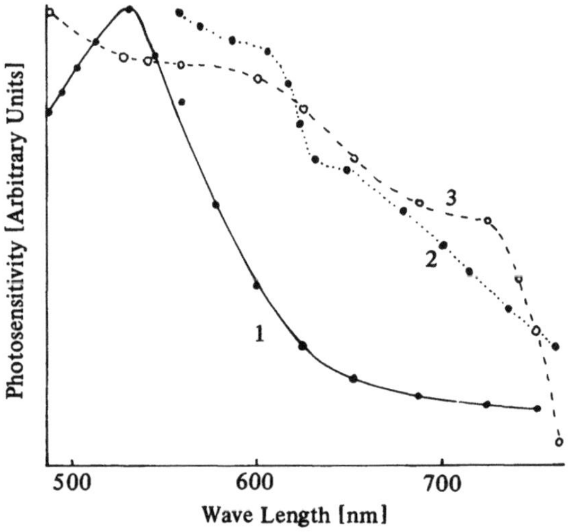

Fig. 8. Spectral photoresponse of
1 CdS
2 CdS + merocyanine A 10 (photoconductivity)
3 CdS/merocyanine A 10 – photocouple

sublimed dye film, 0.1–1 micron in thickness, were placed between a base electrode and a transparent front electrode.

Some of the experimental results obtained with these arrangements can be summarized as follows:

1. On illumination photocurrents up to 10^{-8} ampere and photovoltages up to 300 mV are produced. The fact that series-arrangements produce photovoltaic effects proves that this effect cannot be the result of a Dember effect [17].

Table 1. Photovoltages of CdS/dye systems

CdS + dye	max. photovoltage
Merocyanine A 10	280 mV
Merocyanine FX*	500 mV
Phthalocyanine	210 mV
Pinacyanol	300 mV
Malachite green	30 mV
Crystal violet	30 mV

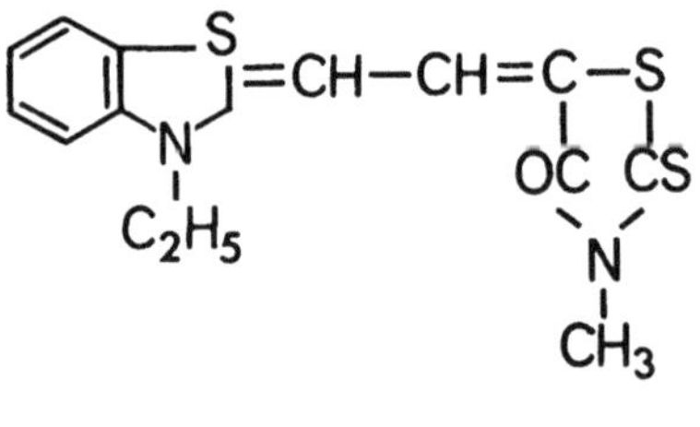

2. The action spectra for photocurrent or photovoltage of the semiconductor dye cells approximately follow the optical absorption spectra (Fig. 8).

3. The photocurrent increases linearly and exponentially, resp., with light intensity (Figs. 9 and 10).

4. The stability of photovoltage and photocurrents proves that there exists steady exchange of electronic charge carriers between the inorganic photoconductor and the organic film. Moreover, we can derive from the photovoltaic effect that the barriers between the dye molecules and the inorganic component are permeable to electronic charge carriers.

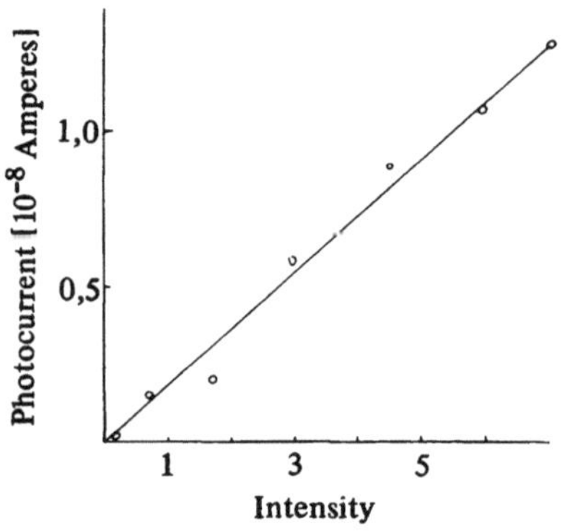

Fig. 9. Photocurrent versus intensity. CdS/merocyanine A 10 – cell (structure of the dye see Fig. 3); λ = 5580 Å

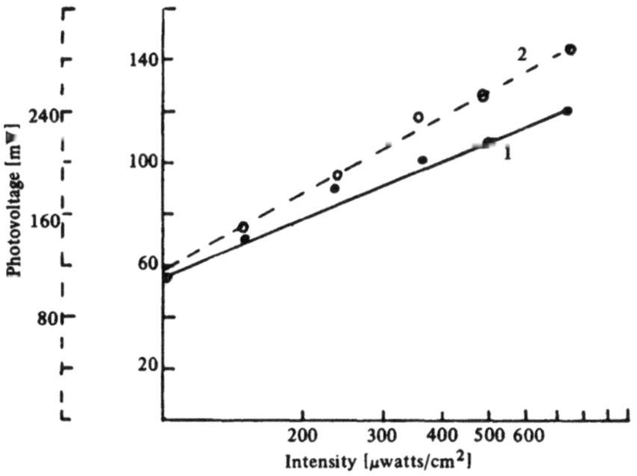

Fig. 10. Photovoltage vs. intensity
1. CdS/merocyanine A 10 – cell
2. AgI/rhodamine B – cell

From the experiments the capability of inorganic-organic junctions for producing photovoltaic effects is proved. The question about the mechanism of this capability, however, remains unsettled, and in this context, it is above all the function of the dyestuff which must be clarified.

4. Some Remarks to the Mechanism of Photoconductivity of Dyes

From a series of experiments it is known that organic dyes are photoconductors. It would be beyond the scope of this paper to discuss the mechanism of photoelectrical effects in organic systems in detail. However, it should be stressed that steady photocurrents are the result of a complicated interaction between the following steps [36]:

a) The direct or indirect generation of electrons and holes in the dark and during illumination;
b) the transport of electronic charge carriers in narrow energy bands with the width of the order of kT at room temperature;
c) the recombination of electrons and holes determined by centres of recombination and traps;
d) the distribution of electrons and holes among available energy states, i.e. between the conduction band, valence band and traps, in accordance with Fermi-statistics.

From these processes results the dependence of the photocurrent on different parameters such as light intensity, electrical field-strength, wavelength and others. Moreover, for the explanation of the behaviour of organic-inorganic junctions the following facts must be taken into special consideration:

1. The quantum yield of the photoconductivity of dye layers is enhanced with decrease in the layer thickness [37, 38], as will be shown in the next table.

Table 2. Effect of dye-thickness on photoconductivity

Dye	Layer thickness	Quantum yield
Pinacyanol	7,500 Å	0.03
	1,800 Å	0.13
	750 Å	0.3
Crystal violet	1,000 Å	0.06
	700 Å	0.13
	300 Å	0.33

From this observation it can be derived that monomolecular layers and small isles of dye aggregates should be highly photosensitive because impurities and other defects are diminished. According to quantum mechanical calculations five molecules are sufficient for the formation of energy bands [39].

2. Insulating dye films can be classified in p- or n-type photoconductors [40, 41]. In many cases this may be the result of an asymmectric trapping which—according to Rose [42]—results from a different number and depth of hole and electron traps. For instance (Fig. 11), with a large number of traps filled with electrons—e.g. from donating centres—the holes of excited electron-hole pairs are rapidly trapped and therefore to a great portion immobile. The electrons, however, are distributed between the conduction band and high-lying shallow traps so that the solid is characterized by an n-type photoconductivity. On the other hand, in the presence of accepting centres it is possible that only low lying traps are filled and that a large portion of empty traps is ready for trapping electrons. This solid is characterized by p-type photoconductivity.

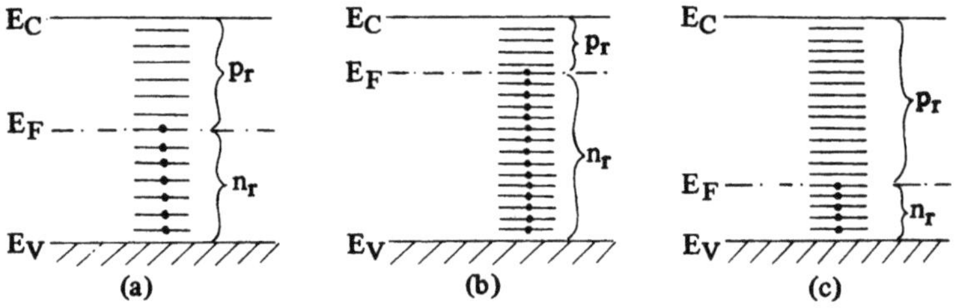

Fig. 11. Dependence of n-type or p-type of organic photoconductors on trapping

(a) Symmetric trapping
(b) Asymmetric trapping: n-type
(c) Asymmetric trapping: p-type
p_r, n_r = empty and filled centres, resp.
E_F = Fermi energy

3. According to the last mentioned classification the steady state Fermi energy of electrons of illuminated n-type dyes should be situated close to the conduction band and that of p-type dyes close to the valence band. In fact, this can be observed: Measured contact potentials against silver, i.e. the thermal work function or steady state Fermi level during illumination, reflect the classification of n-type and p-type dyes. This is shown in table 3 and in the Figure 12 [18, 43].

4. In accordance with the last mentioned theory it is also possible to define Fermi potentials for the non-illuminated dyes which are situated below the Fermi potentials of electrons during illumination [44]. Moreover, by addition of acceptor centres–such as tetracyanethylene or o-chloranile–the Fermi potential of p-type dyes is displaced towards the valence band because of an enlargement of the hole density (Fig. 13). This effect, too, is in agreement with the general semiconductor theory of inorganic solids.

To summarize, it is possible to explain the electrical behaviour of dye layers similar to inorganic

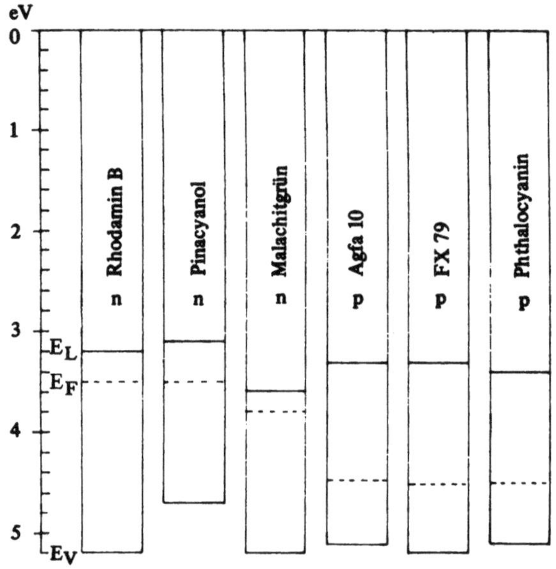

Fig. 12. Energy diagram of n- and p-type dyes

Table 3. Fermi-Energy of Illuminated Dye Films

Dye	Contact potential against Ag [mV]	Type	Thermal work function during illumination
FX 79 (merocyanine)	+ 200	p	4.5
Agfa 10 (merocyanine)	+ 250	p	4.45
Erythrosine	+ 200	p	4.5
Pinacyanol	+1100	n	3.6
Malachite green	+ 700 (+900)	n	4.0 (3.8)
Rhodamine B	+1200	n	3.5
Crystal violet	+ 900	n	3.8
Orthochrom T	+1100	n	3.6
Phenosafranine	+ 800	n	3.9

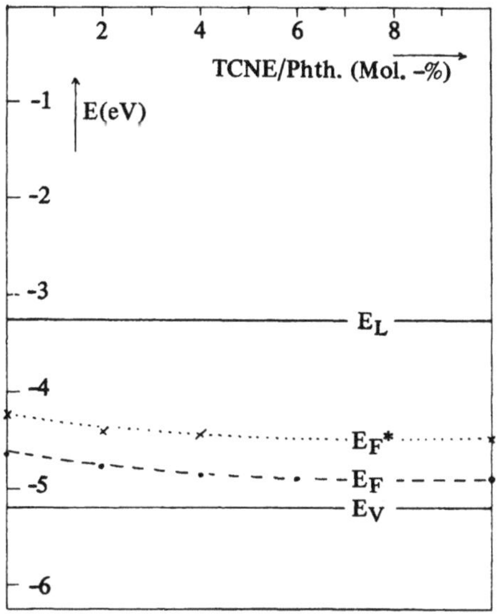

Fig. 13. Shifting of Fermi potentials in Phthalocyanine by doping with tetracyanethylene
(E_F = dark Fermi potential; E_F^* = Fermi potential during illumination)

photoconductors: For instance, we can postulate conduction bands and valence bands, traps, Fermi potentials etc.

5. Energy Levels in Organic Dye Layers

By using the band model for organic dye-photoconductors the discussion of the electrical behaviour or organic-inorganic heterojunctions will be facilitated. Above all it is possible to obtain the disposition of the dye energy levels by using methods which are known in the field of semiconductors.

5.1.Definition and Measurement

The various electric energy levels are indicated in Figure 14. The values are given in electron volts relative to the zero reference energy level at infinity. We have to distinguish the following levels and methods of determination [18]:

1. The macropotential eV_M which results from macroscopic space charges, double layers etc. and which must be taken as the reference level

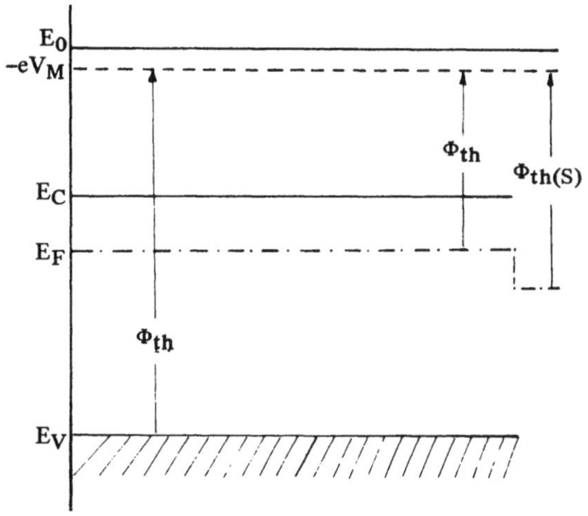

Fig. 14. Scheme of energy levels

for all energy levels. It is important that any displacement of the macropotential which may be, for instance, the result of an electron injection into traps etc. will displace the other levels. Therefore, in junctions with charge interchange there are macropotential differences across the junctions.

2. The position of the valence band which is identical with the ionization potential or the photoelectric work function ϕ_{phot} and which can be obtained from the threshold wave-length for the external photoelectric effect. An example of such a measurement is given for phthalocyanine in the next figure.

3. The position of the conduction band the bottom of which is identical with the electron affinity. This level can be obtained, for example, from the difference between the photoelectric work function and the energy gap ΔE which can be derived from measurements of temperature dependence of dark conduction, or from the spectral dependence of photoconductivity, or from absorption spectra.

4. The location of the Fermi potential E_F below the zero level which can be obtained by the following methods:

Firstly, from measurements of the thermoelectric work function ϕ_{th} which itself can be obtained by the measurement of the contact potential difference between the solid and any metal.

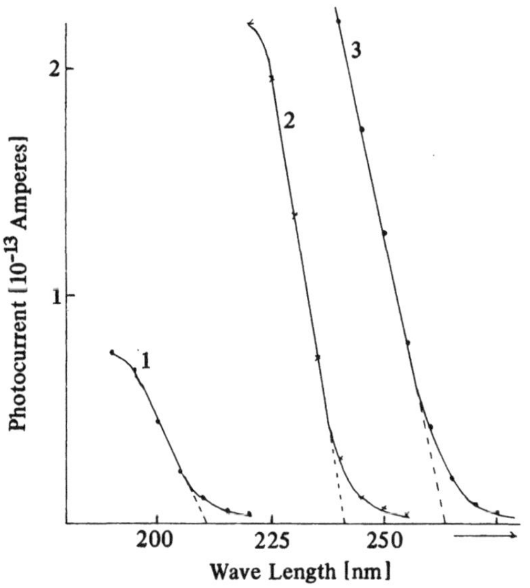

Fig. 15. Measurement of the threshold wave length for the external photoelectrical effect.
2 phthalocyanine (for comparison: 3 Ag; 1 poly-N-vinylcarbazole)

This is possible because the contact potential difference is equal to the difference in the work functions. The methods of determining are, among others, (see [18]):

a) the electron beam retardation method, and

b) the measurement of the external photoelectric effect (i.e. photocurrent vs voltage) because the saturation voltage U_S corresponds to the contact potential; i.e.

$$E_F \approx \phi_{dye} = \phi_{metal} \mp eU_S \ (U_S \lessgtr 0) \ . \qquad (1)$$

With this method the values given in table 3 are obtained [43]. Secondly, measurements of the thermoelectric power Q [mV/°K] can also be used for the determination of the steady state Fermi potential [44] because we have the approximation

$$E_F \approx e \cdot Q \cdot T \ . \qquad (2)$$

With this method the values given in Figure 13 are obtained.

5.2. Results

In the next table the values of the important energy levels of some inorganic semiconductors and some dyes are summarized.

From these results the following facts concerning the position of the energy levels can be derived:

1. According to Scheibes rule [50] the conduction band of many organic layers–i.e. the first excited singlet level–is situated at 3.0 to 3.5 eV below the vacuum level [18, 43]; see also Figure 12.

2. Because of the influence of gases, above all in zinc oxide, the absolute position of the bottom of the conduction band and other levels cannot be given exactly; see [45]. This fact should not be overlooked in discussing the relative positions of the electronic energy levels between dyes and semiconductors, which are given for the example of zinc oxide, cadmium sulfide and dye in the Figure 16. For instance, in clean zinc oxide the conduction band can be shifted up to 3.2 eV.

Table 4. Absolute Position of Energy Levels and the Fermipotential

Substance	Type	E_V[−eV]	E_C[−eV]	E_F[−eV]	Literature
TlI	p	5.6	3.0	5.4	[45, 46]
AgBr	p	6.0	3.5	5.3 (4.6)	[45, 46, 47, 48, 49]
ZnO	n	7.3; 6.8	4.3; 3.7	(4.5) (3.9)	[45, 46, 48, 49]
CdS	n	5.5	3.5	(3.8)	[47]
Phthalocyanine	p	5.2	3.25	4.25*; 4.7	
Pinacyanol	n	5.25	3.13	3.66*	
Malachite green	n	5.5	3.5	4.0*	
Poly-N-vinylcarbazole	p	5.9	2.6	4.7*	

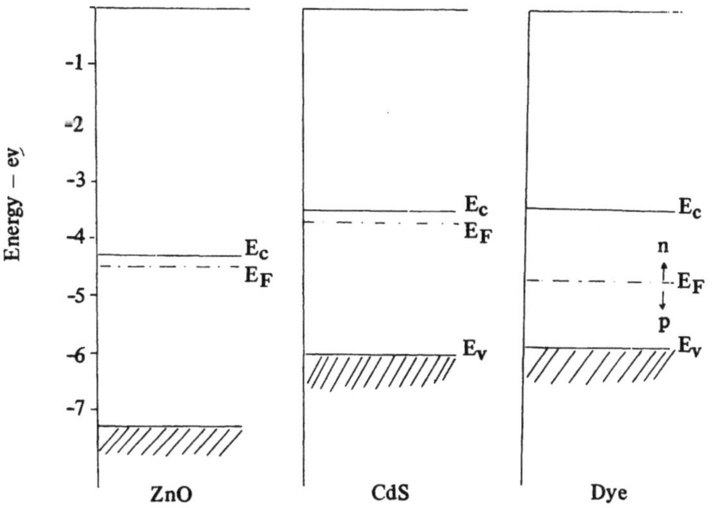

Fig. 16. Absolute position of energy levels and the Fermi potential in ZnO, CdS and dye stuffs

3. According to the p-or n-type of a dye layer the steady state Fermi potential in the dark or during illumination is situated close by the valence or conduction band. From this fact it can be concluded that dye layers contain a high density of trapping levels and other centers which are involved in the distribution equilibrium of electrons and holes.

6. Mechanism of the Electrical Behaviour of Organic-Inorganic Junctions

For describing the photovoltaic effects of dye-semiconductor junctions two cases can be discussed:

Firstly, according to Nelson's hypothesis [29] there is no charge interchange between the dye and the semiconductor and therefore no macropotential difference across the junction in the dark. In this case the photovoltage–during illumination–results from a transfer of electrons from the dye to the semiconductor so that the conduction band of the dye must be situated above the conduction band of the inorganic solid.

From this model we cannot expect any influence on charge retention, or any enhancement of photosensitivity as has been discussed above for zinc oxide binder layers, or any dependence on light intensity.

Secondly, according to models discussed by Grossweiner [31, 32] and by ourselves [33, 34, 18, 51, 52], there is a charge interchange between the dye and the semiconductor in the dark which results in a macropotential difference across the junction.

Grossweiner proposed a Schottky barrier model for zinc oxide/dye couples in which zinc oxide conduction electrons are transferred to trapping levels of the dye in the dark and are returned when the dye is illuminated (Fig. 17). This model was supported by measurements of a resistance increase in zinc oxide films induced by dyeing, as shown in Figure 18.

In our opinion heterojunctions between inorganic and organic films behave substantially in the same way as purely inorganic or organic junctions. These heterojunctions are characterized by an electron transfer in the dark from the system with the higher Fermi potential to the system with the lower Fermi with a consequent build-up of an electrostatic double-layer across the junction as is shown in the simple scheme of Figure 19 and in the energy diagram of Figure 20.

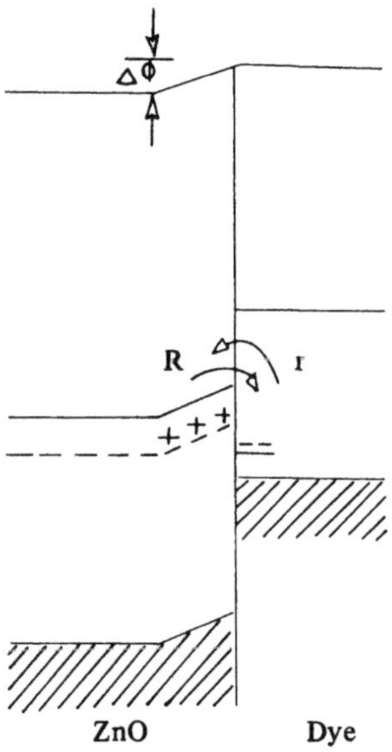

Fig. 17. Schottky barrier according to Grossweiner et al. [31, 32].
R, r = rate constants for thermally-activated processes;
g = hole-electron pairs/cm^3 sec

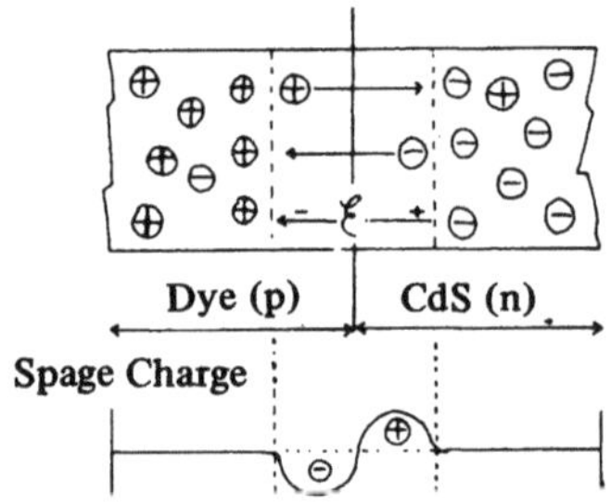

Fig. 19. Simple scheme of space charge barriers in pn-junctions

The build up of a macropotential difference is the consequence of lining up the Fermi potential throughout the system. It is important to take into consideration that a potential difference can be expected in all those cases in which there exists a difference of electron or hole densities between contacted materials, for instance between two n-conductors with various electron concentrations. The electrostatic potential is always more positive in the n-type material or in the region with the greater electron concentration, and more negative in the p-type material or in the region with the greater hole concentration.

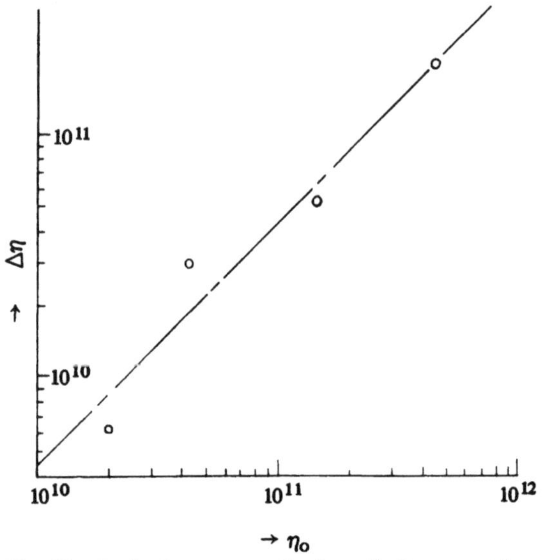

Fig. 18. Dark electron transfer from ZnO to crystal violet (according to Grossweiner et al. [31, 32]).
Δn = number of electrons transferred to the dye
n_0 = initial number of conduction electrons in ZnO

However, the transfer of electrons into a dye can only take place if the difference between the thermal work function of the semiconductor and the electron affinity of the dye is smaller than 1.5 eV or if positive holes in the valence band or deeply situated electron acceptors are present in the dye. The last mentioned case is equivalent to the case discussed by Grossweiner.

As a consequence of the junction potential built-up in the dark various effects can be expected:

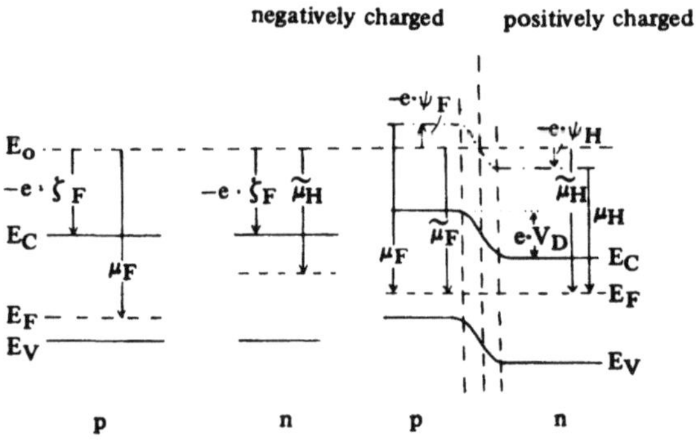

Fig. 20. Formation of a macropotential difference in the dark in inorganic-organic junctions

1. As already described above, the generation of photocurrents and photovoltages as the result of the separation of optically formed electron-hole pairs by the internal electric field. We see that upon illumination the density of free electrons of an n-type semiconductor must be increased when connected with a p-type dye or also with an n-type dye if the difference between the chemical potential of the semiconductor $|\mu_s|$ and that of the dye $|\mu_D|$ is below zero:

$$|\mu_s| - |\mu_D| < 0 \text{ n-type photosensitization} \quad (3)$$

$$|\mu_s| - |\mu_D| > 0 \text{ p-type photosensitization} \quad (4)$$

Table 5 clearly demonstrates that the observed direction of the electron transfer agrees with the direction predicted by the barrier theory.

In our opinion spectral sensitization may be in many cases the result of corresponding dark and photoprocesses.

For instance we can explain zinc oxide or cadmium sulfide photosensitization, i.e. the enhancement of the electron density in the semiconductor by excitation of the dye, because of the characteristic differences between the Fermi potentials of dyes and n-type photoconductors.

2. Assuming that the boundary layer between inorganic and organic systems acts similarly to a pn-junction we can expect an enhancement of the photoelectric sensitivity under the influence of an electrical field. According to this assumption, for example, the photocurrent of a cadmium sulfide/merocyanine couple showed a higher value with a positive charging of the n-type cadmium sulfide than by a negative charging (Fig. 21). This pn-photoeffect may be important in the electrophotographic pn-systems mentioned above.

3. It seems to be possible that the formation of a pn-junction can partly be responsible for a change of the rules of accepting and retaining positive and negative charges by p- and n-type semiconductors. For instance, it can be expected that by combining an n-type semiconductor with a p-type dye positive charges may be retained by the n-type semiconductor because in this case the pn-junctions can be biased in the barrier (reverse) direction.

4. Moreover, it is to be expected that by increasing the hole concentration in the non-il-

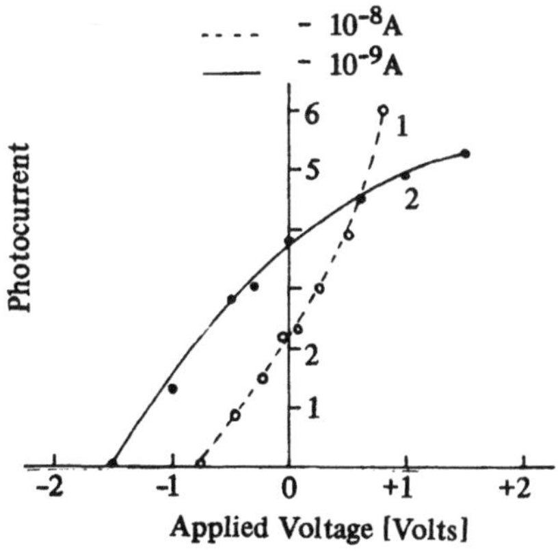

Fig. 21. pn-photoeffect of inorganic-organic model systems under the influence of an external electric field
1. Malachite green (n)/Merocyanine A 10 (p)
2. CdS (n)/Merocyanine A 10 (p)
Light intensity: 10^3 μW/cm^2 (4000–8000Å)

Table 5. Polarity of the Photovoltaic Effect in Inorganic-Organic Photoconductor Systems

Inorganic Photoconductor (I)		Dye		Photo-Charge experiment I	experiment II	pn-theory I	pn-theory II	Literature
CdS	(n)	Phthalocyanine	(p)	–	+	–	+	
CdSe	(n)	Merocyanine	(p)	–	+	–	+	
ZnO	(n)	Eosin-Na	(p)	–	+	–	+	[30]
AgBr	(p)	Rhodamine B	(n)	+	–	+	–	
AgI	(p)	Malachite green	(n)	+	–	+	–	
TlI	(p)	Phenosafranine	(n)	+	–	+	–	[9]
Malachite green	(n)	Merocyanine	(p)	–	+	–	+	

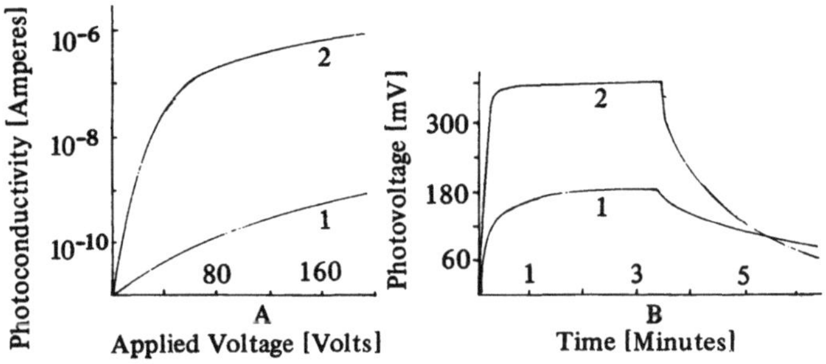

Fig. 22. Model of hypersensitization

A. Photoconductivity
1. Merocyanine A 10
2. Merocyanine A 10 + o-chloranil

B. Photovoltaic effect
1. AgI/merocyanine A 10
2. AgI/merocyanine A 10 + o-chloranil

luminated dye there must be an increased junction potential difference between the n-type semiconductor and the dye and, consequently, an increased photovoltaic effect. This could be observed by doping the dye component of dye/cadmium sulfide couples with electron acceptor molecules (Fig. 22). We believe that the hypersensitization effect observed by Inoue can be explained according to this mechanism.

7. Conclusion

In summary, we see that the experimental facts show that the concept of an junction potential formed in inorganic-organic systems in the dark can give an explanation for many observations made in electrophotographic layers. In this context, however, we should not neglect the influence of chemisorbed oxygen which may also build up a surface barrier layer superimposing the junction potential of inorganic-organic photoconductor systems.

We think that the conception of the formation of electrostatic double layers between dye aggregates and photoconductors–similarly to the generation of boundary layers under the influence of gases–should not be overlooked.

Reference

[1] Young, C. J., and H. G. Greig, RCA Rev. **15,** 471 (1954).

[2] Giaimo, E. C., RCA Rev. **23,** 96 (1962).

[3] Frieser, H., and M. Schlesinger, Phot. Korr. **101,** 69, 133 (1965).

[4] Frieser, H., and M. Schlesinger, Phot. Korr. **8,** Sonderheft (Elektrophotographie) 43 (1966).

[5] Taft, D. D., and S. C. Heidecker, Tappi **50,** A 36 (1967).

[6] Frieser, H., and M. Schlesinger, Phot. Sci. Engng. **12,** 17 (1968).

[7] Inoue, E., H. Kokado and T. Yamaguchi, J. phys. Chem. **69,** 767 (1965).

[8] Hauffe, K., V. Martinez, J. Range and R. Schmidt, Phot. Korr. **104,** 113 (1968).

[9] Terenin, A., and I. Akimov, J. phys. Chem. **69,** 730 (1965).

[10] Comizzoli, R. B., J. Appl. Phys. **41,** 4148 (1970).

[11] Mollwo, E., Z. phys. Chem. **198,** 258 (1951).

[12] Ing, S. W., and Y. S. Chiang, J. Chem. Phys. **46,** 487 (1967).

[13] Ing, S. W., and Y. S. Chiang, J. Chem. Phys. **46,** 478 (1967).

[14] Smith, M., and A. J. Behringer, J. Appl. Phys. **36,** 3475 (1965).

[15] Aftergut, S., J. J. Bartfai and B. C. Wagner, Applied Optics, Suppl. 3., 161 (1969).

[16] Nelson, R. C., J. Opt. Soc. Amer. **51,** 1182 (1961).

[17] Meier, H., and W. Albrecht, Ber. Bunsenges. phys. Chem. **68,** 64 (1964).

[18] Meier, H., Spectral Sensitization, Focal Press, London–New York, 1968.

[19] Shepard, J. W., and B. L. Shely, U.S. Patent 3410767; November 12, 1968.

[20] Brit. Pat. 1137766, 27. Dec. 1968.

[21] Hauffe, K., personal communication.

[22] Hauffe, K., and R. Stechemesser, Phot. Sci. Engng. **11,** 145 (1967).

[23] Kurita, T., Jap. J. Appl. Phys. **5,** 977 (1966).

[24] Inoue, E., T. Yamaguchi and I. Maki, Kogyo Kagaku Zasshi **66,** 428 (1963).

[25] Inoue, E., H. Kokado and T. Yamaguchi, J. Phys. Chem. **69**, 767 (1965).
[26] Meier, H., and W. Albrecht, Z. phys. Chem. N.F. **39**, 249 (1963).
[27] Langston, D. J., TAPPI **49**, 16 (1966).
[28] Nelson, R. C., J. Opt. Soc. Am. **46**, 1016 (1956); J. Chem. Phys. **29**, 388 (1958); J. Phys. Chem. **69**, 714 (1965).
[29] Petruzzella, N., J. Ferrier, M. Schaer and R. C. Nelson, J. Chem. Phys. **50**, 3527 (1969).
[30] Dudkowski, S. J., and L. I. Grossweiner, J. Opt. Soc. Am. **54**, 486 (1964).
[31] Dudkowski, S. J., A. G. Kepka and L. I. Grossweiner, J. Phys. Chem. Solids **28**, 485 (1967).
[32] Grossweiner, L. I., and W. D. Brennen, J. Phot. Science **17**, 189 (1969).
[33] Meier, H., and W. Albrecht, Ber. Bunsenges. phys. Chem. **68**, 64 (1964).
[34] Meier, H., and W. Albrecht, Ber. Bunsenges. phys. Chem. **69**, 160 (1965).
[35] Meier, H., and W. Albrecht, in: Dye Sensitization Symposium Bressanone (W. F. Berg, U. Mazzucato, H. Meier, G. Semerano eds.). Focal Press London–New York, 1970, p. 250–260.
[36] Meier, H., Phot. Kolloquium der ETH Zürich, 1. 7. 1971 (will be published).
[37] Noddack, W., and H. Meier, Z. Elektrochem. **63**, 971 (1959).
[38] Noddack, W., H. Meier and A. Haus, Z. phys. Chem. **212**, 55 (1959).
[39] Sukigara, M., and R. C. Nelson, Mol. Phys. **17**, 387 (1969).
[40] Meier, H., Z. wiss. Photogr., Photophys. and Photochem. **53**, 1 (1958).
[41] Meier, H., Die Photochemie der organischen Farbstoffe, Springer-Verlag, Berlin 1963.
[42] Rose, A., RCA Rev. **12**, 362 (1971).
[43] Meier, H., and W. Albrecht, Ber. Bunsenges. phys. Chem. **69**, 917 (1965).
[44] Meier, H., W. Albrecht and U. Tschirwitz, Ber. Bunsenges. phys. Chem. **75**, 938 (1970).
[45] Terenin, A., and I. Akimov, Z. phys. Chem., Leipzig, **217**, 307 (1961).
[46] Terenin, A., and E. Putzeiko, Proceedings of the International Colloquium Liége, 1959, Pergamon Press, Oxford, 1962, pp. 532–543.
[47] Nelson, R. C., J. Opt. Soc. Am. **46**, 1016 (1956).
[48] Akimov, I., V. M. Bentza, F. I. Vilesov and A. N. Terenin, Phys. stat. sol. **20**, 771 (1967).
[49] Terenin, A. N., I. A. Akimov, V. M. Bentza and F. I. Vilesov in: Dye Sensitization Symposium Bressano (W. F. Berg, U. Mazzucato, H. Meier, G. Semerano, eds.). Focal Press, London–New York, 1970, pp. 187–192.
[50] Scheibe, G., D. Brück and F. Dörr, Ber. dtsch. chem. Ges. **85**, 867 (1952).
[51] Meier, H., Chimica **18**, 170 (1964).
[52] Meier, H., J. phys. Chem. **69**, 719 (1965).

Discussion

Verhille, K.: It certainly is most interesting to describe such systems as ZnO-dye by the band-model as you have just given. Especially the evidence presented for the photocurrent yield as a function of bias voltage in such systems is stricking. ZnO-resin layers which if suitably formulated, are bichargeable show also a lesser photoresponse when charged positively, although the effect is in the order of a factor 2 instead of an order of magnitude.

Meier, H.: The cited examples show that there exists the possibility for a positive charging of zinc oxide after addition of organic substances (see e.g. also T. Kuridu: Jap. J. Appl. Phys. 5 (1966)).

Stockman, D. L.: Have you any explanation for the large difference of the photosensitivity between α- and β-(x-) phthalocyanine?

Meier, H.: In the cases that we have studied there is a correspondence between the absorption spectrum and the photoconductive response, the intensity dependence is linear and this would suggest an extrinsic generation mechanism. Thus it would not be surprising if differences were noted in yield between the α and β forms.

Hauffe, K.: On the basis of our experiments, α-copper phthalocyanine resin layers cannot be charged significantly either positively or negatively. The reason for this behavior seems to be the much larger concentration of oxygen solved in the α-compound causing a large concentration of holes $|e|^{\cdot}$ according to the mechanism

$$O\text{(dissolved in Cu-Ph)} \rightarrow O^{-}\text{(dissolved in Cu-Ph)} + |e|^{\cdot}$$

Therefore, it is supposed that this large concentration of holes is responsible for the fast-dark decay and presumedly also for the large difference of the photosensitivity between α- and β-copper phthalocyanine.

Electrophotographic Properties of Copper Phthalocyanine-Zinc Oxide Systems

K. Hauffe, N. I. Ionescu, A. Meyer-Laack and K. Petrikat

Abstract

The nature of the conductivity in β-copper phthalocyanine and the electron transfer through the interfaces zinc oxide/phtalocyanine and phthalocynanine/chemisorbed species charged either positively or negatively were studied by means of a micro-probe discharge technique. The dark-decay and the speed of discharge under illumination of the following photoconducter systems

Al | In | ZnO Single Crystal | Cu Phthalocyanine Resin

Al | Cu Phthalocynanine Resin

Al | ZnO Powder Resin | Cu Phthalocynanine Resin

Al | ZnO Powder + Cu Phthalocynanine Resin charged up under a DC-operating corona discharge, have been investigated.

It has been found that a positive surface charge on a copper phthalocyanine layer disappeared rapidly when illuminated with light of wavelength ⩾ 550 nm indicating that the singlet excitons generated in the surface region of the phthalocyanine layer are responsible for the fast discharge. Irradiating with light of 370 nm which produces electron-hole pairs in zinc oxide acting as support of the phthalocyanine-resin layer, has no significant influence on the discharge of the positively charged layer system. These results suggest that the generated singlet excitons react with the positive surface charge, the nature of which is not quite understood, whereas the freed positive holes in the phthalocyanine migrate toward the zinc oxide interface and recombine with the free electrons available in zinc oxide due to the barrier-free indium or NESA glass electrode. Further information has been obtained when the surface was negatively charged. Here, only UV-radiation (= 370 nm) was able to increase noticeably the speed of discharge indicating that generation of electron-hole pairs in zinc oxide was needed for the discharge of the negatively charged surface. This result becomes understandable if it is assumed that holes are injected from zinc oxide into phthalocyanine, coupled with rather high mobility in phthalocyanine of these holes which react by recombination with the trapped electrons in the surface.

Based on this study, layers containing only β-copper phthalocyanine resin without and with zinc oxide exhibit exellent electrophotographic behavior and can be used as bichargeable material.

1. The Problem

In order to get more information on the electron transfer through the interface of an inorganic and organic photoconductor, it is desirable to measure the dark and photocurrent through the interface zinc oxide/copper phthalocyanine by means of the electrophotographic charging and discharging technique. Since zinc oxide is a well-known n-type photoconductor a detailed introduction of its electrical properties can be omitted. However, a short introductory discussion on the electrical properties of copper phthalocyanine is necessary for the understanding of the conduction and charge-transfer mechanism.

Copper phthalocyanine—written below in the abbreviation Cu-Ph—is an industrially important blue pigment which resists bleaching. It is available in the α or β modification depending on the gas pressure and the temperature of the support during precipitation. An x-ray amorphous modification appears if the evaporated Cu-Ph is precipitated on a support the temperature of which is $< 30°$ C [1]. A slightly chlorinated product is also available which was employed for some measurements. The mechanism of the electrical conductivity of Cu-Ph, particularly of the β modification is presently not completely understood. The study of the electrical properties is complicated because of its very low conductivity of $< 10^{-14}\,ohm^{-1}\,cm^{-1}$. In addition, the study is further complicated by a significant solubility of oxygen which acts as trapping center for electrons in phthalocyanine. As justifiably supposed, the very small intrinsic conductivity is changed to a p-type conductivity of several orders of magnitude higher due to the presence of the electron acceptor oxygen within the crystal and on its surface [2–6]. The strong change in the surface conductivity demonstrates dramatically the electronic interchange of the adsorbed gases with Cu-Ph [7, 8]. Electron spin-resonance measurements did not give any indication of a charge-transfer complex. Only oxygen as a "free radical" could be detected [9].

Due to its significant photoconductive behavior, Cu-Ph seems to be a suitable material for electrophotographic layers*). In the beginning it was believed that the photoconduction should be caused by a direct excitation from the ground state into the conduction band [10]. Later however, it was found that it is caused by a singlet-exciton excitation [11]. These electrically neutral species may dissociate under various experimental conditions and act as charge carriers. Recently, the photo-effect was studied particularly in metal-free phthalocyanine [12] for the purpose of examining the generation and the transport of charge carriers. The exponential increase of the quantum yield with temperature forces us to assume that the singlet excitons generated by illumination (~1.8 eV) either dissociate thermally or react with the adsorbed species. Therefore, the charge carrier production occurs obviously either by dissociation due to phonons or by a surface reaction. The energy of singlet excitons has been determined from the fluorescence spectrum to be 1.77 eV [4]. Since β-Cu-Ph exhibits superior electrophotographic behavior in comparison to α-Cu-Ph, generally only β-Cu-Ph was used in this study. The trap density depends to a great extent on the annealing temperature and atmosphere—as for instance oxygen or hydrogen; it was determined for β-Cu-Ph to be about $6 \times 10^{15}\,cm^{-3}$ [13]. A notably high concentration of traps ($10^{19} - 10^{20}\,cm^{-3}$) located 0.38 eV below the edge of the very narrow conduction band, was found by Barbe and Westgate [21]. In this connection a paper by Devaux and Schott [22] must also be quoted. Worth mentioning are also the relatively small mobilities of electrons and holes with $u_- = 3 \times 10^{-3}$ and $u_+ = 2 \times 10^{-2}\,cm^2/volt\text{-}sec$ [13]. The surprisingly small quantum yield for charge-carrier generation, $< 10^{-4}$ at 25° C [12], may be due to a fast recombination the mechanism of which is at present not quite clear. In very pure crystals, the recombination rate is governed by a bimolecular rate law [18, 19]. But in doped crystals, the rate of recombination follows a monomolecular rate law. In the case of oxygen-doped crystals, charge carriers are generated by a reaction of singlet excitons with oxygen atoms with a quantum yield of nearly 1. The influence of oxygen on the photoconductivity was reported in several papers [8, 2, 14].

*) Because of the blue color of Cu-Ph, it is particularly recommended for electrophotographic offset printing plates and for reusable purposes.

Worth mentioning is the infrared photoconduction at about 1100 nm which is increased after an annealing in oxygen and has practically disappeared after a hydrogen treatment [6]. Since the infrared photoconduction was detected only in Cu-Ph and not in metal-free phthalocyanine, the assumption seems to be obvious that the singlet-triplet absorption responsible for the infrared photoconduction, must be due to the spin-orbital coupling of the copper ion in Cu-Ph [6, 15]. This has earlier been pointed out as a possible singlet-triplet interaction by Terenin [8].

The influence of the surface states on the photoconduction is significant. The depth of the surface states is about 0.3–0.6 eV [16]. In spite of the dependence of the photoconductivity on the atmosphere, the action spectrum remains unchanged [17]. According to our present knowledge, we may assume that the rather large concentration of surface states ($> 10^{11}\,cm^{-2}$) and of trapping centres caused by the presence of oxygen, is mainly responsible for the noticeable p-type conductivity. Also the possibility of the appearance of space-charge limited currents must be considered for the evaluation of the conduction mechanism as was pointed out by Sussman [20].

On the basis of this short review on the electrical properties of Cu-Ph and under consideration of the study of its structure by Honigmann et al. [23], an attempt will be made to elucidate the mechanism of the charge-transfer through the interface Cu-Ph/ZnO and the charge transport through Cu-Ph. Therefore in one case a Cu-Ph resin layer coated on a ZnO resin layer and vice versa and in the other case on a ZnO single crystal was employed.

In the beginning, a few experiments on Cu-Ph evaporated on aluminum and on Cu-Ph resin layers coated on aluminum will be reported because these results are rather informative for the main problem in this paper. The two choices of support–either aluminum or zinc oxide–permitted further information on the electron-transfer mechanism to be obtained.

2. Experimental

For the study of the electrophotographic properties of Cu-Ph and of the charge transfer through the interface Cu-Ph/metal and Cu-Ph/ZnO, suitable resin toluene dispersions have been coated either on aluminum or on zinc oxide single crystals. Additionally, Cu-Ph resin layers have been coated also on zinc oxide resin layers as shall be dealt with using a few selected examples. The resins used were silicone resins from Wacker Chemie, the RP 665 from Monsanto and Sinolac® 600S from G. M. Langer & Co. The $(000\bar{1})$ face of the ZnO crystal had an indium layer evaporated on it to guarantee a barrier-free contact and was provided with a silver paste electrode. While the surface of the crystal amounted to about 0.2 cm^2, the geometric surface of the Cu-Ph resin layers on metal or on a ZnO resin layer was several cm^2. This fact forced us to employ two different measuring systems.

For the measurements with single crystals coated with suitable Cu-Ph resin layers, the following measuring device as represented in Figure 1, was employed [25]. In a small measuring chamber consisting of a brass block (I) with a revolving gold plated brass cylinder (II) and a sample holder (III), the coated zinc oxide crystal can be brought selectively either under the corona discharge (1) or under the translucent probe (5) at a distance of 0.5 mm. The chamber could be evacuated and flushed with various gases at different pressures. Air at 1 atm was used. The probe was a thin glass disk covered with evaporated gold and transparent in the wave-length region employed. The conductive surface of the probe facing the coated surface of the crystal was connected to the input terminal of a Keithley vibrating electrometer.

The Cu-Ph supporting ZnO crystal is charged under a corona operating with a dc high-voltage device of ± 12 kV. After the charging it is swung under the probe. We employed gold as a standard for the reference probe. Through the transparent probe, light of various wave lengths was irradiated on to the charged sample. In our

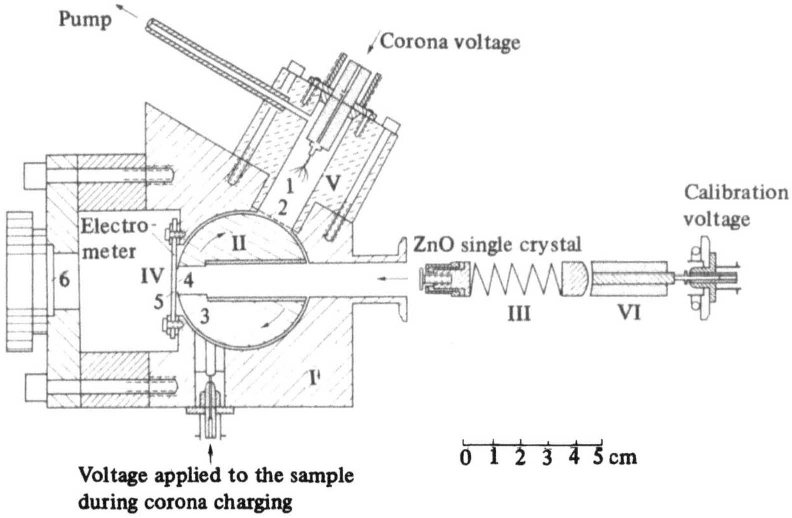

Fig. 1. Schematic representation of the experimental set-up for charging and discharging measurements of small single crystals.
I - a brass block as chamber, II - the movable cylinder with the ZnO sample, III and VI - the sample holder with the voltage device, IV - the translucent probe and V - the housing with the corona 1 and the screening grid 2

charging technique [25], the screened grid (2) was grounded and the rear of the sample was connected with a calibration voltage of ± 300 volts. By this technique, satisfactorily reproducible surface voltages were obtained.

For the charging and discharging experiments with layers of larger areas, another measuring unit was employed, represented in Figure 2. Also here, the measuring chamber can be evacuated and filled with gases at various pressures. By means of a revolving turntable (4) supporting the sample and the standard as reference, the sample can be brought at will either under the corona (5) for charging or under the probe (2) which is transparent for the employed wave lengths irradiated via a surface mirror (1). The measuring technique was reported in detail elsewhere [24].

Threefold sublimated Cu-Ph in its β modification and also a β-Cu-Ph sample supplied from the laboratory of the BASF*) with an average

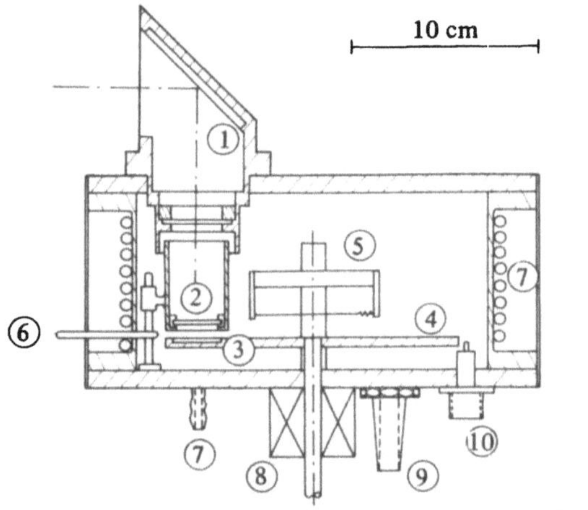

Fig. 2. Schematic drawing of the measuring unit for charging and discharging measurements of larger areas of layers.
1 - surface mirror, 2 - probe, 3 - sample, 4 - rotating table, 5 - corona, 6 - light conductor, 7 - thermostat, 8 - ball-bearing and packing, 9 - vacuum and 10 - electrical connection

chlorine content of 0.6% were mainly employed for the experiments dealt with in this report. A few experiments have also been carried out with α-Cu-Ph and with strongly chlorinated copper phthalocyanines. Generally, equal weight parts

*) We appreciated very much the supply of refined β-copper phthalocyanine by Dr. B. Honigmann of BASF.

(1:1) but also 2:1 weight parts of Cu-Ph and of silicone or Sinolac or RP 665 resin in toluene were dispersed by means of an ultrasonic rod. Often a further addition of toluene had to be used in order to ease brushing for the coating. It has been found that a small addition of methanol improved the adhesion of the layer on aluminium.

3. Experimental Results

3.1. Copper Phthalocyanine Layers Evaporated on Aluminum

Neither a positive nor a negative charging was possible with layers of Cu-Ph evaporated between 10^{-3} and 10^{-4} torr at 500–550° C on an aluminum foil at room temperature which was heated up to 40–50° C during the precipitation. A subsequent annealing of these layers at 10^{-4} torr between 200 and 220° C for about 1 hour resulted in a small improvement with respect to positive charging (about + 12 volts). No negative charging, however, was detectable.

If the experiment was repeated under the above mentioned conditions with the exception that now the aluminum support was heated up to 200–220° C, then layers were obtained which could be charged up positively from 300 to 500 volts and negatively from 60 to 100 volts. A subsequent annealing at 120° C in air of 1 atm decreased the ability for charge acceptance which could be recovered by annealing in nitrogen of 10^{-2} torr. In all cases, however, the dark-decay was so fast, that a discharge under illumination was not of interest.

3.2. Copper Phthalocyanine-Resin Layers

The first experiments have been carried out with Cu-Ph silicone resin layers with Cu-Ph: resin weight ratios of 2:1, 2:3, 1:2 and 2:5. Generally, both the ability to charge positively and negatively increased with increasing resin content.

The highest discharge speed under illumination with the bright light of a xenon high-pressure lamp was obtained with a layer composed of 2:1 parts by weight of Cu-Ph to resin. As can be seen from Figure 3, the surface voltage of 310 volts is decreased to 40 volts in 0.1 sec after the beginning of illumination. However, after 3 seconds a residual voltage of 20 volts still remains. It is also worth mentioning that layers with a higher resin content (1:2.5) which could be charged up to 800–1000 volts, exhibited a high speed of discharge under illumination. The remaining residual voltage of about 100 volts, however, is very high.

As with evaporated layers, an oxygen treatment of these Cu-Ph resin layers at 250° C caused a decrease of both the negative and positive charging level. The initial value of the charge level could be regained by a vacuum annealing at 10^{-2} torr nitrogen. A treatment with hydrogen of 400 torr had no further influence on the electrophotographic behavior of these layers. The annealing temperature in these experiments was also 250° C. No influence was detected at an oxygen pressure <30 torr. A longer storage time of these layers in air resulted in an increase of the positive and in a decrease of the negative charge acceptance.

After these more general measurements, layers in the weight ratio 2:1 of phthalocyanine and silicone resin, listed in Table 1, were produced

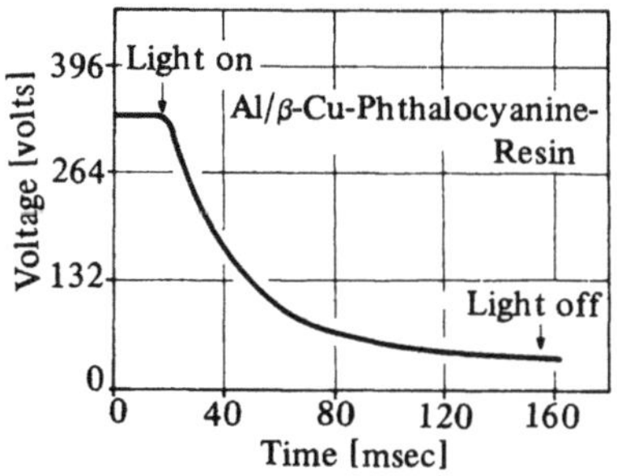

Fig. 3. The occurence of a discharge of a positively charged β-copper phthalocyanine resin layer (2 : 1) under illumination with the unfiltered light of a xenon high-pressure lamp

Table 1. Copper Phthalocyanine and Chlorinated Copper Phthalocyanine Powder Supplied from BASF

Sample No	Compound	Specific Surface m^2/g
1	β-Cu-Ph	68
2	β-Cu-Ph	15
3	α-Cu-Ph	69
4	α-Cu-Ph	37
5	α-Cl_{16}Cu-Ph	73
6	α-Cl_{16}Cu-Ph	29

and their electrophotographic behavior was studied after 1 day storage in the dark at 1 atm air. As may be seen from Figures 4 and 5, the fine grained β-Cu-Ph embedded in silicone resin showed the highest surface potential at both polarities (+ 290 and – 240 volts) and the fastest light-decay when illuminated via an edge filter with λ ⩾ 420 nm. The larger-grained β-Cu-Ph exhibited a significantly lower sensitivity, as can be seen from curve 2 in the Figures 4 and 5. The α-modification of Cu-Ph could not be charged up either positively or negatively. Obviously, the structure of Cu-Ph should have a

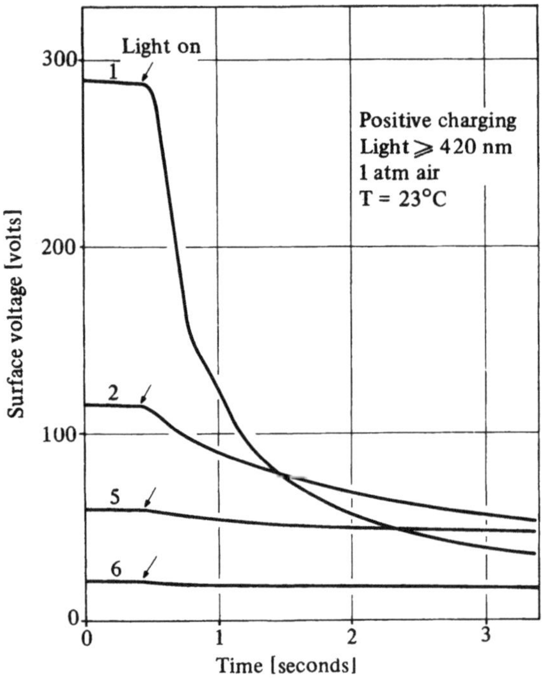

Fig. 4. The light-decay of positively charged β-Cu-Ph resin layers (curves 1 and 2) and α-Cl_{12}Cu-Ph resin layers (curves 5 and 6); see Table 1

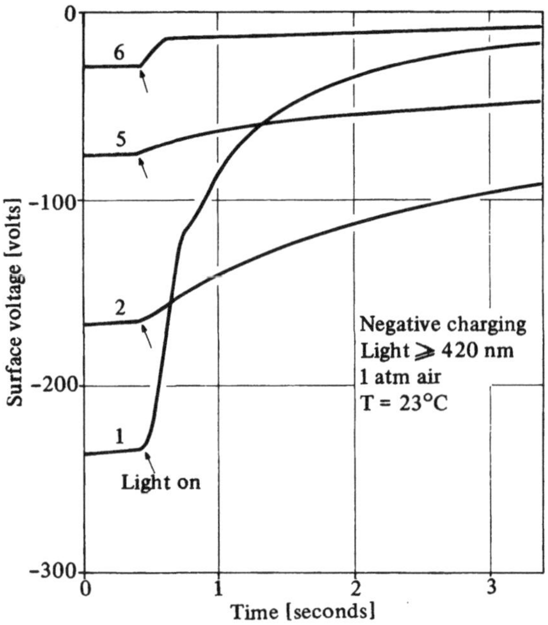

Fig. 5. Corresponding measurements with negatively charged layers as in Figure 4

large influence on its electrophotographic properties. As demonstrated by the curves 5 and 6 in the Figures 4 and 5, the chlorinated products exhibited only poor electrophotographic behavior.

A rather large decrease of the initial charge is observed with a 40 micron thick β-Cu-Ph silicone resin layer (2:1) with increasing temperature between 20 and 50° C (Fig. 6). The dark-decay, however, is roughly independent of the temperature. A similar result was obtained with a 30 micron layer of β-Cu-Ph embedded in Sinolac resin in the weight ratio 1:2. In analogy to the discharge experiments in Figure 4 with silicone as resin, the speed of the discharge of a 25 micron thick β-Cu-Ph Sinolac layer (1:2) at 23° C was determined at an illumination of $\lambda < 1300$ nm (curve 1) and between $420 < \lambda < 700$ nm (curve 2 in Fig. 7).

Informative for the elucidation of the conduction mechanism in Cu-Ph is also polarity and extent of the photovoltage of a β-Cu-Ph silicone resin layer (1:1) coated on aluminum. As can be seen in Figure 8, the surface potential,

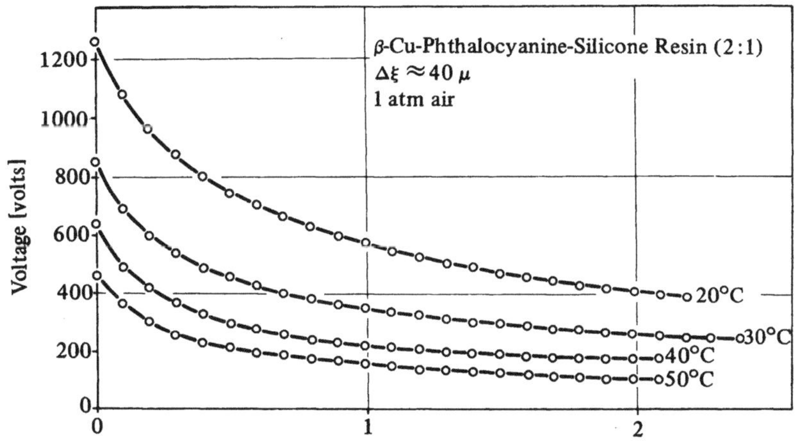

Fig. 6. Dark-decay curves of a β-Cu-Ph silicone resin layer (2:1) at various temperatures

measured versus a gold layer standard, was in the dark −130 mV if the phthalocyanine used was sublimed 3 times in vacuo of 10^{-4} torr nitrogen at 250° C. The illumination with light between 420 and 650 nm caused a significant increase to −260 mV (curve 1 in Fig. 8). When the light was switched off a fast return toward positive direction was found but without attaining the initial value. If the layer was annealed about 2 hours at 120° C in 1 atm air, the surface potential in the dark became −210 mV, but under illumination it exhibited exactly the same value of −260 mV as before annealing (curve 2 in Fig. 8). The subsequent dark period caused a return only to −240 mV indicating the irreversibility of the electron trapping in the surface presumedly due to oxygen chemisorption during the illumination period. A similar behavior was obtained in studies of the negative photovoltage with β-Cu-Ph layers evaporated on aluminum.

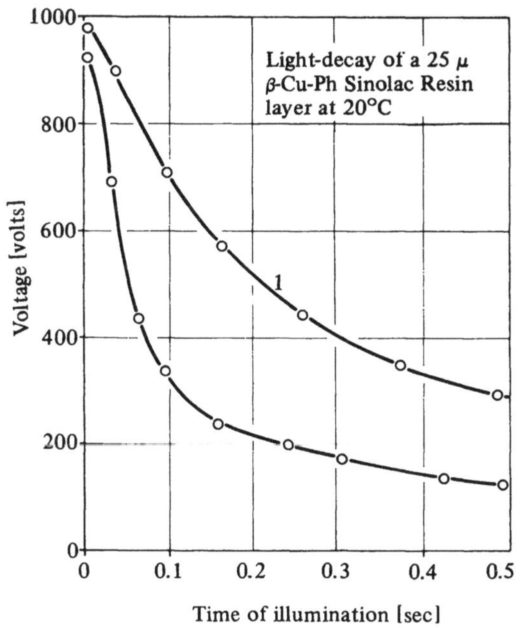

Fig. 7. Light-decay curves of a β-Cu-Ph Sinolac resin layer (1:2) at 20° C in air with light of λ < 1300 nm (curve 1) and of 420 < λ < 700 nm.

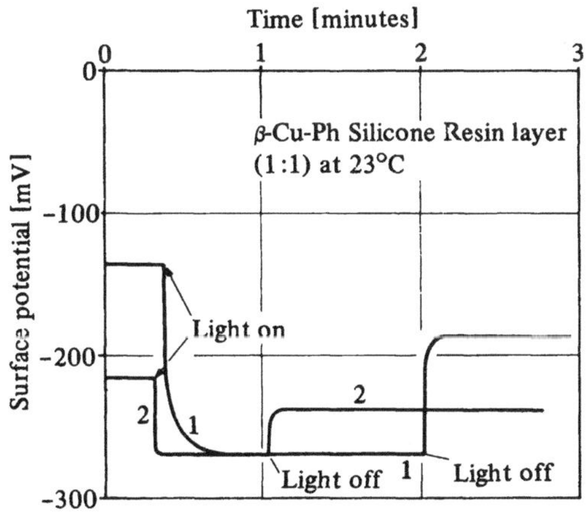

Fig. 8. The change of the surface potential due to illumination of a β-Cu-Ph silicone resin layer (1:1) in air at 23° C; curve 1: a layer before and curve 2: a layer after annealing in oxygen

3.3. Copper Phthalocyanine-Resin Layers on Zinc Oxide

In order to compare charging and discharging experiments in the dark and under illumination with Cu-Ph resin layers on aluminum as support, layers were coated on a photoconductive support with a barrier for the entrance of electrons in order to study the charge transfer through the interface Cu-Ph/inorganic photoconductor, such as zinc oxide. For these experiments, single crystals of pure zinc oxide were used. Two characteristic discharge curves with a positively and negatively charged Cu-Ph silicone resin layer (2:3) coated on an undoped ZnO crystal are represented in Figure 9. Curve 1 demonstrates the dark-decay and the speed of the discharge during illumination with the bright light of a mercury high-pressure lamp (HBO-500 watts) after a positive charging. The corresponding dark and light-decay after a negative charging is represented by curve 2 in Figure 9. An identical discharge curve after a positive corona charging with a larger time resolution (Fig. 10) demonstrates the fast voltage decay under illumination. We also determined the influence of the resin content on the initial potential, the speed of discharge under illumination and on the remaining residual surface potential. In Figure 11 the resin content is as follows: curves 1 and 4 (1:3), 2 and 5 (2:3) and 3 and 6 (1:1). With decreasing resin content it is observed that after a positive charging the initial potential is larger and after a negative charging smaller. The speed of the light-decay is faster and that of the residual surface potential is smaller.

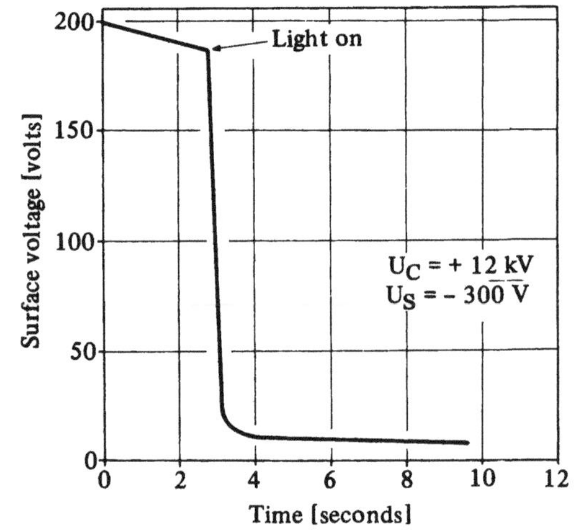

Fig. 10. Typical light-decay curve of a positively charged β-Cu-Ph silicone resin layer (1:1) coated on a ZnO crystal under the same bright light as in Fig. 9

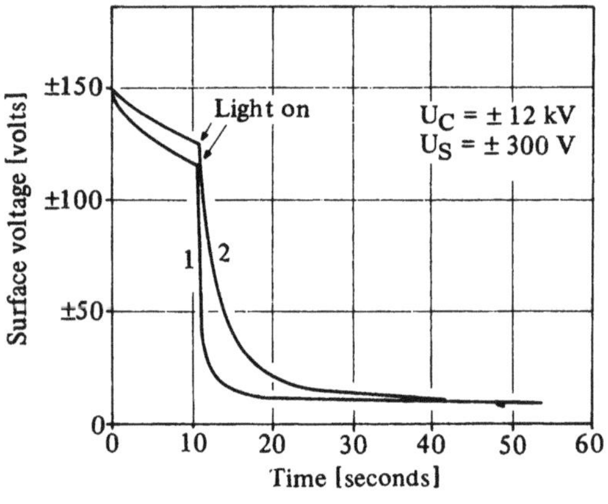

Fig. 9. Discharge under illumination with the bright light of a 500 watts mercury high-pressure lamp of a corona charged ß-Cu-Ph silicone resin layer (2:3) coated on a ZnO crystal; 1. after positive and 2. after negative charging

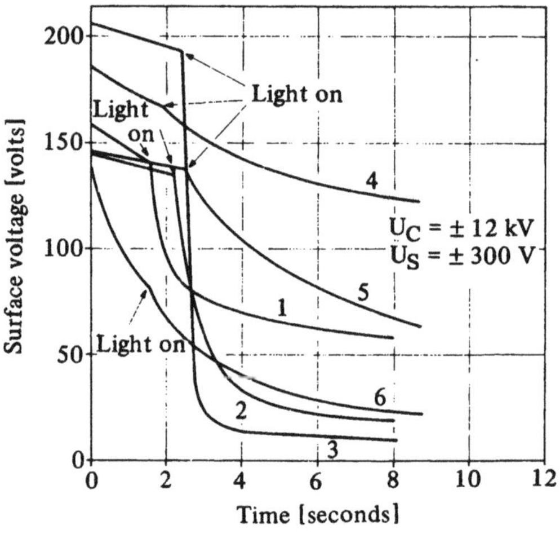

Fig. 11. Discharge curves of positively and negatively charged Cu-Ph silicone resin layers with various resin ratios under illumination with light of $\lambda \geqslant 520$ nm. 1. positive and 1:3, 2. positive and 2:3, 3. positve and 1:1, 4. negative and 1:3, 5. negative and 2:3, 6. negative and 1:1

For the first studies, the described discharge experiments were performed employing mostly the unfiltered light of a 500 watts mercury high-pressure lamp. Since β-Cu-Ph exhibits two maxima in its optical absorption spectrum, one at 550 and the other at 800 nm, it was obvious that one should investigate the sensitizing ability in the two optical maxima. In order to get more direct information on the electron transfer through the interface ZnO/Cu-Ph, a zinc oxide crystal was brushed with β-Cu-Ph crystallites so that a practically invisible distribution of very small Cu-Ph particles was available on the surface. This procedure was necessary because no solvent for Cu-Ph coating on zinc oxide is known. As can be clearly seen from Figure 12, the rate of discharge of the negatively charged ZnO single crystal with the fine β-Cu-Ph particles*) as sensitizer was increased only under illumination with light of 502, 550 and ⩾550 nm (see curves 2, 3 and 4 in Fig. 12) while an illumination with an 800 and ⩾780 nm light had no influence on the discharge (curve 5 is identical to curve 1 for the dark-decay).

The influence of the wave-length region on the speed of discharge of a Cu-Ph resin coated ZnO single crystal after both positive and negative charging can be seen from Figure 13. The layer used had a weight ratio of Cu-Ph and silicone resin of 1:1. A significant increase of the discharge rate after positive charging was obtained by illumination with light through an edge filter of ⩾520 (curve 1), ⩾450 nm (curve 2) and with full light (curve 7). Irradiation with a 370 nm light resulted in a slower discharge (curve 3). A negative charging with this layer was characterized by a noticeable dark-decay. Illumination through the edge filters had no influence on the speed of the discharge (curve 4 and 5). Only the full light of the mercury high-pressure lamp caused an appreciable effect on the discharge after negative charging (curve 6).

Additional information about the mechanism of the charging and discharging of Cu-Ph resin layers was obtained when the layer was coated on a ZnO crystal which was doped with Li_2O. As known [25], an incorporation of Li_2O into the lattice of zinc oxide, according to

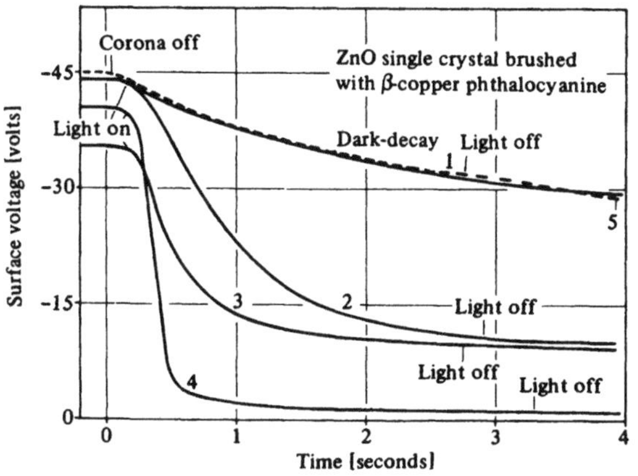

Fig. 12. The dark- and light-decay of a negatively charged ZnO crystal brushed with Cu-Ph crystals from which traces remaining on the surface operate as sensitizer.
1: dark-decay, 2: λ = 502 nm, 3: λ = 545, 4: λ ⩾ 550 nm and 5: λ = 808 and λ ⩾ 780 nm light

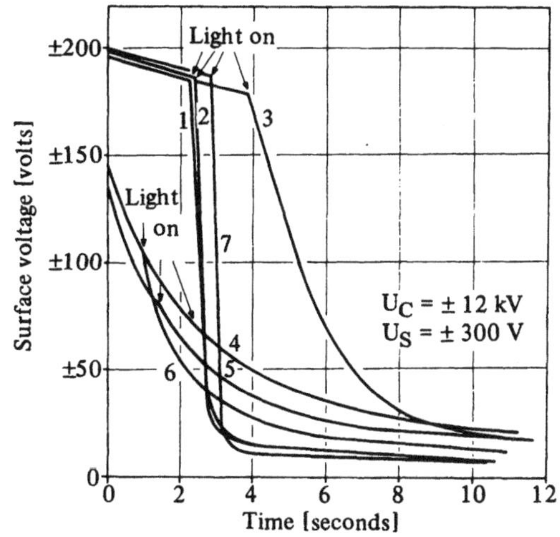

Fig. 13. The dark- and light-decay of a positively and negatively charged ß-Cu-Ph silicone resin layer (1:1) on a ZnO crystal at various wave lengths.
1. positive, λ ⩾ 520 nm, 2. positive, λ ⩾ 440 nm, 3. positive, λ ⩾ 372 nm, 4. positive, λ ⩾ 520 nm, 5. negative, λ ⩾ 440 nm, 6. negative and 7. positive with the full light of a 500 watts mercury high pressure lamp

*) The very fine β-Cu-Ph particles act as "pigment sensitizer" in contrast to the usual sensitizers which are adsorbed as molecules.

$$Li_2O + e' = 2\,Li\,|Zn|' + Zn^{\cdot} + ZnO \qquad (3.1)$$

decreases the concentration of free electrons e′ and increases the concentration of donors $Zn^{\cdot}$. Beside the generation of singly charged donors which have trapped an electron, also doubly charged donors $Zn^{\cdot\cdot}$ may be present. Such a doped ZnO crystal as support which is highly insulating, causes significant changes in the electrophotographic behavior of the Cu-Ph silicone resin layers. The fast dark-decay of a positively charged layer and the small height of its surface voltage (only 65–70 volts) is surprising. These experiments will be dealt with in detail elsewhere.

3.4. Copper Phthalocyanine Resin Layers in Contact with Zinc Oxide Resin Layers

In order to approach the systems used in electrophotography, we repeated similar charging and discharging experiments with the following double-layer systems:

Aluminum | ZnO + Resin | Cu-Ph + Resin (I)

Aluminum | Cu-Ph + Resin | ZnO + Resin (II)

Aluminum | ZnO + Cu-Ph + Resin (III)

Here, the Cu-Ph resin layer with a weight ratio 1:2 was either coated on a zinc oxide powder resin layer at a weight ratio of 6:1, or coated on aluminum while the ZnO resin layer now represented the external layer. The layer system (I) could be charged up positively to about 400–500 volts because of its rather thick coatings (20 and 50 microns) accompanied by a slow dark-decay (Fig. 14). The speed of the discharge under illumination with the full light of the lamp (curve 1 in Fig. 14) was satisfactory while the discharge with the light through an interference filter of 365 nm (curve 2) was rather slow. The illumination with light between $590 < \lambda < 1300$ nm, however, was also very effective for the discharge (curve 3). Because of the large difference in the light intensities, these results are of qualitative value. As may be concluded from curve 3, no

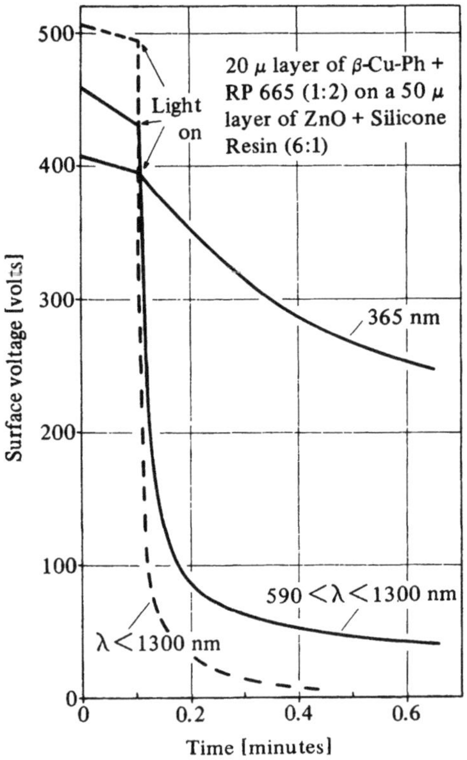

Fig. 14. Discharge curves under various illumination conditions of a 20 μ β-Cu-Ph RP665 resin layer (1:2) on a 50 μ ZnO silicone resin layer (6:1)

light with shorter wave lenghts, particularly at 380 nm, is necessary and, therefore, no generation of electron-hole pairs in the single ZnO grains is necessary for the charge transport through the ZnO resin layer.

Identical curves for the dark-decay and the discharge under illumination by means of various wave-length regions were obtained with the same Cu-Ph resin layer coated on a ZnO crystal after a positive (Fig. 15) and a negative charging (Fig. 16). As can be seen from the curves 3 and 4 in Figure 15 and from the curves 2 and 5 in Figure 16, the shorter infrared region of the incident light seems to be very important for the discharge.

Further valuable information was obtained by discharge measurements of a positively and negatively charged layer system (II) at 365 nm and by illumination through an edge filter of $\lambda \geqslant 590$ nm (Figs. 17 and 18). As demonstrated unequivocally

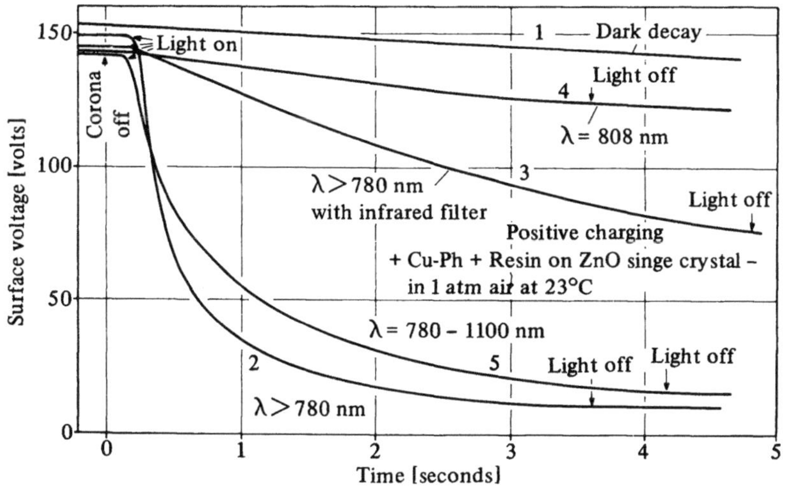

Fig. 15. Dark- and light-decay curves of a positively charged β-Cu-Ph resin layer coated on a ZnO crystal.

1: dark-decay, 2: λ ≥ 780 nm (edge filter),
3: λ ≥ 780 nm with an infrared filter, 4: λ = 808 nm and 5: λ = 780 - 1100 nm

for the discharge of the layer sequence: Al | Cu-Ph + Resin | ZnO + Resin | Air, the shorter wave length of about < 380 nm of the incident light seems to be very necessary for a fast discharge. In contrast to the results mentioned above, illumination through an edge filter of λ ≥ 590 nm exhibited only a poor efficiency.

Finally, zinc oxide powder (either Fotosiegel A or Azofax 661) together with β-Cu-Ph was dispersed in a silicone resin toluene solution,

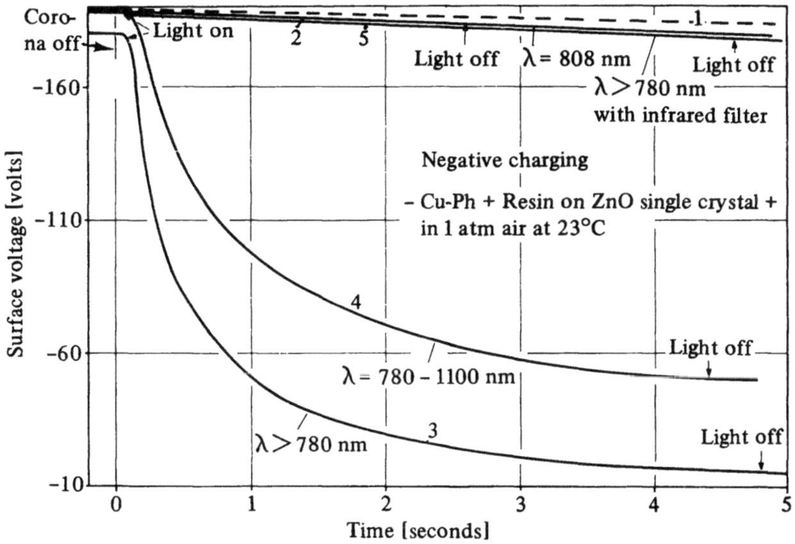

Fig. 16. Identical experiments as in Figure 15 after a negative charging.

1: dark-decay, 2: λ = 808 nm, 3: λ ≥ 780 nm,
4: λ = 780 - 1100 nm, 5: λ ≥ 780 nm with an infrared filter

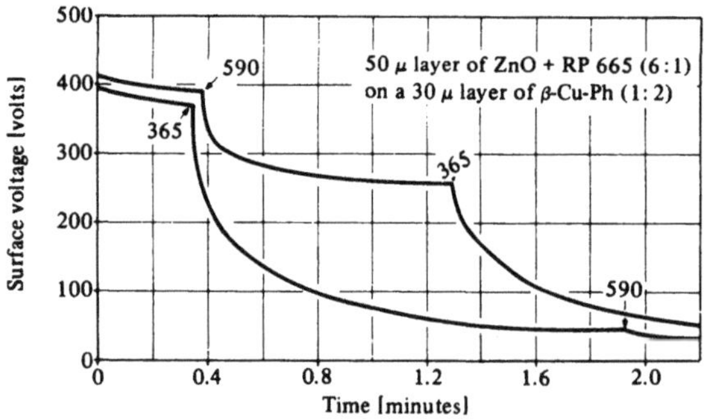

Fig. 17. Dark- and light-decay of a positively charged 50 μ ZnO RP665 resin layer (6:1) on a 30 μ β-Cu-Ph RP665 resin layer (1:2).
Illumination with 365 nm (interference filter) and $\geqslant$ 590 nm (edge filter)

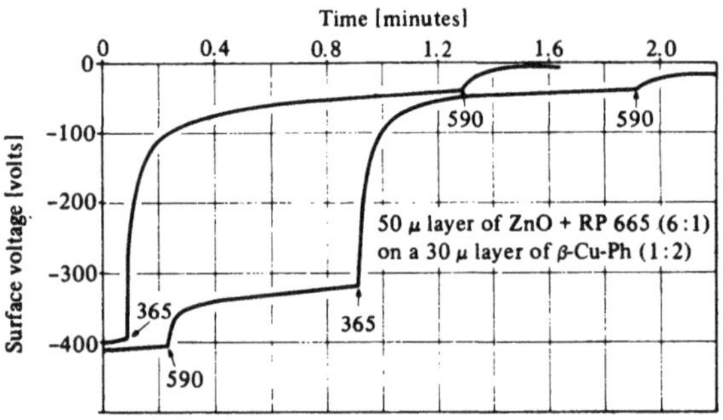

Fig. 18. Dark- and light-decay of the same layer as in Figure 17 after a negative charging

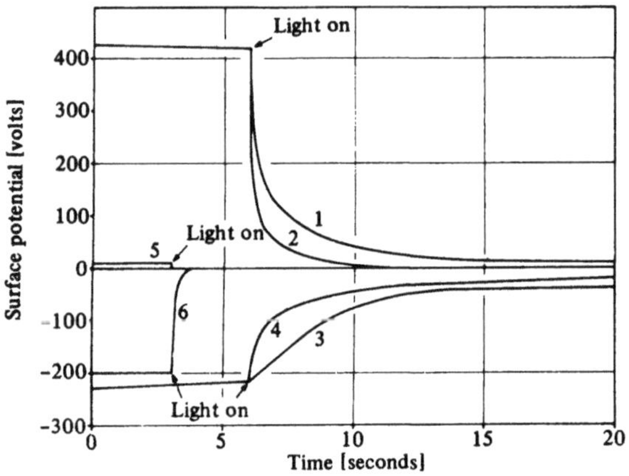

Fig. 19. Discharge curves under illumination of a positively charged (curves 1 and 2) and a negatively charged (curves 3 and 4) ZnO β-Cu-Ph silicone resin layer (3:2:1). For comparison identical charging and discharging curves with ZnO silicone resin layers (3:1) are represented by the curves 5 and 6

coated on an aluminium foil and annealed at 100° C for about 1 hour. The weight ratio of the mixture of ZnO : Cu-Ph : resin was 3 : 2 : 1. As can be seen from Figure 19, this layer can be charged positively and negatively. The dark-decay was rather small. The initial potential after a positive charging was 420 volts, roughly twice as high as the initial potential after a negative charging. For the layer composition used, the speed of the discharge of the positively charged layer was much larger than the speed after negative charging. This fact may be seen from the curves 1 to 4 in Figure 19. But this is only an arbitrary example. Without doubt the speed of discharge should be significantly increased if the composition of the layers is changed. For comparison, the positive and negative charging and discharging of an analogous zinc oxide silicone resin layer are plotted versus time (curve 5 and 6). As can be seen, a positive charging of this layer is not possible.

4. Discussion

In the subsequent discussion, we tentatively assume on the basis of the present results that the Cu-Ph grains embedded in the resin are preferably p-conducting pigments. An n-type conductivity, however, should not be excluded. The assumption is based on the experimental facts that firstly the mobility of holes is about one order of magnitude larger than the mobility of electrons and secondly the concentration of holes is much larger due to the oxygen up-take than the concentration of electrons. As was demonstrated by the experimental results, the investigation of the discharge of the various layer systems:

Al | Cu-Ph + Resin

Al | ZnO Single Crystal | Cu-Ph + Resin

Al | ZnO Powder + Resin | Cu-Ph + Resin

Al | Cu-Ph + Resin | ZnO Powder + Resin

Al | ZnO Powder + Cu-Ph + Resin

has promoted the understanding of the mechanism of the discharge and the charge transfer through the interface ZnO/Cu-Ph

The fact that p-type conductivity is dominant in Cu-Ph crystals was found by the increase of the conductivity with increasing oxygen pressure, presumedly, due to additional generation of holes ⊕ (Cu-Ph) according to

$$O_2(\text{gas}) \rightleftharpoons O_2^-(\text{ads}) + \oplus^{\cdot}(\text{Cu-Ph}) \quad (4.1)$$

near the surface and in the bulk of the crystals with dissolved oxygen

$$O(\text{Cu-Ph}) \rightleftharpoons O^-(\text{Cu-Ph}) + \oplus\,(\text{Cu-Ph}) \quad (4.2)$$

The hole mechanism is also in agreement with the decrease of the conductivity after admission of NH_3 due to annihilation of holes according to [2]:

$$NH_3(\text{gas}) + \oplus\,(\text{Cu-Ph}) \rightarrow NH_3^+\,(\text{ads}) \quad (4.3)$$

or in combination with a surface reaction, such as

$$NH_3(\text{gas}) + O_2^-\,(\text{ads}) + \oplus\,(\text{Cu-Ph}) \rightarrow NH_3 \cdot O_2\,(\text{ads}) \quad (4.4)$$

As further evidence for the dominant p-type character of Cu-Ph, the negative photovoltage of a Cu-Ph layer under illumination may be quoted, as shown in Figure 8. Such a deflection toward negative values becomes understandable if the following reaction is assumed between the physisorbed oxygen molecules and the light-generated singlet exitons ⊖ ~ ⊕ (Cu-Ph) in the phthalocyanine crystals:

$$O_2^x(\text{ads}) + \ominus\!\sim\!\oplus\,(\text{Cu-Ph}) \rightarrow O_2^-(\text{ads}) + \oplus\,(\text{Cu-Ph}) \quad (4.5)$$

resulting in an increase both of negatively charged oxygen species adsorbed on the surface of the Cu-Ph crystallites and of holes in the space-charge layer.

While the description of the electronic disorder of the combination Indium | ZnO single crystal by means of the band model does not in principle cause any difficulty, an equivalent discussion in the case of a copper phthalocyanine resin layer is scarcely possible. The reason for this dilemma may be seen in the fact that a direct application of the band model to organic compound semiconductors is generally not allowed. Furthermore, the mechanism of the electron transfer through the sequence of the Cu-Ph grains embedded in the resin is largely unknown at present. In layers with a rather high content of Cu-Ph, it may be assumed that the charge flux through the layer occurs mainly through chains of Cu-Ph particles available in the layer.

For the mechanism of the positive charging of the layer, we tentatively propose that positively charged oxygen species arriving at the surface of the phthalocyanine crystallites may inject holes into Cu-Ph in the ground state according to

$$O_2^+(\text{gas}) \rightarrow O_2^x(\text{ads}) + \oplus\,(\text{Cu-Ph}) \quad (4.6)$$

under subsequent desorption

$$O_2^x(\text{ads}) \rightarrow O_2(\text{gas}).$$

These holes migrate into the bulk of the grains which are located near the surface of the Cu-Ph resin layer. Because of the large electric field between ZnO and Cu-Ph resin layer, it is supposed that the positive charges present in the single Cu-Ph grains are distributed broadly through the entire Cu-Ph resin layer similar to space charges in a homogeneous semiconductor with the largest concentration near the surface. This is schematically represented in Figure 20 and is possible only if the transfer of holes from Cu-Ph grain to Cu-Ph grain is strongly hindered so that a significant flux of holes toward the negative countercharge is to a great extent prevented. The countercharge may be assumed in the zinc oxide located roughly at the interface ZnO/Cu-Ph.

The negative charging of the layer which will be dealt with below, should be caused by a negative surface charge mainly represented by $O_2^-(\text{ads})$ and by a positive space-charge in zinc oxide. We assume that the dark-decay both of a positively and of a negatively charged Cu-Ph layer–evapo-

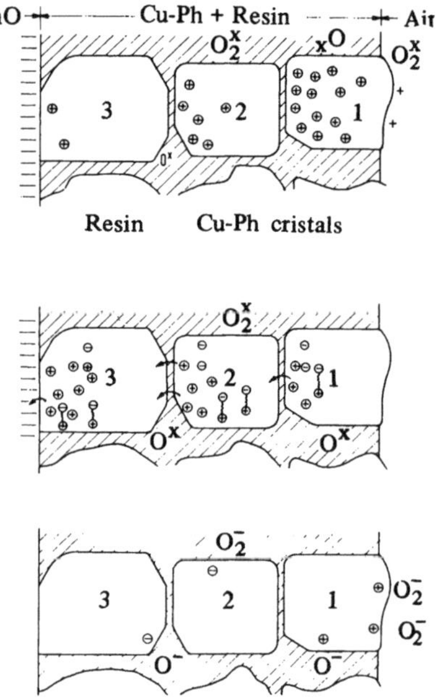

Fig. 20. Schematic representation of a sequence of positively charged Cu-Ph crystallites embedded in a resin coated on zinc oxide ($\ominus \sim \oplus$ = singlet exciton):
a) after a positive charging
b) during the discharge under illumination
c) after the light-decay.
(A positively charged layer has only physisorbed oxygen species O_2^x or O^x (ads); after the discharge, however, negatively charged oxygen species are chemisorbed)

rated on aluminum or coated with a suitable resin on aluminum or zinc oxide—may occur if holes can pass the boundaries from grain to grain. Such a "boundary hopping" mechanism seems to be much more favored in an evaporated Cu-Ph layer than in a layer where the Cu-Ph crystallites are embedded in a resin. This is incidentally further evidence of the important role of the resin on the electrical behavior of Cu-Ph resin layers.

According to the schematic representation of the distribution of the positive charges in the Cu-Ph resin layer and under the assumption of a dominant hole transport, these positive charges are neutralized by the generated singlet excitons $\ominus\!\sim\!\oplus$ with the energy released exciting the remaining hole. These excited holes can pass the grain boundary toward the neighboring crystal:

$$\ominus\!\sim\!\oplus(1) + \oplus(1) \rightarrow \oplus(2) \qquad (4.7a)$$

and

$$\ominus\!\sim\!\oplus(2) + \oplus(2) \rightarrow \oplus(3) \qquad (4.7b)$$

As soon as the holes have arrived at the interface Cu-Ph/ZnO, a recombination across the interface may occur:

$$\ominus\!\sim\!\oplus(3) + \oplus(3) + e'(ZnO) \rightarrow \text{Zero} \qquad (4.7c)$$

Here the numbers refer to the single crystallites, illustrated in Figure 20.

As may be seen from Figure 20 and the sequence of the reaction steps (4.7), the transfer of the holes through the grain boundaries should be favored by the generated energy due to the reaction of the singlet excitons with the positive charges in the crystallites. As soon as a hole has entered crystal 3 and has arrived at the interface Cu-Ph/ZnO, then a recombination according to (4.7c) may occur. Because of the slow dark-decay, the recombination step

$$\oplus(3) + e'(ZnO) \rightarrow \text{Zero} \qquad (4.7d)$$

should occur more slowly than step (4.7c).

A similar mechanism may be tentatively proposed for the discharge of a negatively charged layer. In this case, the majority of the negative charges may be assumed to be located in the surface. Under the additional assumption of a preferential hole transport, the discharge begins from the interface ZnO/Cu-Ph and ends at the interface Cu-Ph/Air, as schematically represented in Figure 21 according to:

$$\ominus\!\sim\!\oplus(3) \rightarrow e'(ZnO) + \oplus(3) \qquad (4.8a)$$

$$\ominus\!\sim\!\oplus(3) + \oplus(3) \rightarrow \oplus(2) \qquad (4.8b)$$

$$\ominus\!\sim\!\oplus(2) + \oplus(2) \rightarrow \oplus(1) \qquad (4.8c)$$

and finally

$$\ominus\!\sim\!\oplus(1) + \oplus(1) + O_2^-(ads) \rightarrow O_2(gas) \qquad (4.8d)$$

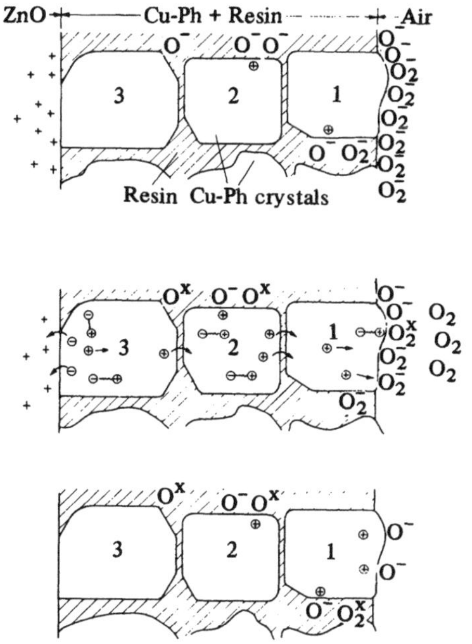

Fig. 21. Schematic representation of a sequence of negatively charged Cu-Ph crystallites embedded in a resin coated on zinc oxide.
a) just after a negative charging
b) during the discharge under illumination
c) after the light-decay

For the occurrence of this discharge mechanism, it is necessary for the resin to be transparent for the light necessary for the illumination of the Cu-Ph crystallites at the interface Cu-Ph/ZnO. On the basis of the experiments described in section 3, we are forced to assume two different mechanisms of discharge of the double layer under illumination. In the first case, under illumination with light of $\geqslant$ 550 nm, it may be tentatively proposed that a discharge caused by a charge flux through the Cu-Ph resin layer is accompanied by a recombination process such as (4.7c). The discharge of a positively or negatively charged layer under illumination with light of $\lambda \geqslant 780$ nm, however, should not occur by an electron recombination process but only by a flux of positive charges toward the interface Cu-Ph/ZnO as mentioned above. These conclusions seem to be obvious if it is considered that Cu-Ph pigments on a ZnO surface do not sensitize the discharge under illumination with light of $\geqslant$ 780 nm while a layer of Cu-Ph resin coated on zinc oxide demonstrates a fast discharge as can be seen from curve 1 in Figure 14, curve 2 in Figure 15 and curve 3 in Figure 16. Since the height of the charging strongly depends on the temperature (Fig. 6), therefore, at present a significant contribution of thermic energy to the discharge in the wave-length region up to 1300 nm cannot be excluded. Preliminary experiments to measure the temperature increase during the long wave-length illumination by means of a micro-thermocouple, resulted only in a small increase of the temperature. But further experiments are in progress.

Acknowledgement

We are grateful to Dr. S. R. Morrison for stimulating discussions. The financial support by the Deutsche Forschungsgemeinschaft for the experimental equipment and by the Alexander von Humboldt-Stiftung for the salary of one of us (N.I.I.) is appreciated.

References

[1] Karaser, F., and J. C. Decius, J. Am. Chem. Soc. **34,** 4716 (1952).
[2] Kaufhold, J., and K. Hauffe, Ber. Bunsenges. phys. Chem. **69,** 169 (1965).
[3] Heilmeier, G. H., and L. A. Zanoni, J. Phys. Chem. Solids **25,** 603 (1964).
[4] Assour, J. M., and S. E. Harrison, J. Am. Chem. Soc. **87,** 651 (1965).
[5] Day, P., and R. J. P. Williams, J. Chem. Phys. **42,** 4049 (1965).
[6] Harrison, S. E., J. Chem. Phys. **50,** 4739 (1969).
[7] Wartanjan, A. T., and I. A. Karpowitsch, Z. phys. Chem. UdSSR **27,** 178 (1953).
[8] Putzejko, E. K., and A. N. Terenin, Z. phys. Chem. UdSSR **30,** 1019 (1956).
[9] Assour, J. M., and S. E. Harrison, J. phys. Chem. **68,** 872 (1964).
[10] Heilmeier, G. H., G. Warfield, and S. E. Harrison, J. Appl. Phys. **34,** 2278 (1963).
[11] Harrison, S. E., and K. H. Ludewig, J. Chem. Phys. **45,** 347 (1966).
[12] Usov, N. N., and V. A. Benderskii, Phys. stat. solidi **37,** 535 (1970).
[13] Hamann, C., Phys. stat. solidi **26,** 311 (1968).
[14] Gamo, K., K. Masuda, and J. Yamaguchi, J. Phys. Soc. Japan **25,** 431 (1968).

[15] Hamann, C., Wiss. Z. Techn. Hochschule Karl-Marx-Stadt **10**, 265 (1968).
[16] Devaux, P., and M. Schott, Phys. stat. solidi **27**, 201 (1968).
[17] Devaux, P., M. Lazerges, and M. Schott, Phys. stat. solidi **4**, 43 (1964).
[18] Heilmeier, G., and G. Warfield, J. Chem. Phys. **38**, 169 (1963).
[19] Devaux, P., and J. Mas, Solid State Comm. **7**, 1095 (1969).
[20] Sussman, A., J. Appl. Phys. **38**, 2738, 2748 (1967)
[21] Barbe, D. F., and C. R. Westgate, Solid State Comm. **7**, 563 (1969).
[22] Devaux, P., and M. Schott, Phys. stat. solidi **20**, 301 (1967).
[23] Honigmann, B., H.-U. Lenne, und R. Schrödel, Z. Kristallogr. **122**, 185 (1965).
[24] Stechemesser, R., Diss. Göttingen 1967.
[25] Schmidt, R., Diss. Göttingen 1970.
Hauffe, K., and R. Schmidt, Phys. stat. sol. (a) **3**, 173 (1970).
[26] Hauffe, K., and A. L. Vierk, Z. Phys. Chem. **196**, 160 (1950).

Discussion

Verhille, K.: The wave-lenth dependence of ZnO single crystals brushed with phthalocyanine or coated with resin-phthalocyanine is entering different as you have shown. Do you have only closer information on this?

Hauffe, K.: On the basis of these results, we may tentatively assume that for the real sensitization of the electron transfer through the interface ZnO/copper phthalocyanine light of 550 nm is necessary. For the charge transport in the phthalocyanine resin layer, however, light of $\geqslant$780 nm seems to be already sufficient. A more detailed study is in progress.

Berg, W. F.: Have you done any experiments on additivity of radiation of different wavelength as a function of level of irradiance and temperature? Such experiments might, in analogy to the silver-halide case, help to elucidate the mechanism involved.

Hauffe, K.: Such experiments have been carried out partly and are under progress (doctor thesis of K. Petrikat).

Phthalocyanine-Binder Photoreceptors for Xerography

J. W. Weigl, J. Mammino, G. L. Whittaker, R. W. Radler and J. F. Byrne*

Abstract

A useful new class of organic photoconductor coatings for xerography is described. They consist of microcrystalline dispersions of metal-free phthalocyanine in suitable binders. Their spectral response extends from 350 to 850 nm, their white light sensitivity is equal to that of selenium. The coatings are flexible enough to be fabricated into coated belts and papers, as well as drums, and they may be re-used for many thousands of cycles. The mobile carriers are holes; the photoresponse and sensitometry appear to be determined largely by the nature of the dielectric barriers between phthalocyanine particles. Some major effects of crystal form and binder composition will be discussed, in relation to the general nature of charge carrier generation and transport in organic photoconductor systems.

1. Introduction

This paper describes a new class of fast, reusable photoconductor-binder coatings based upon metal-free phthalocyanine. These exhibit spectral response extending from 350 to 850 nm, tungsten light sensitivity equal to that of selenium, and sufficient mechanical flexibility to be fabricated into xerographic belts as well as drums. A number of years ago it was discovered in our laboratories [1] that resin-binder coatings containing relatively low concentrations of phthalocyanine pigment exhibited surprisingly high xerographic photosensitivity. Since a wide variety of binders could be used, it became possible to devise useful photoreceptors which combine the high photosensitivity of phthalocyanine with the desirable electrical and mechanical properties of the resin binder. This discovery was used to devise flexible, reusable photoconductive coatings for application in transfer xerography.

*Engineering Physics Laboratory, Battelle Memorial Institute, Columbus, Ohio

2. Materials

2.1. Phthalocyanine Pigments

Since the coatings under consideration consist of a phthalocyanine-pigmented resin layer on a suitable conductive base, one must control three sets of materials: the pigment, the binder, and the conductive substrate. The purity and crystal form of the pigment are among the most critical parameters.*

Two crystal forms of phthalocyanine have been known hitherto [2]: the relatively amorphous, metastable α-form, which results from sulfuric acid precipitation, and the highly crystalline β-form, which may be formed from the α-polymorph by thermal or solvent recrystallization.

*The pigment used was of commercial origin, or especially synthesized [5], purified by exhaustive solvent extraction, sulfuric acid solution, reprecipitation, ammonia neutralization, and a de-ionized water rinse.

In the course of this work, a third crystal form was discovered, which turned out to be considerably more sensitive in binder plates than the others. This polymorph, which was named x-phthalocyanine, is also crystalline and stable, and readily distinguishable from the other crystal forms by its distinctive X-ray scattering [3] and infrared absorption patterns [4], as well as a strongly shifted electronic absorption spectrum [4]. It may be prepared from the α-form, by prolonged dry impact milling, by salt milling under special conditions [3], or preferably, by solvent recrystallization in the presence of nucleating x-seed crystals [6].

Various metal phthalocyanines, peripherally substituted derivatives, and related macrocycles were investigated. All, including CuPc, were found to be substantially inferior to x-H_2Pc in xerographic performance, due variously to inadequate charge acceptance or low photoresponse in binder coatings.

2.2. Dielectric Binder

A thermosetting epoxy-phenolic polymer blend was used for most of the experiments.* This material was selected on the basis of its excellent dielectric properties, its ability to disperse the pigment effectively, and to produce coatings with high photosensitivity. Coatings produced from it possess desirable mechanical properties, including high gloss, excellent abrasion resistance, and good adhesion to flexible belt bases. Due care was taken to exlude or immobilize ionizable residues, such as plasticizers and cross-linking catalysts.

2.3. Resin-Binder Dispersion

The manner in which the pigment is dispersed in the binder is of considerable importance. For the coating to have useful photosensitivity in epoxy enamels such as those described, over 20 weight percent of the pigment should be present in the x-crystal form. If x-phthalocyanine is simply dispersed in the binder resin, coated, and cured, one forms plates with sensitometric characteristics, which will be described below as being of "type 1". Alternatively, one may allow α-pigment to recrystallize in the binder with 5% each of x- and β-seed crystals, until an 80% β-plus 20% x-H_2Pc mixture is formed. When such a mixed-grain suspension is coated and cured, the resultant plates exhibit significantly different ("type 2") characteristics. Plates prepared from 80 : 20 mixtures of separately crystallized β- and x-pigments behave simply as very insensitive type 1 coatings, and are quite unsatisfactory in performance when compared to type 2 plates of the same nominal composition.

The differences in electrical behavior, described below, show that there are subtle interactions between the pigment and the binder, which are not yet well understood and undoubtedly involve barriers to charge transport between adjacent H_2Pc particles. Electron micrographs of thin sections of type 1 coatings (Fig. 1) indicate that the crystallites are acicular, typically 0.1 – 1 μ long and $\lesssim$ 0.1 μ in crossection. They are deposited in such a way as to provide contorted, but apparently continuous, paths for carrier transport, interrupted by a number of thin resin barrier. The characteristically in type 2 coatings appear to be characteristically longer (0.2 – 5 μ) and thicker ($\lesssim$ 0.5 μ) so that carrier transport across the coating may require the crossing of fewer barriers. The electrical performance of type 2 coatings is consistent with the idea that barriers are less dominant than in type 1 layers.

---

*Composition of Epoxy-Phenolic A ("EPA") Binder Solution

	% by Weight
Epon-1007 (Shell Chemical Company)	35.5
Methylon-75201 (General Electric Company)	20.0
Uformite-F240 (Rohm & Haas Chemical Compony)	4.5
Methyl Isobutyl Ketone	30.0
Methyl Ethyl Ketone	10.0

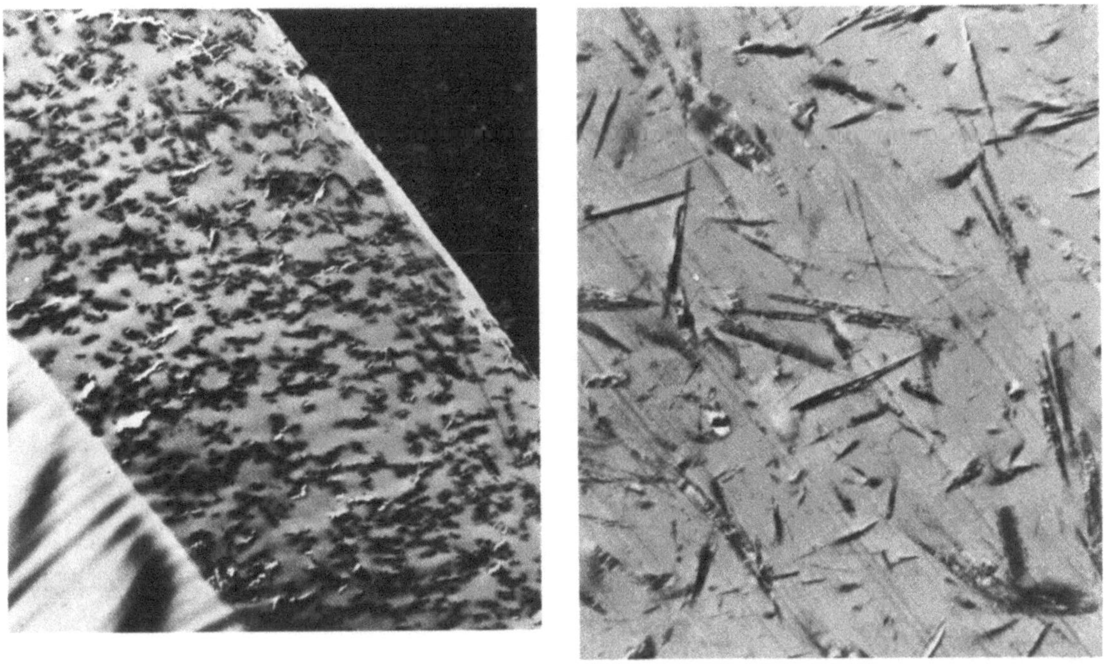

a) Type 1 b) Type 2

Fig. 1. Transmission electron micrographs (~10,000 X) of crossections through phthalocyanine-binder coatings

2.4. Conductive Base and Coating Process

In order to perform properly in the xerographic process, the photosensitive enamel must be coated on a conductive base, which forms a blocking contact to the photoconductor [7]. Interfaces of copper, gold, or their alloys, including brass, were found to be undesirable, because the plate voltage produced under constant charging conditions gradually decreased during prolonged cycling. Satisfactory bases for positively charged coatings include aluminum (with its natural oxide coating), cold-rolled and stainless steel, nickel, and tin oxide-coated ("NESA") glass.

The viscosity of coating slurries may be adjusted, as required, to permit the coating of flat plates by Mayer bar or Bird applicators, or of drums and belts by means of an automatically programmed spray gun. After drying in a dust free hood, the coating is heat-cured to produce a hard, smooth dark blue enamel, $10-30\mu$ in thickness. The photoelectric properties of the resulting xerographic plates have been found to be stable in storage over periods of many months.

3. Performance

3.1. Charging Characteristics

Positive charge was applied to test plates and measured by means of an integrating electrometer connected to the base. Sector-shaped samples, $7.0\,cm^2$ in area, were rotated at 10 RPS under a scorotron [8] and a potential measuring probe connected to an oscilloscope (Fig. 2). The integrated charge (Q) applied to the coating was plotted against surface potential (V), as measured by the probe.

The dark charging characteristics of type 1 plates are found to be quite strongly affected by prior

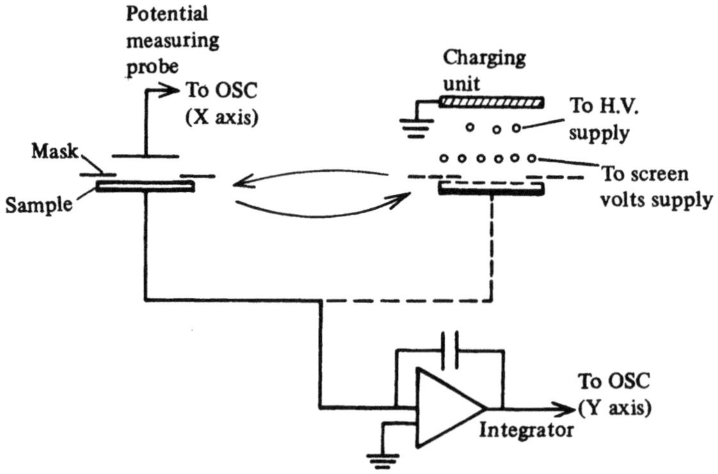

Fig. 2. Schematic diagram of apparatus used to measure charging curves (See text)

illumination. "Rested" plates, which have been stored in the dark for 12 hours at room temperature, or for a few minutes at 100° C, produce linear Q/V plots with a slope of the order of 160 pfd/cm^2. These values are found to be consistent in each case with the capacitance computed from their thickness and dielectric constant (3.2 ± 0.2). (Fig. 3a, curves A and B). Pre-exposure with green light, which is absorbed fairly uniformly throughout the thickness of the plate, greatly increases the amount of charge required to produce a given plate potential. The initial portion of this curve may be represented by $Q\,\alpha V^{1/2}$, which is generally characteristic of plates with trapped space charge [9] (Fig. 3a, curve C). Thus, the extra positive charge required after pre-illumination may serve to neutralize some negative space charge, which has been trapped in the binder. The density of this charge is of the order of 6×10^{14} electrons/cm^3, or 10^{12} electrons/cm^2 of surface area. When the corona polarity is now reversed (Fig. 3a, curve D), the negative charge required to return the plate to zero

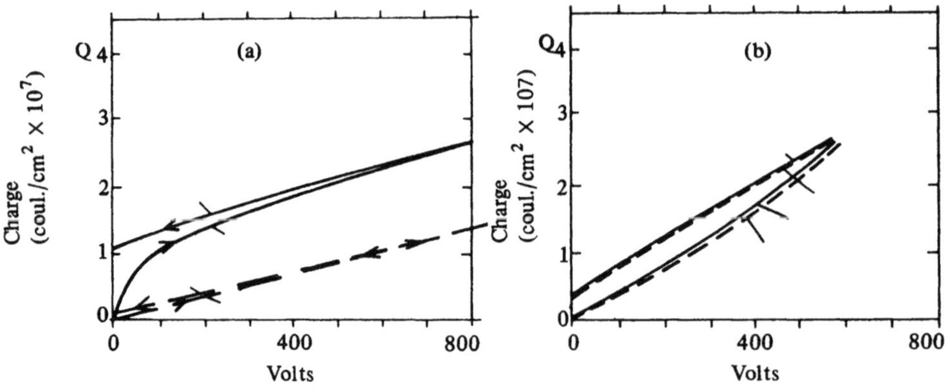

a) Type 1, b) Type 2.

Fig. 3. Typical Q-V characteristics of phthalocyanine binder coatings for positive charge
------ dark rested, ________ pre-exposed.
A, C – positive corona charging sequence B, D – subsequent negative corona discharge sequence

potential is only the much smaller amount corresponding approximately to the normal capacitive charge. This shows that the photoinduced depletion charge is swept out in the first charge cycle and does not re-enter the coating when the polarity is reversed.

As may be seen from Figure 3b, the effect of pre-exposure upon the charging current of "type 2" plates is very much smaller, and the differences between pre-illuminated and rested samples are often negligible. These observations will be used below in an attempt to account for the nature of an interesting induction phenomenon observed in "type 1" plates.

3.2. Dark Discharge

Figure 4 shows typical dark-discharge curves, and Figure 5, an Arrhenius plot for the initial dark-decay rate of typical "type 1" coatings ($E \cong 0.5$ e. V.). In general, as noted below (Fig. 8), as pigmentation is increased, the charge acceptance of the coatings decreases. Pigment exposed at the surface appears to act as an efficient channel for the dark injection of corona charge into the bulk of the plate; this problem can be avoided by the use of effective film forming binders or thin protective overcoatings. The dielectric strength of these coatings exceeds 100 v/μ; when appropriate

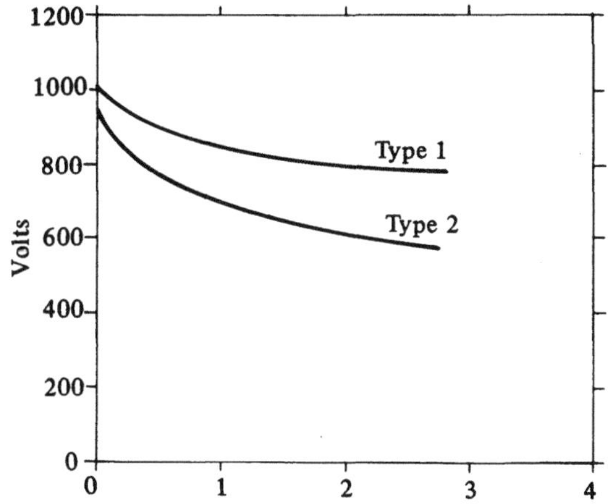

Fig. 4. Typical dark discharge curves for two kinds of phthalocyanine plates

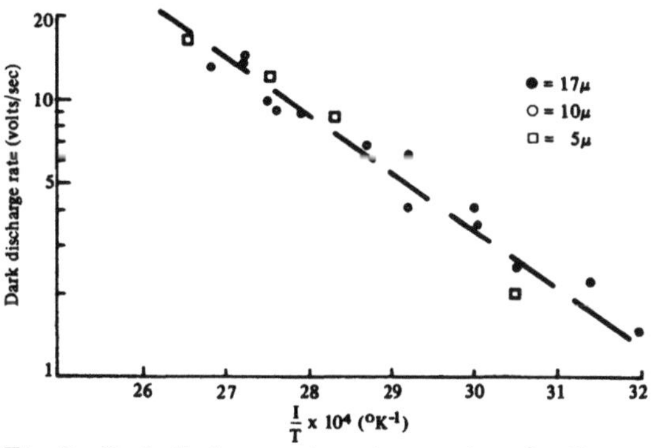

Fig. 5. Dark discharge rate vs temperature for three Type 1 plates, all charged to initial potential (V_0) of 275V. The slope corresponds to an activation energy (of the form E/kT) of $\sim$0.5 eV

blocking electrodes are used as base, charge acceptance is limited only by the field at which localized breakdown regions begin to appear in the coating.

3.3. Photo-induced Discharge Characteristics ("PIDC's")

The photosensitivity of the coatings was measured by the well-known PIDC method [10]: the layers were charged, then exposed under a transparent-probe electrometer [11].

Figure 6 shows the very striking differences between the photoinduced discharge characteristics of type 1 and 2 plates. Type 1 plates require a certain minimal integrated exposure (irradiance x time) which is roughly independent of irradiance. During this "induction period," little discharge takes place; then, quite suddenly,

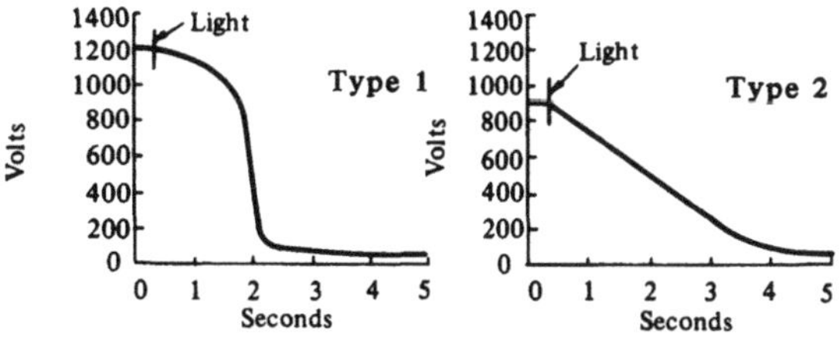

Fig. 6. Typical PIDC for Type 1 and Type 2 plates. Irradiance was approximately 1.5 $\mu w/cm^2$, using a tungsten lamp with filter passing wavelengths from 650 to 800 nm

the layer discharges with great rapidity, as though some non-destructive photo-induced breakdown phenomenon had occurred. The magnitude of the induction exposure determines the effective photosensitivity of "type 1" plates. It is strongly affected by crystal form, being far greater for α- than for β- or x-H_2Pc (Fig. 7).*

As Figurc 7 indicates, the photosensitivity of positively charged phthalocyanine-binder plates for strongly absorbed red radiation is > 8 x greater than for negatively charged plates, indicating that holes are the more mobile carriers. All results reported from here on refer to positively charged plates.

Photosensitivity is strongly dependent upon purity, concentration, and crystal form of the

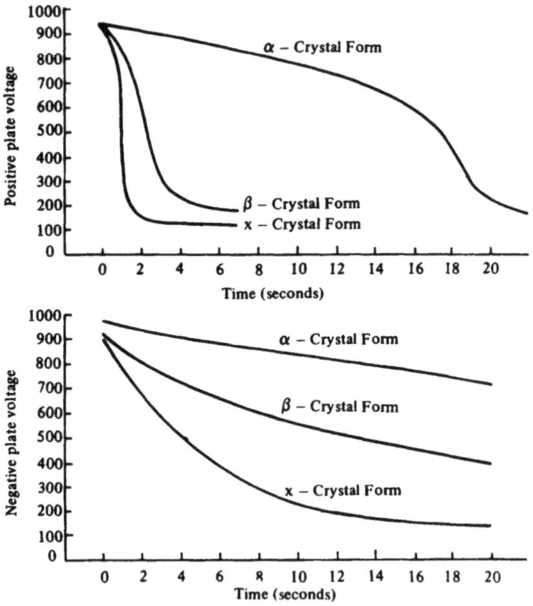

Fig. 7. PIDC's of 25μ thick α -, β -, and x-phthalocyanine plates, especially made to show the effect of crystal from on photodischarge. Top: positive charge, bottom: negative charge. Tungsten illumination (0.2 f. c. at 2800° K)

*Phthalocyanine dispersions in certain photoconductive binders, such as polyvinyl carbazole, do not display this induction phenomenon. Their properties are described in a separate publication [12]."Type 2" coatings exhibit a much smaller induction effect (see below).

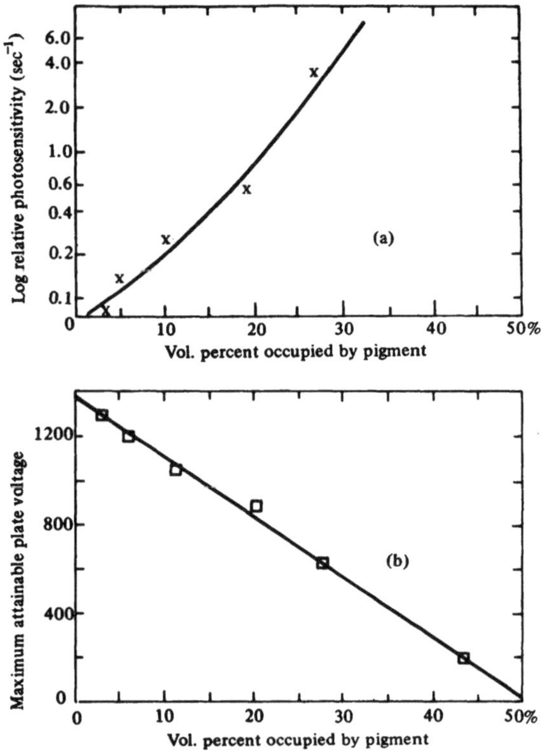

Fig. 8. Effect of pigment concentration upon (a) photosensitivity and (b) charge acceptance (bottom diagram) of a typical series of "type 1" plates

pigment, as well as on the chemical constitution of the binder. The effect of pigment concentration upon photosensitivity and charge acceptance of the coatings is illustrated in Figure 8. Photosensitivity rises *gradually* over the range from 5 to about 30 volume percent. There is no indication of a sudden increase of sensitivity at some point where universal particle-particle contact is established, as has been observed in previously known pigment-binder systems [13, 14, 15].On the contrary: direct particle contact is damaging to good xerographic performance. As the pigment concentration is raised and the particles begin to contact each other without intervening resin barriers, the charge acceptance of the plates drops progressively and fatigue effects increase. The weight ratio of one part pigment to six parts resin (11 volume percent pigment) is near the top of a broad optimum range, and much of the

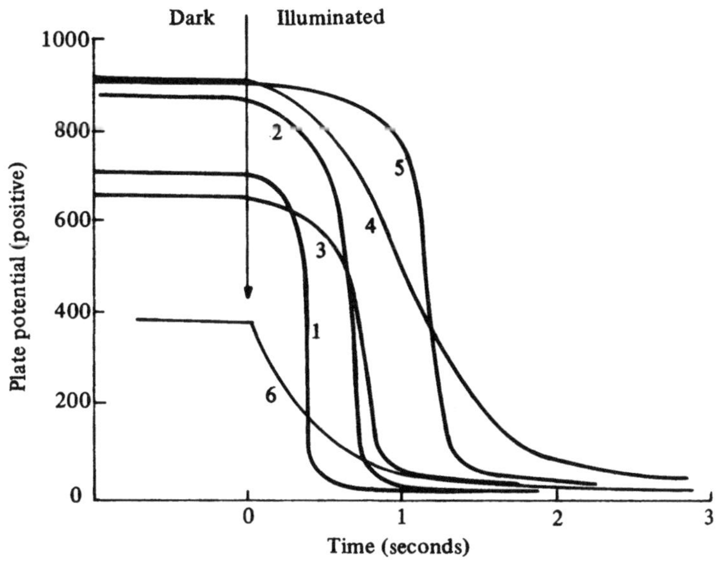

Fig. 9. Effect of resin binder on PIDC of aluminium base "type 1" phthalocyanine-binder plates, all at 11 vol. % pigmentation. Tungsten illumination. (0.2 f. c. at 2800° K)

remaining discussion will be concerned with the properties of coatings of this composition.

In addition to providing mechanical strength, the binder plays an essential role as dielectric: it separates contiguous pigment particles by thin insulating barriers, which prevent dark discharge without significantly impeding the injection of photo-excited holes and their transport from particle to particle. Figure 9 shows photo-induced xerographic discharge curves (PIDC's) for a group of typical binder coatings, using the same x-H_2Pc pigment concentration. Clearly the chemical structure of the binder, and its ability to produce good pigment dispersions, are factors affecting its utility for H_2Pc-coatings.

3.4. Factors Affecting Induction Exposure

For a given "type 1" plate, the induction period is found to be approximately inverse to irradiance over a 100 x range (Fig. 10) [16]. For a given pair of pigment and binder materials, the induction exposure is approximately proportional to (concentration)$^{-1/3}$ (Fig. 11), suggesting that it is related to the mean spacing between crystallites. At constant composition, the induction period is found to be proportional to plate potential (Fig. 12); at constant composition and constant field (constant surface charge density), it rises with increasing coating thickness (Fig. 13).

The nature of this induction effect is not clearly understood. It is cumulative depending upon the total number of quanta absorbed, more or less independently of the illumination and timing. The spacing and thickness dependence, mentioned above, suggests that it is related to carrier trapping at multiple interfaces between pigment particles.

The top of Figure 14 shows that a partial exposure of "type 1" plates, just sufficient to overcome the induction period, leads to continued fast decay of potential, even in the dark. (See, e.g., the curves for 0.58 and 0.8 relative intensity). This "post-flash discharge" phenomenon makes it very difficult to control middle-tones when "type 1" plates are exposed by flash illumination. The induction period results in unusually high image contrast and low

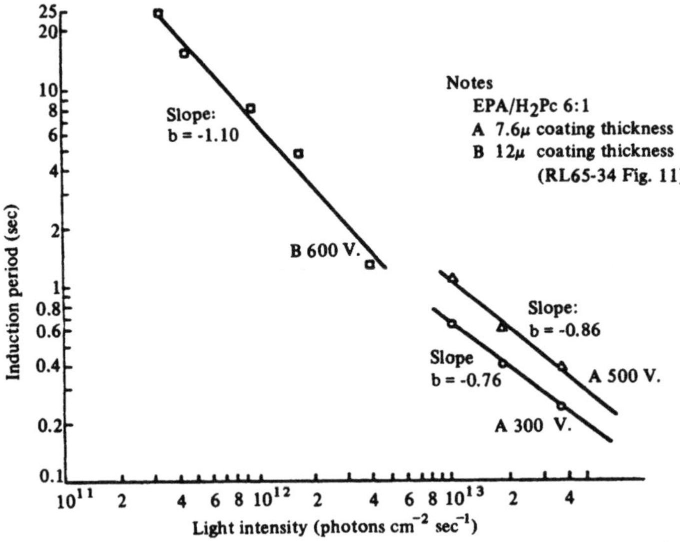

Fig. 10. Reciprocity plot for induction exposure. induction period (illumination)^b, where b ranges from -0.76 to -1.10 for different samples. (Reciprocity requires $b = -1.0$)

exposure latitude. On the other hand, "type 2" plates (Fig. 14, bottom) exhibit "conventional" photodischarge and sensitometric characteristics, similar to those of selenium [15, 17] and may be used for applications which require rapid charge stabilization after flash exposure, as well as reasonable exposure latitude.

3.5. Spectral Response, Reciprocity Law

The overall photosensitivity of type 1 and type 2 plates of equal pigmentation is about equal when both are measured with 2800° K tungsten light. However, the spectral response curves (Fig. 15) differ significantly because the x-polymorph, which is present in much higher concentration in type 1 coatings, contributes to a strong peak at about 800 nm, while the α- and β-forms absorb mainly between 670 and 740 nm [4]. They are least responsive to $\lambda \approx 500$ nm radiation, due to weak absorption in this region. At ~ 40 V/μ, an average of ~ 0.1 carrier is transported across the coating per incident photon of red light in 11 vol. % x-H_2Pc-EPA coatings. Practical tungsten light exposures are ~ 0.2 fc sec., or ~ 2.5 ergs/cm^2.

Figure 16 shows that the reciprocity relationship is obeyed quite well by both types of plates over a wide range of exposure times, i. e., from about 30 ms to 15 seconds.

3.6. Cycling Properties

Although the dark discharge rate of properly prepared phthalocyanine enamels is only slightly affected by pre-illumination in the absence of field, it does tend to increase gradually in the course of cyclic charging and photodischarging. This effect can be minimized by

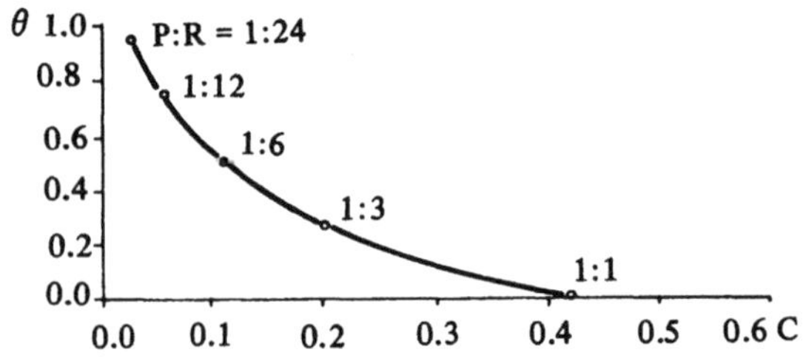

Fig. 11. Induction period (Θ, in sec.) vs volume concentration of pigment (C). Data points are shown for a series of pigment:resin ratios. Type 1 coating 30 μ thick.
Illumination = 2×10^{13} phot/cm^2 sec at λ 620 nm; initial potential 500 V (Ref. 16). $\Theta \propto C^{-1/3}$

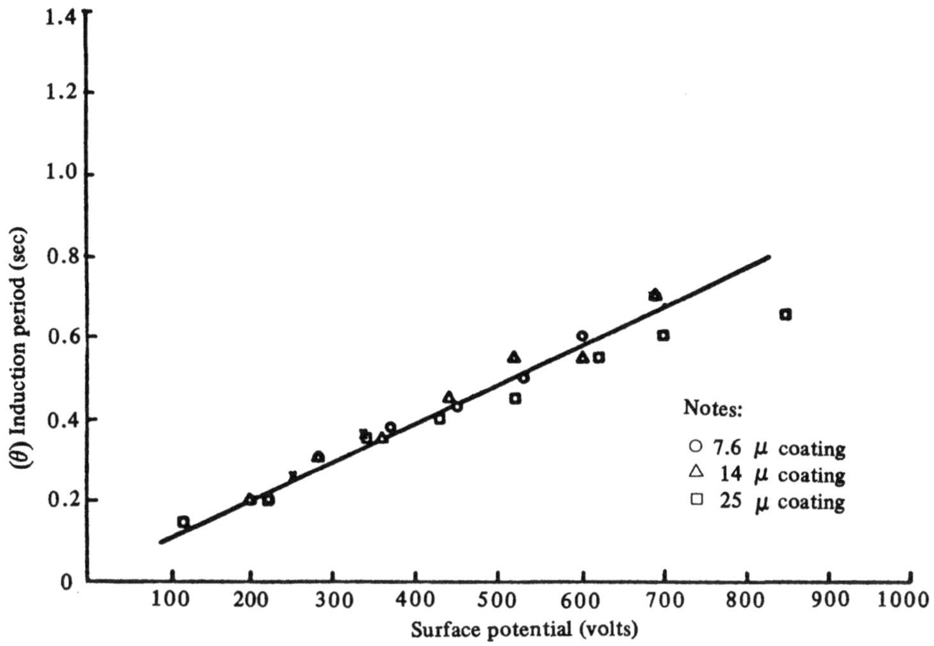

Fig. 12. Induction period vs plate potential for a series of 11 vol. % coatings of type 1. (Illumination same as in Fig. 11)

avoiding contact of the photoreceptor surface with certain types of contaminants, and is accordingly thought to be due to a gradual increase in the dark injection of carriers from corona ions.

Electrical cycling tests were performed with a rotary scanner in which the sample was

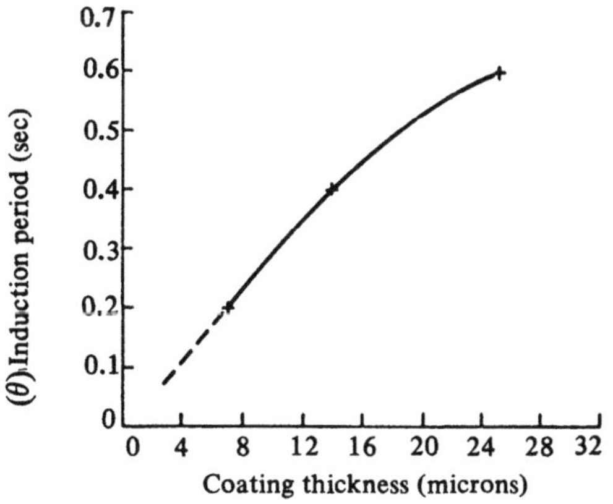

Fig. 13. Induction period vs coating thickness at constant applied field (2.8×10^5 v/cm) for a series of 11 vol. % coatings of type 1. (Illumination same as in Fig. 11)

sequentially cycled under a corotron, an exposure lamp with a transparent probe electrometer, and a row of tungsten "erasure" lamps, similar to those used in conventional xerographic machines [18]. Figure 17 shows representative results for type 1 and type 2 plates at a number of exposure levels, obtained from PIDC curves taken at various intervals during cycling. Ideally, all lines should be horizontal and equally spaced. In fact, the intermediate exposure curves shift rapidly, although the zero- and high-exposure potentials remain quite constant. With high contrast development, this shift of the intermediate exposure curves manifests itself as a progressive increase in the effective photosensitivity during the first dozen cycles, after which a steady sensitivity level is maintained—typically at about twice the initial value—as long as the plate is cycled, over at least some hundreds of cycles. This sensitivity fatigue may be relaxed by prolonged dark storage at room temperature, or by heating the plates to 100° C for about 2 minutes. Therefore, careful exposure control is required when "type 1" phthalocyanine plates are cycled in intermittent use.

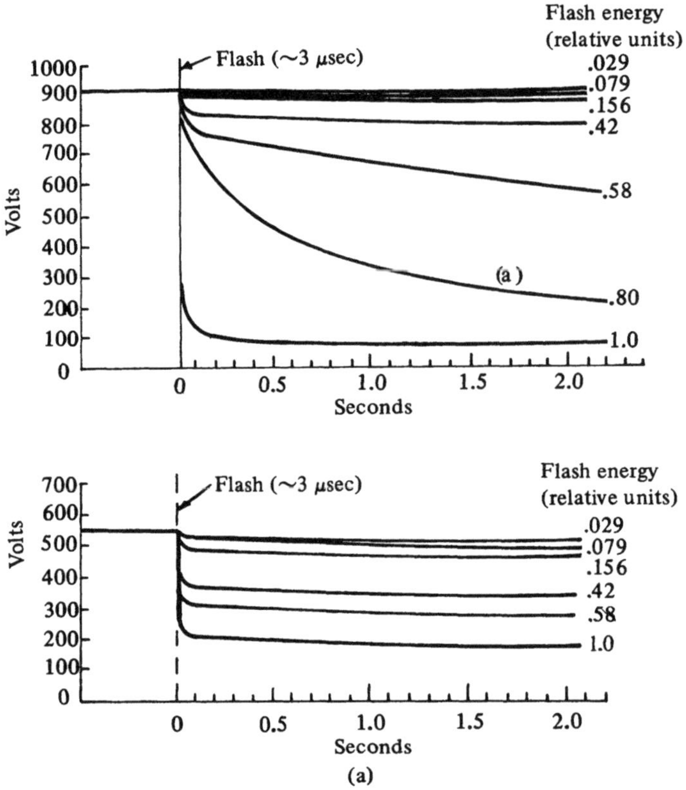

Fig. 14. Post-flash discharge of type 1 and type 2 coatings (Upper and lower diagram, respectively). (See text)

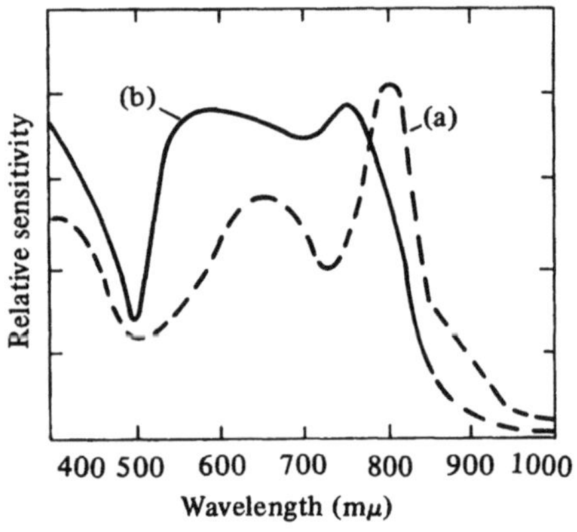

Fig. 15. Spectral sensitivity of (a) type 1 and (b) Type 2 plates. Sensitivity is defined as the reciprocal time to discharge plate from 300 v to 100 v. (Illumination: 5×10^{12} phot/cm^2 sec, at each of 15 wavelengths)

The combined effects of post-flash discharge and sensitivity fatigue make type 1 plates unreliable in long term cyclic use, except for high contrast imaging. Type 2 plates offer much better sensitometric stability.

3.7. Imaging Properties

Neither the mechanical nor the triboelectric surface properties of these coatings have as yet been properly adapted to specific machine requirements. Long image-cycling tests were carried out in a Xerox copier. About 50,000 copies of satisfactory quality were made on a single phthalocyanine enamel coated drum in a machine, equipped with standard cool white lamps and a lens with variable aperture. Resolution was limited by the optics and the development system, rather than by the photore-

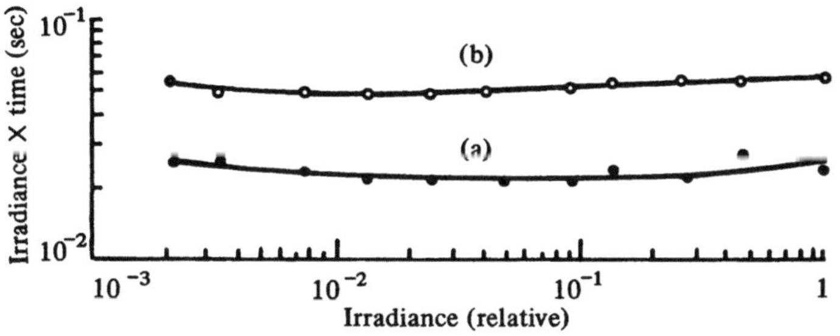

Fig. 16. Reciprocity relationship plot for (a) type 1 and (b) type 2 plates. (The time is that required to discharge 200 volts from the initial potential. A tungsten lamp with filters passing λ 600 – 650 nm is used.)

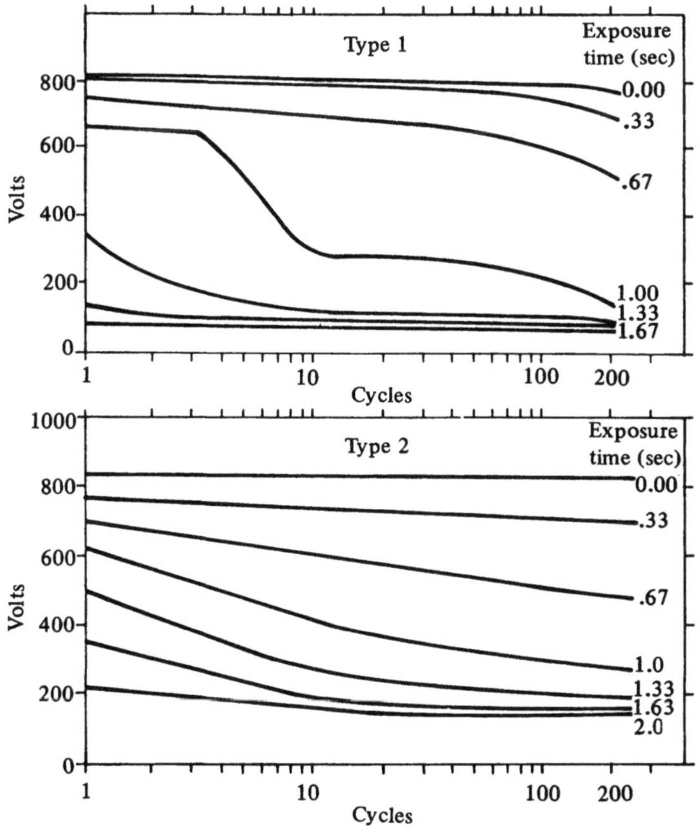

Fig. 17. Cycling performance of phthalocyanine binder plates. Each line measures the photodischarge produced by a given exposure, at the stated imaging cycle. Stable performance produces a family of straight, evenly spaced lines. Curvature of midrange exposure lines indicates a progressive increase in sensitivity to middle tones as plate is cycled

ceptor itself. In separate tests with high-resolution liquid developers, phthalocyanine-binder coatings have demonstrated resolution capabilities in exess of 180 lp/mm [19].

4. Discussion

The outstanding novelty of this system lies in the fact that excellent photosensitivity can be achieved at only 10 to 15% mean free volume concentration of phthalocyanine pigment–even though the latter is randomly dispersed through the binder. Previously known photoconductor pigment-binder systems, such as those based upon CdS, CdS/Se, ZnO, monoclinic selenium [13, 15], and a wide variety of organic compounds [14],exhibit a threshold for photosensitivity at a photoconductor loading of the order of 40% mean free volume, and require pigmentation of 60 – 70 volume percent for optimal sensitivity. This has generally been attributed to a need for universal particle-particle contact to allow effective carrier transport through these materials. As a result, the electrical and mechanical properties of conventional binder photoreceptors have been largely determined by the pigment rather than by the binder.

In the phthalocyanine-binder system, on the other hand, pigmentation is sufficiently low that direct particle-particle contact may not be essential. In fact, as might be expected from the relatively high dark conductivity of phthalocyanine pigments [2,20],if the concentration of such a layer is raised to the point where interparticle contact becomes prevalent, its charge acceptance drops rapidly and the coating becomes unable to sustain the high fields required for xerographic imaging.

The photoresponse of phthalocyanine-binder coatings is much higher when these are illuminated from the positive than from the negative side. This indicates that the majority carriers are holes, which are photogenerated near the illuminated surface and injected in the prevailing strong field across the resin barriers between a series of contiguous pigment particles. To account for the high gain observed under high field conditions, one must assume that the average hole can pass through a distance comparable to the plate thickness without being trapped.

The induction phenomenon observed in "type 1" phthalocyanine plates cannot be accounted for by a simple trapping model, such as has been invoked for the case of zinc oxide binder layers [21], but appears to be related to the role of the minority carriers. The charging experiments described above indicate that negative charge is trapped in the resin interstices of type 1 plates when the pigment is illuminated. Once this charge reaches a critical density, it appears to facilitate very efficient hole transport through the coating, resulting in the observed sigmoid PIDC. The mechanism for this is not understood at present; it may involve intense microfields between adjacent particles. Type 2 plates exhibit neither negative space charge nor induction phenomena; evidently in these, photoinjected holes may pass through the layer more or less unimpeded. Electron micrographs suggest that the crystallites in type 2 layers may be consistently longer and the barriers thinner (as well as less numerous per passage) than in type 1 coatings, and that the barriers may therefore play a less critical role in this case.*

5. Summary

In summary, phthalocyanine-binder coatings are produced by dispersing particles of a relatively conductive phthalocyanine pigment in a mech-

*During the course of this investigation, Bartfai et al. [22] described the use of dilute CuPc dispersions as panchromatic media for photoplastic recording. These coatings were reported to have poor cycling properties as well as low photosensitivity: the data of their Figure 8 and Figure 11 indicate a gain of the order of 10^{-3} and exposure requirements of $\gtrsim$ 100 ergs/cm^2 The poor cycling properties reported for these coatings may be related to the use of a very large volume ratio of thermoplastic binder to phthalocyanine; their low photosensitivity may be related to this, as well as to the use of CuPc instead of the more sensitive x-H_2Pc.

anically strong dielectric binder, in sufficient concentration to assure that the particles are separated by very thin resin shells. These composite coatings behave as sensitive photoconductive insulators with high sensitivity from 400 – 800 nm, and fair behaviour in cyclic xerography. The key parameter appears to be not so much the intrinsic photoconductivity of the pigment, as its ability to inject photo-excited positive charge carriers through thin dielectric barriers between adjacent particles.

Since a wide range of binder resins may be employed, one may use phthalocyanine as pigment for a variety of special-purpose photoconductive coatings, whose chemical, mechanical, electrical, and surface properties may be determined essentially by those of the binder.

Acknowledgements

The authors gratefully acknowledge the assistance of many of their colleagues on various aspects of this work, particularly: Mrs. G. D. Jvirblis, Messrs. P. F. Kurz, H. R. Carrington, F. L. Moore, J. Bergfjord, C. Bielan, E. C. Faucz, G. Fekete, and Drs. P. L. Gerace and M. Lardon. Finally, we wish to thank Dr. John Dessauer, Dr. Fred Schwertz and Mr. Robert Gundlach for their support and encouragement at the inception of this research.

References

[1] Gt. Brit. Pat. 1,116,553 (U. S. filing date 6/15/64).

[2] Moser, F. H., and A. L. Thomas, Phthalocyanine Compounds (N. Y., Reinhold Publ. Co., 1963).

[3] Byrne, J. F., and P. F. Kurz, U. S. Pat. 3,357,989 (U. S. filing date 10/29/65, issued 12/12/67).

[4] Sharp, J. H., and M. Lardon, J. Phys. Chem. **72,** 3230 (1968).

[5] Brach, P. J., S. J. Grammatica, O. A. Ossanna, and L. Weinberger, J. Heterocyclic Chem. 7, 1403 (1970).

[6] Radler, R. W., U. S. Pat. 3,594,163 (U. S. filing date 7/21/66, issued 7/20/71)

[7] Hope, W. D., and M. Levy, "Photoreceptor Base Materials," Chapter IV, Xerography and Related Processes, edited by J. H. Dessauer and H. E Clark, The Focal Press, New York, 1965.

[8] Vyverberg, R. G., "Charging Photoconductive Surfaces", Xerography and Related Processes, op. cit., Chapter VII.

[9] Amick, J. A., RCA Review, **20,** 770 (1959).

[10] Regensburger, P. J., J. Appl. Phys. **35,** 1863 (1964).

[11] Nielsen, P., Rev. Sci. Inst. **38,** 1486 (1967).

[12] Hackett, C. F., J. Chem. Phys. **55,** 3178 (1971).

[13] Wood, C., "Xerographic Properties of Photoconductor Binder Layers", Xerography and Related Processess, op. cit., Chapter V.

[14] Smith, M., and J. W. Weigl, "Organic Photoconductors", Xerography and Related Processes, op. cit., Chapter VI.

[15] Schaffert, R. M., Electrophotography, Focal Press, London & N. Y., 1965, pp. 275-84.

[16] Lardon, M., and E. C. Faucz, unpublished.

[17] a) Li, H. T., and P. J. Regensburger, J. Appl. Phys. **34,** 1730 (1963).
b) Tabak, M. D., and P. J. Warter, Phys. Rev. **173,** 899 (1968).

[18] Robertson, D., unpublished.

[19] Sankus, J. G., personal communication.

[20] a) Assour, J. M., and S. E. Harrison, J. Phys. and Chem. of Solids **26,** 670 (1965).
b) Harrison, S. E., and K. H. Ludewig, J. Chem. Phys. **45,** 343 (1966).

[21] Arneth, R., and B. Lorenz, Reprographie, **3,** 199 (1963).

[22] Bartfai, J. J., V. Ozarow and J. Gaynor, Phot. Sci. & Engrg. **10,** 60 (1966).

Discussion

Arneth, R.: 1) Why can't you explain the "induction period" of type I layers by trapping?
2) You mentioned if you evaluate the steep slope of light decay curve you obtain quantum efficiencies above 1. Can you give figures?

Weigl, J. W.: 1) What I meant to say was—no traps of lifetime greater than the photodischarge time could be involved, since all positive carriers are drawn out in every photo-induced discharge.
2) It is meaningless to calculate quantum efficiencies from the steep portion of the PIDC,

since all photons absorbed during the induction period must be counted, as well. To mention an extreme case: one can pre-illuminate a charged coating, to the induction exposure, then turn off the light, just as the fast discharge commences. The discharge now continues, even though no further photons are absorbed. (See Figure 14) The real quantum efficiencies are ~ 0,1–0,2 under the most favorable conditions.

Rochlitz, J.: 1) Have you observed a dependence of photoresponse upon particle size?
2) What do you think is the reason of the lack of S-shape in the case of the Phthalocyanine/PVK system?

Weigl, J. W.: 1) No. The reasons are: a) It was too difficult to grow mono-disperse particles of β- and x-H_2Pc of stepped sizes for use in "type 1" coatings.
b) It was not possible to separate the effects of particle size from those of particle contacts, as formed, e. g. in "type 2" plates.

2) The properties of the binder which appear to control the shape of the PIDC are its ability to transport and trap injected carriers.

Rochlitz, J.: Did you achieve the 50.000 copying-cycles using a special barrier-layer between the Al-substrate and the Phthalocyanine-binder-layer?

Weigl, J. W.: No special interface was required; the natural aluminum oxide on the drum was sufficient.

The Photoconductivity of Metal Chelates

H. H. Kosche

Abstract

The behaviour of organic metal chelates as photoconductors is interpreted on the basis of experimental results. The influence of the metal-containing chelate application and the chelate structure of photoconductive layers and the sensitization of photoconductors by chelates is discussed.

1. Introduction

The organic metal chelate compounds represent a particular group of compounds which have been discussed originally by Werner [1], Weinland [2] and Pfeiffer [3] and recently by Martell and Kalvin [4]. Metal chelates with a more simple configuration possess a metal atom in the center, surrounded by the organic chelate formers. At a suitable composition, these compounds are good photoconductors.

When these photoconductive chelates are used with suitable binders or react with poly-isocyanates to photoconductive resins, then photoconductive layers are available exhibiting electrophotographic properties which can be employed in reprography and for offset printing plates using electrostatically chargeable toner material [5].

In the present paper, the photoconductivity of organic metal chelate compounds is described as a function of the chelated metal atom and it is pointed out which metals are particularly suitable to produce photoconductive chelates. The mechanism of the photoconductivity is tentatively discussed. Subsequently, the sensitization of zinc oxide is discussed on the basis of a chelate formation. Donor-acceptor complexes of sensitizers are pointed out. By means of the electron configuration of the chelates operating as sensitizers on zinc oxide, a mechanism of electron injection from the sensitizer into the ZnO crystal is outlined. Additionally, the charge transport through the binder is interpreted as a function of hydrogen bridges polarized in the electric field. Also the photoconductivity of the resins belonging to the group of polyurethane or polyurea detected by the author [6] can be understood only by the effect of hydrogen bridges of the mesomeric or tautomeric states of these compounds.

Also the bichargeability found with metal oxides which have been covered on the surface with photoconductors containing groups of isocyanates, is caused by a chelate formation [7].

By this method for example, a negatively and positively chargeable zinc oxide-resin layer can be obtained.

The experimental results dealt with in this paper are based on the studies carried out between 1960 and 1961. A similar behavior of the photoconductivity of some metal chelates, as for instance N, N-di-2-hydroxy-benylidene-dianisidine, of 8-hydroxyquinolines [9] and their derivatives as well as sulphur containing aromatic substituted heterocyclic compounds.

2. Experimental

The production of the various metal chelates has been described extensively in the literature. Only few of them are employed as analytical reagents for test reaction both of metals and organic compounds. As was demonstrated by Pfeiffer [10], the resistant and brillant dye lacquers are organic metal chelates which are good photoconductors if they are highly purified and no ionogenic groups are present.

For the measurement of the photoconductivity, layers of a thickness of 15 microns consisting of 3 parts by weight of the photoconductor and one part by weight of the binder, have been prepared. The photoconductive chelate pigments had an average grain size of 5 microns. Soluble chelates have been solved together with the binder and cast as layers. As binder for chelates roughly all resins can be used which have been employed for zinc oxide layers. Cellulose ester and vinyl-copolymers, as for instance Gelvar 202 are preferred [12]. The light sensitivity of these photoconductive layers has been measured in the spectral range of both UV fluorescence lamps and tungsten light.

3. Grouping of the chelate formers

Because of the better lucidity, the organic chelate compounds have been grouped by their chemical structure and steric pattern. Metal chelates have been produced with metals which have been chosen according to their position in the periodic systems of elements. Particularly extensive investigations have been carried out on the metal chelates with the transition elements as well as with lead and bismuth. Simultaneously, the light absorption of the solid photoconductors has been studied.

In the following Tables I–V and Figures 1–5, special compounds are represented, the structure and the photo-conductivity of which are especially of interest. In the footnotes of the formulas, the investigated metals are quoted. (Figs. 1–5).

Table 1 shows the 1st. Group: **Oximes, aromatic o-nitroso-hydroxy compounds and hydroxane acids.**

Class of Compounds	Active Group	Assumed complex ring	Formula	Examples
Monoxime of α-diketones and aromatic o-nitroso-hydroxy compounds as isonitrosoketones	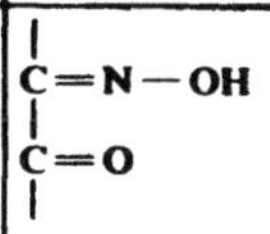	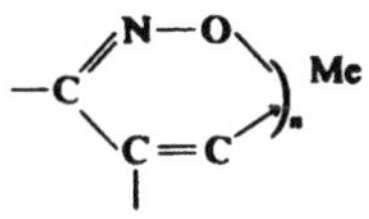	1,2	Complex salts of the α-nitroso-β-naphtol of isatin-β-oxime (s. Schultz-Farbstofftabellen, 7 ed. 1. vol. p. 3-7 "Quinone-oxime dyes")
Dioximes	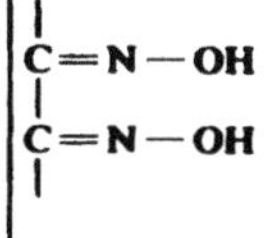	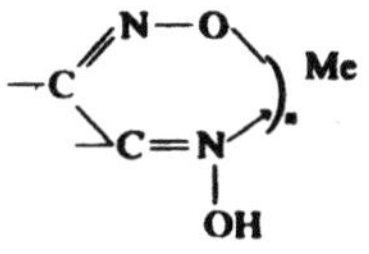	3,4	Complex salts of the diacetyldioxime, benzildioxime
Ketoxime	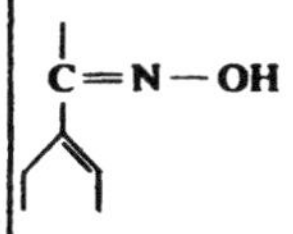	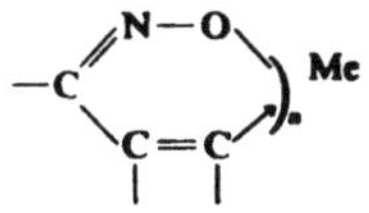		Complex salts of the benzo-phenonoxime
S-atoms containing oximes	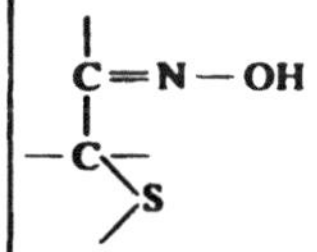	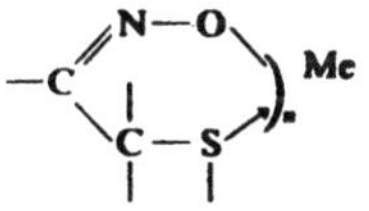	6	Complex salts of the oxime of the phenyl-α-thienyl-ketone
N-atoms containing oximes	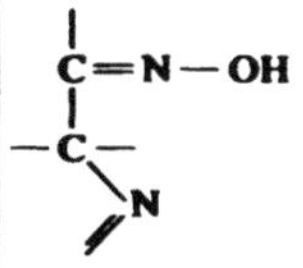	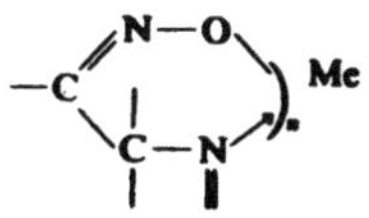	7	Complex salts of the oxime of the phenyl-α-pyridyl-ketone
N-nitroso compounds of the guanidines in the oxime form	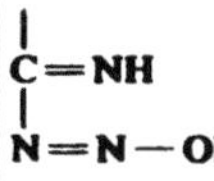	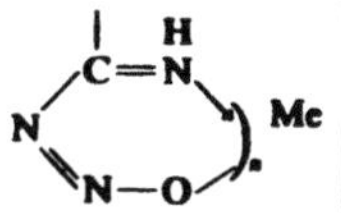	8	Complex salts of the nitroso-guanidine
Nitroso-arylhydroxyl-amine in the oxime form	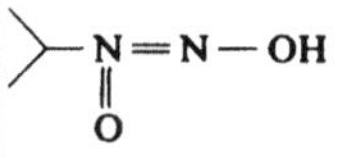	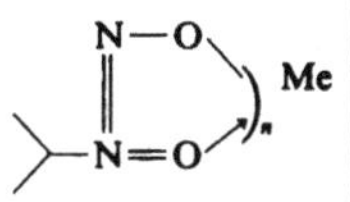	9	Complex salts of the dimethyl-amino-nitroso-phenylhydroxyl-amine and the nitroso-phenyl-hydroxylamine
Hydroxame acids	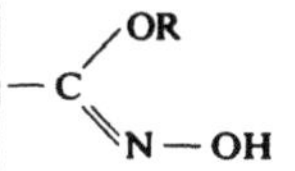 R = H, Alkyl oder Acyl	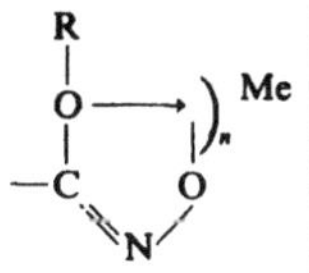	10	Complex salts of the hydroxamine cinnamic acids

Fig. 1 The following formulas are examples for special chelates of the 1st group.

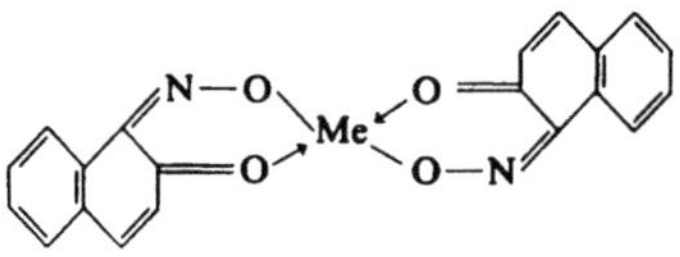

(1)

Me = Pb, UO_2, Cu

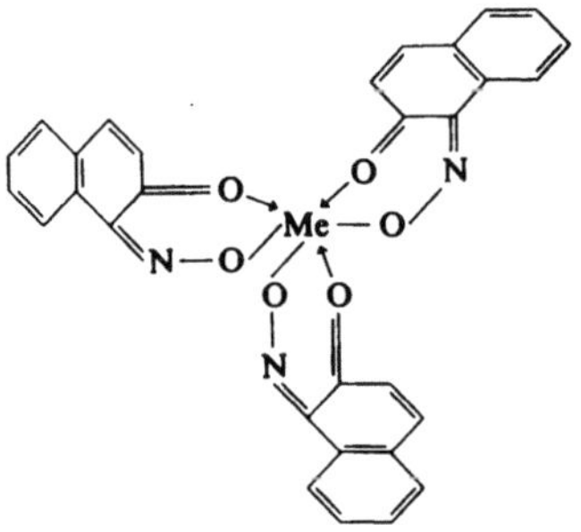

(2)

Me = Co, Fe

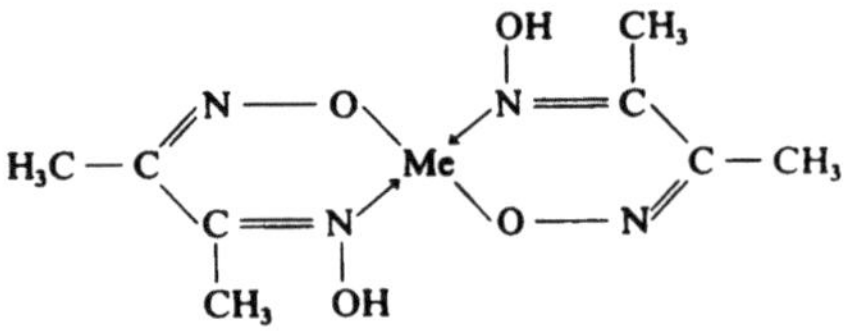

(3)

Me = Ni

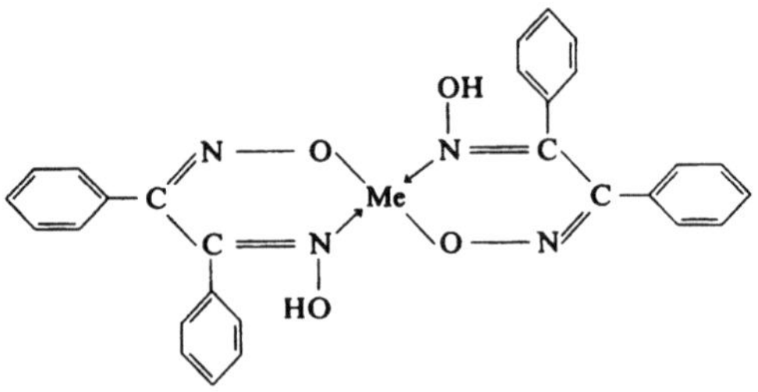

(4)

Me = Ni

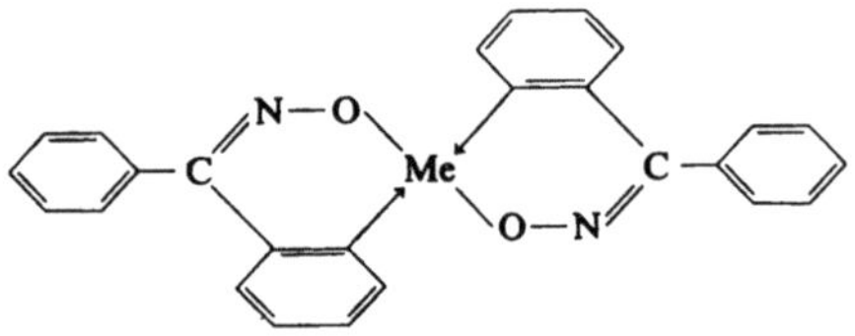

(5)

Me = Pb

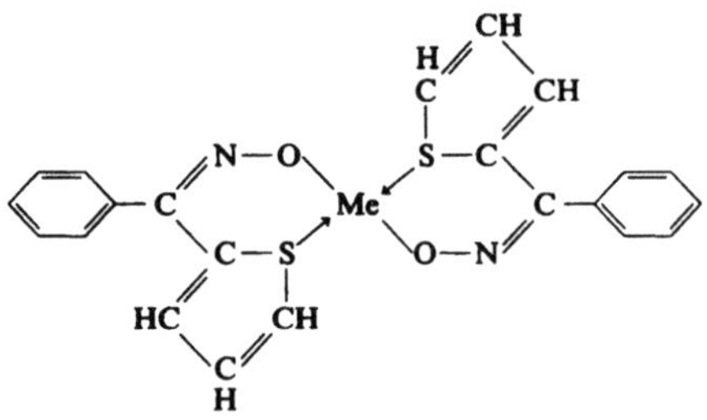

(6)

Me = Pb

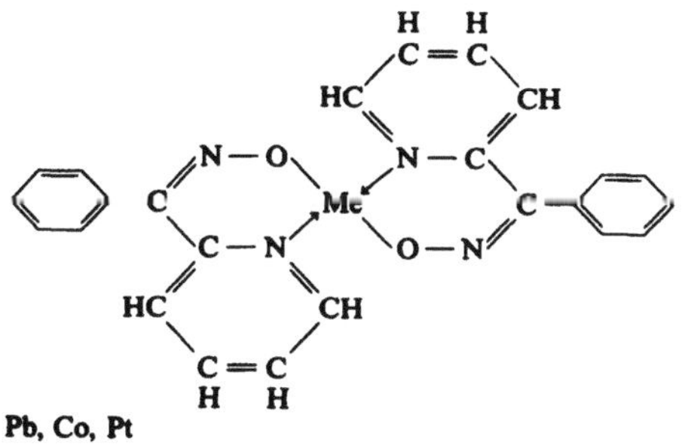

(7)

Me = Ni, Pb, Co, Pt

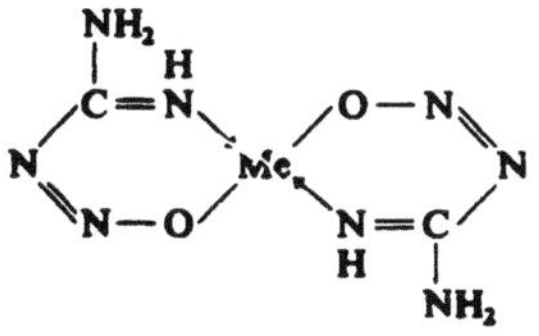

(8)

Me = Pd

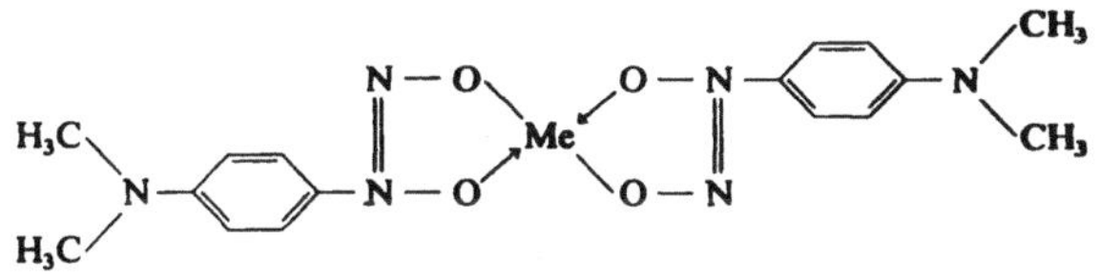

(9)

Me = Ni, Pb, Co, Fe, Ba, Cu

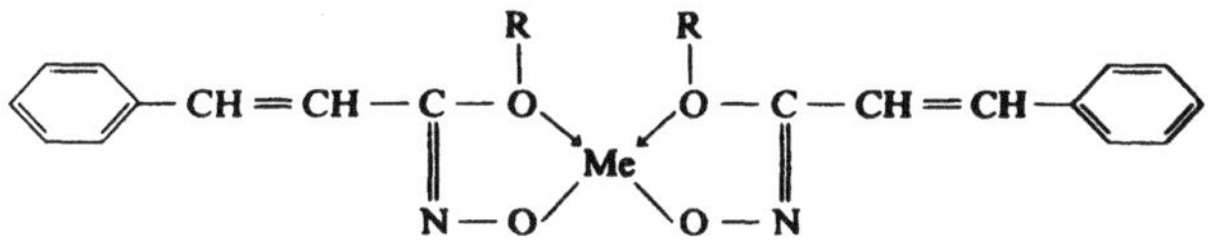

(10)

Me = Cu, Ni, Co, Cd

Table 2 shows the 2nd group: **Arylthiocarbazones, mono- and dihydrazides, biguanides, amidines, oxyamidines, azocompounds and Formazans.**

Class of Compounds	Functional Group	Assumed complex ring	Formula	Examples
Arylthiocarbazones, prim. and sec. arylthiocarbazonates	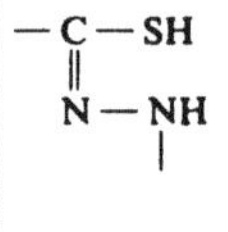	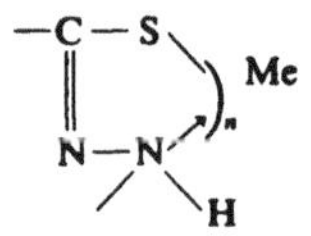	11	Complex salts of diphenyl thiocarbazone (Ag-complex salt), di-p-tolythiocarbazone, di-2,4-dimethylphenylthiocarbazone,, di-α-naphtylthiocarbazone, di-β-naphtylthiocarbazone and of di-p-biphenylthiocarbazone
Biguanide, also substituted	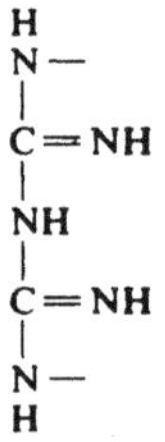		12	Complex salt of o-tolylbiguanide
Aromatic or heterocyclic ring systems containing azo compounds which carry at least one substitute of X = OH, SH, COOH or SO_3 in o- or α- position to the azogroup	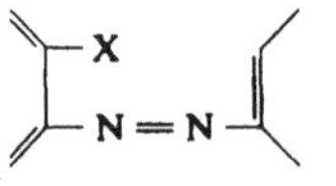	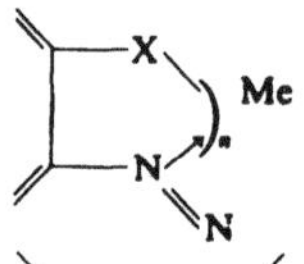	13	Complex salt of 1–(2-pyridylazo)–2-naphthol
	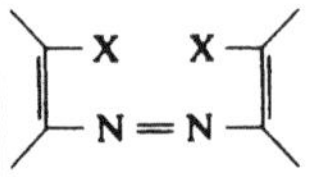	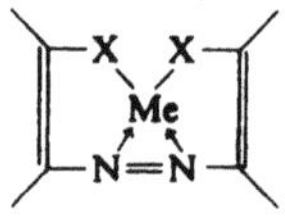	14	Complex salts of 2′,2′ dioxyazobenzene, dyes or metal lacquers of the azo series for example 2-oxynaphthyl-1-(4′-nitro-1′-azobenzene)
	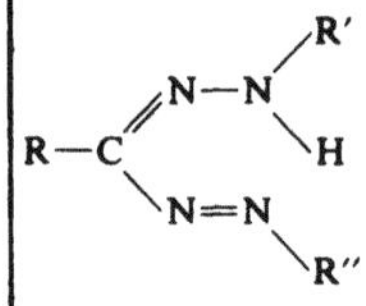	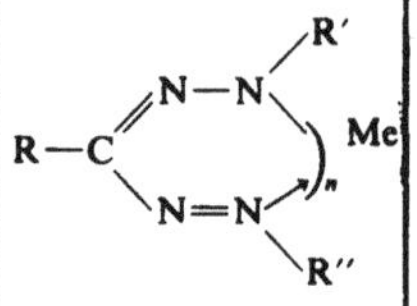	15	Complex salts of Lithol red R (Schultz-Farbstofftabellen 7th edition 1931 Nr. 219)
Formazans with at least one group of–OH″, –SH or –COOH in Aryl in ortho-position to the Formazan chain.	R = Alkyl, Carboxyalkyl, Carboxyaryl, R′, R″ = Aryl			Cu- or Ni- complex salts of triphenylformazan-2′ carboxylic acid

Fig. 2. A selection of compounds of the 2nd group and the chelate-forming positions is shown.

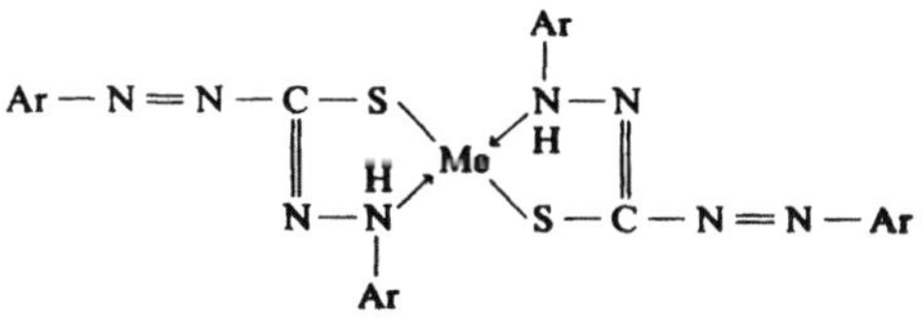

(11)

Me = Dithizionmetals after G. Iwantscheff, "The Dithizon and its Application in the Micro and Traces Analysis", Weinheim (1958);

Ar =

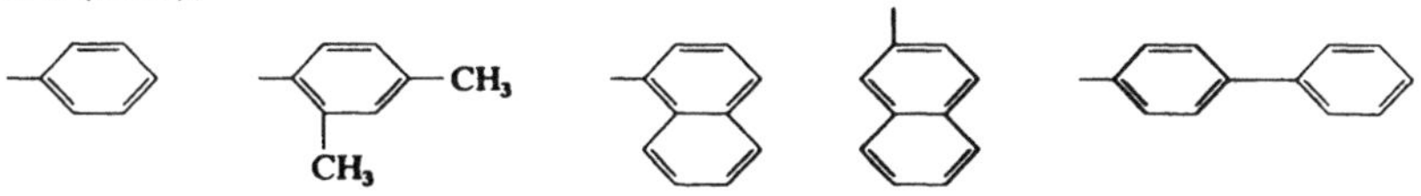

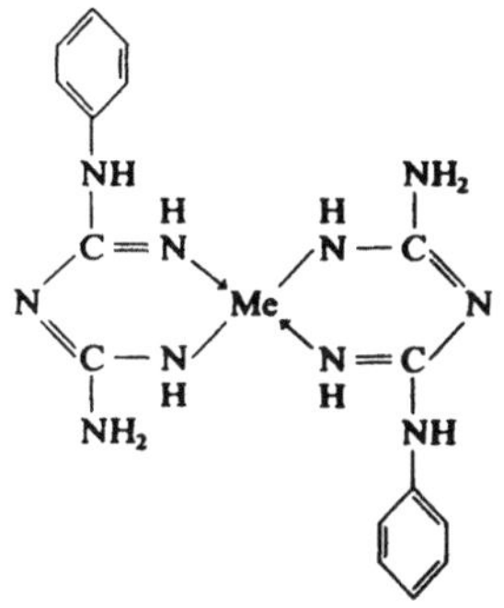

(12)

Me = Cu

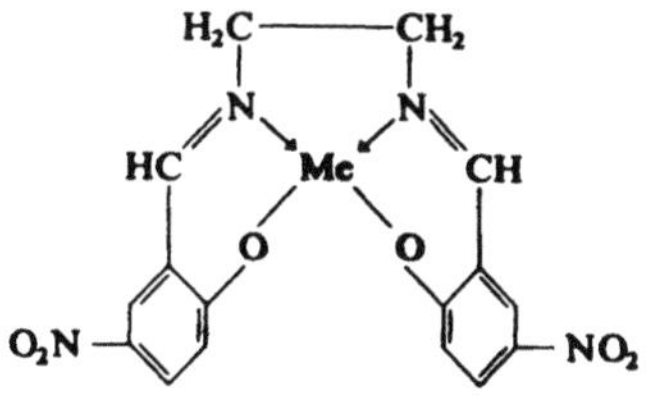

(13)

Me = Cu, Co, Ni, Pb, Fe, Cd

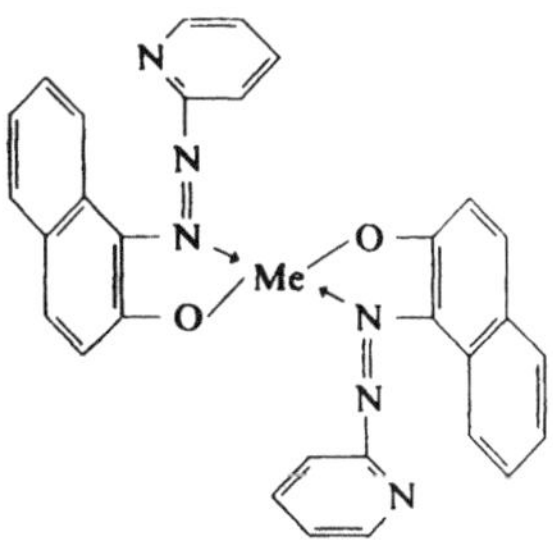

(14)

Me = Zn, Cu, Cd, Co, Pb, Fe, Bi, Ba

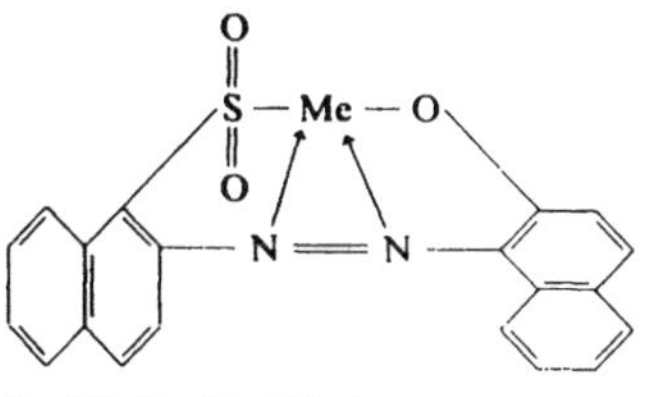

(15)

Me = Ba, Ca, Ni, Cu, Pb, Cd, Co

Table 3 shows the 3rd group: **Amino acids, pyridine, quinoxaline, quinazoline, quinazine, quinoline compounds, hydroxy acridines, acridones, oxyquinacridines, indoles, and murexides**

Class of Compounds	Functional Group	Assumed complex ring	Formula	Examples
α-Amino acids	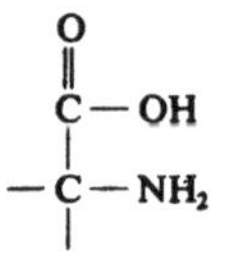	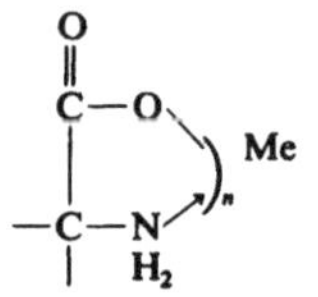	16	Complex salts of glycine or of histidine
Aromatic or heterocyclic o-(1,2)-amino carboxylic acids, also substituted or heterocyclic carboxylic acids containing a carboxylic group in the α-position to nitrogen or–as to quinoline–in the 8-position	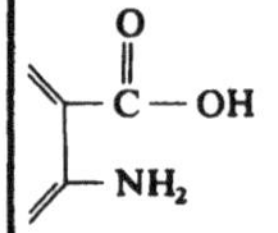 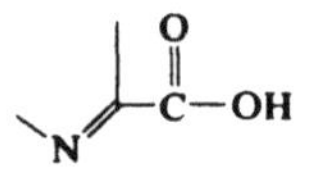	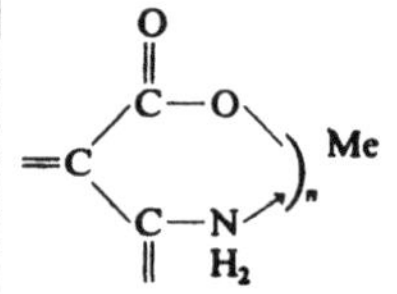 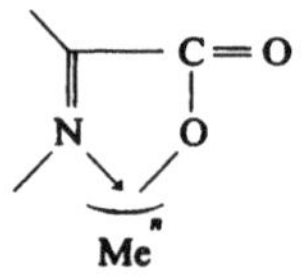 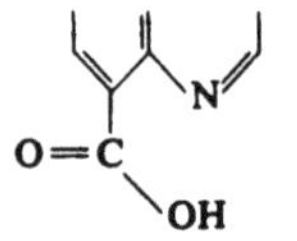	17	Complex salts of anthranilic acid, of o-amino naphthalene carboxylic acid, of α-pyridine carboxylic acid, of quinoline-8-carboxylic acid
Quinolines, resp. quinaldines containing an -SH- group in the 8-position	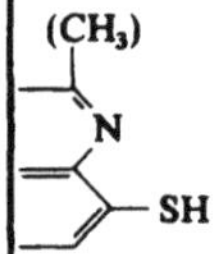	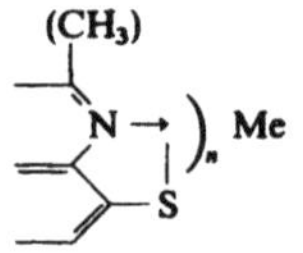	18 19 20	Complex salts of 8-mercapto-quinaldine, of 5,7-dibromo-8-mercaptoquinoline, of 4-styrene-8-mercaptoquinoline complex salt of the 4-oxyacridine
Acridines containing an -OH- or an -SH- group in the 4- and/or the 5-position	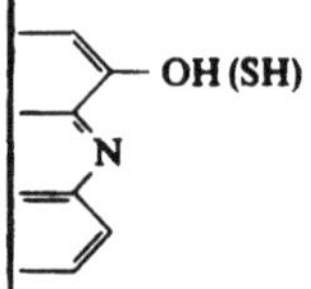	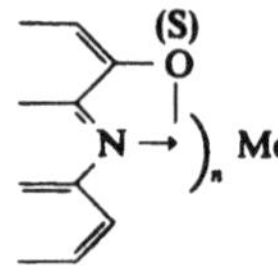	21	
Indoles having an =O-, =NH-, =S- group in the β-position	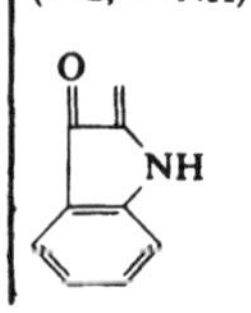	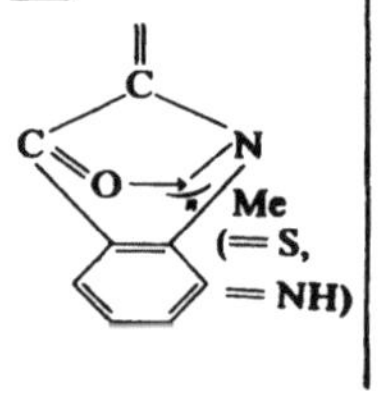	22	Complex salts of isatin

Class of Compounds	Functional Group	Assumed complex ring	Formula	Examples
Compounds with a pyridine or a pyrazine ring having, as to the pyridine ring in the α-position one, as to the pyrazine ring in the 2,3-position two aldehyde, keto, carboxylic acid carboxylic acid imide, or carboxylic acid hydrazide groups, reacting eventually after hydration in the pseudoacidoform	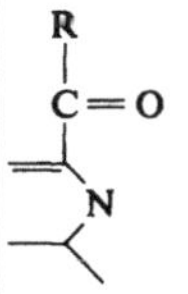 (R = Alkyl, Aryl oder Oxalkyl)	hydratisiert 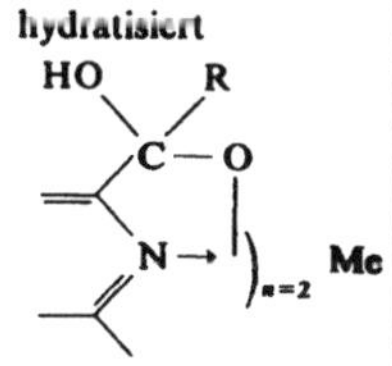	23	Complexes of quinoline-2-carboxylic acid imide, of quinoline-2-carboxylic acid hydrazide, of pyrazine-2,3-dicarboxylic imide or hydrazide especially when using copper
	(=N—N<) 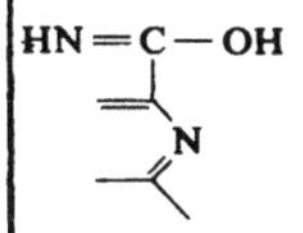	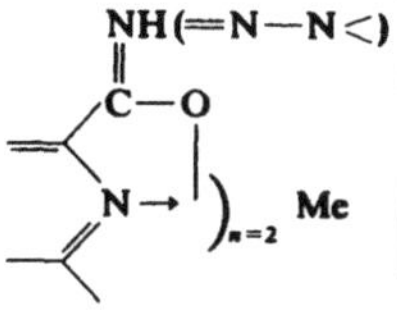		

Fig. 3. A selection of compounds of the 3rd group and the chelate-forming positions is shown.

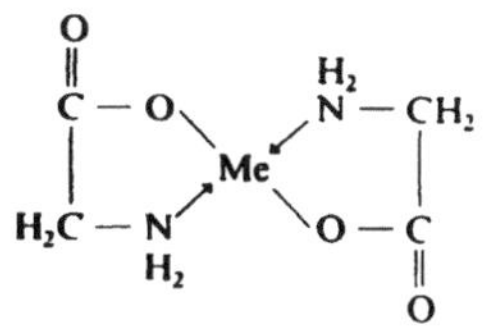

Me = Cu, Au (16)

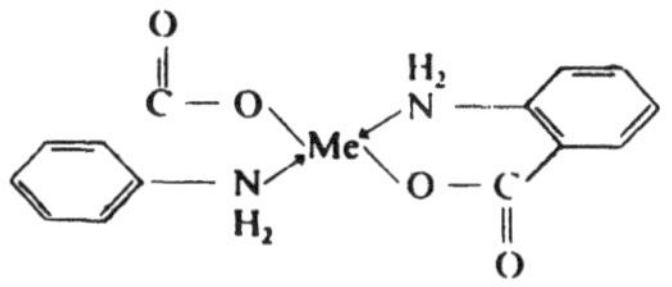

Me = Zn, Cd, Cu, Ni, Co, Pb, Hg (17)

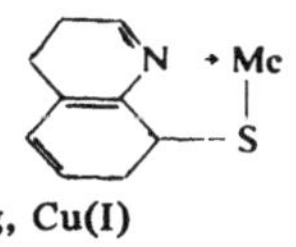

Me = Ag, Cu(I) (18)

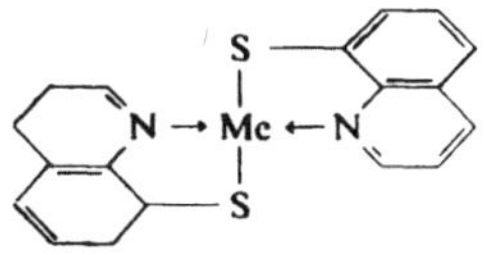

Me = Ba, Zn, Cd, Pb, Cu(II), Ni, Sn, Al, Mn, Fe (19)

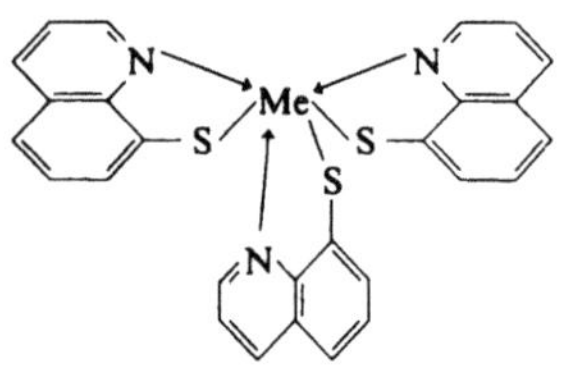

(20)

Me = Al, Cr(III), Bi, Fe(III)

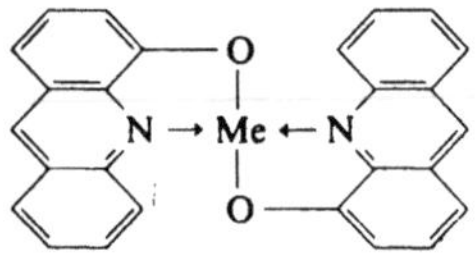

(21)

Me = Zn, Cd, Pb, Ni, Cu; Me^{3} = Ringsystems analogous to formula 27 with Co(III), Al, Fe(III), Cr(III)

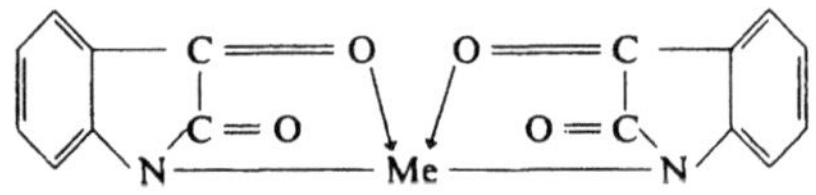

(22)

Me = Pb, Ni, Cu

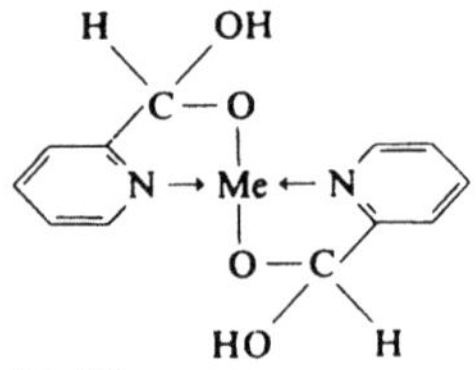

(23)

Me = Cu, Co, Ni

Table 4 shows the 4th group: **Hydroxy ketones or their enols, naphtoquinones, anthraquinones, flavones.**

Class of Compounds	Functional Group	Assumed Complex ring	Formula	Examples
β-Hydroxy Ketones	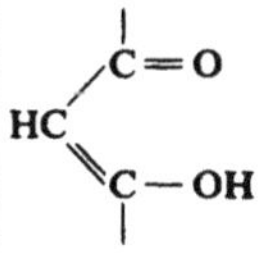	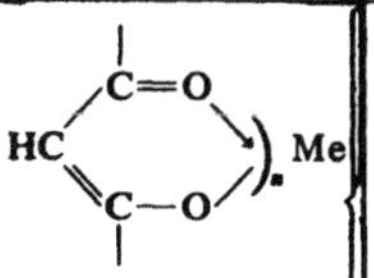	24 25	Complex salts of acetic acid ester, of acetic acid-4-nitro-anilide, of acetic acid-4-chloranilide
Bis-β-diketones at aromatic or heterocyclic rings in enole form	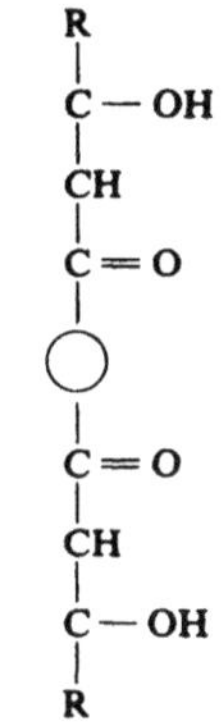 R = Alkyl		26	Complex salts of terephthalyl-1,4-diacetone
Quinones, also naphthenquinones, especially polynuclear, which have an substituted X, especially -OH and -SH in neighborhood to an = O-group, flavones	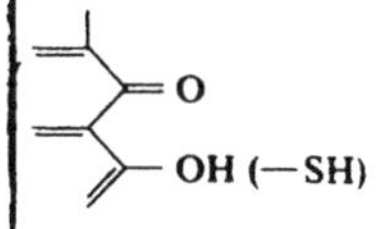	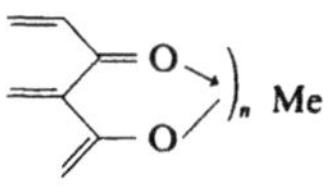	27 28 29	Complexes of 1,2-dihydroxy-anthraquinone (alizarin), of alizarin orange, alizarin blue, alizarin bordeaux red and quinizarin under chain-forming complex-building with Me = 2, under spacious threedimensional with Me = 3, complexes of morin

Fig. 4. A selection of compounds of the 4th group is shown.

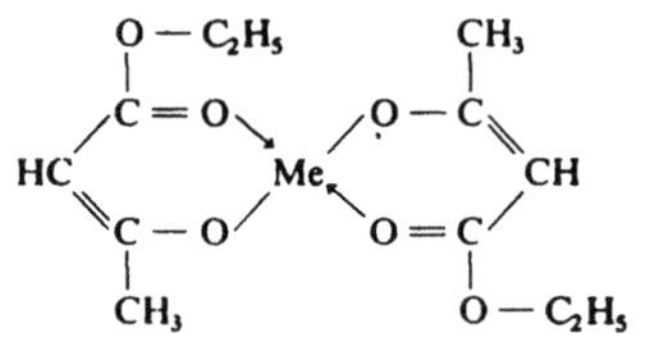

(24)

Mc = Cu

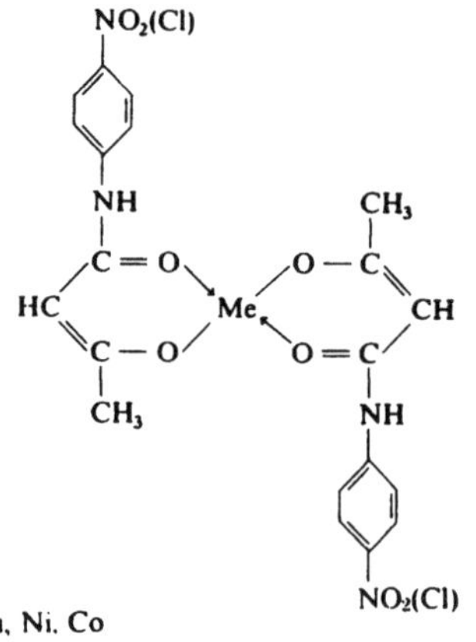

(25)

Mc = Cu, Ni, Co

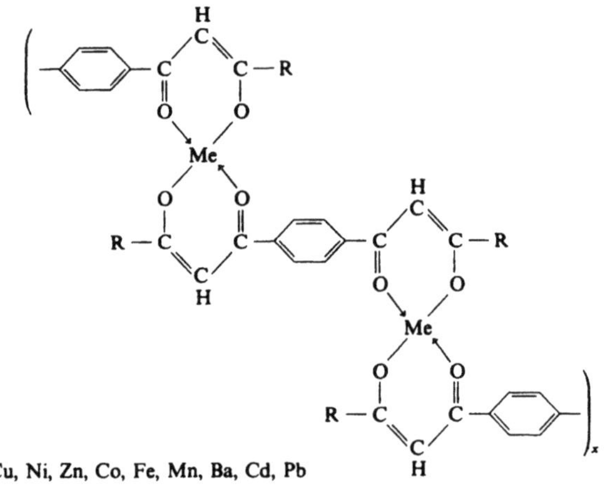

(26)

Me = Cu, Ni, Zn, Co, Fe, Mn, Ba, Cd, Pb

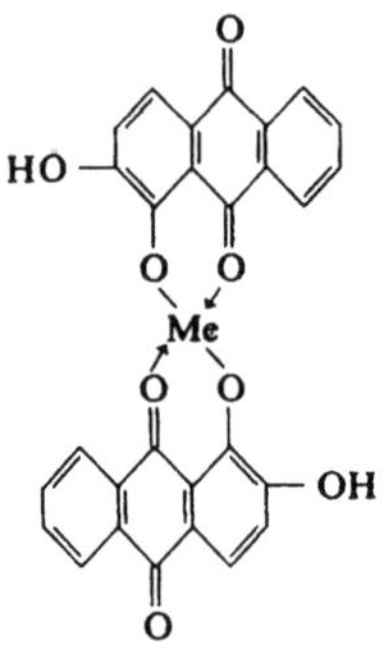

(27)

Me = Ni, Mn, Co, Be, Pb, Cd, for Me(III) = analogous to 35

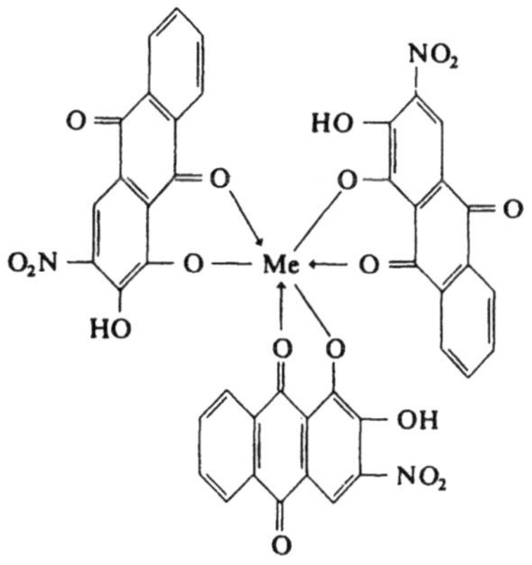

(28)

Me = Al, Cr, Fe, W, for Me(II) = analogous to 31

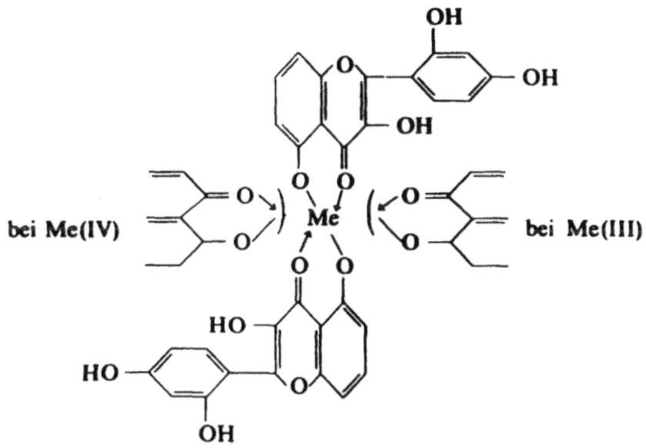

(29)

Me(II) = Pb, Cd, Ni, Mn, Be, Zn, Cu, Me(III) = Al, Cr, W, Me(IV) = Zr, Sn

Table 5 shows the 5th group: **Imidazoles, thiazoles, thiocyanure acids**

Class of Compounds	Functional Group	Assumed Complex ring	Formula	Examples
Substituted imidazoles with a substitutent X in 2-position (X = -OH, -SH, -COOH, or $-SO_3H$), which are or orthocondensated with aromatic ring systems at the C-atoms 4 and 5 or have heterocyclic substitutents	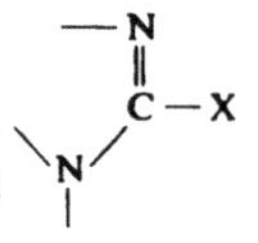	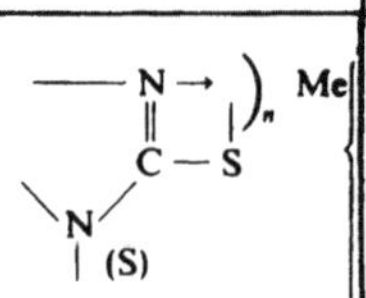	30	complex salts of 2-mercaptobenz-imide azole
Thiazoles corresponding to the preceding part	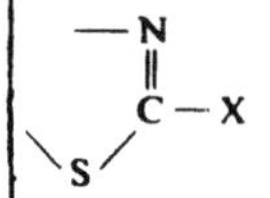 X = —OH, —SH, —COOH oder $—SO_3H$		31	Complex salts of 2-mercaptobenz-thiazole
Thiocyanure acids, especially trithiocyanure acids, but also those which contain 1 or SH-groups and moreover 2 or 1 OH-groups: technical mixtures by reaction of cyanure chloride with sodium hydrogen sulfide	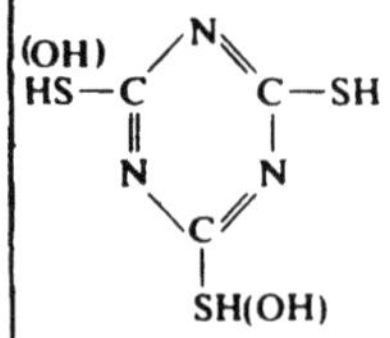	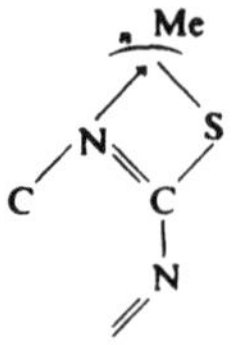		Complex salts of the thiocyanure acids with Ba, Mn, Co, Ni, Pb, Cd, Cu, partly in sterical arrangement

Fig. 5. A selection of compounds of the 5th group is shown.

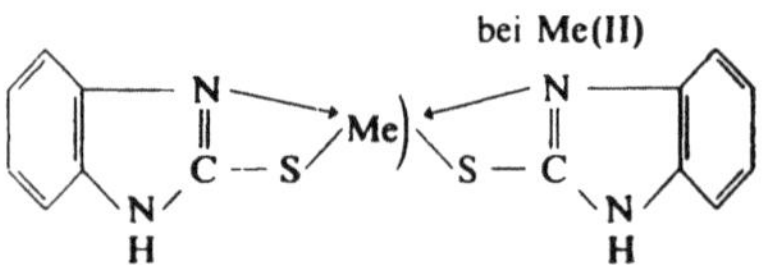

(30)

Me(I) = Ag, Me(II) = Zn, Cd, Pb, Cu

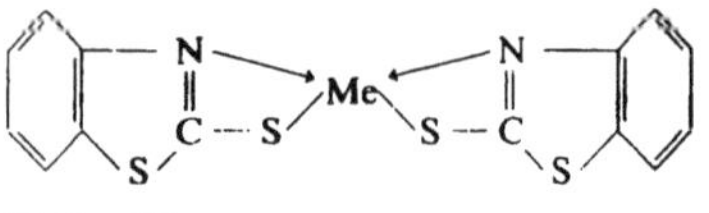

(31)

Me = Pb, Tl, Cu, Cd

The preceding studies have been carried out to find photoconductive chelates from the various classes of compounds. From these experiments, the structure which is responsible for the photoconductivity, has been elucidated. On the basis of our knowledge on the steric structure of the chelates, the influence of the metallic central atom on the photoconductivity has been investigated. On this point of view, the synthesis has been carried out of metal chelates of 2-mercapto-benzo-thiazole and of 8-hydroxyquinoline with various metals which were coated with the above mentioned resin. The chargeability, the dark-decay and the light sensitivity with UV light of these layers have been tested by the Dyn-tester and the Victoreen apparatus. The wave-length dependence and the rate of the discharge represented in the figures 6–11.

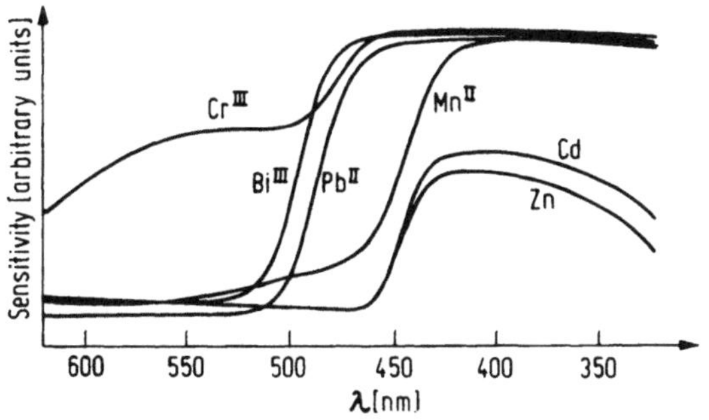

Fig. 6. Action spectra of 8-hydroxyquinoline-metalchelates

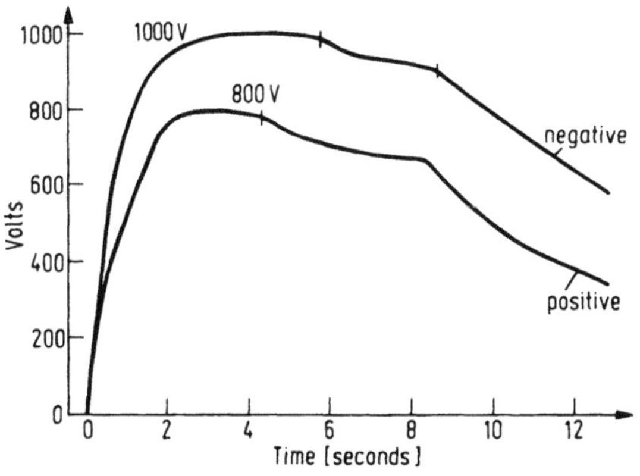

Fig. 7. Charge acceptance, dark-decay and tungsten light sensitivity of Zn^{II}-mercapto-benzthiazole

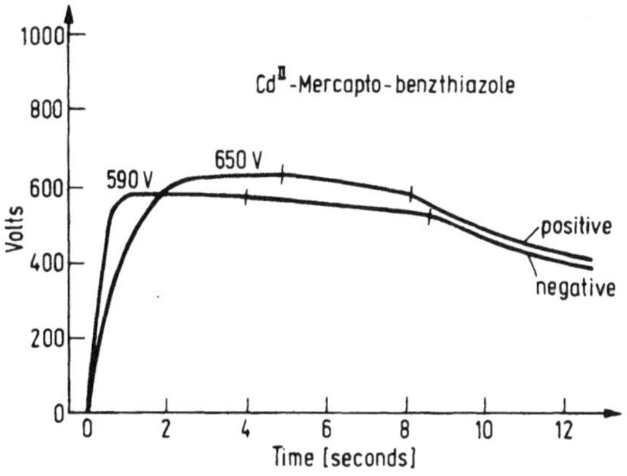

Fig. 8. Charge acceptance, dark-decay and tungsten light sensitivity of Cd^{II}-mercapto-benzthiazole

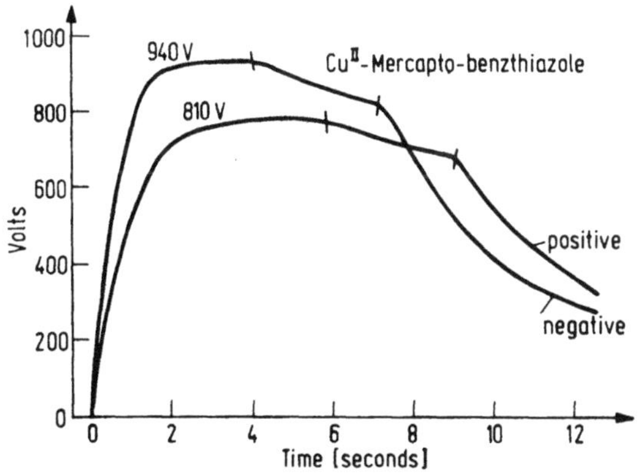

Fig. 9. Charge acceptance, dark-decay and tungsten light sensitivity of Cu^{II}-mercapto-benzthiazole

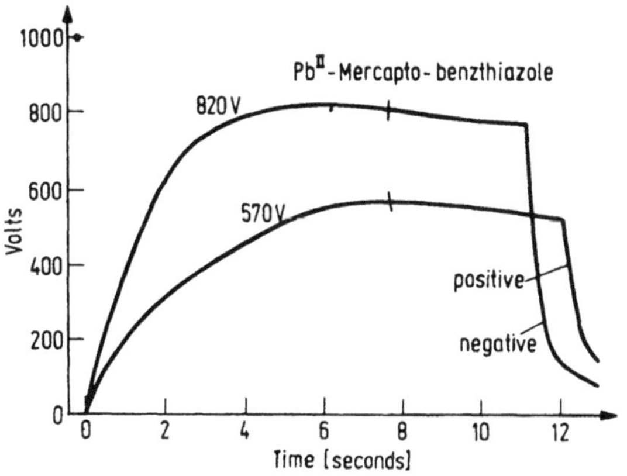

Fig. 10. Charge acceptance, dark-decay and tungsten light sensitivity of Pb^{II}-mercapto-benzthiazole

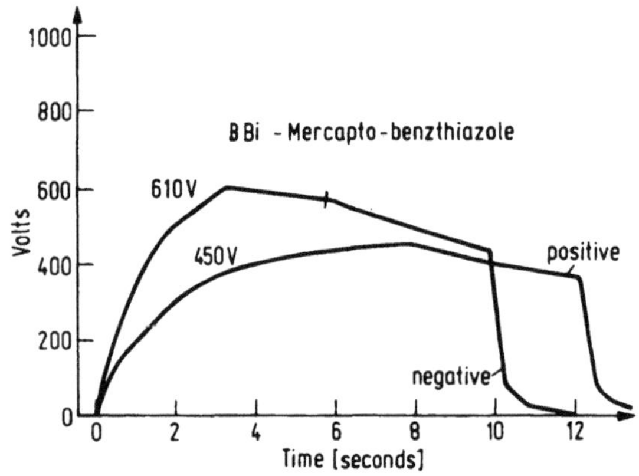

Fig. 11. Charge acceptance, dark-decay and tungsten light sensivity of Bi^{III}-mercapto-benzthiazole

4. Discussion

The action spectra indicate that the optimum light absorption of the chelates increases with increasing atomic number of the metals. Simultaneously, the maximum is shifted to longer wave lengths. The spectral shift of the light absorption of the 3-point fixed chromium complex of the 8-hydroxyquinolines is in agreement with the model of Löhr, Arneth and Winkelmann [16]. They have discussed that in organic donor-acceptor complexes the increase of the electron affinity of the acceptor causes a shift to longer wave lengths. In the periodic system adjacent elements yields roughly the same course of the absorption spectrum. The reason for this behavior may be suggested in the same structure of the compounds with regard to the arrangement and the occupation of the electron levels. With the exception of bismuth and chromium, all metals of the chelates have the coordination number 4 forming with two organic chelate formers molecules in a plane or in a tetrahedron structure. These chelates exhibit their action maximum in the UV region.

Layers of 2-mercapto-benzothiazole and Gelvar 202 in the weight ratio 3 : 1 can be charged both positively and negatively under a corona. However, the positive charge acceptance is smaller than the negative charge acceptance. Particularly striking is the deviation at a Ni (II) chelate. Furthermore, the light sensitivity of positively charged layers is smaller than that of negatively charged layers. In additon, the light sensitivity increases with increasing atomic number. The chelates are especially sensitive in the UV region.

If the organic chelate formers posses additionally electromeric substituents which can takc up or give off electrons, then it is obvious that an electron shift between the over-compensated metal atom and the organic electromeric substituents can readily occur when electric fields are present or light energy interfer. Therefore, we may assume that an electron exchange occurs dependent on the state of the electrophotographic layer.

Since the aromatic system due to its oscillating π-electrons represents the state of a one-dimensional electron gas, it exhibits a large electronic conductivity. Thus it is assumed that at an incorporation of a suitable metal, a large electron exchange between the metal and the electromeric substituents may be preferred. Due to the rather long distance caused by the aromatic system, such mesomeric limiting structures from dipoles with strong moments may be expected. Under consideration of this fact, chelates can be produced with high photoconductivity. In agreement with this concept, we expected a significant photoconductivity in metal chelates with aromatic derivatives, reported by Schlesinger [13].

The course of the positive and negative charging of nickel-2-diacetyldioxime and of copper-2-betain and their discharge under illumination with UV light are represented in Figures 12 and 13. At that time, it was surprising and for the further understanding of the photoconductivity in the metal chelates important that such simply composed compounds are photoconductive.

Figure 14 represents the formula of diacetyldioxime (I) and of betain (II) and their chelate derivatives. If the representation of a 5-point fixed chelate ring is selected, then

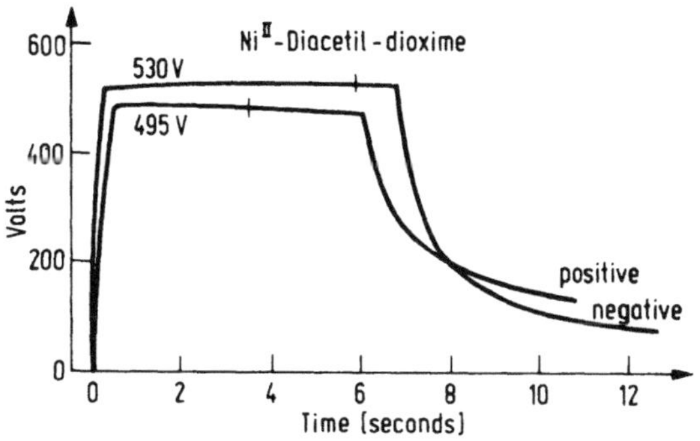

Fig. 12. Charge acceptance, dark-decay and tungsten light sensitivity of Ni^{II}-diacetildioxime

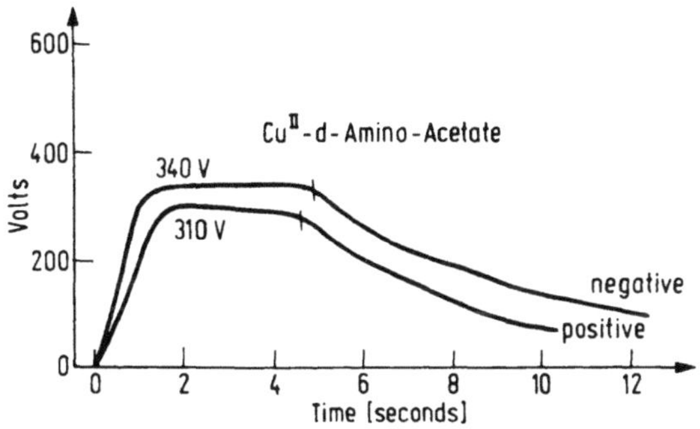

Fig. 13. Charge acceptance, dark-decay and tungsten light sensitivity of Cu^{II}-d-aminoacetate

according to IIA besides a chelate formation with a metal atom also chelates with two additional hydrogen atoms become possible. If need be, these hydrogen bridges could make understandable the high speed of charge acceptance by the chelates since due to the

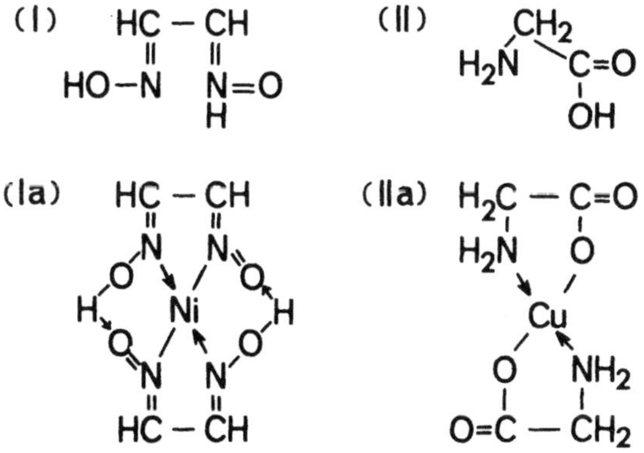

Fig. 14. Diacetildioxime (I) and its Ni^{II}-chelate (Ia), α-amino-acetate (II) and its Cu^{II}-chelate (IIa)

interchange of the nitrogen atoms with the metal atoms an additional polarization appears. Formula IIA indicates the structure of the chelate of copper betain. This compound exhibits only a weak photoconductivity, but experimentally confirmed. The two mentioned compounds are typical representatives of the so-called penetration complexes which fill up the inner electron levels of the transition metal ions with donor elements.

Copper-2-betain has a very simple organic structure without a noticeable conductivity since in the copper-2-chelate of glycol the probability of the electron transfer via the CH_2 group is very small. Therefore, the photoconductivity must be expected due to an electron transfer between the metal ion and the organic donor.

4.1. Photoconductivity of inner molecular donor-acceptor complexes

On the basis of our present results, we are forced to assume that the transition metals form a photoelectrically effective donor-acceptor complex if electrons from the donors of the chelate former are incorporated in their inner electron shells. Due to the incorporation of electron pairs, the metal ion obtains an excess of negative charge which results in a reduction of the binding energy of electrons in the upper shell. These electrons, however, can be easily excited and shifted by smaller energy than the energy necessary for the excitation in the intrinsic region. These electrons in the upper level are responsible for the charge transfer if the chelate polarized in the electric field due to corona charging is excited by illumination. Depending on the structure, the metal chelates can form one or several donor-acceptor complexes. But the increase of the number of donor-acceptor complexes results necessarily in an increase of mobile excited electrons in the molecule which are available for the formation of polar structures at electrical excitation. A full efficiency of charge-transfer processes, however, may be expected only in such donor-acceptor complexes in which metal ions and donors form real covalent bonds.

The ions of the alkali metals are useless because of their strong bonds forming with the donor. Covalent bonds appear particularly with the transition elements. A high electron affinity is exhibited by the noble metals, followed by Ag, Hg and in part also by copper. Ni, Co, Fe and Mn exhibit a weaker electron affinity but still too strong covalent bonds. On the other hand, metal chelates with zinc and cadmium demonstrate already an appreciable polar bond mechanism.

As donors preferably non-metals of the 5. and 6. group of the periodic system and also halogens come into question. Particularly preferred is nitrogen since it is mainly responsible for the formation of covalent bonds. The appearance of covalent bonds during the chelate formation can be prevented by a steric hindrance (due to special bond angles) since the spatial arrangements of the atoms is determined by the angles of bonds which can be deduced from the electron levels of the metal ion supplied from donors. If these steric requirements are not fulfilled then the metal ion is bound polar under decrease or loss of the photoconductivity. These compounds cannot be used as photoconductors in electrophotographic layers.

If a complete covalent bond is established then one might expect a quadratic plane or a tetrahedral configuration for a metal atom with a coordination number 4 and an octahedral configuration for a metal atom with a coordination number 6. By no means it is necessary that the formers of the organic chelates form with the metal ion ring systems with a covalent bond. It is often more likely that the chelate formers make their ionic group available to the one metal ion and their donor group available to another metal ion. In addition, metal chelates are known which obtain donors only from the chelate former. These chelates arise of the type of ethylene diamine. In this connection, it is also worth mentioning that macromolecular species of organic metal chelates are available which exhibit an appreciable photoconductivity.

4.2. Sensitization experiments of photoconductive metal chelates

During the investigation of the photoconductivity, it has been obviously attempted to sensitize the various chelates with organic dyes employed in zinc oxide electrophotography. It has been found, however, that organic metal chelates, the coordinative bonds of which are saturated cannot be sensitized by the usually employed sensitizers which contain a chelate-forming group. Thus such coordinatively saturated chelates become photoconductive only during illumination in the intrinsic energy region. However, a significant sensitization was detected when the sensitizer as chelate former has been introduced into the metal chelate under formation of a coordinatively saturated mixed metal chelate. For their production, partial chelates have been prepared, the valencies of which were saturated by chelate formers on the one side and by ionogenic groups on the other side. Preferentially by alkali, the ionogenic group was substituted by a chelate-forming sensitizer. As particularly exchangeable components for the ionogenic groups, the insoluble salts of halogens, rhodanide, cyanide and of hydroxyl and acidic ions can be used.

It has been originally found by us that many chelate formers, as for instance morin, are also significant sensitizers for the photoconductivity of zinc oxide-resin layers. The examination of the sensitizers for zinc oxide has indicated indeed that the most effective sensitizers, technically used in zinc oxide-resin layers often demonstrate a structure similar to that of a metal chelate. Here, we may mention the derivatives of sulphophthalein, eosin, of azo compounds with H-acid substituents in σ position, alizarines etc.

4.3. Sensitization of ZnO crystals by chelate-type sensitizers

On the basis of the discussion in the last chapter under consideration of the chelate mechanism, the attempt is made to describe the sensitization of the electron-transfer through a ZnO resin layer from the special point of view of a chelate-induced sensitization. The previous studies make it obvious that the sensitization should occur on preferred spots of a zinc oxide crystal where a chelate formation has been established. By various papers [14], the presence of hydroxyl groups on the zinc oxide surface has been confirmed. In addition, also other impurities, such as carbonates and sulphide ions, are available on the surface influencing the adsorption of the sensitizers. Furthermore, excess zinc ions and free electrons in the bulk and on the surface of the zinc oxide crystallites have to be considered. In the case of eosin and the higher halides of the sulphophthaleines, it could be suggested that the halides act as electron donors and the polar binding occurs between the carboxyl or hydroxyl group. We have to forego the concept that the chelate formation at the surface of a photoconductive crystal must be fixed in a two-point configuration which, generally, is the rule of adsorbed metal chelate molecules. Probably, the adsorption of a sensitizer may occur at the edge of a crystallite as a chelate in a one-point configuration.

If a chelate formation with a zinc ion in interstitial position on the surface does occur during adsorption of a sensitizer, then the charge density surrounding the zinc ion is enlarged by two electrons. We tentatively assume that this large charge density near the metal atom of the sensitizing chelate molecule determines the transfer energy of the sensitizer. By the injection of the donor electrons towards the zinc ions reacting with the dye molecules, energetically favorable levels for an electron injection into the conduction band of zinc oxide have been supplied. Therefore, only partial amounts of energy is necessary for both the transfer and the flux of electrons. This partial amount of energy is enough in order to cause a transfer of an electron from the excited sensitizer via the donor-acceptor complex generated by a reaction between a chelate former and a metal ion. Hauffe et al. [15] were able to demonstrate by means of photocurrent measurements that morin (an excellent chelate former) and azo compounds can exhibit significant sensitization for the electron transfer through charged zinc oxide single crystals. Special binary senitizer systems, such as rhodamine B and an azo dye, exhibit a higher sensitizing ability than the single dyes. It may be supposed that mixtures of various sensitizers can become particularly effective for zinc oxide if the following requirements are fulfilled by the sensitizers:

1. Type A has to be a strong chelate former with zinc ions and must possess additional acceptor groups.
2. Type B should have both donor and acceptor groups in the conductive organic framework.
3. Type C should possess only a donor group and an optical absorption for the sensitization.

Sensitizers of the type A form a chelate with zinc ions available on the surface of the ZnO crystals and act additonally as acceptor whereby between the A and B type molecules donor-acceptor complexes are formed. Since sensitizers of the B type possess additional acceptor groups, they can form further donor-acceptor complexes with compounds of the C type. But molecules of the C type can only react with molecules of the A type. It is required that molecules of type A and B must belong to chemical compounds suitable for a charge transfer. This means that the chemical compounds must consist of aromatic, conjugated unsaturated or of heterocyclic compounds with aromatic properties. Molecules of the B type can be precipitated in several layers one after another.

Since the donor-acceptor complexes decrease the energy necessary for the charge transfer, they are useful as sensitizers in order to increase the sensitivity under illumination with a discharge lamp employed in technical units. On the basis of this mechanism, however, it is not necessary that a compound of type A itself must be sensitizer. Generally, also chelate formers with any sensitizing properties may be considered which at least posses acceptor properties and an electronically conductive structure between the acceptor and the chelate-forming donor. It is supposed that chelate formers of this type exist in special binders of electrophotographic layers.

These explanations have practical consequences for the sensitization of zinc oxide resin coatings. If an equilibrium may be assumed for a reaction of a chelate-forming sensitizer with zinc ions on the surface of ZnO crystallites, then the stronger bound sensitizer displaces the sensitizer with the weaker bond. By this simple displacement mechanism, the decrease of the sensitization of the single dyes in a binary or ternary mixture becomes understandable. The same seems to be valid for donor-acceptor complexes of the sensitizer types B and C. Besides this mechanism, however, also a mutual electron interchange of the adsorbed sensitizer types A, B, and C may be responsible for both the increase and the decrease of the sensitizing efficiency, as could be demonstrated by adsorption and electrochemical cell technique experiments.

On the basis of this discussion, the dependence of the sensitization on the structure of the binder becomes also understandable. Furthermore, we believe to understand why sensitizers are particularly effective when they are dissolved whereby their mutual donor-acceptor action is terminated. Since silver belongs to the complex-forming transition elements, the same mechanism might be expected for the sensitization of silver halide crystals. The experimental fact is well known that in many cases the sensitizers with their increasing sensitivity in mixtures are effective for both silver halide and zinc oxide.

If the chemical structure of the resins used for electrophotographic layers is considered then it can be recognized that all the useful resins contain groups of hydrogen chelates or at least hydrogen bridges. There exist resins which are able to form chelates with zinc ions available on the surface of the ZnO crystallites under the same conditions as described for the sensitizers. Since the chelate formation is always connected with the existence of excess zinc ions and electrons as consequence of the donor-acceptor complex, an influence of the lattice-defect structure and of the conductivity cannot be excluded. Under this point of view, a first indication is available for the understanding of the phenomenon why certain polar resins used as binders increase the photoconductivity of zinc oxide. If we consider in addition the interference of the electric field during the corona discharge which results because of the polarization of the binder in an increase of the conductivity, then a preorientation of the hydrogen chelates due to polarization may be assumed.

Finally, it is of interest to know that the electrophotographic properties of zinc oxide was detected with zinc oxide–silicone resin layers. Such silicone resins, however, represent model substances with hydrogen chelates.

References

[1] A. Werner, Neuere Anschauungen auf dem Gebiet der anorganischen Chemie, neu bearbeitet von P. Pfeiffer, Vieweg (1923)

[2] R. Weinland, Einführung in die Chemie der Komplexverbindungen, Enke (1924)

[3] P. Pfeiffer, Organische Molekülverbindungen, Enke (1922)

[4] S. E. Martell und M. Kalvin, Die Chemie der Metallchelatverbindungen, Verl. Chemie (1958)

[5] H. H. Kosche DBP 1 301 249 (Renker GmbH Düren)

[6] H. H. Kosche DBP 1 296 001 (Renker GmbH Düren)

[7] H. H. Kosche DBP 1 278 241 (Renker GmbH Düren)

[8] H. Schlesinger DBP 1 192 044 (Kalle A. G. Wiesbaden)

[9] H. Schlesinger DBP 1 111 015 (Kalle A. G. Wiesbaden)
[10] H. Schlesinger DBP 1 113 360 (Kalle A. G. Wiesbaden)
[11] P. Pfeiffer, Chem. Ztg. 59, 205-12 (1935)
[12] Monsanto, Plastic Prod. and Resin Div. St. Louis/Miss.
[13] Victoreen Inst. Comp. Cleveland Ohio
[14] Gmelius Handbuch der anorg. Chemie. Bd. Zink System Nr. 32, 8. Aufl. S. 131
[15] K. Hauffe, H. Pusch, J. Range und D. Rein: in diesem Buch, S. 219ff.
[16] B. Löhr, R. Arneth.und D. Winkelmann: in diesem Buch, S. 178ff.

High Sensitivity Electrophotographic Development

R. B. Lewis and H. M. Stark

Abstract

Certain laboratory and commercial applications of xerography require high resolution and high sensitivity development methods. In these cases it is common to use either powder cloud or liquid electrophoretic development. Here we report new work and review some older results which provide a theoretical and experimental picture of the mechanisms, properties, and limitations of these development systems in these applications. Many of the features of the two systems are common and comparisons between them are useful.

In the case of powder cloud development we describe a typical device and list some nominal values of powder cloud properties: number density, charge, mobility, and covering power. A mobility model predicts approximate development rates, and calculated field patterns determine the structure of the developed image.

In the case of electrophoretic development evidence shows that, for solid areas, the initial development rate is proportional to the initial electric field, and to the conductivity of the developer. A simple equivalent circuit accounts for these features of development, but as development proceeds, space charge effects and toner depletion cause departure from the ideal model.

A good measure of sensitivity is the smallest potential step yielding a perceptible line image. We give experimental results for both systems and show how the results relate to a theoretical calculation based on computed field patterns.

1. Introduction

The primary motivation for this work has been the development of X-ray-exposed selenium for medical diagnostic purposes. High sensitivity demands, it seems to us, that the method be carrierless. This restricts discussion to powder cloud and electrophoretic techniques. We suppose also that the requirements of wide process latitude and good visual detection of detail call for non-absolute, edge-enhanced development that must stop short of image charge saturation. The plan of discussion is to describe first in some detail the mechanisms in powder cloud development, then to review mechanisms in electrophoretic development, and throughout to contrast the two whenever it appears that comparisons may be useful.

2. Powder Cloud Development

Figure 1 shows the hardware that we have used. The box itself is about 3″ deep, metal, and the plate to be developed rests face down on top. The cloud itself is introduced at the bottom, under a baffle, by an aspirator-type powder cloud generator which works in a series of puffs. This allows the cloud to quiet down between puffs but promotes high turbulence for mixing. There are vents on the side of the box but they are not shown. Throughout this discussion it will be assumed that, while moving air controls toner motion throughout most of the box, near the plate surface that motion is damped and electric fields dominate. The

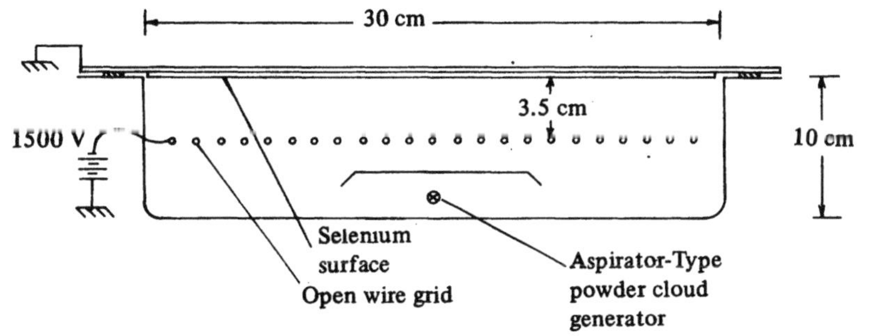

Fig. 1. Schematic of a powder cloud development box with the selenium plate in place

cloud itself contains, as made, charges of both signs. The grid serves to separate the two components of the cloud and drives, in this case, the negative component toward the plate. It provides what we will call a bias field of about 400 volts per centimeter. Notice that we do not bias out background but rather deliver toner to almost every part of the plate so that fringe fields can sort out the particles and detect information nearly everywhere. The grid serves also as a distant development electrode, but this is essentially open box development.

Estimates of powder cloud properties are shown in Table 1.

The only unusual measurement is that of powder cloud mass density. It was made by light absorption and will be commented upon later. Stokes mobility is computed for an average particle and is about 1% of that of an ion in air. The toner itself is a fairly standard, insulating, pigmented thermoplastic. To interpret certain results, it is necessary to have the information in Figure 2. The important thing is that over our range of measurement, transmission optical density is linear in deposited mass. The reflection results are about what one might expect.

It is probably enough to understand two things about any development system, speed and

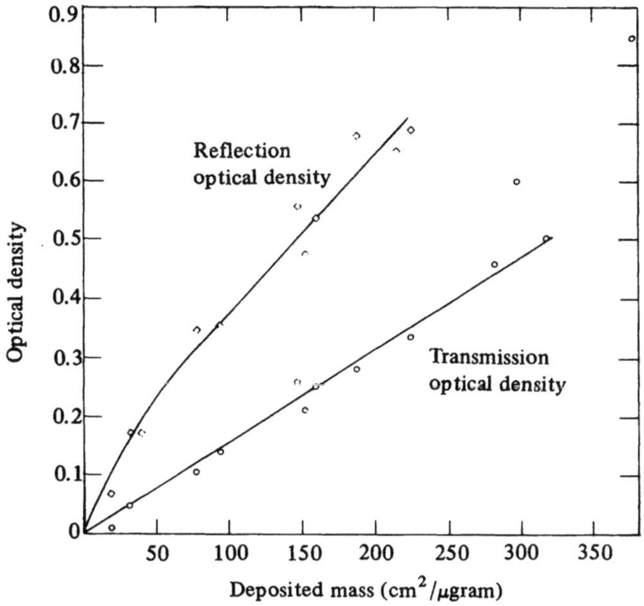

Fig. 2. Covering power measurements for powder cloud toner–taken with red light

Table 1

Number Mean Diameter	4μ	Coulter Counter
Cloud Mass Density	$\approx 6\mu gm/cm^3$	Light Absorption
Developed Q/M	$3-5\mu coul/gm$	Electrometer and Balance
Stokes Mobility	10^{-2} cm^2/volt-sec	Computed

sensitivity. Although speed is not really a primary consideration, it is worth some examination. Figures 3 and 4 show some measurements taken by light absorption. In this apparatus a laser beam passed through the box about a centimeter under the plate and into a linear detector mounted outside. The scale of mass per unit volume is the result of a separate calibration based on known toner masses deposited on microscope cover slips. Of course, the measurement is an average of density along the laser beam. Now suppose that each toner particle moves singly with the computed Stoke's mobility, that the electric field is about 400 V/cm, and that there are about 3 centimeters to clear out. If follows that the clearing time should be about one second. In both traces each of the spikes marks a puff of toner. Comparison between the case of no bias field and that of 600 volts/cm shows a density halving time of from 2 to 3 seconds, in line with our estimates.

If one is struggling to reduce development time, there is only one thing to do, and that is to increase the powder cloud number density. There is really very little to be done about the mobility. Such an increase, however, will promote the recombination of oppositely charged particles. Consider the case of two oppositely charged particles coming together under their own fields. If their separation is r, then

$$-\frac{dr}{dt} = \frac{q_1\mu_2}{r^2} + \frac{q_2\mu_1}{r^2}$$

$\mu \equiv$ mobility

$\tau \equiv$ time to recombine

$r_0 \equiv$ initial separation

therefore,

$$\tau \cong \frac{r_0^3}{6q\mu}$$

assuming $r_0 \gg$ particle size and particles identical but for sign.

Such recombination tends to remove developable toner or to reduce its developability, and in the limit of strong recombination, the total toner extracted by a unit field would be approximately

(number density) (mobility) (toner survival time) $\equiv N\mu\tau$,

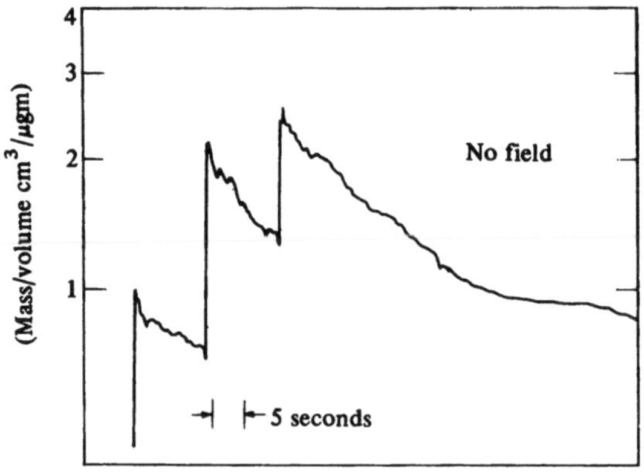

Fig. 3. Mass density in the powder cloud box versus time. Time increases to the right and there is no bias field present

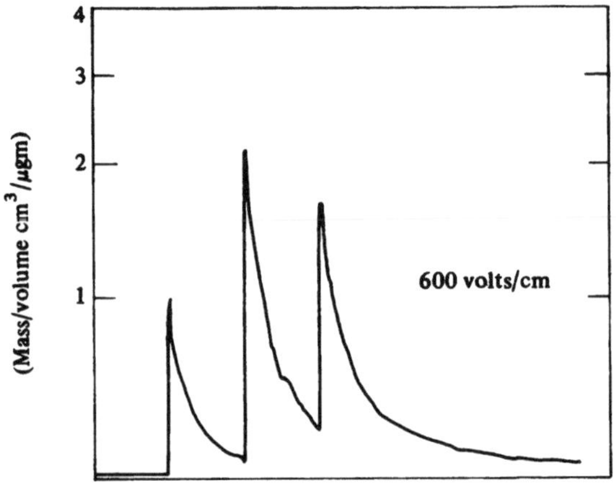

Fig. 4. The same measurement as in Figure 3 but with a strong bias field applied

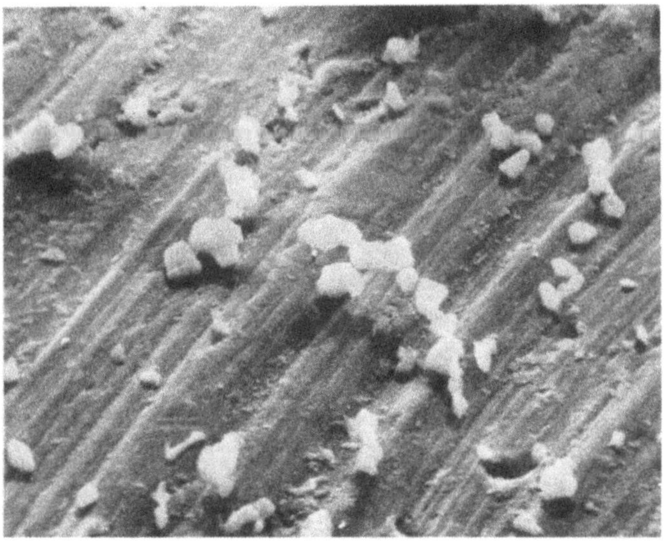

Fig. 5. Scanning electron microphotograph of toner particles developed onto a sample stub. The toner is nominally 4 microns in diameter. Particle recombination is evident

but r_0 varies as $N^{-1/3}$ so, beyond some aerosol number density, total development per toner burst will saturate. A scanning electron microphotograph (Fig. 5) of deposited toner does indicate some recombination. Large associations like that in the center of the picture are not particularly rare. They seem to be the result of stringing toner particles like beads along lines of the bias field.

In considering the question of image structure and formation we will retain the mobility model, either having in mind a rather dilute cloud or supposing that where necessary we can think in terms of the mobility of agglomerates. The model electrostatic image will be a step function in plate surface charge density, and this will be specified in terms of floating potentials as is the custom in xerography. Figure 6 shows such a step image together with the electric fields over its surface. These were computed by John O'Brien following the method of Neugebauer [1]. The solution is a good approximation compounded of the charge step in free space and a superimposed bias field between the grid and the photoreceptor substrate. Thus, reflections of the charge distribution in the grid plane (a distant development electrode) have been ignored. The first thing to notice is that every field line terminating on the plate surface either arches back to the plate or reaches away into the depths of the box. These lines are possible toner trajectories and we assert that those that arch back are never occupied by toner because they do not penetrate very far above the plate. This leads to a white gap at the image border and this gap extends a little bit into the step itself. The other lines do have access to a region of uniform toner density, the air motion randomized interior of the development box. Now the rate of toner delivery to any element of the plate surface is in general proportional to the local quantity (Normal Electric Field) x (Toner Number Density). Number density is 0 in the arch, and we how argue that, under some reasonable approximations, it is everywhere else constant; so that, outside the arch, deposited mass is linear in *only* the normal field. Argument:

1. $\bar{v} = \mu \bar{E}$ — mobility motion

2. $\mathrm{div}(\bar{v}N) + \frac{\partial N}{\partial t} = 0$ — conservation of toner

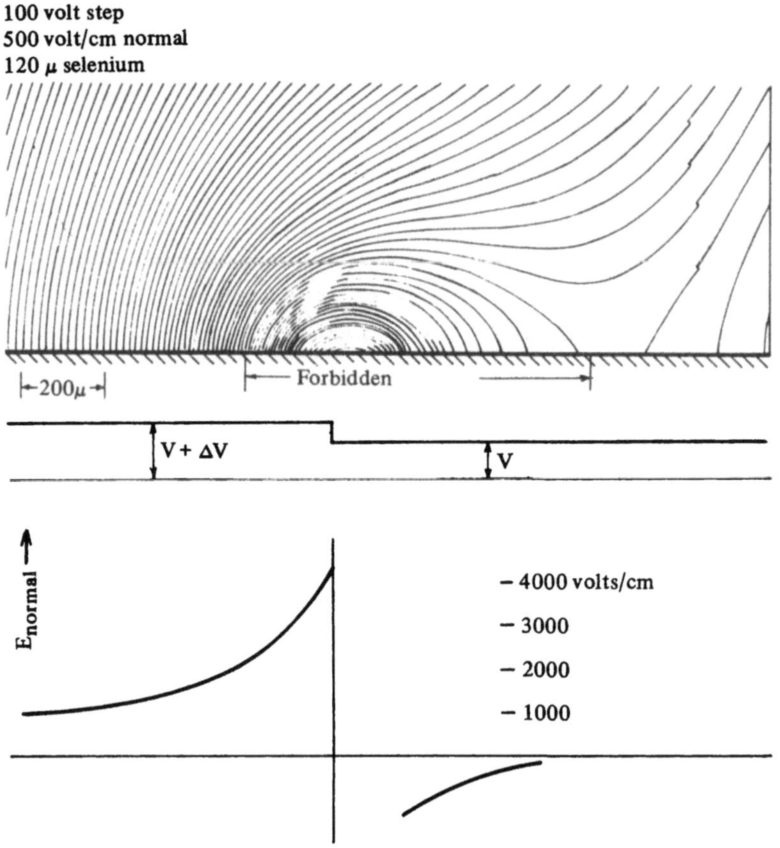

Fig. 6. The electric field pattern over a step in plate charge density and below a plot of the normal component of electric field

3. $\operatorname{div} \bar{E} = 0$ — no space charge in the box volume, an approximation.

4. $\frac{dN}{dt} = \frac{\partial N}{\partial t} + \bar{v} \cdot \operatorname{grad} N$ — the rate of change of toner number density in a frame moving with a particular particle.

5. $\frac{dN}{dt} = -\mu\bar{E} \cdot \operatorname{grad} N - \mu N \operatorname{div} \bar{E} + \mu\bar{E} \cdot \operatorname{grad} N \equiv 0$

The last equation follows from the preceding ones and means that in steady state the toner number density is preserved along field lines; it never changes from the uniform distribution deep in the box.

Figure 7 shows a simple apparatus designed to assess the magnitude of the electric field perturbations due to the toner cloud itself. A grounded aluminium plate replaces the selenium-coated one and carries a charge-sensing electrode 10 cm long and 0.1 cm wide. This electrode is

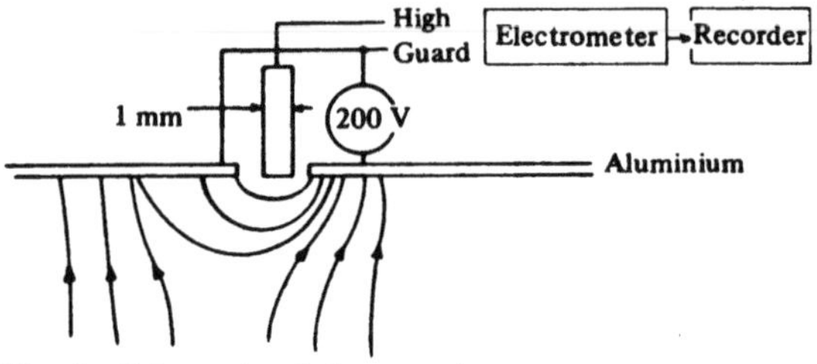

Fig. 7. Schematic of the experiment to measure toner-cloud-induced changes in the electric field in the powder cloud box

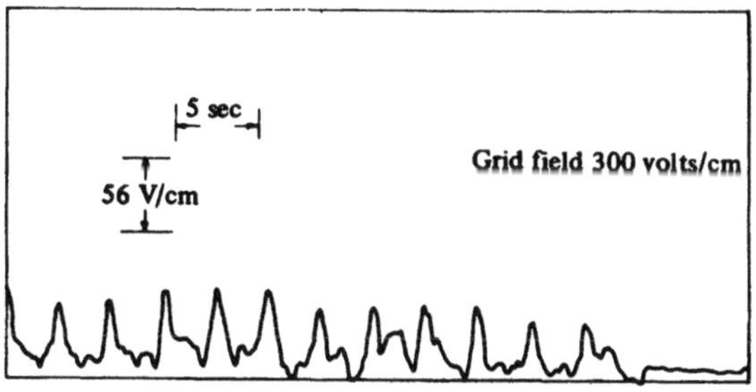

Fig. 8. An output trace of the apparatus of Figure 7. Time increases to the left. Each field spike corresponds to the introduction of a toner burst and each is about 20% of the field already in the box

virtually grounded and guarded in the usual way and senses field changes by the charge flows they induce. The device may be calibrated by switching in various values of grid bias. In order to keep toner from depositing directly on the sensing electrode it was buried inside the small field arch provided by a 200 volt step. The effects of that arch do not extend very far into the box so the device is sensitive to most of what happens. We observe in Figure 8 that, at each puff, there occurs a transient, cloud-caused field change about 20% as large as the bias field. In passing, please notice that its time duration is again about what might be expected on the basis of toner mobility.

It should be possible to predict the structure of real images. Their most prominent feature is the white edge described above so we examine that first. The Xeroradiography group (Xerox, Pasadena) X-ray-exposed several charged plates through a step wedge, measured surface potentials on each side of the steps, developed them, then transferred and fixed the resulting images. From these images we read, by reflection densitometry, the widths of the white gaps. Table 2 contains the data.

Table 2

Point	Bias	V_1	V_2	(White gap/micron)
1.	750	546	296	2750 ± 500
2.	750	296	78	4670 ± 1000
3.	750	78	13	1500 ± 500
4.	1000	543	303	2250 ± 500
5.	1000	303	90	2720 ± 250
6.	1000	90	10	1250 ± 250
7.	1500	534	287	1750 ± 250
8.	1500	287	83	1800 ± 250
9.	1500	83	11	650 ± 250

The analytical problem is to find the linear dimension of the field arch given plate potentials and biases. A fairly direct analysis shows that this dimension is a function of $\Delta V/V_{\text{applied}}$, where

$\Delta V \equiv |V_2 - V_1|$ V_1, V_2 potentials on either side of the step

$$V_{\text{applied}} \equiv V_{\text{average}} + |V\ \text{bias}|$$

$$V_{\text{average}} \equiv \frac{(V_1 + V_2)}{2}$$

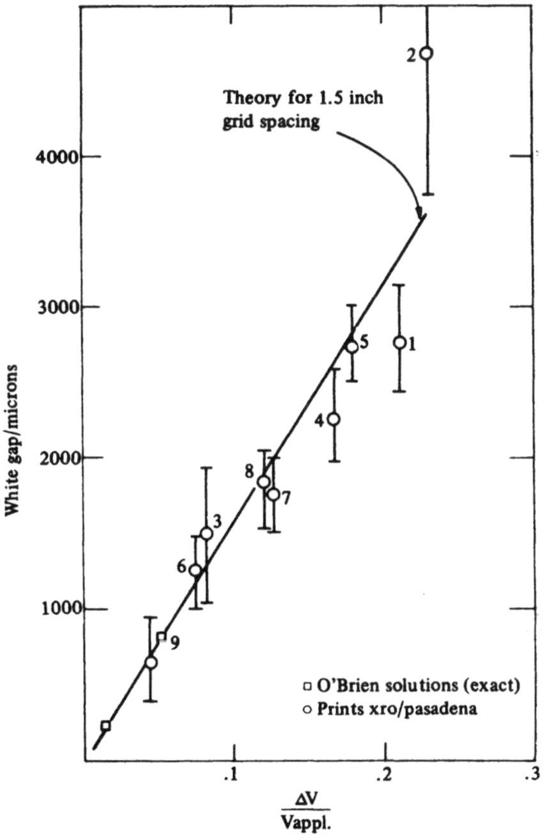

Fig. 9. Measured values of the white gaps at the edges of charge steps together with the results expected from a solution to the electrostatics problem

Figure 9 displays experimental points and theoretical results. The line is an approximate analytic solution and there are two points measured from O'Brien's more precise computer results. The experimental error bars reflect a lack of edge sharpness in the images.

To further examine the details of edge development we prepared carefully controlled charge steps on a split electrode plate. Two adjacent plane electrodes were formed by cutting a straight insulating gap (width $< 40\mu$) in the conducting side of a sheet of aluminized mylar. This was bonded to a rigid substrate and overcoated with plain mylar of thickness 60 microns. The method of use was first to place known potentials on the electrodes, then to neutralize the mylar surface with alcohol and an alpha emitter, and then finally to remove the potential sources and tie the electrodes to ground. The resulting surface charge distribution is a good approximation to a step, and it is electrically smooth and of known magnitude. Toner developed onto these images was totally stripped on transparent tape, mounted in immersion oil between glass plates, then scanned by a transmission densitometer. The results are the line traces of Figures 10, 11, and 12. The points in these figures are theoretical; they are calculated values of normal electric field at the plate surface fitted to the data according to the following procedure: first recall that optical density is proportional to deposited mass per unit area. If, according to the theory just given, deposited mass per unit area is proportional to normal electric field (except in white gaps) then it follows that transmission density should be linear in normal electric field. The constant of proportionality connecting them can be set by the broad area response away from the edge, thus determining the vertical scale. We have set the theoretical origin by the symmetry of the density trace; and, of course, the horizontal scale is distance and that is set by the ratio arm of the densitometer. So far as the edge response itself is concerned, this fit is without adjustable constants. The theoretical traces are generally slimmer than the experimental results but agreement is sufficiently good to support our contention that the central ideas are quantitatively understood whatever other effects must be added as perturbations. We remark that the electrical effects of deposited toner have been ignored. Separate calculations indicate that development under these conditions is just beginning to saturate fine detail. Also, the three volt edge is definitely useful and, in fact, one volt edges can be detected visually.

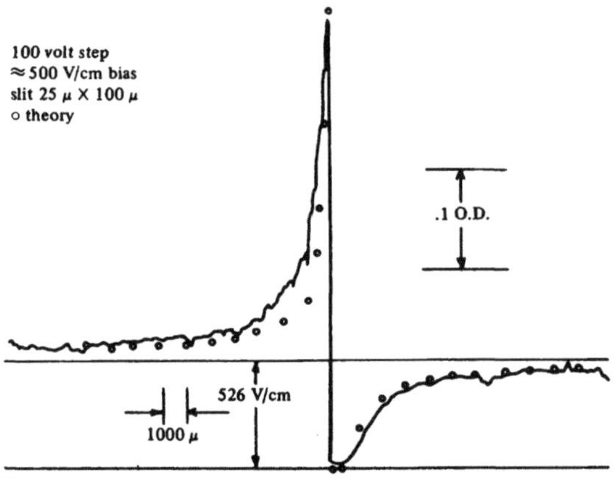

Fig. 10. A density trace across a 100 volt step and points fitted according to theory

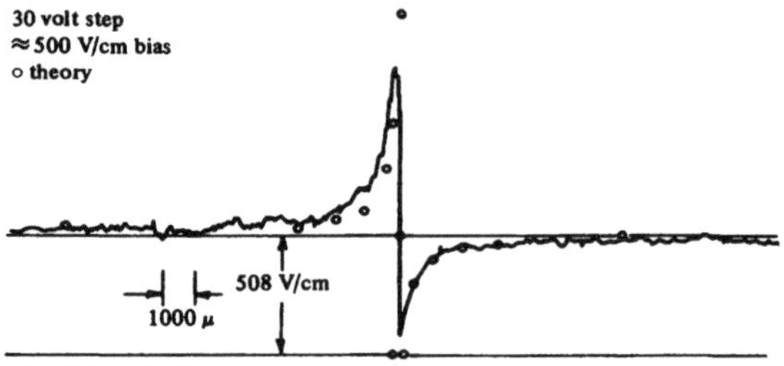

Fig. 11. A density trace across a 30 volt step and points fitted according to theory

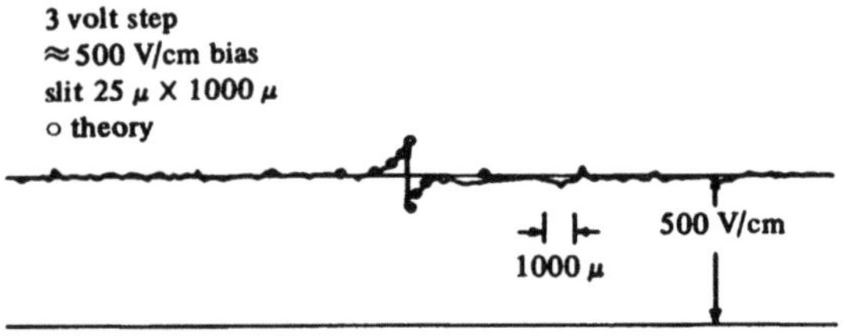

Fig. 12. A density trace across a 3 volt step and points fitted according to theory

3. Electrophoretic Development

The method was suggested by Metcalf [2] and has been reviewed by Claus and Mayer [3], and by Schaffert [4]. It is, of course, used in several commercial processes. In these applications development is taken to charge saturation and kinetics are not of central interest. For us they are, and most of what follows deals with the kinetics of solid area development. A typical experimental development cell is diagrammed in Figure 13. The photoconductor is simulated by mylar cemented to a conductive glass substrate. Ordinarily the cell is charged, assembled, then immersed in the developer for a time t. During development, either the current or charge flowing between the development electrode and photoconductor substrate is measured; then, afterwards, both the residual charge on the dielectric and the transmission optical density. Over a wide range of conditions it has been shown that floating potential measurements of photoreceptor discharge are consistent with externally measured charge flow [5].

The liquid developer used in these experiments was prepared in our laboratory by ball milling together carbon black and a control resin [3]. The resultant paste-like mixture was then dispersed in Isopar G, a non-conducting (10^{16} Ω-cm) hydrocarbon fluid. The purpose of the control resin is to adjust the charge on the particles and the rate of flocculation of the mixture. In our developer, this agent was chosen

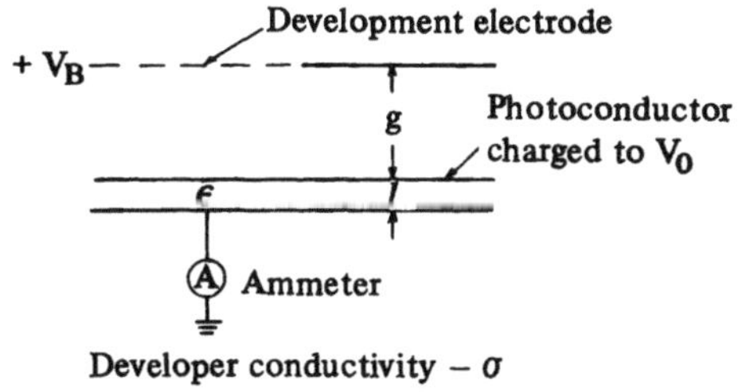

Fig. 13. Typical development configuration showing the development region and the charged photoconductor

to ensure that the carbon particles (toner) were negatively charged and that the dispersions were stable for a minimum of several weeks. In general, 1% by volume of the carbon black and control resin mixture was dispersed in the Isopar and the resulting conductivity was about $3.0 \times 10^{-12} (\Omega\text{-cm})^{-1}$. Table 3 lists some of the important properties of the developer.

These numbers are typical of electrophoretic developers and a comparison with powder cloud parameters is useful. First, liquid developers contain smaller particles and so are capable of higher resolution. Their toner mass density is greater so there is available much more material close to the electrostatic image. This speed advantage in suspended mass is augmented by higher covering power and almost entirely offset by a much smaller particle mobility. Rate of optical density deposition for a unit field is given by the product:

(mass density) (charge/mass) (mobility) (1/deposited charge for O.D. = 1);

Table 3

Developer Properties	Liquid	Cloud
Particle Diameter	$\sim .1\,\mu$	$4\,\mu$
Toner Density	~ 1 mg/cm^3	$6\,\mu$gm/cm^3
Developed Q/M	$\sim 300-500\,\mu$c/gm	$3-5\,\mu$c/gm
Stokes Mobility	$\sim 10^{-5}$cm^2/volt-sec	10^{-2}cm^2/volt-sec
Deposited Charge for Transmission O.D. of 0.5	$5 - 10$ nc/cm^2	1.2 nc/cm^2
Inherent Speed	$\sim 3 \times 10^{-4} \frac{\text{OD units}}{\text{sec}} / \frac{\text{volt}}{\text{cm}}$	$\sim 2 \times 10^{-4} \frac{\text{OD units}}{\text{sec}} / \frac{\text{volt}}{\text{cm}}$
Negative Charge Density	4×10^{-7}coul/cm^3	2×10^{-11}coul/cm^3

and for these examples both developers have about the same inherent speed. Their saturating tendencies are different, however; note that, for the same developed optical density (in this case transmission), it is necessary to deposit several times more charge in the liquid case. The entry "negative charge density" is what would be measured if counter charges were removed. Its low value in the powder cloud is why space charge is not important there.

Electrical measurements are easier than weighings, so covering data is presented that way in Figure 14. These results are for a wide variety of values of initial potential, development time, and photoreceptor thickness; and they are consistent with a single, linear relationship. For our purposes it is not necessary to know what fraction of this current is carried by pigment particles and what remainder by unobserved ions. It is evidently the case that these fractions are fixed.

Developer conductivity is a central idea and requires careful discussion. The measurements were made in a cell identical to the development cell with the photoconductor replaced with a metal electrode. The effect of allowing the current to flow for extended periods of time is shown in Figure 15 where it is seen that the current falls by almost a factor of two within .5 sec after the voltage is applied. It is useful to define, for the initial decay, a time constant given by the initial rate of decay divided by the initial current. Note that this time constant is not dependent on agitation and toner depletion. By using pulse techniques [10], we found this time dependence of the conductivity to hold down to times less than 1 millisecond. Field dependent [6] effects on conductivity have also been observed; however, at ordinary development fields, conductivity is substantially constant. Finally we show in Figure 16 the dependence of low field initial conductivity upon toner concentration. The observ-

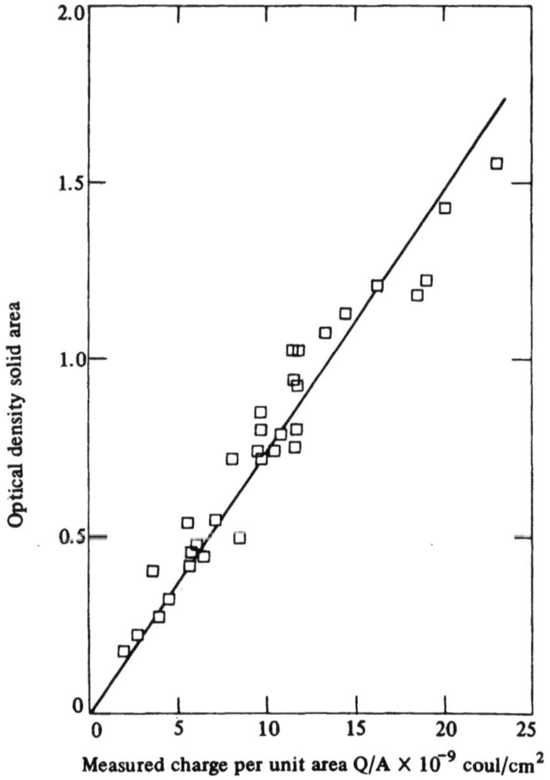

Fig. 14. Optical density of developed image vs. measured charge/area for 100 μ thick dielectric

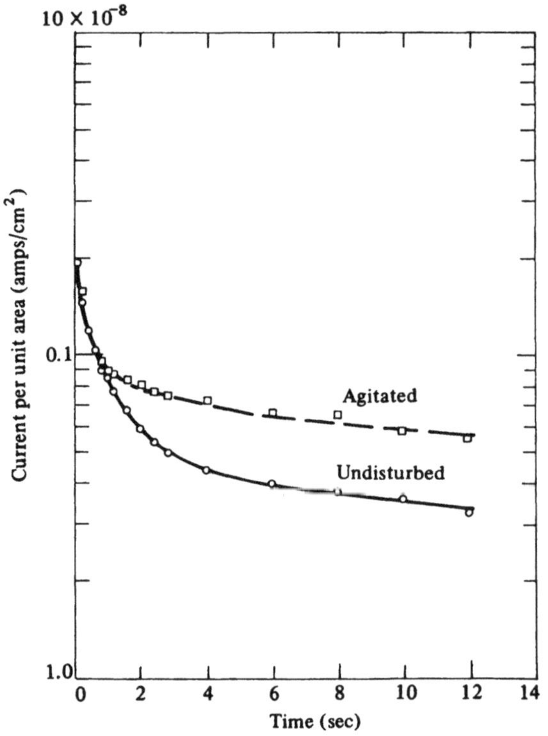

Fig. 15. Current vs. time with the applied voltage held constant. The effect of agitation is shown

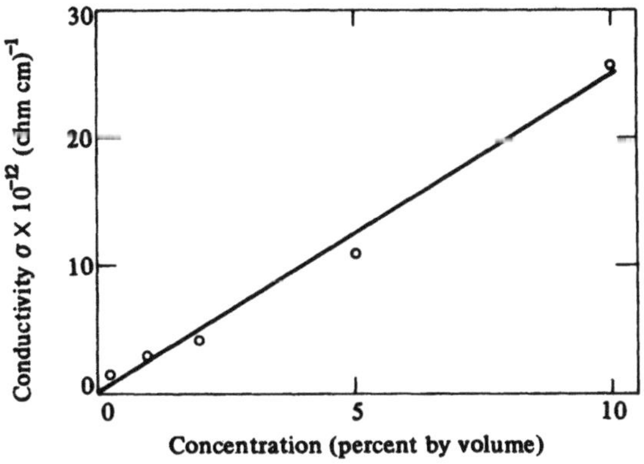

Fig. 16. Conductivity vs. concentration of toner

ed proportionality between concentration and conductivity supports the postulated ohmic behavior of the developer.

We now discuss the results of development experiments with particular attention to the rate at which development proceeds in time. In view of the linear relationship between charge and optical density, it is sufficient to measure just the current through the development electrode. As development proceeds, the charge on the photoconductor is neutralized and the field in the development region is reduced in proportion to the amount of neutralization. The process is equivalent to the discharge of a capacitor (the photoconductor) through a resistor (the developer). For the usual situation, the photoreceptor is much thinner than the distance to the development electrode and the electrical effects of the toner layer thickness may be neglected. Under these conditions, the field in the development region is easily determined and the electric current is given by

$$I = I_0 e^{-t/\tau}, \text{ where} \tag{1}$$

$$I_0 \equiv \frac{(V_0 - V_B)\sigma}{l}$$

$$\tau \equiv \frac{\epsilon}{\sigma} \frac{g}{l}$$

The time constant, τ, is determined by the conductivity of the developer, the capacitance per unit area of the photoconductor, and the development electrode spacing. Equation 1 then describes the rate at which the density increases for a given charge on the photoconductor. Thus,

$$D = D_{max}(1 - e^{-t/\tau}), \tag{2}$$

where D_{max} is the density for complete neutralization.

When the development electrode spacing approaches infinity or when the photoconductor thickness goes to zero, the development time constant approaches infinity. This is because in either case the fields due to the image distribution remain in the photoconductor. Thus, in the limit of no development electrode, broad area development does not occur at all. In this regard, it is important to define carefully what is meant by broad area development. Solid areas can fill in from their edges where fields do extend out of the photoconductor. Once the charge at the edge has been neutralized, field lines move further into the interior and development proceeds into the central areas, giving the impression of broad area development. Clearly these results cannot apply to the case of line copy.

Results of this kind have been reported in the literature by many workers and have been, in part, summarized by Schaffert [4]. In the case of electrophoretic development, Anfilov and Fridkin [7] obtained a relation of the form of Equation 2. They did not take account of the effect of the development electrode and the fact that the photoconductors is neutral, hence their results do not apply to the usual liquid development situation. More recently, Dahlquist and Brodie [6] reported experimental results in which they attribute the development time to either a dielectric relaxation or a diffusion limitation. However, they did not take into account the effect of the photoconductor thickness on the electric fields in the development region due to the charge distribution in the photoconductor. As a result, we find it difficult to interpret their experimental results pertaining to development rate. Kurita [8] obtained a relation similar to Equation 2 following a line of reasoning similar to ours.

A typical decay curve is shown in Figure 17 in which log current is plotted versus time. To obtain this curve the raw data was smoothed near t = 0. This was necessary because initial turbulence in the fluid caused up to 30% fluctuation in the current and this fluctuation continued for 1 to 2 seconds after immersion of the sample. The measured current is always within 50% of the value predicted from geometry and developer conductivity. The decay curve does not follow the straight line implied by the simple, lumped parameter model; but rather the rate of discharge (the slope of the curve) is always decreasing. The decreasing rate of decay is caused by the decrease in the conductivity of the ink itself.

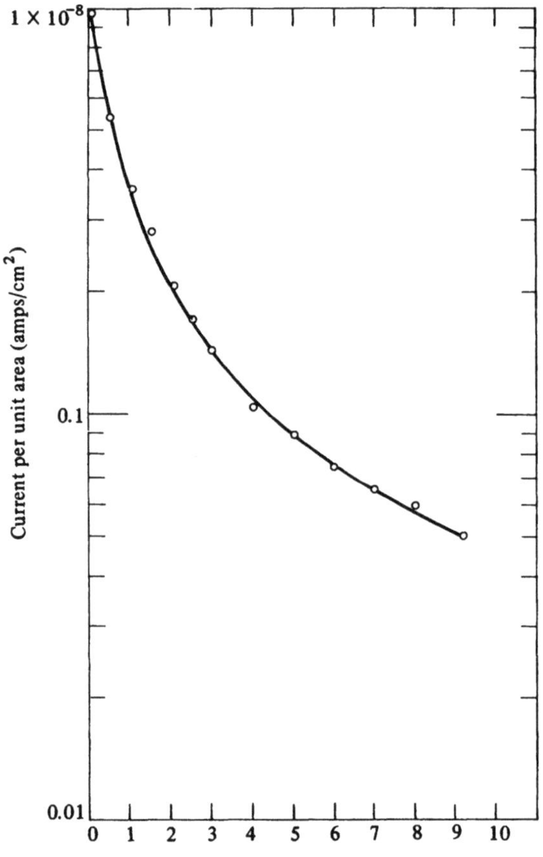

Fig. 17. Decay curve showing current/area vs. time during development of 100 μ thick dielectric

We show in Figure 18 the initial slope of the decay curve plotted as a function of (1/g). The scatter in the points is due to the initial turbulence in the developing fluid. The best fit straight line has a slope of 0.11 sec^{-1} which, if interpreted in terms of Equation 1, indicates a conductivity of 3.3 x $10^{-12}(\Omega\text{-cm})^{-1}$. Measured values of the conductivity of this fluid varied between 2.7 x 10^{-12} and 4 x $10^{-12}(\Omega\text{-cm})^{-1}$. The non-zero intercept of $1/\tau$ = 0.178 sec^{-1} cannot be understood in terms of the simple model. To account for the intercept, the time constant associated with the conductivity itself must be included. It can be shown that the reciprocals of the time constants add [5]. From conductivity measurements the reciprocal of the time constant for this developer varied between 0.16 and 0.21 sec^{-1} compared to the value obtained from the development experiments of 0.18 sec^{-1}. This completes our review of the kinetics of broad area electrophoretic development. It is self-consistent and works by translating any effects of developer depletion or space charge into an equivalent conductivity. It serves as a basis for the characterization of the developer and has sorted out a number of apparent difficulties in the interpretation of data.

The sensitometry of edge development depends strongly on kinetics and, although our experi-

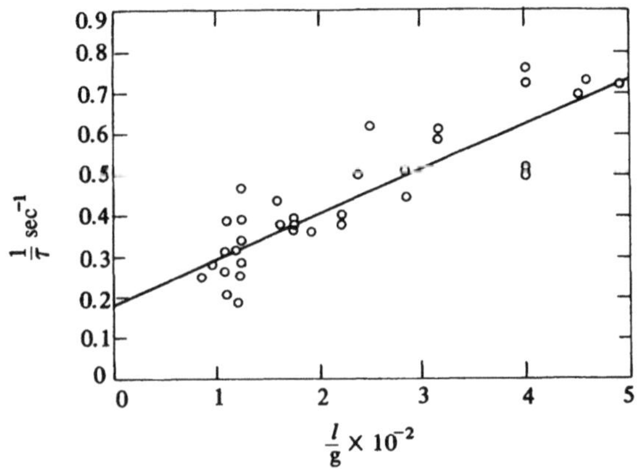

Fig. 18. Initial slope of decay curve vs. 1/g

mental program is still quite incomplete, we have preliminary remarks that are perhaps of some value. First, electrical perturbation by the developer charge is a more serious problem than in powder cloud. Where it is due to toner deposited on the image it is not difficult to quantify, but where it operates as space charge it creates apparently severe theoretical problems. Although the same warning could have been delivered in the case of powder cloud it is not safe to ignore space charge effects in these liquids. Some of the necessary models have been worked out and are found in the literature of plasma studies and the ionic conduction of fluids [9]. In any case, the higher charge deposition per unit optical density of this electrophoretic developer has made it unattractive for our purposes. The situation is illustrated by developing a 100 volt edge in a field geometry similar to that used in the powder cloud work. In this case the electrode spacing was 1 cm and a 500 volt/cm bias field drove toner toward the plate. To slow development the developer was diluted to 25% of standard concentration, and the computed electrical time constant for this developer in this geometry was about 5 seconds. A 2 second development (Fig. 19) does show edge structure, while a 20 second development does not. In the latter there is no white edge and the response profile approximates the profile of the image

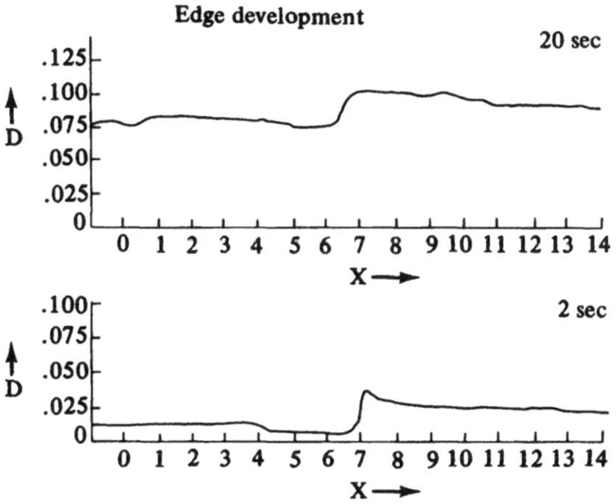

Fig. 19. Optical density as a function of position for 100 volt liquid-developed edges

charge density. Both images are smooth and of very low density. In this experiment liquid kinetics have been obscured because measurable images are saturated. Our mylar-coated plates had capacities corresponding to thick selenium but there is no reason why higher capacity plates should not be used for study purposes, even in the absence of a lower-charged electrophoretic developer.

Speaking practically on the basis of visual comparison, and under the general conditions we have described, powder cloud development can reveal 1 volt edges while liquid development is not useful unless the edge exceeds 5 volts. We have ascribed this difference to image neutralization by the liquid developer. It is by no means certain, however, that lower-charged liquid developers cannot be made. Finally, recall Metcalf's "chargeless xerography" in which images are made on initially uncharged zinc oxide paper by long electrophoretic development. It is known that entire continuous tone pictures can be compressed into potential ranges that lie below 1 volt. There is no inconsistency. Metcalf's images are the result of carrier diffusion over comparatively short distances; and, while the resulting potentials are small, the charges may be quite large. In medical xeroradiography, radiation to the patient sets a limit on the number of charge pairs available for development, so it is charge sensitivity that is required, even though for reasons of latitude and perception we cannot approach broad area charge neutralization.

4. Concluding Remarks

Rather special requirements have motivated a quantitative study of high sensitivity edge development. The response is directly related to field-controlled initial development rates and this non-equilibrium problem is rich in the physics of development mechanisms. The two systems cited should be regarded as typical examples and not optimum for any particular case. Our comparison of them is incomplete and we have suppressed questions of space charge and the motion of the suspending media—both of which are essentially collective problems. It seems to us that the single particle, mobility-controlled model that we have employed must always work in either system at sufficiently low toner density. Evidently practical powder cloud processes are already sufficiently dilute; and in them image formation is understood and structure can be predicted from plate charge distributions through their determination of plate normal fields. We expect that this program can be applied further to the case of liquid development. It is underway and rests now on a treatment of broad area development and the resulting elucidation of developer properties. It has been demonstrated that, with special techniques, edge response is a convenient indicator of the inner workings of the developer. We leave for another occasion the construction of a description based on sinusoidal inputs and the complex question of the visual perception of information processed in this way.

The design of the powder cloud apparatus and process is not ours: it was evolved by many Xerox researchers in Rochester and Pasadena whom we thank for equipment and for detailed discussions. One of us (R.B.L.) appreciates the field computations of John O'Brien, discussions with John Weigl on chargeless xerography, the advice of Paul Roetling on densitometric techniques, and particularly the experimental skill of E. W. Connors who worked out all the powder cloud techniques and experiments. Howard M. Stark thanks P. K. Watson for helpful conversations, and R. S. Menchel for his careful experimental work in the study of liquid development.

References

[1] Neugebauer, H. E. J., Xerography and Related Processes, J. H. Dessauer and H. E. Clark, Eds., (Focal Press, London, 1965) Chap. VIII.

[2] Metcalf, K. A., J. Sci. Instrum. **32**, 79 (1955).

[3] Claus, C. J., and Mayer, E. F., Xerography and Related Processes. (Focal Press, London, 1965) Chap. XII.

[4] Schaffert, R. M., Electrophotography, (Focal Press, London, 1965) p. 373.

[5] Stark, H. M., and Menchel, R. S., J. Appl. Phys. **41**, 2905 (1970).

[6] Dahlquist, J. A., and Brodie, I., J. Appl. Phys. **40**, 3103 (1969).

[7] Anfilov, I. V., and Fridkin, V. M., Zhur. Nauch. I. Priklad. Fot I. Kinematog **5**, 367 (1960). In Russian.

[8] Kurita, T., Denshi Shahin (Electrophotography) **3**, 26 (1961).

[9] Demetraides, A., in a panese and Hill, J. H., J. Appl. Phys. **36**, 747 (1965).

[10] Watson, P. K., and Sharbaugh, A. H., Rev. Sci. Instrum. **35**, 1310 (1964).

Discussion

Lange, P.: 1) What type of grid was used in the experiment?

2) Did you use a spherical type toner (spray dried material) or a ground milled type toner?

Lewis, R. B.: 1) The grid is an array of fine steel wires spaced about 5 mm a part.
2) The toner is made by standard methods which lead to widely size distributed angular particles, not spheres.

Matulionis, A.: I would like to know how close is the experimental ability to develop 1 volt edge from the theoretical limit for this type of developing technique.

Lewis, R. B.: Because of the edge-enhancement it is not possible to apply directly the usual theories of human visual perception. It should be possible to modify them, however, though we have not yet succeeded in doing so.

Stotz, S.: Did you measure the conductivity of the liquid developer after centrifuging the toner particles?

Have you any idea about the relation between the total conductivity and the conductivity without toner particles?

Stark, H. M.: Upon electrical removal of the toner the conductivity of the fluid returned to approximately the conductivity of the original fluid in which the toner was dispersed. This was on the order of $10^{16}\,\Omega$-cm. We were able to move up and down on the toner concentration conductivity curve by these methods.

Pell. E. M.: What determines the optimum spacing between grid and photoconductor? (It presumably affects the degree of solid area versus edge development).

Lewis, R. B.: Grid spacing has been determined empirically. If it is too close there will be enhanced solid area development and that will bury detail in areas of high density. The maximum distance is probably determined by space requirements, or perhaps eventually by the requirement that fringing effects at the plate edges be small.

Particle Charge and Stability of Liquid Developers in Electrophotography

S. Stotz

Abstract

The application of the DLVO theory on the stability of lyophobic colloids to liquid developers yields a relationship between the size and minimum specific charge of the toner particles in a stable suspension. A stability diagram is presented from which the necessary increase in specific charge of the particles required by a reduction in their size can be seen.

Variable charging of toner particles is accomplished by oleophilic ionogen substances. A model system was investigated consisting of colloidal carbon black suspended in liquid isoparaffines with additions of Ca petroleum sulphonate. The mean specific charge of the toner particles was determined by electrophoretic deposition of a weighable mass on an insulating plastic foil.

The contribution of the different charged species (negative carbon black particles, both positive and negative Ca petroleum sulphonate micelles) to the total conductivity as well as their influence on the determined quantity of the specific charge was evaluated. In this way systematic deviations at lower carbon black concentrations could be corrected. In particular with a constant Ca petroleum sulphonate concentration in the liquid phase of about 4.3 x 10^{-5} k mole/m^3, the corrected specific charge of carbon black particles was found to be about 0.07 C/kg over a wide range of carbon black concentrations (0.5–15^0/oo).

1. Introduction

A liquid developer, in electrophotography, is a colloidal suspension of electrically charged pigment particles, the toner, in an insulating liquid [1, 2]. The concentration of pigment is in the range of 0.1 to 1^0/$_{oo}$, and the actual particle size is about 0.1–1 μm. The charging of the particles can in principle be performed by contact electrification, just as in dry powder development. Yet K. A. Metcalfe [3, 4], a pioneer of liquid development, apparently overestimated the significance of Coehn's rule [5] (of two dielectrics in contact, the one with the higher dielectric constant becomes positively charged) in these systems. An unambiguous and higher charging of the toner particles can be achieved by the addition of oil-soluble ionogenic substances (surfactants), e.g. metallic salts of organic acids with sufficiently long aliphatic chains. By predominant adsorption of one ionic species the particles receive a net charge, the amount of which can be regulated simply by changing the additive concentration. This is a very convenient method of adapting the sensitivity of the toner (deposited mass per surface charge) to a particular purpose. The polarity is controlled by the appropriate choice of the surfactant. For example, a suspension of carbon black in liquid isoparaffins becomes negatively charged by calcium petroleum sulphonate and positively charged by calcium diisopropylsalicylate.

The electrical charge of the toner particles is also essential for the stability (resistance to agglomeration). If in the course of increasing the sensitivity the particle charge is decreased, further agglomeration to greater particles will occur. Just as in silver halide photography, so also with liquid development in electrophotography the rule is: the higher the sensitivity, the lower the resolution. Application of the DLVO (Derjaguin, Landau, Verwey, Overbeek) theory [6, 7, 8, 9] concerning the stability of lyophobic colloids to liquid developers gives a relationship between the size and the minimum specific charge (charge per unit mass) of the toner particles in a stable suspension. These magnitudes determine the stability of a liquid developer system.

2. Stability of Suspensions in Nonpolar Media

According to the DLVO theory the stability of lyophobic colloids is determined by two factors*:

1. the electrostatic repulsion of the diffuse electrical double layer around the particles,
2. the van der Waals-London attraction.

If the total potential energy of two particles

$$V_{tot} = V_R + V_A \qquad (1)$$

is plotted as a function of the distance between them (Fig. 1), the suspension is in practice considered to be stable if $V_{max} \geqslant 10$ kT. The DLVO theory was developed in connection with aqueous systems, in which its essential points have been confirmed. But its formalism is applicable also to systems of lower dielectric constant, and then it can lead to even simpler results [9].

In practice, however, the existence and kind of ions in nonpolar media (dielectric constant 2–5), a necessary condition for the charging of suspended particles, was a matter of debate for a long time. Movements of particles in electric fields frequently proved to be due to dielectrophoresis [10], i.e. caused by the dielectric polarisation of uncharged particles in inhomogeneous fields. This plays an important role in practical development processes, particularly in dry developers. Very thorough investigations of carbon black suspension in benzene (dielectric constant 2.5) have, however, enabled van der Minne and Hermanie [11] to demonstrate genuine electrophoretic movement of the particles: the velocity increased linearly with the field strength and changed in sign on reversal of the field. Neither behaviour is found in dielectrophoresis. The authors further observed a relationship between the electric charge of the particles (calculated from the magnitude of the velocity) and the stability of the suspension.

In the meantime quite a number of oil-soluble ionogenic substances, e.g. petroleum sulphonates, octoates, and salicylates have become available; in 10^{-3} molar solution in hydrocarbons these have conductivities of the order of 10^{-8} to 10^{-10} ohm^{-1} m^{-1}. The appreciable degree of dissociation of these materials is due to their ability to form micelles (colloid aggregates), which are able to incorporate lyophobic ions and screen them from the nonpolar solvent. The experimental result, mentioned above that carbon black if suspended with petroleum sulphonates in hydrocarbons becomes negatively charged, can therefore be explained simply by the adsorption of negative $SO_3C_nH_{2n+1}$ ions on the carbon black, with incorporation of the positive ions in micelles in the solution.

The energy of respulsion of two spherical particles bearing charges of equal sign is expressed by [8, 9]

$$V_R = \frac{4\pi\epsilon_0\epsilon a^2\psi_0^2}{H + 2a}\exp(-\kappa H) \qquad (2)$$

in which

ϵ_0 = permittivity of vacuum
ϵ = relative dielectric constant of the solvent,
a = particle radius,

* Repulsion on account of steric hindrance, for example by uncharged, adsorbed macromolecules, a topic of recent discussions, is not taken into account in this theory, though it can in certain circumstances be included.

ψ_0 = potential at the surface of the particles,
H = separation of the surfaces,
κ = reciprocal thickness of the electrical double layer.

The condition (2) which is valid for low surface potentials and $\kappa a \ll 1$, are amply met in hydrocarbons with $\epsilon < 5$. The thickness of the electrical double layer is of the order of 10 μm.

The energy of attraction due to the van der Waals forces of two spherical particles is given by [12]

$$V_A = -\frac{A}{6}\left(\frac{2}{s^2-4} + \frac{2}{s^2} + \ln\frac{s^2-4}{s^2}\right) \qquad (3)$$

The Hamaker constant, A, is of the order of 10^{-19} J; some variation in its value having little effect on the following results. The dimensionless quantity s is the quotient of centre-to-centre distance, R, and particle radius, a.

Given the particle radius, a, and the surface potential, ψ_0, the total energy is simply a function of the surface separation, H. The curve in Figure 1 has a distinctly asymmetrical maximum, because the van der Waals attraction becomes effective only at short range, i.e. at a distance of the order of magnitude of the radius of the particles. For the calculation of the activation energy, V_{max}, it is therefore possible to neglect the exponential term in (2), since $\kappa a \ll 1$ implies $\kappa H_{max} \ll 1$. Hence one obtains the simple Coulomb repulsion of two spheres:

$$V_R = \frac{4\pi\epsilon_0\epsilon a^2\psi_0^2}{H + 2a} \qquad (2a)$$

V_{max} as a function of ψ_0 and a is shown in Figure 2. The condition $V_{max} \geqslant 10$ kT attaches to each radius a minimum surface potential. The minimum charge of the particles can be calculated with adequate accuracy from

$$q_{min} = 4\pi\epsilon_0\epsilon a\psi_{0\,min} \qquad (4)$$

the capacitance of the electrical double layer in such nonpolar media being about the same as the capacitance of a sphere of radius a.

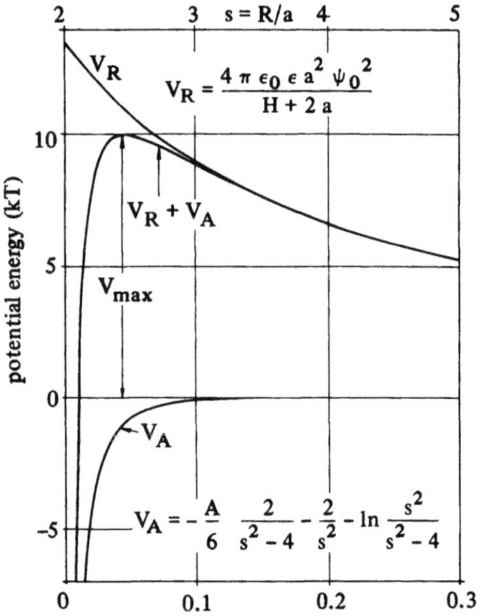

Fig. 1. Potential energy of two spherical particles, i.e. the sum of the electrostatic repulsion, V_R, and the van der Waals attraction, V_A, as a function of the surface separation H = R − 2a or s = (R/a). $\epsilon = 2$; a = 0.1 μm; ψ_0 = 70 mV; A = 10^{-19} J.

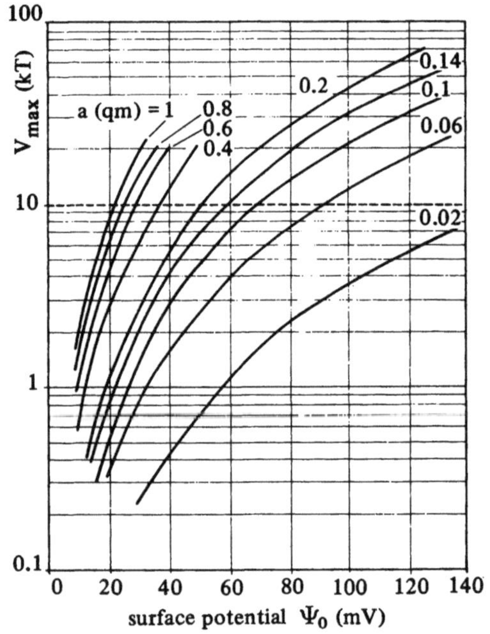

Fig. 2. Maximum of potential energy of two particles as a function of surface potential, ψ_0, and particle radius, a.

In Figure 3 the minimum specific charge of a stable carbon black suspension in liquid isoparaffin is plotted as a function of the particle radius. The values of practical developer suspensions, i.e. those stable against coagulation, must lie above the curve. A higher resolution, i.e. a smaller particle size, therefore implies a greater specific charge, and hence, a lower sensitivity. If on the other hand, we start with toner particles having a radius of 0.1 μm and a specific charge of 0.2 C/kg, and if the latter is diminished, for instance by a change in the chemical composition of the system, the particles will agglomerate until a stable state is reached again.

Usually stability is discussed in connection with the zeta potential, which in these nonpolar systems can as a first approximation be equated with the above surface potential. The values of the zeta potential quoted for such practical systems are generally smaller than 150 mV [13, 14]. To permit a comparison with our representation, Figure 3 shows a set of curves for given

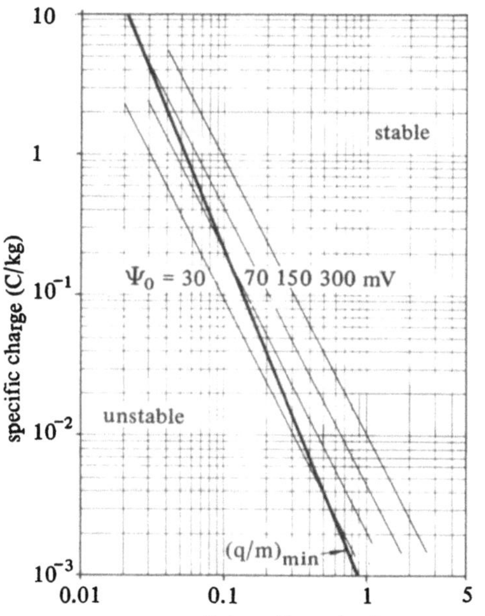

Fig. 3. Minimum specific charge (V_{max} = 10 kT) of a stable carbon black suspension as a function of particle radius. The graph shows also a set of curves for a constant surface potential (zeta potential) of 30, 70, 150, and 300 mV.

values of the surface potential, ψ_0. Obviously real, stable systems (zeta potential 40–150 mV) meet the stability condition of this theoretical treatment in a rather narrow region.

3. Determination of the Mean Specific Charge of Suspended Particles

The charge of suspended toner particles can be determined by measuring their velocity in an applied electric field. But the microelectrophoretic method, frequently used in aqueous systems, is rather difficult and tedious in systems of small electrical conductivity and low dielectric constant. Moreover in the case of electrophoretic deposition one is more interested in the average toner particle charge, rather than in the charge of individual particles. The method used by us is explained in Figure 4. A cylindrical chamber is divided by an insulating foil. The suspension under investigation is poured into the two compartments thus formed. As soon as a d.c. voltage is applied to the metal electrodes on either side of the foil, the toner particles as well as the corresponding opposite charges begin to migrate in the electric field. They can both be discharged at the metal electrodes but not at the insulating foil, so that the latter becomes charged as the experiment progresses. Deposition ceases when the potential difference between each electrode and the foil surface facing it has vanished, in other words when the voltage across the foil is equal to the applied voltage. The charge on the foil is calculated from the applied voltage and the known foil capacity, and the deposited mass is determined by weighing. The value of the specific charge thus determined is, in actual practice, the value in which one is interested for the calculation of given amounts of precipitate. But it does not represent the specific charge of the toner in all cases. For instance, if charging is performed by the calcium petroleum sulphonate mentioned previously, then in addition to the negatively charged toner particles and the positive countercharges there are also negatively charged sulphonate micelles in the solution. These micellar ions, which bear charges of the same sign, are deposited together with the toner particles, i.e. the latter carry a smaller amount of charge than calculated. The true specific charge must therefore be estimated by means of an extrapolation method, in which the carbon black concentration is increased while the sulphonate concentration in the solvent phase is held constant. In this way the interference due to equally charged sulphonate micelles becomes increasingly depressed: the apparent specific charge, which requires correction, approaches a limiting value, the true specific charge.

4. Experimental Part

4.1. Materials, and Preparation of the Suspensions

Use was made of Shellsol T, a solvent containing practically no aromatics, manufactured by Deutsche Shell Chemie. Composition: 92% isoparaffins, 8% naphthenes; b.p. 188–210° C; specific resistance 5 x 10^{15} ohm cm.

The carbon black used, Spezialschwarz 4 (Degussa, Frankfurt) is a carbon black pigment which after manufacture is slightly oxidised, in order to obtain increased adsorption power with respect to polar molecules. Mean diameter of the primary particles 25 nm, BET surface 180 m^2/g; density 1.8 g/cm^3. Prior to the experiments the carbon black was heated to 100° C to remove adsorbed water.

For the ionogenic surfactant a neutral calcium petroleum sulphonate (total basic number 1–3), obtained from Orogil-Mineralöl-Additives Vertriebsgesellschaft mbH, Frankfurt, was used as a 60% solution in mineral oil. The composition corresponded approximately to $Ca(SO_3C_{32}H_{65})_2$, equivalent to a mean molecular weight of about 1100.

Dispersion took place in two steps. First of all a 10% stock dispersion was prepared in a ball mill; milling time 65 hours. From this the various suspensions were prepared by dilution with Shellsol T and with further additions of calcium petroleum sulphonate. In this step an ultrasonic vibrator was used as an aid to dispersion. For reasons of reproducibility the suspensions were either tested 20 minutes after preparation or treated ultrasonically, for 5 minutes, 20 minutes before the test. To determine the concentration of sulphonate in the solution the particles of carbon black were centrifuged in stainless steel tubes on a laboratory centrifuge for up to two hours at 13,000 r.p.m., depending on the stability of the suspensions.

4.2. Determination of the Specific Charge

The testing circuit (Fig. 4) consists of a direct voltage mains unit, a sensitive ammeter, and the measuring cell, circuited in series. The cell was manufactured from polyvinyl chloride (Trovidur). For the insulating foil we used a polycarbonate foil (Makrofol) with a thickness of 10 μm; specific insulation resistance 10^{17} ohm cm, relative dielectric constant 2.8, dielectric strength 180 kV/mm. The diameter of the chamber, and therefore the diameter of the precipitate, was 9 cm. With an applied voltage of 700 V the calculated charge on the foil is then 1.1 x 10^{-5} C. The experiment was stopped as soon as the charging current had fallen to a hundredth of its initial value (after 3–5 min). In the case of suspension of lower conductivity (equivalent as a rule to a lower specific charge of the pigment particles) a longer duration of the experiment can be avoided by a reduction of the foil capa-

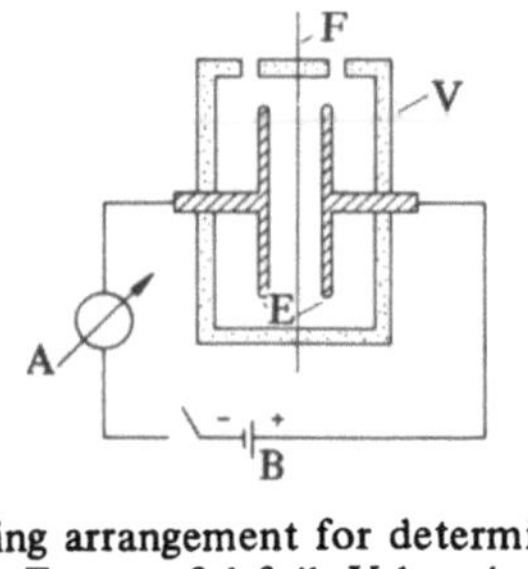

Fig. 4. Measuring arrangement for determining the specific charge. F macrofol foil; V housing of Trovidur; E brass electrodes; A ammeter; B mains supply unit.

citance (greater thickness of the foil). Short experimental times are desirable, above all, if sedimentation would lead to changes in the concentration of the suspensions.

4.3. Determination of the Electrical Conductivity

The individual units of the test circuit (Fig. 5) were screened from external stray fields. The design of the test chamber (cylindrical capacitor) can be seen from Figure 6.

If a conducting medium occupies the space between the electrodes of a cylindrical capacitor, the electrode voltage U(V), current strength I(A), and conductivity $\sigma\,(\mathrm{ohm}^{-1}\,\mathrm{m}^{-1})$ are related by

$$\sigma = K\frac{I}{U} \qquad (5)$$

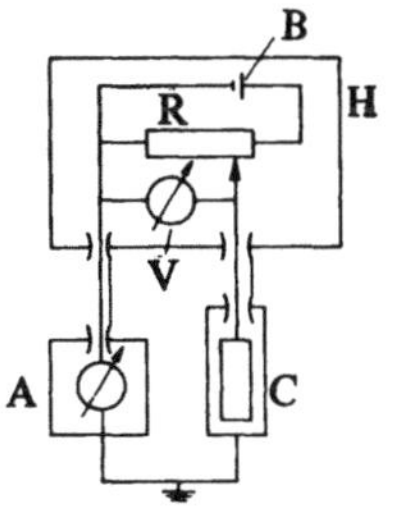

Fig. 5. Measuring arrangement for determining the conductivity H metallic housing; B battery, R variable resistor; V voltmeter; A ammeter; C measuring cell

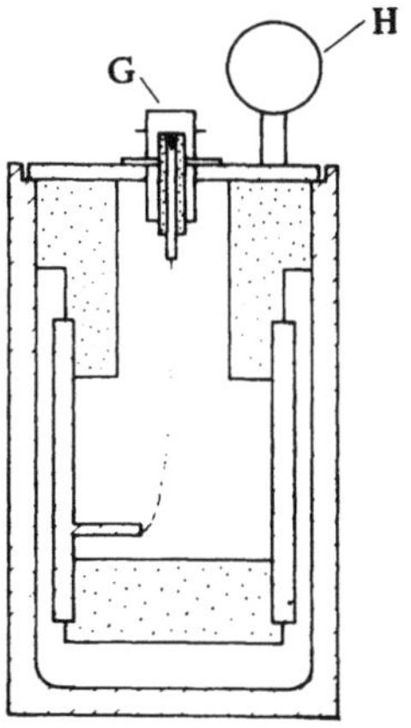

Fig. 6. Measuring cell
steel; Lupolen, G insulated grommet; H carrying handle.

with the cell constant

$$K = \frac{\ln(d_2/d_1)}{2\pi l} = 0.214\ \mathrm{m}^{-1} \qquad (6)$$

in which

d_2 is the outer diameter, 5 cm,
d_1 is the inner diameter, 4.4 cm,
l is the length of the inner electrode, 9.5 cm.

We used the shortest possible duration of the test and the lowest possible applied voltage (5 seconds and 1 volt) in order to restrict deposition at the electrodes to a minimum. To avoid possible polarisation of the electrodes, we measured twice in either direction of the current and then calculated the mean value. Duplicate measurements with direct voltage pulses (600 V, 50 ms) and recordings with an oscilloscope yielded similar results.

5. Results

Figure 7 shows the conductivity, in Shellsol T, of the calcium petroleum sulphonate used by us. The values are reproducible and they remain constant for longer periods. Measurement of the conductivity thus represents a convenient method of determining the concentration of sulphonate.

Next the specific charge of the carbon black particles was determined, their concentration being in the range of 0.5 to $15\,^0/_{00}$. Here the composition of the suspensions was chosen so that, after equlibrium had set in and the carbon black had been centrifuged, the solution had a constant concentration of sulphonate (4.3×10^{-5} kmole/m^3, corresponding to a conductivtity of $2 \times 10^{-10}\ \mathrm{ohm}^{-1}\mathrm{m}^{-1}$).

The points plotted in the graph of Figure 8 show the specific charge determined in these suspensions as a function of the concentration of the carbon black. The apparent increase of the specific charge at lower concentration reveals that, with carbon black concentrations of less

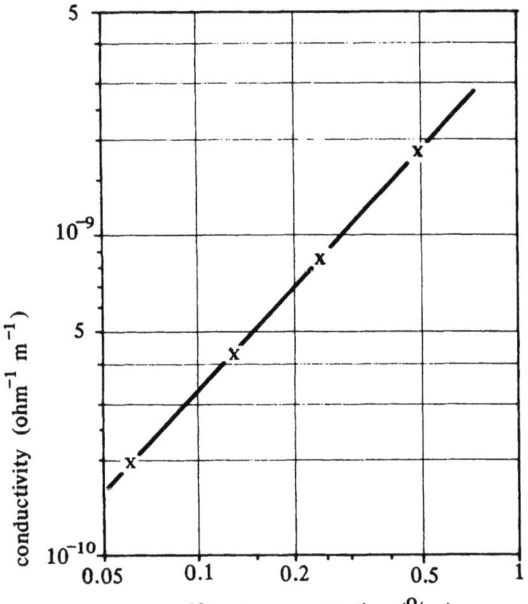

Fig. 7. Electrical conductivity of a calcium petroleum sulphonate solution in Shellsol T as a function of the concentration of sulphonate.

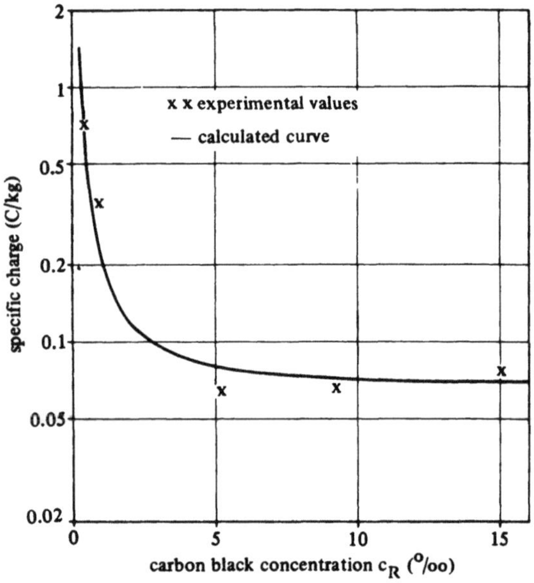

Fig. 8. Specific charge of carbon black as a function of the carbon black concentration at a constant sulphonate concentration in the solution of 4.3×10^{-5} kmol m^{-3}. (–) calculated curve (cf. Section 6.3).

than 5°/oo, the contribution of the similarly charged sulphonate micelles cannot be neglected. Only above 5°/oo is the standard value of about 0.07 C/kg obtained.

6. Discussion

6.1. Estimation of the Degree of Dissociation of Calcium Petroleum Sulphonate in Shellsol T

According to Peri [15] the number of calcium petroleum sulphonate micelles in n-decane is estimated to be about 40 in the concentration range used by us. For simplicity we assume spherical micelles, although the real shape is open to discussion. The radius r of the micelles is found to be about 2.5 nm if the size of the SO_3 tetrahedra and the length of the aliphatic chains are taken into consideration. The same result is obtained if the radius, r, is calculated by means of the diffusion coefficient, given also by Peri, of 6.25×10^{-11} m^2/s and with the relation

$$D = \frac{kT}{6\pi\eta r} \tag{7}$$

in which η is the viscosity of the solvent.

The following estimation of the number of charge carriers in calcium petroleum sulphonate solutions, therefore assumes both positive and negative charge carriers to be singly ionised, spherical micelles with radius of 2.5 nm.

The total conductivity is made up of the contributions of positive and negative charge carriers:

$$\sigma_L = \overset{+}{\sigma}_L + \overset{-}{\sigma}_L \tag{8}$$

For each type of charge carrier the general relations apply:

$$\overset{+}{\sigma}_L = \overset{+}{n}_L \, e u^+$$

$$\overset{-}{\sigma}_L = \overset{-}{n}_L \, e u^- \tag{9}$$

in which

$\overset{+}{n}_L$, $\overset{-}{n}_L$ is the number of positive or negative charge carriers per unit volume,
e is the elementary charge,
u^+, u^- represents the mobility of the positive or negative charge carriers.

The number of ions, their size, and therefore, their mobility are the same for both kinds of charge carriers.

With $\overset{+}{n}_L = \overset{-}{n}_L = \overset{\pm}{n}_L$ and $u^+ = u^- = u$ we obtain from (8) and (9)

$$\sigma_L = 2\overset{\pm}{n}_L e u \tag{10}$$

The mobility is calculated from Stokes' law:

$$u = \frac{e}{6\pi\eta r} \tag{11}$$

Substitution of (11) in (10), and rearrangement, leads to

$$\overset{\pm}{n}_L = \frac{\sigma_L}{2eu} = \frac{3\pi\sigma_L\eta r}{e^2} \tag{12}$$

With the numerical values

$\sigma_L = 2 \times 10^{-10}\ \text{ohm}^{-1}\text{m}^{-1}$
$\eta = 1.4 \times 10^{-3}\text{m}^{-1}\ \text{kg s}^{-1}$
$r = 2.5 \times 10^{-9}\ \text{m}$
$e = 1.6 \times 10^{-19}\ \text{C}$

we obtain

$$\overset{\pm}{n}_L = 2.6 \times 10^{17}\text{m}^{-3}$$

The number of neutral molecules in our tests is

$$n_L = 2.6 \times 10^{22}\text{m}^{-3}$$

so that the degree of dissociation is

$$\alpha = \frac{\overset{\pm}{n}_L}{n_L} = 1.0 \times 10^{-5} \tag{13}$$

The experimental fact (Fig. 7) that the electrical conductivity of pure sulphonate solutions is proportional to the sulphonate concentration can be explained as follows. The micellar ions are formed as a result of 2nd order reaction between two neutral micelles. Applying the Law of Mass Action to such a mechanism

$$K_D = \frac{\overset{+}{n}_L\overset{-}{n}_L}{n_L^2} = \frac{(\overset{\pm}{n}_L)^2}{n_L^2} = \alpha^2 = 1.0 \times 10^{-10} \tag{14}$$

This confirms the observed linear dependence between σ_L n_L, while the Law of Mass Action applied to the dissociation of a single molecule gives

$$K'_D = \frac{\overset{+}{n}_L\overset{-}{n}_L}{n_L} = \frac{(\overset{\pm}{n}_L)^2}{n_L} \tag{15}$$

i.e. σ_L varying as $(n_L)^{1/2}$. However, exact knowledge of the mechanism of charge carrier creation is here not necessary because in the calculation in section 6.2 only the product

$$\overset{+}{n}_L\overset{-}{n}_L = K_D n_L^2 = K'_D n_L = (\overset{\pm}{n}_L)^2 = 6.8 \times 10^{34}\text{m}^{-6} \tag{16}$$

is involved.

6.2. A simple Model for Calculating the Concentration of Charge Carriers as a Function of the Carbon Black Concentration

The mean diameter of the actual particles in the developer (agglomerates of primary particles of carbon black), as determined by electron microscopy, is about 0.3 μm in the composition range under investigation, irrespective of the carbon black concentration. In the following calculation we shall adopt the simplification of assuming spherical particles having this diameter. The negative charge of the particles is due to the adsorbed negative SO_3R-ions. Their number per unit volume, $\overset{-}{n}_{ad}$, as a function of the carbon black concentration, c_R (in parts by weight), is

$$\overset{-}{n}_{ad} = \frac{(q/m)_E\rho_S c_R}{e} = Bc_R \tag{17}$$

with

$$B = \frac{(q/m)_E \rho_S}{e} = 3.33 \times 10^{20} \mathrm{m}^{-3}$$

$(q/m)_E$ = limiting value of the specific charge, from Figure 8

ρ_S = density of Shellsol T.

In addition to (17) there is the condition of electrical neutrality

$$n_L^+ = n_{ad}^- + n_L^- \tag{18}$$

and the law of mass action, applied to the sulphonate ions n_L^+ and n_L^- in the solution

$$\frac{n_L^+ \, n_L^-}{n_L^2} = K_D \,. \tag{19}$$

The dissociation constant, K_D, and the number of neutral molecules are known from Section 6.1.

The three unknowns, n_L^+, n_L^-, and n_{ad}^-, can be calculated as a function of the carbon black concentration by means of (17), (18), and (19). Thus n_L^+, for example, is obtained from

$$n_L^+ = \frac{B\,c_R}{2}\left[1 + \left(1 + \frac{4\,K_D n_L^2}{B^2 c_R^2}\right)^{1/2}\right] \tag{20}$$

With rising concentration of the carbon black the square root term converges toward 1, i.e. n_L^+ converges toward $B\,c_R = n_{ad}^-$ (Fig. 9). The number of negative micelles then becomes negligibly small.

The individual conductivities are calculated from the charge carrier concentrations by means of (9). The asymptotic values are, for high carbon black concentrations:

$$\sigma_L^+(\mathrm{ohm}^{-1}\mathrm{m}^{-1}) = 1.3 \times 10^{-7} c_R \tag{21}$$

$$\sigma_L^-(\mathrm{ohm}^{-1}\mathrm{m}^{-1}) = \frac{7.9 \times 10^{-14}}{c_R} \tag{22}$$

After appropriate rearrangement, the contribution of the carbon black particles is given, in the entire concentration range, by

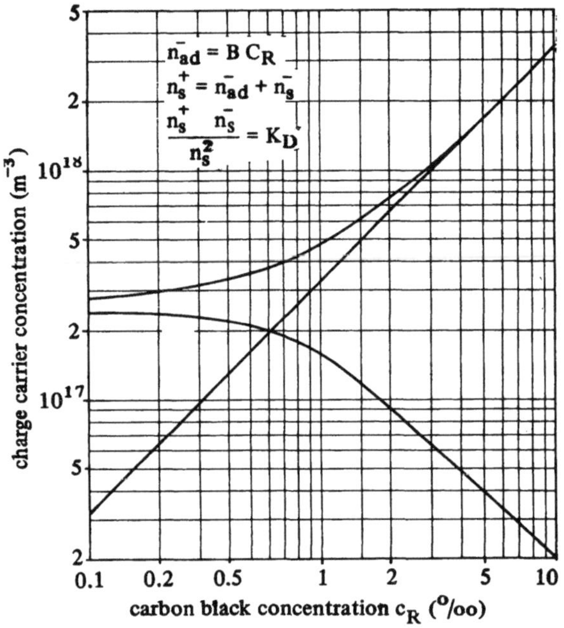

Fig. 9. Number of charge carriers in a carbon black suspension as a function of carbon black concentration, calculated from Eqs. (17)–(19) with
$B = 3.3 \times 10^{20}\ \mathrm{m}^{-3}$
$n_s = 2.6 \times 10^{22}\ \mathrm{m}^{-3}$
$K_D = 1.0 \times 10^{-10}$

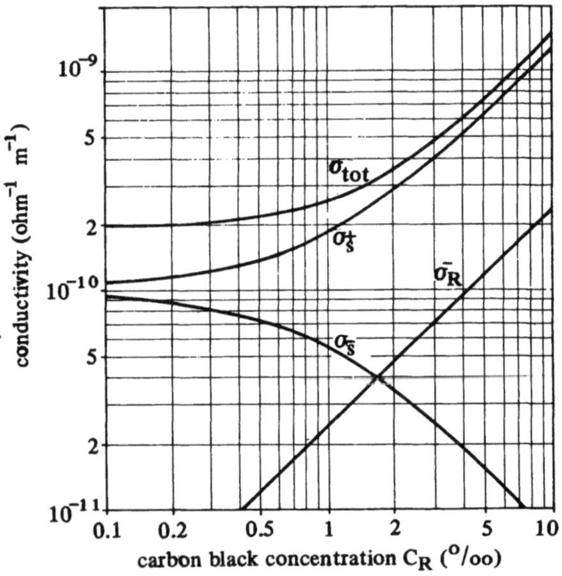

Fig. 10. Total conductivity and individual conductivities in a carbon black suspension as a function of carbon black concentration, calculated from the charge carrier concentrations in Fig. 9

$$\bar{\sigma_R} = \frac{(q/m)_E^2\ \varrho_S\ \varrho_R\ d_R^2\ c_R}{18\eta} \qquad (23)$$

$$\bar{\sigma_R}(ohm^{-1}m^{-1}) = 2.4 \times 10^{-8}\ c_R$$

in which ρ_R = density of carbon black
d_R = size of carbon black particles.

In Figure 10 the calculated individual conductivities as well as the total conductivity are plotted as a function of the concentration of the carbon black (at a constant concentration of sulphonate in the solution). At 1.5 ‰ the two negative contributions are about equal, and from 5 ‰ upward the contribution of the carbon black is greater by more than one order of magnitude than that of the negative sulphonate micelles.

6.3. Correction of the Experimentally Determined Specific Charge

Once the individual conductivities are known it becomes possible to correct the apparent specific charge. All that is necessary is to multiply the values determined at the various carbon black concentrations by the factor $\bar{\sigma_R}/(\bar{\sigma_R} + \bar{\sigma_L})$. The contributions to the charging of the foil are in fact directly proportional to the electrical conductivities. In view of the simplicity of the model and the limited accuracy of measurements the corrected specific charges show good correspondance (Table 1).

Table 1. Measured and corrected specific charge as a function of the concentration of carbon black

$$(q/m)_{corr.} = (q/m)_{exp} \times \frac{\bar{\sigma_R}}{\bar{\sigma_R} + \bar{\sigma_L}}$$

c_R ‰	$(q/m)_{exp}$ C/kg	$\frac{\bar{\sigma_R}}{\bar{\sigma_R} + \bar{\sigma_L}}$	$(q/m)_{corr.}$ C/kg
0.5	0.7	0.14	0.098
1	0.31	0.30	0.093
5	0.07	0.89	0.062
9	0.069	0.96	0.066
15	0.077	1.0	0.077

While the apparent values show deviations of as much as one order of magnitude, the corrected values fluctuate only by ±30%.

Conversely it is possible, given a true specific charge of 0.07 C/kg, to calculate the specific charge expected to be measured in our arrangement. The fully drawn curve in Figure 8 was obtained in this way.

7. Conclusions

The present method for determining the specific charge of suspended toner particles will in any case supply the numerical value required for calculating the electrophoretic deposition. Furthermore it is possible in special systems to determine the true specific charge of the toner particles by means of an appropriate procedure.

From the pair of values here found, viz. specific charge = 0.07 C/kg, radius = 0.15 μm, the surface potential is calculated to be 53 mV. The values lie just below the stability curve of Figure 3. The slight instability to be expected from this is in agreement with experimental observations. The system under notice is furthermore suitable for performing elementary model calculations, thus providing a good insight into the properties of similar suspensions.

Acknowledgements

The author would like to thank H. J. Göttig and D. Wädow for assistance in the experiments, Dr. N. Hansen for fruitful discussions and H. G. E. Wallace for careful attention to the manuscript.

References

[1] Dessauer, J. H., and H. E. Clark (Editors), "Xerography and Related Processes", The Focal Press, London, New York 1965.
[2] Schaffert, R. M., "Electrophotography", The Focal Press, London, New York 1965.

[3] Metcalfe, K. A., J. Sci. Instr. **32,** 74 (1955).

[4] Metcalfe, K. A., R. J. Wright, J. Sci. Instr. **33,** 194 (1956).

[5] Coehn, A., Ergebnisse der exakten Naturwissenschaften **1,** 175 (1922).

[6] Derjaguin, B. V., and L. Landau, Acta Physicochim URSS **14,** 633 (1941).

[7] Verwey, E. J. W., and J. Th. G. Overbeek, "Theory of the Stability of Lyophobic Colloids" Elsevier, Amsterdam 1948.

[8] Kruyt, H. R., (Editor) Colloid Science, vol. I Elsevier, Amsterdam 1952.

[9] Lyklema, J., Adv. Colloid Interf. Sci. **2,** 65 (1968).

[10] Pohl, H. A., and W. F. Pickard (Editors), "Dielectrophoretic and Electrophoretic Deposition" The Electrochemical Society, New York 1969.

[11] van der Minne, J. L., and P. H. J. Hermanie, J. Colloid. Sci. **7,** 600 (1952); ibid. **8,** 38 (1953).

[12] Hamaker, H. C., Physica **4,** 1058 (1937).

[13] McGown, D. N. L., G. D. Parfitt and E. Willis, J. Colloid Sci. **20,** 650 (1965).

[14] Lewis, K. E., G. P. Parfitt, Trans. Faraday Soc. **62,** 1652 (1966).

[15] Peri, J. B., American Chemical Society, Atlantic City Meeting, Sept. 17–21, 1956.

Sensitized Photolysis on Semiconductors as Analogy to the Conventional Photographic Process

V. M. Fridkin and A. D. Sablin-Javorski

Abstract

A study of physical development on semiconductors was made. It was demonstrated that illumination of the semiconductor in the intrinsic region of photosensitivity resulted in a reduction of silver ions adsorbed on the surface of the semiconductor. After this, the latent image can be developed by physical developers. This method was taken as a principle of infrared photography on semiconductors with narrow width of the forbidden band. As an example a method of infrared photography on PbS layers has been discussed.

1. Introduction

Usually the main reason for the development of non-silver photography is believed to be exhaustion of the supply of silver; this is a vital factor in the near future. In our opinion there is another reason due to the logic of the evolution of photographic science. The discovery of photography in 1839 had taken place long before the creation of modern solid state theory and in particular long before the discovery of photoconductivity. The investigations in solid state physics and mainly Gudden and Pohl's researches into the photoelectric properties of crystals for the first time revealed the connection between latent-image formation and photoconductivity of silver-halide crystals.

In the thirties the evolution of ion-crystal and band theory resulted in the Gurney-Mott theory of the photographic image involving the two-stage mechanism of latent-image formation. The first (electronic) stage of the process includes electron-hole pair formation as the result of light absorption, the electron being localized at a deep trapping level. At the second (ionic) stage the captured electron is neutralized by an interstitial silver ion to form a neutral silver atom. The latter acts by deepening the electron trap resulting in an increase in size and activity of the latent-image center.

The evolution of the theory of halide-silver photography and the deeper understanding of its mechanism led to attempts to look for some analogues of the conventional photographic process. One of these is electrophotography, discovered by Carlsson in 1938. That uses directly the photoconductivity of semiconductors with wide forbidden bands. So from the point of view of conventional photography, electrophotography is a one-stage process; its main deficiency is its relatively low light-sensitivity. However from the analysis of the Gurney-Mott theory it seems appropriate to conclude that a two-stage photographic process is not the characteristic of the silver-halides only, but should occur with a wide range of ionic crystals and semiconductors. Thus some analogues of conventional photography other than electrophotography may yet be revealed. Today it is very difficult to predict whether photographic science will evolve in ways other than halide-silver photography.

The present paper is devoted to one of the analogues of conventional photography which we designate as *sensitized photolysis on semiconductors* (I).

2. Sensitized photolysis on semiconductors with wide and narrow forbidden bands

It is known that latent-image formation in AgX is due to the interaction of electrons and holes with interstitial ions and cation vacancies. However, a two-stage mechanism of latent-image formation seems to be realized for any semiconductor if there are some metal ions situated on its surface. In such a case the ions adsorbed on the surface act the part of the interstitial silver ions in the silver halide lattice, and the electronic stage of the photographic process depends on the intrinsic and/or extrinsic photoconductivity of the semiconductor. As in the case of AgX this new photographic process will be of value whenever the process of photolysis predominates over the process of recombination of the photoexcited carriers. This condition requires efficient hole capture at the surface on the levels acting as chemical sensitizers. It is essential that the spectral sensitivity of the process depends on the energy gap of the semiconductor (in the case of the intrinsic photoconductor) and it may be greatly varied over the range from narrow- to wide-band semiconductors. For narrow-band semiconductors the photographic effect appears to be limited by background fog owing to the reduction of the metal ions by the equilibrium carriers. For wide-band semiconductors the electronic and ionic stages may be separated for a time. At the electronic stage the deep surface levels are filled. The dark regression of the latent image at this stage is slower, the deeper the level and the lower the temperature. At the ionic stage latent-image formation in the form of neutral metal particles takes place; the latent image then has to be developed by methods of physical development. For narrow-band semiconductors both stages must proceed simultaneously. It is characteristic of the new process that the non-equilibrium electrons reducing the metal ions may be either the majority or the minority carriers.

The above idea was realized in some independent investigations [1 – 7]. In the publications [1, 3 – 7] such a technique was used for infrared photography on semiconductors with narrow forbidden band. In what follows, this method will be designated as *sensitized photolysis on semiconductors.*

The purpose of the experiment to be described was to investigate the process of photoreduction of silver ion provided from aqueous silver-nitrate solution on the surface of some semiconductors and dielectrics (silica, PbI_2, PbS, CdS, HgI_2, SbSI). For some of the above substances the spectral distribution of the effect was revealed i. e. reduction of the silver occurred in the region of intrinsic photosensitivity of the semiconductor. The experimental procedure was as follows. The plate of the semiconductor under examination was placed into a beaker containing the silver-nitrate solution. Through the bottom of the beaker the image was projected on to the surface of the sample in a suitable spectral region. Having been exposed the sample was taken out of the solution to be developed by means of physical development. In some experiments exposure was carried out directly under the layer of physical developer. In other experiments a gelatin layer was coated over the surface and the image processed and developed. The image was revealed to be negative, providing evidence for photostimulated reduction of silver. In Figure 1 the curves of spectral distribution of the photographic effect are shown for an HgI_2 polycrystalline layer and two dielectrics: mica and polymethylmetacrylate.

Optical density of the print after deducting the fog value is plotted against wave length. The spectral distribution for HgI_2 (Fig. 1, c) reveals the region of sensitization with a maximum at 520 nm characteristic of the spectral distribution of photoconductivity.

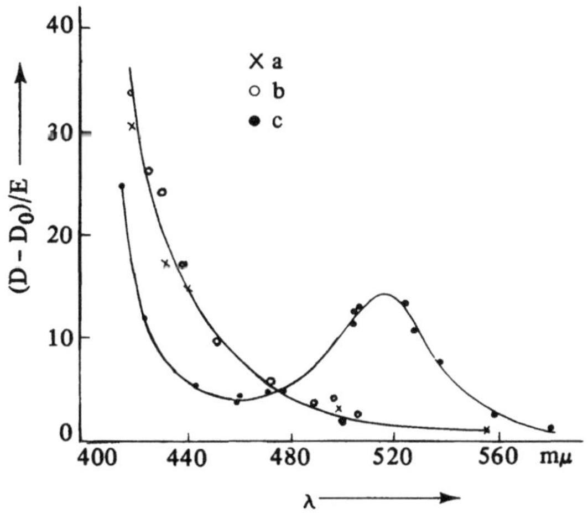

Fig. 1. Spectral distribution of the optical density produced by a constant amount of light energy for mica (a), polymethylmetacrylate (b), HgI_2 polycrystalline layers (c)

Distinct sensitization was observed also for polycrystalline CdS- and PbI_2- layers. As for dielectrics, the spectral sensitivity is situated in the blue region (Fig. 1, a, b) and does not depend on the dielectric itself. Thus the sensitivity of the dielectric does not appear to be intrinsic and is perhaps due to the formation of a layer of AgCl or of silver oxide on the surface of the dielectric being in contact with the solution. This suggestion is proved by the absence of a photographic effect on the dielectrics when the electronic and ionic stages are separated for some time.
Such a result was observed in a PbS-layer revealing the maximum of the photoconductivity at $\lambda = 2{,}5\ \mu m$. P-type PbS layers were investigated, which had been prepared by evaporation *in vacuo* or by chemical precipitation through reduction of the lead complex by thiourea. The photographic effect on PbS layers takes place at $\lambda > 1{,}8\ \mu m$, the sample being lit through a Ge-filter. To obtain a dense image an exposure time $\sim 10^2$ sec of energy flux $\sim 10^{-4}$ Watt/cm^2 were required. The photographic sensitivity correlates with the photoelectric one as is seen from the spectral distribution of photoconductivity. The mechanism of latent-image formation does not exclude in principle the formation of Ag_2S. However, the investigation of the spectral distribution of photoconductivity and photographic effect in Ag_2S shows the photoconductivity in Ag_2S to be limited at $\lambda \cong 1{,}4\ \mu m$.

The process of reduction of methylene blue in contact with the PbS layer was studied, the latter being exposed through a Ge-filter. The latent image in the dye layer was developed by a physical developer. The process is an analogue of the reduction of methylen blue in AgX-emulsions. In the above case reduction is effected by intrinsic excitation of the narrow-band semiconductor.

It is of interest that for some semiconductors (e. g. SbSI) a reversal of the image takes place: having exposed the semiconductor in the region of its intrinsic photosensitivity a positive image is formed.

3. Sensitized photolysis on Ag_2S layers

A more detailed investigation was carried out in polycrystalline Ag_2S layers. In order to study the two-stage mechanism of latent-image formation we investigated the temperature dependence of the characteristic curves, sensitivity and curves of constant density.

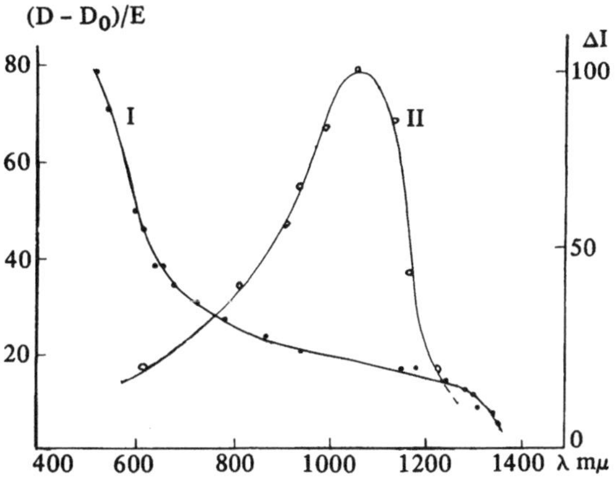

Fig. 2. Spectral distribution of the optical denstiy (I) and photoconductivity (II) produced by a constant amount of light energy for Ag_2S. Photoconductivity data are taken from (9)

The Ag_2S layers were prepared by chemical precipitation from aqueous solution of silver complexes with thiourea on mica or paper. The layers were thermoactivated. Directly before exposure the layers were moistened in an $AgNO_3$-solution containing such sensitizing additions as hydroquinone, metal and citric acid. Then the layer to be exposed was placed into a beaker, the necessary temperature being maintained. After exposure the layer was developed in a physical developer at 0° C, washed, and fixed in acid thiourea solution.

In Figure 2 the spectral distributions of the photographic effect (curve I) and of the photoconductivity (curve II) for Ag_2S are shown at room temperature. As mentioned above the long wave-length limit of photoconductivity for Ag_2S lies at $\lambda \sim 1{,}4\ \mu m$ corresponding to $E_g \cong 0{,}9$ eV [8].

The long wave limit of sensitized photolysis is seen from Figure 2 to coincide with the long wave limit of intrinsic photoconductivity of the semiconductor. However, by contrast to HgI_2, the spectral distribution of the photographic effect for Ag_2S does not coincide with the spectral distribution of its photoconductivity. Different from that for photoconductivity, the curve for the photographic effect does not reveal any maximum, but it should be noted that in the short-wave region the sharp increase

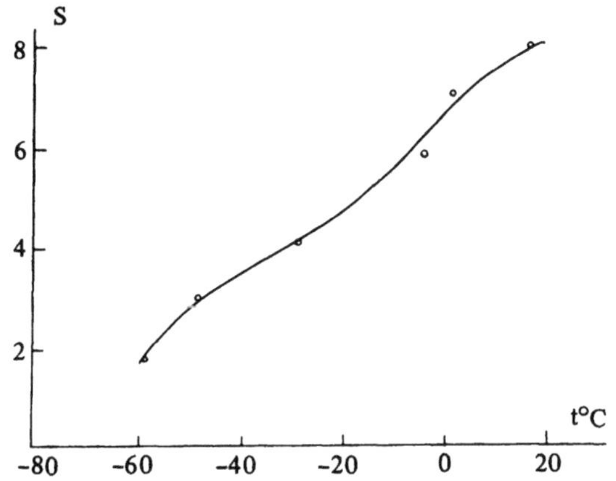

Fig. 4. Temperature dependence of the photosensitivity for Ag_2S

in photosensitivity corresponds to a sharp decrease in photoconductivity.

Characteristic curves of the dependence of D-D_0 (D_0 = the fog value) plotted against log exposure time (lg H) for a series of temperatures (−15° C : + 60° C) are shown in Figure 3. Figure 4 shows the temperature dependence of sensitivity S determined as $S = 1/H$ for $D - D_0 = 0{,}5$. Photosensitivity is seen from these data to decrease with decreasing temperature. Curves of equal density in Figure 5 are analogous to the right branch of that for the conventional photographic process (the region of low light intensity). At low temperatures there appears a region in the curves where the reciprocity law is fulfilled. With higher temperature this region is displaced to larger values of log t.

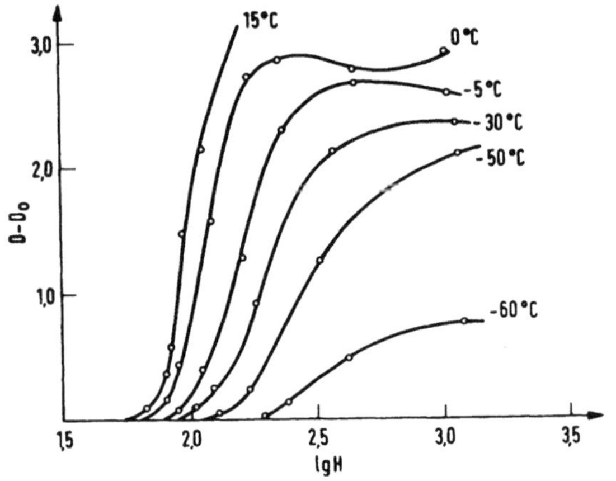

Fig. 3. Characteristic curves for Ag_2S

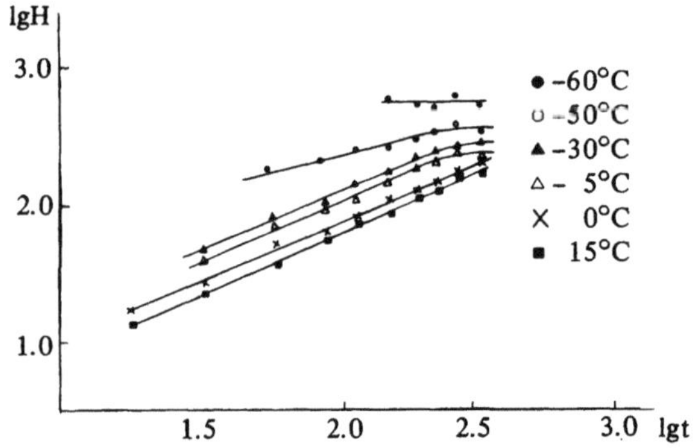

Fig. 5. Curves of constant density for Ag_2S

Such a temperature dependence is known to be characteristic of the two-stage photography process in AgX.

Finally it is to be noted that not only the curves of constant density but also some other results demonstrate the role of the ionic stage in latent-image formation in Ag_2S. For instance if a layer has not been moistened with $AgNO_3$ solution, on exposure and development, no photographic effect is observed. This means the impossibility of separating for a time the electronic and ionic stages in Ag_2S even at low temperatures. Thus the ionic stage proceeds simultaneously with the electronic one even at low temperatures when $AgNO_3$ forms a film of ice on the Ag_2S surface. Thus sensitized photolysis is possible if two solid phases are in contact, one of them being a semiconductor-sensitizer, another a crystal with some ions as extrinsic or intrinsic defects. This suggestion reveals a new possibility of realizing sensitized photolysis.

4. The mechanism of sensitized photolysis on semiconductors

As mentioned above the efficiency of sensitized photolysis on semiconductors is higher the lower the probability of electron-hole recombination as in the case of silver-halide emulsions. Thus effective photolysis involves the essential modification of surface recombination. Adsorbed metal ions prove to be effective means of electron trapping, the holes being captured in some other levels. These levels may well be

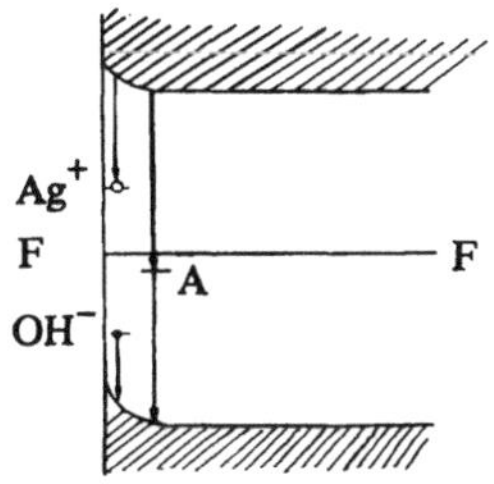

Fig. 6. Mechanism of sensitized photolysis on the surface of a semiconductor

the hydroxyl ions or the molecules of an oxidiser adsorbed on the surface of the semiconductor. Thus the problem of chemical sensitization of the layer for the process considered is the same as that for silver-halide photography though the nature of the hole trapping differs for the various cases. In Figure 6 the electron and hole levels of recombination under conditions of sensitized photolysis is schematically shown. The quantum yield approximates unity if both kinds of traps become more efficient as compared to the intrinsic surface trap A. If surface recombination through center A is more effective than the photolysis, the spectral distributions of photocurrent and photographic effect will coincide, revealing a maximum in the region of the absorption edge. In the opposite case if recombination through the surface center A is insignificant as compared with photolysis, then the increase in photographic sensitivity corresponds to the short-wave decrease of the photocurrent. These two alternatives seem to occur fo HgI_2 (Fig. 1) and Ag_2S (Fig. 2) respectively.

In the above case of sensitized photolysis the reciprocity failure, the curves of constant density and the temperature dependence of photosensitivity should be similar to those characteristic of AgX layers. That was proved in particular for low-intensity reciprocity failure (Fig. 5). In the high-intensity and low-temperature range the reciprocity failure observed would be due to the inertia of the ionic stage.

In connection with that it is of interest to note both the smooth nature of the temperature dependence of photosensitivity and any peculiarities in the curves (Figs. 3 – 5) while passing from the fluid to the solid phase of $AgNO_3$ on decreasing the temperature during exposure. Thus the covering of the surface of the semiconductor by Ag^+ ions seems to be sufficient to provide for the further physical development of the latent-image centers. Hence adsorption and desorption of silver ions during latent-image formation is negligible.

Our examination disregards possible chemical reactions on the surface of the semiconductor. However, in particular the reversal of image may be due to some chemical reactions. For instance the transition from negative to positive may be affected by the efficient capture of the holes by semiconductor anions (e. g. halide ion) and their further interaction with silver. The investigation of the new photographic process is of interest not only in itself but for studying the nature of the latent-image centers in silver halide emulsions. If in both cases the centers are the same (as is likely to be so), then the investigation of centers on the surface of well-known semiconductors such as Ge and Si has to be accomplished by means of conventional methods such as the kinetics of photoconductivity, field effects etc. In this way the parameters of the silver centers and their evolution in the process of latent-image formation may be determined.

5. Acknowledgement

The authors wish to thank Prof. Dr. W. F. Berg for his kind suggestion to present this paper at the Colloquium dedicated to the jubilee of Prof. Dr. John Eggert.

References

[1] Sablin-Javorski, A. D, V. M. Fridkin, Journ. of Scientific and Applied Photography and Kinematography, **14**, 458, (1969).

[2] SPSE Ann. Conf. Boston, Massachusetts, June, Preprints p. 10-14, (1968).

[3] Sablin-Javorski, A. D., V. M. Fridkin, L. M. Belyaev, Krystallographya, **15**, 591, (1970).

[4] Paritski, L. G., S. M. Ryvkin, Journ. of Scient. and Appl. Photography and Kinematography, **15**, 184, (1970).

[5] Sablin-Javorski, A. D., L. M. Belyaev, V. M. Fridkin, DAN, **193**, 1283, (1970).

[6] Sablin-Javorski, A. D., L. M. Belyaev, V. M. Fridkin, Intern. Congr. on Photogr. Sci. July-Aug. 1970, Moscow, Preprints E., **94**, (1970).

[7] Belyakov, L. V., D. N. Goryatchev, Zh. G. Dholyan, L. G. Paritski, S. M. Ryvkin, Intern. Congr. of Photogr. Sci. July-August (1970), Moscow, Preprints E, **90**, (1970).

[8] Bube, R., "Photoconductivity of Solids", J. Wiley and Sons, New York (1960).

[9] Miseliuk E. G., Murtens, Izv. Akad. Nauk USSR, ser. fiz., **16**, 115, (1952).

Discussion

Schaffert, R.: I want to ask whether anyone has an opinion as to whether this can be considered an electrophotographic process. I am willing to accept it as such. However, the people at Itek Corp. do not classify it as an electrophotographic process. In comparing this type of process with "chargeless electrophotography" it seems to me the only difference is that one uses electrophorectic development, while the other uses chemical development.

Berg, W. F. (Fridkin): I would also accept this as an electrophotographic process and would even classify silver-halide systems under that heading. This is surely a question of semantics. Silver-halide photography has an electronic step immediately followed by a chemical one—is this enough to make a basic distinction?

Computer Simulation of the Electrophotographic Process in a Simple Model

James E. Lu Valle

Abstract

A computer simulation of the electrophotographic process has been made. The simple model used in this work assumed that the corona charging or sensitization step always gave a uniformly charged photoconductor. In the exposure step, a pulse exposure of one microsecond was used; it was assumed that the surface voltage decay involved a simple RC circuit of capacitor and resistance in series. In the development step (toning) it was assumed that each toner was a single particle size with all particles carrying the same charge. Four particles sizes were used, 0.2, 0.5, 1.0 and 2.0 microns in diameter with charge densities of 1, 3, 10, 30, 100 or 1000 charges per particle. In the transfer (printing) step, a transfer function was introduced in order to obtain a tonal output similar to the tonal input. This simple model has suggested several different approaches for a more sophisticated model.

1. Introduction

Many phenomena are observed in electrophotographic copiers which are little understood. For instance, tailing, shadowing, offsetting, background, etc. are evanescent effects, here today and gone tomorrow; unfortunately they frequently return for another visit. If these problems arise on a copy machine in the field, it is always the fault of the toner or the photoconductor (at least this is the usual consensus). If they occur during the design of a new copier, a three-cornered exchange takes place between machine, toner and photoconductor designers, each group blaming the other two groups for the problem.

To understand these phenomena, factorial experiments are required where several variables and their interactions are studied simultaneously. A copy machine is not the vehicle for such experiments as too many parameters are uncontrollable. A simple system may be set up for experimental purposes (usually only one phase of the copier system is simplified and the remaining phases ignored) and observations made on this system. When the data are given to the machine, toner and photoconductor designers, they usually say "What does this mean? How do we use the data to improve our copier? Your system is virtually unrelated to our complex copier system."

Clearly another approach is needed. The advent of modern large scale computers and interactive computer systems (in which the computer and the user converse through a time-sharing terminal) has made it feasible to simulate models of complex systems. To achieve this, however, the model must be carefully prepared and the system clearly specified. During the problem definition, simulation, programming, debugging, and validation stages an interactive computer system is invaluable. Simulation of this type is a stepwise process. A simple model is simulated and validated and then a more complex model is simulated and validated. If the overall system is quite complicated, it is possible to break it down into several subsystems and to prepare simulation models of each subsystem. After validation of the discrete subsystems, they are combined into the overall system model. This is then validated. Figure 1 is a simple flow chart for system simulation. If the entire system is simulated, boxes E, F and G are unnecessary.

It was decided to simulate a simple model of an electrophotographic system, E.P.S. The overall E.P.S. was separated into five subsystems as shown in Figure 2; namely, sensitization (corona charging), latent image formation (exposure), development (toning), printing (toner transfer), and fixing (fusing). These five subsystems are present in the Xerographic process. In the Electrofax process the printing step is combined with the development step as the toner is not transferred.

In this simulation, the APL programming language [1] and specifically the APL\360 subset [2] has been used because programs written in APL are more powerful and much shorter than programs written in FORTRAN or PL1. It is not unusual for the FORTRAN program to be ten times as long as the APL program. As variables in APL may be scalers, vectors, matrices and higher order arrays, it is frequently possible to write APL programs without time-consuming iterative loops which are necessary in complex FORTRAN programs.

2. The simulation model

How much of the overall E.P.S. should be simulated in the first models? If a complete subsystem were to be excluded, it should be one that could be included at a later date without materially affecting the simulation of the other subsystems. In the simulation without this subsystem, the assumption should be made

System simulation flow chart

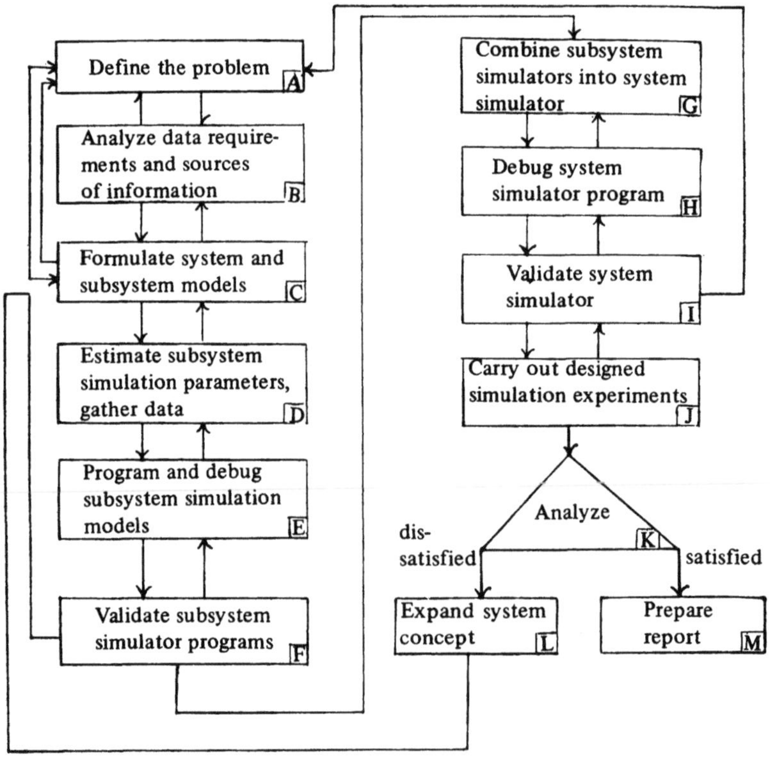

Fig. 1. Flow chart of simulation process

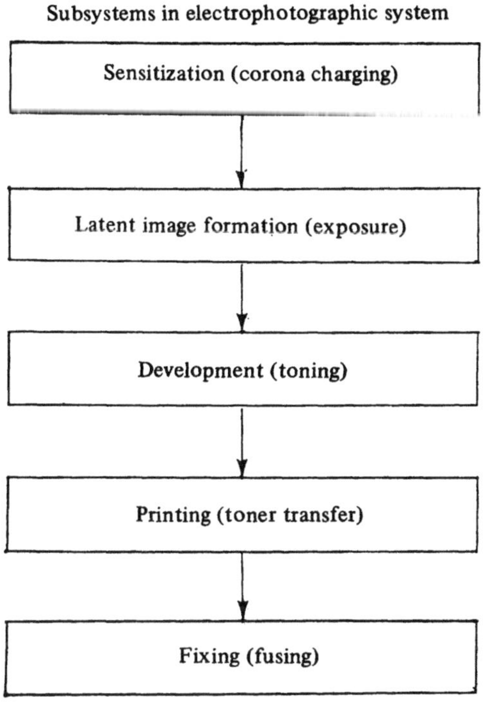

Fig. 2. Flow chart of electrophotographic system

Random discharge matrix

50	50	50	50	500	50	50	50	500	50
50	50	50	50	50	50	50	50	50	50
50	50	50	50	50	50	500	50	50	50
50	50	50	50	50	50	50	50	50	50
50	50	50	50	50	50	50	50	50	50
50	50	50	50	50	50	50	30	50	50
500	50	50	50	50	50	50	50	50	50
50	50	50	50	50	50	50	50	50	50
50	50	50	50	50	50	50	50	50	50
50	50	50	500	50	50	50	50	50	50

Randiscm

90

Random discharge matrix

50	500	50	500	50	500	50	50	50	50
50	50	50	50	50	50	50	50	50	50
50	50	500	50	50	50	50	50	50	50
50	500	50	50	50	50	500	50	50	50
50	50	50	50	50	50	50	50	50	50
50	50	50	50	50	50	50	50	50	50
50	50	50	50	500	50	50	50	50	50
50	50	50	50	500	50	500	50	50	50
50	50	50	50	50	50	50	50	50	50
50	50	50	50	50	500	50	50	50	50

Randiscm

51

Random discharge matrix

500	500	500	500	500	50	50	500	500	500
50	500	500	50	50	500	50	50	500	500
50	500	500	50	50	50	50	500	50	500
500	500	500	50	500	50	50	50	50	500
500	500	50	50	500	500	500	50	50	500
50	500	50	50	50	500	50	500	50	500
500	500	50	500	50	500	50	500	50	50
50	50	50	500	50	500	500	500	50	50
50	500	500	50	50	50	500	50	50	50
50	50	500	500	50	50	50	500	500	500

Fig. 3. Randomly discharged surfaces

that the omitted subsystem operates perfectly on its input to give a perfect output.

Sensitization (Corona Charging), the first subsystem of Figure 2 involves the photoconductor, the rate at which the photoconductor moves past the corona, the distance between the corona and the photoconductor, the voltage applied to the corona, the sign of the voltage applied to the corona, and the composition of the ambient air [3–7]. If sensitization is not to be included in the model, it must be assumed that the photoconductor is capable of accepting a uniform charge over its surface and that the ambient composition is always constant. As we must eventually include sensitization in the model, it should be possible to simulate a randomly charged surface with various discontinuites. Figure 3 is an example of the simulation of randomly discharged surfaces of 0.01 $(\text{micron})^2$ in area. It appears that there will be no problem in randomly charging or discharging the photoconductor surface. The decision was made to exclude the sensitization subsystem in the first simulation models of the E.P.S.

Latent Image Formation (Exposure), the second subsystem in the E.P.S. must be included in the simulation model. The light intensity, the duration of exposure, the sensitivity to light, the resistivity and the capacitance of the photoconductor as well as the retention of charge at the deposited site on the surface of the photoconductor are all involved in latent image formation.

The modulated photodischarge of amorphous selenium has been widely investigated [8–16]. A series of papers on the field dependent photoinjection process in amorphous selenium by Batra et al have clarified many of the phenomena involved [16]. This work also appears to be applicable to some organic photoconductors.

The modulated photodischarge of zinc oxide single crystals and zinc oxide-binder layers has been studied by several investigators [17–29]. However, the system is much more complicated than amorphous selenium [21]. The necessity of dye sensitization [23–25] in order to utilize light in the visible range and the use of an insulating binder yields a very complex system. The nature of the binder, the binder/ZnO ratio, and the nature of the sensitizing dyes all drastically affect the properties of the composite photoconductor. The discharge process is not yet clearly understood in ZnO.

Organic photoconductors are becoming important in this field [30–34]. They also frequently must be sensitized to visible light. The assumption shall be made that the discharge process is similar to that in amorphous selenium.

As the initial simulation model should be quite general, the assumption was made that the discharge process was exponential with an RC time constant that remained constant. The initial work has been done with a dimensionless exponent; i. e., time is measured in time constant units. The light striking the photoconductor is modulated by the luminance of the original which is controlled by its reflection optical densities. The exposure light was a one microsecond flash sufficiently intense to discharge the photoconductor when the original was a perfect white reflector.

The work on field dependent photo-injection of charges in selenium [11, 16] led to the decision to use a pulse exposure of one microsecond in the model work. This permits the assumption that all of the charge is injected at constant field.

In ZnO, many of the observed phenomena, such as high background voltages are eliminated if the exposure is properly adjusted. For instance, the rapid decay following cessation of exposure disappears if the exposure time is lengthened or the intensity of the exposure lamp increased. If however the ZnO-binder layer is imperfect, i. e., does not form a good photoconductor, it should not be used. The light exposure for a good photoconductor may be adjusted to give very low background regardless of the initial A.S.V. Increasing the light intensity will decrease the required exposure time to the desired time interval.

Let the apparent surface voltage on the sensitized unexposed photoconductor be V_0. A point, P(1), on the original with reflection density D will cause the voltage V(D) at the image point PI(1) on the photoconductor at time τ after the microsecond flash exposure.

$$V(D) = V_0 \ (1 - (1 - e^{-\tau}) \ e^{-D}) \qquad (1)$$

The modulation term e^{-D} arises as follows:

$$D = \text{Log}(I_0/I_R)$$

where I_0 is the incident light and I_R is the reflected light.

Thus

$$I_R = I_0 \ e^{-D} \qquad (3)$$

as I_0 is constant in these experiments, it has been set equal to unity. If both sides of Eqn. 3 are multiplied by the exposure time and logarithms taken, the luminance of the original is given by

$$\text{log luminance} = \log I_R t = (\log I_0 t) - D \qquad (4)$$
$$= \text{constant} - D$$

For convenience, the constant has been set at 3.

$$\text{log luminance} = 3 - D \qquad (5)$$

The contrast voltage, CV, at the image point PI(1) will be given by the difference between V(D) and the voltage generated at an associated background point V(B).

$$CV(D) = V(D) - V(B) \qquad (6)$$

The associated background voltage was taken as that for D = 0.0 in these calculations. The response of the exposure simulation model may be examined in two ways. A plot of contrast voltage for several reflection densities against time should give a family of exponential curves with the highest contrast voltage corresponding to the highest reflection density (at all times, τ), i. e., the lowest luminosity. If the contrast voltages for the several luminosities are plotted against the log luminance, an exponential curve should be observed. Thus the response of the photoconductor is such that a linear input becomes an exponential output.

Development (Toning), the third subsystem in the E.P.S., must be included in the simulation model. It is a function of retention of charge on the photoconductor, the toner particles (size, charge per particle, optical properties, shape), the capacitance of the photoconductor, the presence or absence of a development electrode, the spacing between the development electrode and the photoconductor, and the bias potential of the development electrode. The problems that arise are different in brush development, cascade development, aerosol development and liquid development [35].

The development process has not been as extensively investigated [35–43] as the latent image formation process. The development process is quite complex, and the experimental data are difficult to interpret. The work of several investigators has shown that charge neutralization is seldom achieved [38–43]. In a copier, the exposed photoconductor spends a short limited time in the developer so development is never complete.

It was assumed that the deposition of developer follows an exponential law [40, 41]. That is depletion of the developer takes place during development even in the case of a circulating developer.

$$D = D_M (1-\exp(-t/\omega)) \qquad (7)$$

where D is the density obtained at time t and D_M is density obtained at neutralization. The calculation of D_M was made with the assumption that toner is deposited in hexagonal closest packing layers until the charge per unit area is neutralized. A nonfringing field has been assumed between the electrode and the surface of the photoconductor. The time constant (ω) is a function of the conductivity of the developer, the capacitance per unit area of the photoconductor and the development electrode spacing [41]. In the initial computer simulation, a dimensionless exponent has been used with time taken in units of the time constant. It was assumed that the toner particles are spheres and that the spheres pack in hexagonal closest packing. In the simulations it was also assumed that complete classification has been utilized; i. e., each calculation is carried out for one particle size over a range of charge density. The number of layers of toner or the toner thickness is proportional to the mass of toner deposited which is proportional to the optical density; hence, the toner layer thickness will be used as a measure of the optical density.

In the utilization of Eqn. 7, the effect of toner charge density and the ratio of charge to mass of the toner particles has not been included. Both the charge density and the charge to mass ratio will have very definite effects on the rate of deposition. For the same size toner particles, increase in charge density should increase the attainable resolution [37]. However when the charge density becomes so high that only a single incomplete layer is laid down, the resolution will decrease rapidly with decrease in the coverage per unit area of the photoconductor by the toner particles.

Development should take place in such a manner that a plot of the number of layers of toner deposited or the thickness of the deposited layer against the contrast voltage at development should show a linear relationship. The relationship between exposure and the thickness of the toner layer should be exponential, the latter relationship arises from the exponential decay to the contrast voltage on the photoconductor and because the charge per unit area is directly proportional to the contrast voltage. (Actually to the absolute voltage). Since a contrast voltage taken at a time equal to 6 time constants was used, it differs but little from the absolute voltage.

Printing (Toner Transfer), the fourth subsystem of the E.P.S., must be included. It has had little or no experimental or theoretical investigation [44]. However, one would not expect a thin layer of toner to transfer as easily as a thick layer. If the toner charge is not injected into the photoconductor, the toner layer should be held quite strongly. Two possible functions have been used to calculate the toner transfer function. Let the maximum thickness of toner on the photoconductor be TM. It may be that the toner of thickness T would transfer with difficulty proportional to $(T/TM)^2$ due to the fall-off in coulomb attraction with increasing thickness of the layer. It is possible that toner of thickness T may transfer with difficulty proportional to $\exp(T/TM) \times (\exp -1)$. The second exponential is used to normalize the transfer function. In a thick toner layer, provided the particles are not too large, there will be coulombic repulsion between the particles which will also make them tend to transfer more easily than the particles in a thin toner layer. These transfer functions will be discussed in the section on the results of simulation.

Either function will tend to make a plot of toner thickness vs. log exposure less exponential and more nearly linear—a linear plot would give tonal reproduction of the original used in this study. Both functions have been tested and the second function (coulomb attraction fall-off) appears to give more nearly a linear relationship between toner layer thickness and log exposure.

Fixing (Fusing), the last subsystem in the E.P.S., has not been treated separately in the simulation to be discussed. In the computations on toner thickness, the spheres are reduced to equivalent cylinders retaining hexagonal closest packing and it is assumed that fusing or pressure causes the particles to spread to fill the 9% of cavities in each layer (the reduced layer). In a third or fourth round the fixing step will be treated separately.

If the E.P.S. has done a good job of reproducing the original, a tonal reproduction plot [45] will verify the mode. This will be the measure of the effectiveness of the simulation model.

The simulation model of the E.P.S. consists of three subsystems:

Latent Image Formation (Exposure)
Development (Toning)
Printing (Toner Transfer)

It has been assumed that sensitization by corona charging is perfect and that the photoconductor is perfect in its ability to retain a charge on its surface. Fixing has been implicitly included in the development step which is fine for the "Electrofax" process. In the Xerographic process it should be utilized after the transfer step.

3. Results of the simulation

In the initial simulations, the charge decay was allowed to continue to a time equal to six times the time constant of the photoconductor. The original was a step wedge on a perfect reflector with reflection optical densities of 0.0, 0.05, 0.14, 0.28, 0.42, 0.56, 0.70, 0.84, 0.98 and 1.12. Each step was separated from the next step by a border at optical density zero. Figure 4 is a plot of the contrast voltage against time. The model exhibits a curve for latent image formation for each step of the step wedge. The step of 0.0 reflection optical density has disappeared because it was used for the background voltage. Figure 5 is a plot of the contrast voltage (for times of 1, 2, 3, 4, 5 and 6 time constant units) against the log luminance of the original. All of the curves exhibit an exponential relationship. Thus the photoconductor response is nonlinear.

A linear discharge with variation in light intensity for zinc oxide has been discussed by Amick [21]. He postulates successive trapping and detrapping as the carriers are carried through the photoconductor. Examination of his data show that for any given exposure, the charge decays exponentially and that the linear discharge was exhibited at best by only two experimental layers for a light exposure between 10^{11} and 10^{12} photons per square centimeter.

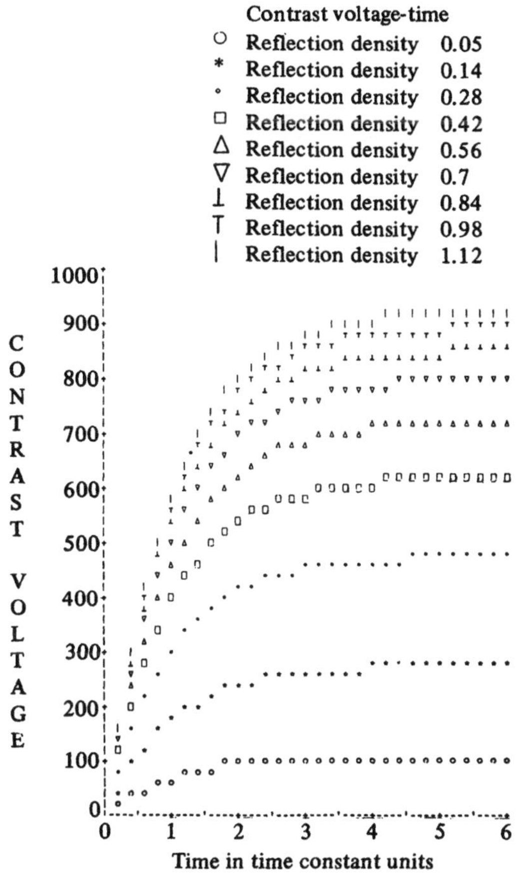

Fig. 4. Change in contrast voltage with time for nine exposure levels

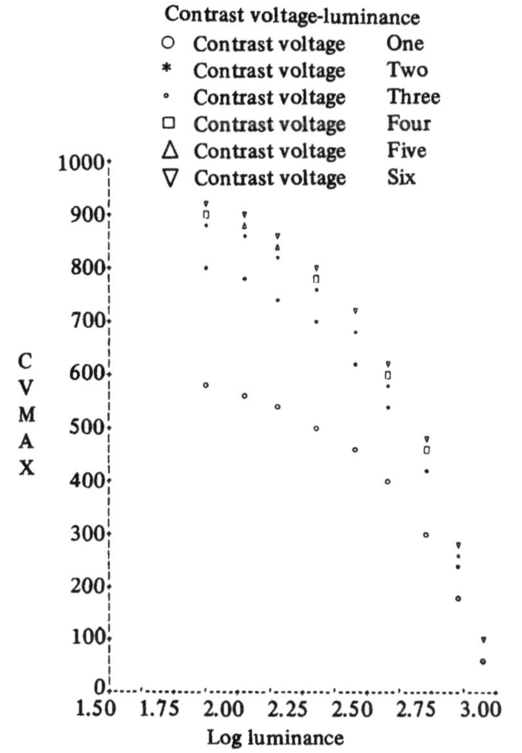

Fig. 5. Change in contrast voltage with exposure at one to six time constant units

Seki, Batra, Gill, Kanazawa and Schechtman [16e] use a model for selenium in which the equivalent circuit is a capacitance in parallel with a two terminal element whose current voltage characteristic is directly related to the field dependent photoinjection efficiency. The model used in these calculations has a simple RC equivalent circuit—the exposure was a microsecond flash in order to photo-inject the charges at constant field. It was assumed that the light exposure was a one microsecond flash with an intensity sufficient to completely discharge the photoconductor if the original were a perfectly white reflector. Figure 6 is a plot of the reflection density against the log luminance and as might be expected, it is a straight line. Figure 6 is the reference, it gives the details of the tonal response of the original. If the simulation is good, its output should be very similar to Figure 6, namely a linear curve of toner optical density against log exposure. Neugebauer [46] has discussed the electric fields arising from electrostatic images. Streifer and Stark [37] have discussed the problems in toning as the resolution is increased. The fringing effects they have discussed have been neglected in this first simulation. All of the calculations in this paper assume the presence of a closely spaced development electrode. The toner deposition was originally carried to neutralization but this was obviously unrealistic so Eqn. 7 was used to limit deposition. After examination of the output an upper and lower limit were placed on the toner deposition. If the single layer had less than twenty percent coverage, it was set to zero coverage except for the background calculations where the actual number of toner layers deposited were used. The upper

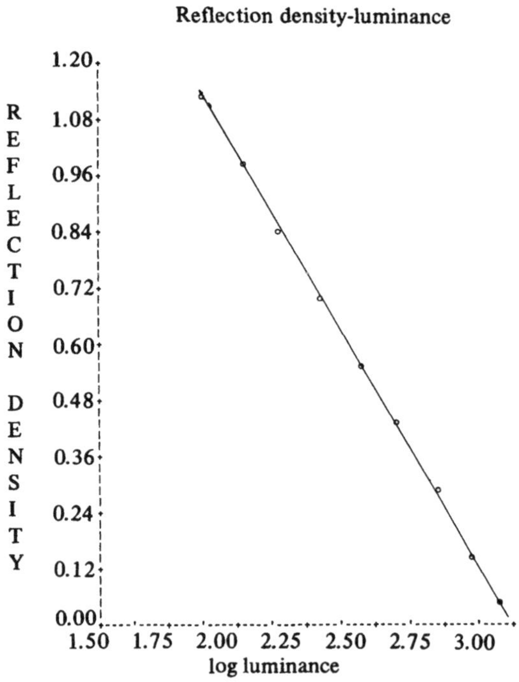

Fig. 6. The reflection density-luminance relationship of the original

limit for toner deposition on the photoconductor was placed at twentyfive microns. This upper limit should probably be given not as thickness but as a certain number of layers of toner; however, this leads to optical density problems; therefore, a layer thickness of twenty-five microns was assumed. This limitation was placed on the reduced layers. (It was assumed that during fixation, the particles become compressed into equivalent cylinders and spread sufficiently to cover all vacant interstices (that is the layer thickness was reduced).

All calculations on toner deposition were made simultaneously for four sizes of spherical particles, 2, 1, 0.5 and 0.2 microns in diameter, for all nine latent images formed in the exposure step and for charge densities, charge per particle, of 1, 3, 10, 30, 100 or 1000 electron charges. An array of nine six by four matrices were generated by these calculations. By simple transposition of the matrices, it is possible to present the data in the following forms; (In all cases, the matrix data is plotted a column at a time).

The data may be presented as nine six by four matrices or nine four by six matrices; in this case, *each matrix represents a constant exposure level,* the six rows or six columns represent the six levels of charge density and the four rows or four columns represent the four particle sizes.

The data may also be presented as six nine by four matrices or six four by nine matrices; in this case, *each matrix represents a constant charge per particle* and the nine rows or nine columns the nine exposure levels whereas the four columns or four rows represent the four particle sizes.

The data may be also presented as four nine by six or four six by nine matrices; in this case, *each matrix represents a constant particle size* and the nine rows or nine columns, the nine exposure levels whereas the six columns or six rows represent the six levels of charge density.

The six different presentations include in each case 216 data points. For the purposes of this paper the 4–9–6 matrix was chosen, where each matrix (or plot) will represent a constant particle size, the nine rows of each matrix will therefore represent the nine exposure levels and the six columns will represent the six levels of charge density.

The photoconductor was set to have a capacitance of 3, 30, or 300 picofarads per square centimeter. The time for deposition of toner ranged from 0.0001 to 1 time constant unit. The choice of capacitance and time constant is not really as wide as it may appear. For instance a capacitance of 300 picofarads and a deposition time of 0.0001 gives virtually the same numerical values for toner deposition as 30 picofarads and 0.001 or 3 picofarads and 0.01. For illustrative purposes a choice of 300 picofarads and time of 0.02 units was chosen; except in Figure 12 where for scaling purposes, the capacitance was maintained at 300 picofarads and the time reduced to 0.001 and 0.0003 units.

Figure 7a, b, c are plots of the logrithm of the number of background toner layers against the logarithm of the charge per particle for the four particle sizes used in the simulation. The background was calculated for the original at a reflection optical density of 0.05 which is rather low. The data indicate that the particle size and the charge per particle play an important role in determining the background density of the photoconductor. Note that contrary to what might be expected, the background layer is heaviest for the large particles and least for the small particles. As was expected, the background layer decreases with increase in charge density for any given particle size. The data of Figure 7a indicate that a development time of 0.02 units is too long for a 300 picofarad capacitance photoconductor. The background is much too heavy.

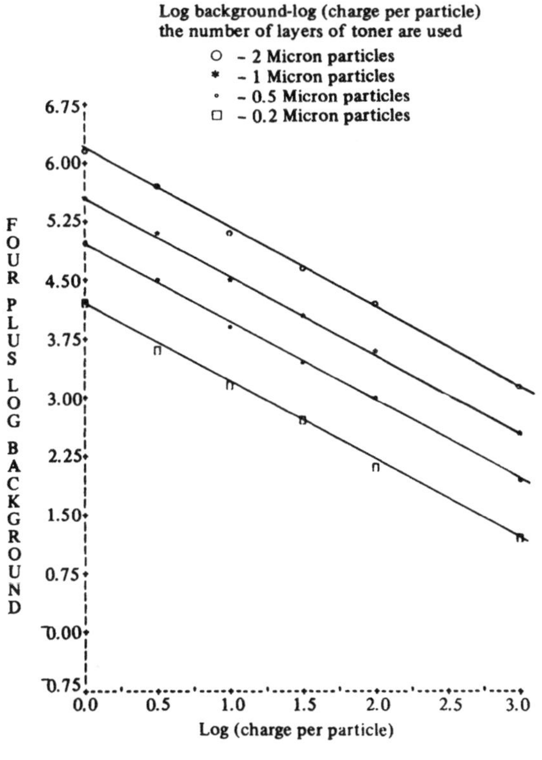

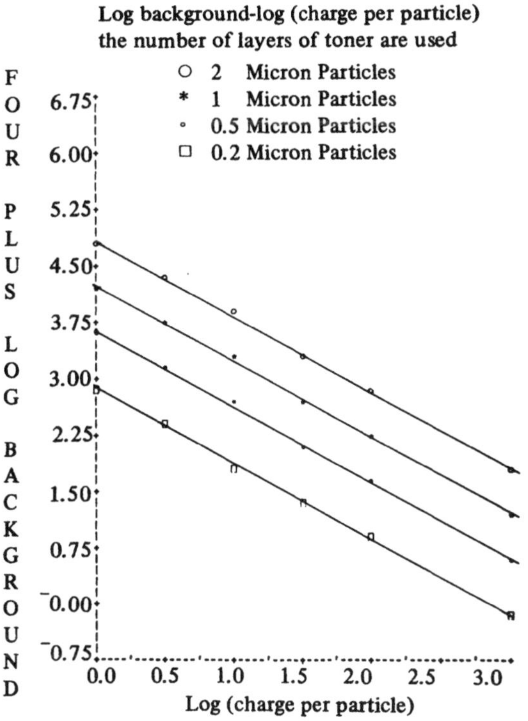

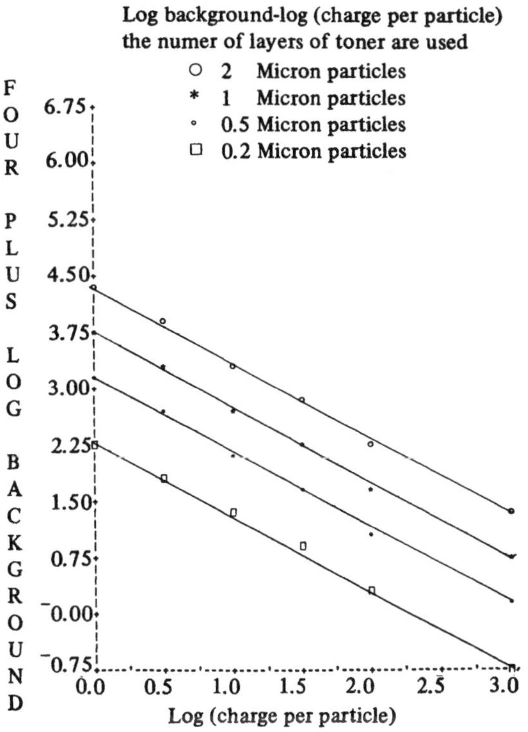

Fig. 7. The relationship between the background and the charge density for four different size toner particles at (a) 0.02 time units, (b) 0.001 time units and (c) 0.0003 time units for deposition of toner

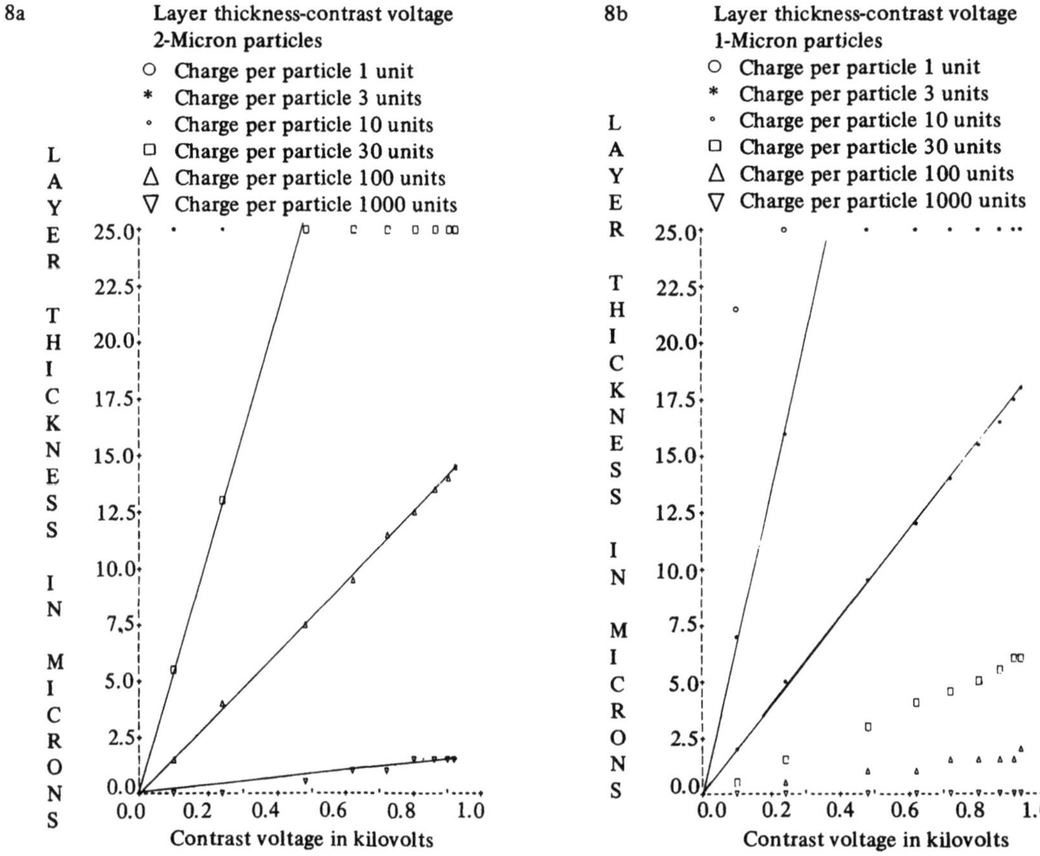

Fig. 8. The relationship between the toner layer thickness and the contrast voltage for (a) 2 micron particles, (b) 1 micron particle, (c) 0.5 micron particle and (d) 0.2 micron particle at 0.02 time units

Figures 8a, b, c, d are plots of toner layer thickness against the contrast voltage for the four particle sizes. Each curve is for a different charge density. Figure 8a shows that the two micron particles at charge densities of one, three and ten charges per particle have already deposited a layer thicker than 25 microns. Figure 8b shows that one micron particles at charge densities of one and three charges per particle have deposited layers thicker than twenty-five microns except in areas of very low luminosity. Figure 8c shows that the 0.5 micron particles at unit charge density have not quite deposited a layer twenty-five microns thick. Figure 8d shows that the 0.2 micron particles at unit charge density have barely deposited a layer 1.5 microns thick. A layer of 0.2 micron particles contains 6.25 times as many particles as a layer of 0.5 micron particles. The layers are 2.5 times thinner for the 0.2 micron particles than for the 0.5 micron particles; hence, the observed nonlinear fall-off in the layer thickness from 0.5 micron to 0.2 micron particles. Examination of the data for the four particle sizes show that the thickness of the layer deposited at a given time falls off rapidly with increase in charge density. All of the figures show that there is a linear relationship between the contrast voltage (and hence the charge per unit area on the photoconductor) and the thickness of the deposited toner layer. The fall-off in the thickness of the layer with higher charge densities imply that it will be more difficult to obtain good covering power and high optical densities when the toner particles have high charge density.

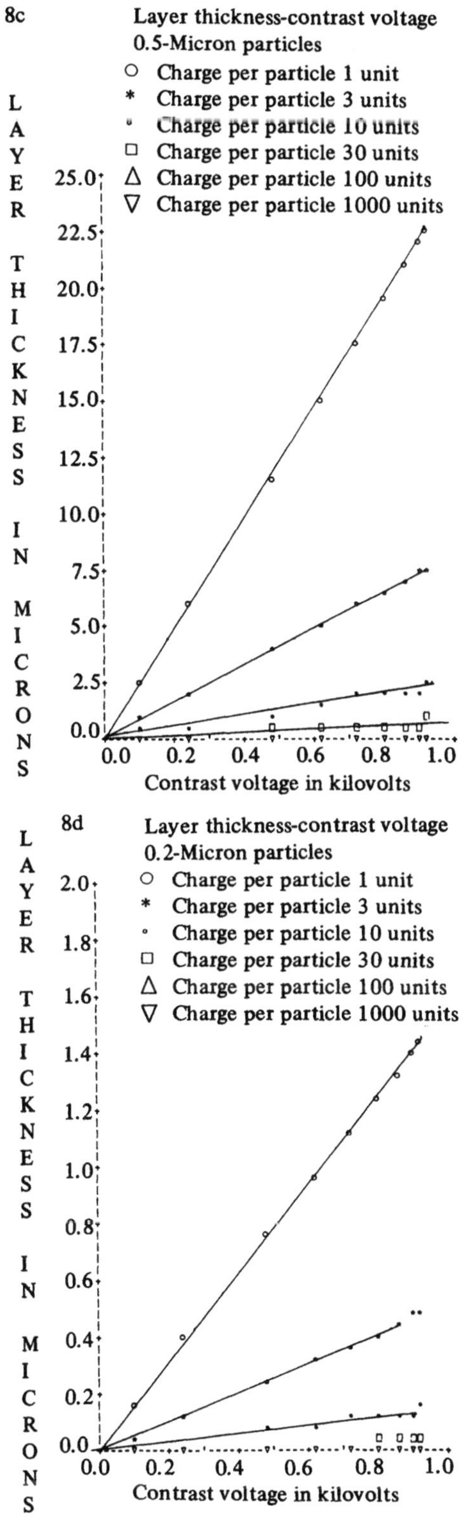

In Figures 9a, b, c, d the toner layer thickness has been plotted against the log exposure; the same deposition conditions were used as for Figures 8a, b, c, d. The thickness of the toner layer decreases exponentially with increase in exposure. The size of the particle and the charge density on the particle does not alter this relationship. When we compare Figure 5 and Figure 9, it becomes clear that the exponential relationship between toner layer thickness and the log exposure arises because of the exponential relationship between the contrast voltage and the log luminance of the original. The linear relationship between the reflection optical densities on the original and the log luminance of the original was converted by the

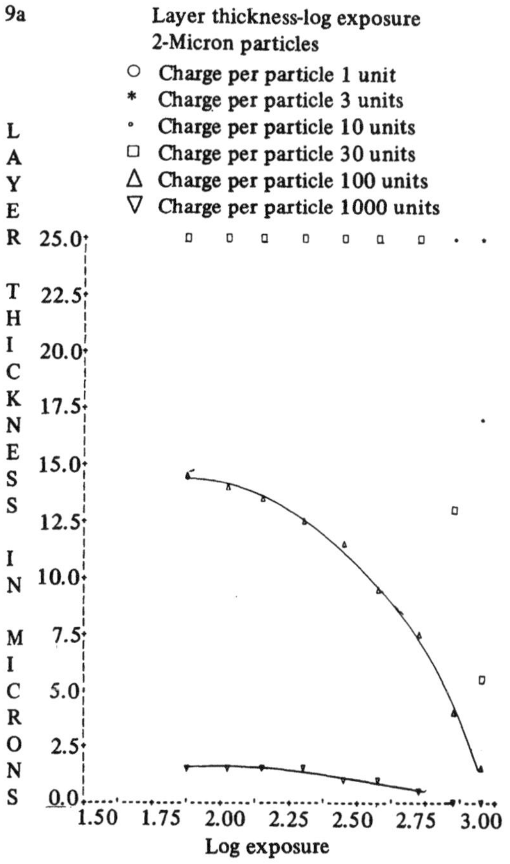

Fig. 9. The relationship between the toner layer thickness the exposure for (a) 2 micron particles, (b) 1 micron particle, (c) 0.5 micron particle and (d) 0.2 micron particle at 0.02 time units

9b

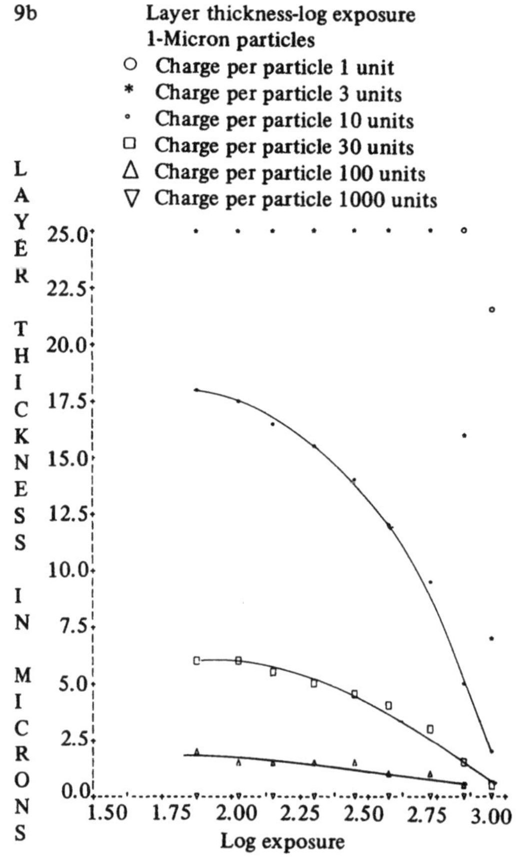

9c

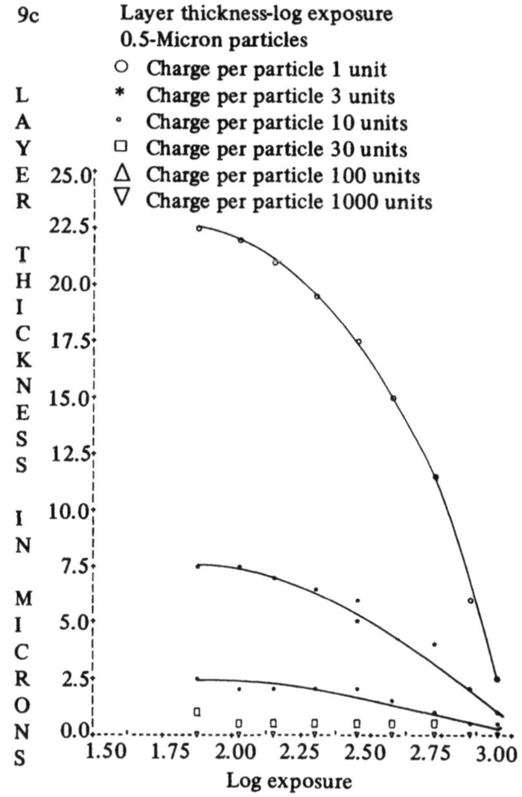

9d

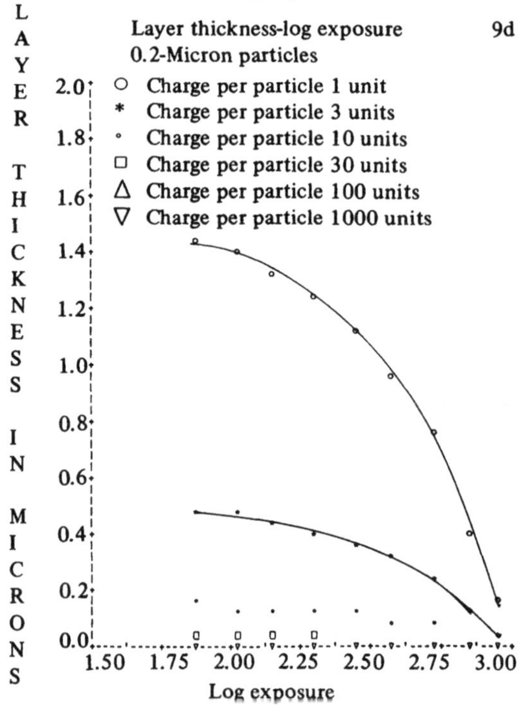

photoconductor to an exponential relationship and the latter relationship has been preserved in the development step.

If it is assumed that the charge on the toner particle is not injected into the photoconductor, the toner particles are bound to the photoconductor by coulombic forces. Under these circumstances a thin layer of toner should be held much more tenaciously than a thick layer. The ease of transfer of the toner should vary with $(T/TM)^2$ where T is the thickness of the toner layer to be transferred and TM is the maximum thickness of toner on the photoconductor. Then the amount transferred will be $T \times (T/TM)^2 \times f$ where f is a factor less than unity since not all of the thickest layer will be transferred. In the calculations f was used at 0.5.

It also appeared possible that the ease of transfer might follow an exponential law and vary with

(exp (T/TM)) x (exp −1)

where (exp−1) normalizes the transfer function. Under these circumstances the amount transferred will be

T x (exp (T/TM)) x (exp −1) x 0.5

Both of the toner transfer functions were tested. Figures 10a, b, c, d give the results when the normalized exponential transfer function was used. The toner deposit is not quite linearly related to the log exposure. Figures 11a, b, c, d give the results when the coulomb fall-off transfer function was used. The toner deposit is almost linearly related to the log exposure. It appears that a coulomb law fall-off transfer function is somewhat better than the exponential transfer function. A transfer function to account for coulomb repulsion among the particles in a thick layer has not been introduced. This would tend to remove the remaining nonlinearity in the toner thickness-log exposure relationship. Figures 12a, b are the plots of the thickness of the transferred toner layer against log exposure for the same capacitance as Figure 11 but for times of 0.001 and 0.0003 units. That is for one-twentieth and

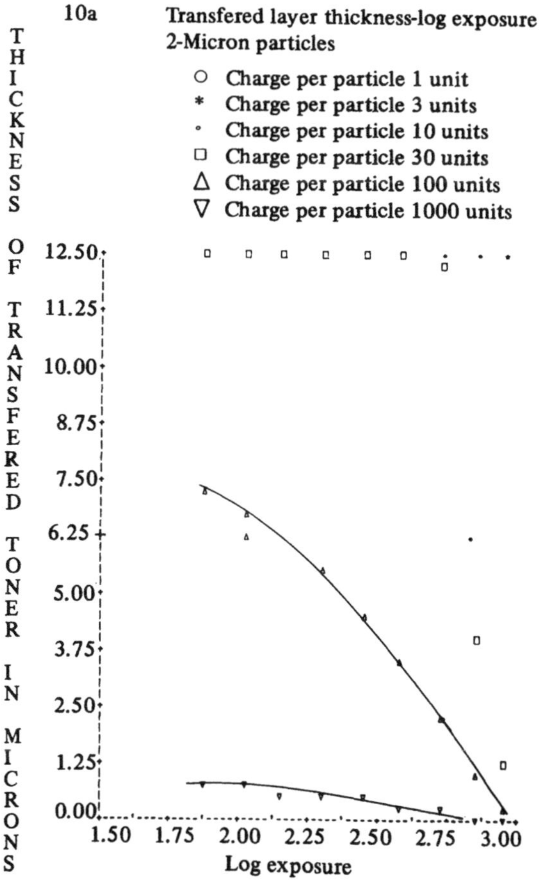

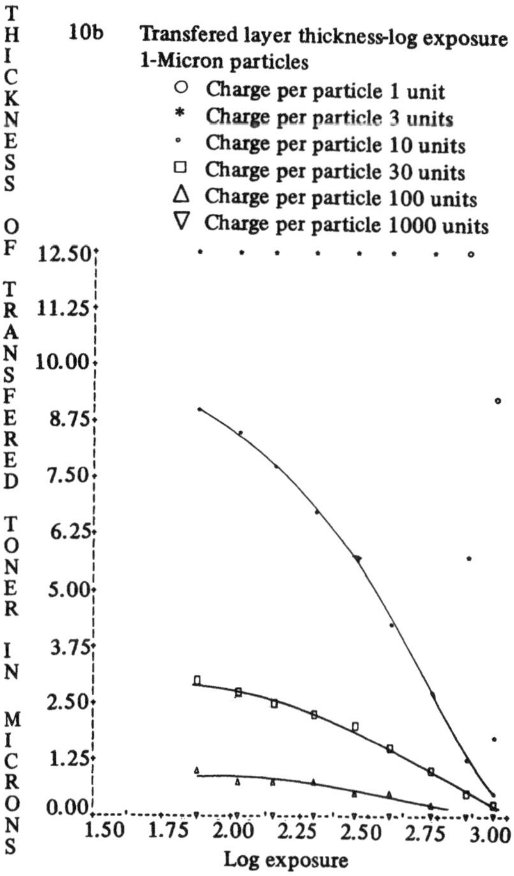

Fig. 10. The toner layer transferred under the exponential transfer function of (a) 2 micron particles, (b) 1 micron particle, (c) 0.5 micron particle and (d) 0.2 micron particle for a toner deposition time of 0.02 time units

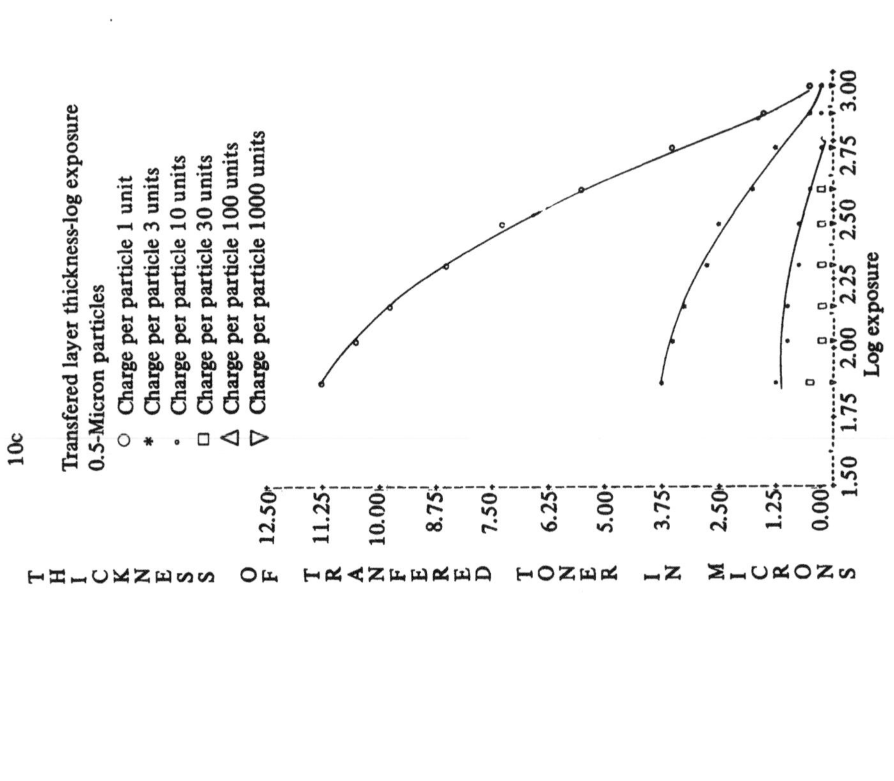
10c
Transfered layer thickness-log exposure
0.5-Micron particles
○ Charge per particle 1 unit
* Charge per particle 3 units
∘ Charge per particle 10 units
□ Charge per particle 30 units
△ Charge per particle 100 units
▽ Charge per particle 1000 units
THICKNESS OF TRANSFERRED TONER IN MICRONS
12.50 11.25 10.00 8.75 7.50 6.25 5.00 3.75 2.50 1.25 0.00
1.50 1.75 2.00 2.25 2.50 2.75 3.00
Log exposure

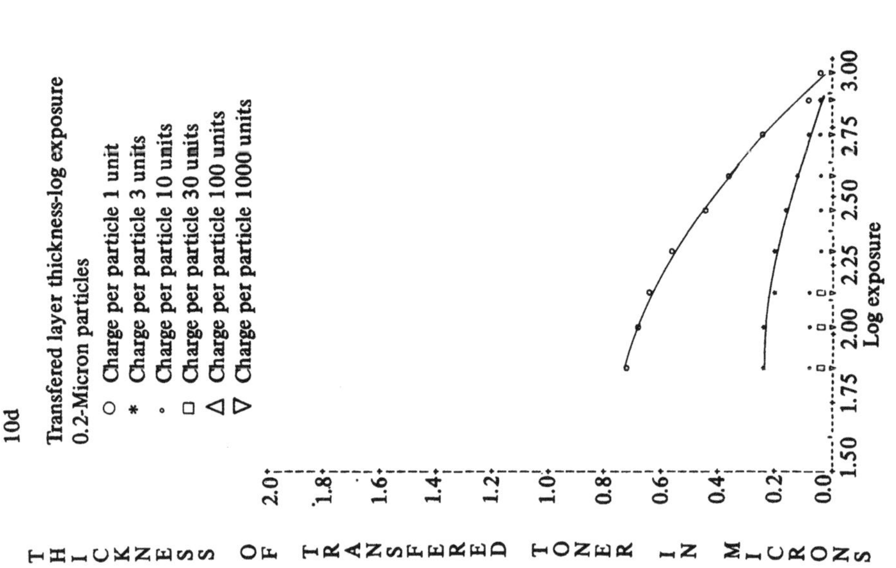
10d
Transfered layer thickness-log exposure
0.2-Micron particles
○ Charge per particle 1 unit
* Charge per particle 3 units
∘ Charge per particle 10 units
□ Charge per particle 30 units
△ Charge per particle 100 units
▽ Charge per particle 1000 units
THICKNESS OF TRANSFERRED TONER IN MICRONS
2.0 1.8 1.6 1.4 1.2 1.0 0.8 0.6 0.4 0.2 0.0
1.50 1.75 2.00 2.25 2.50 2.75 3.00
Log exposure

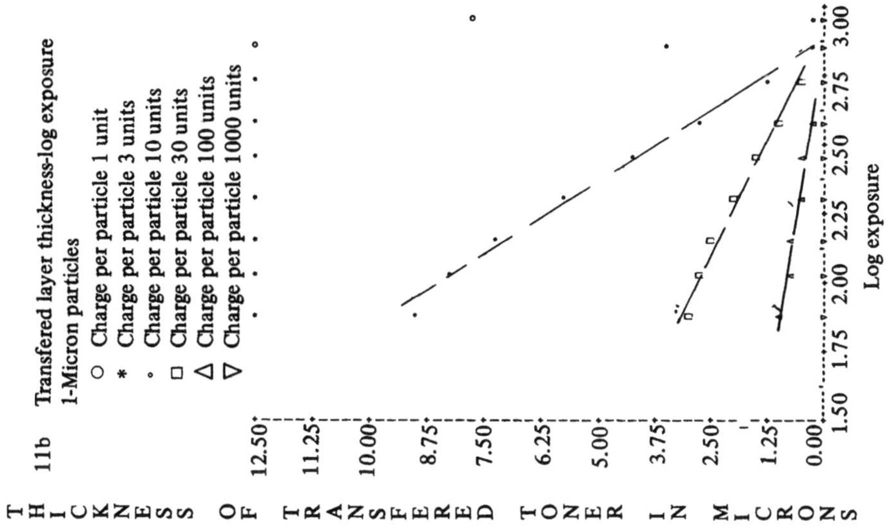

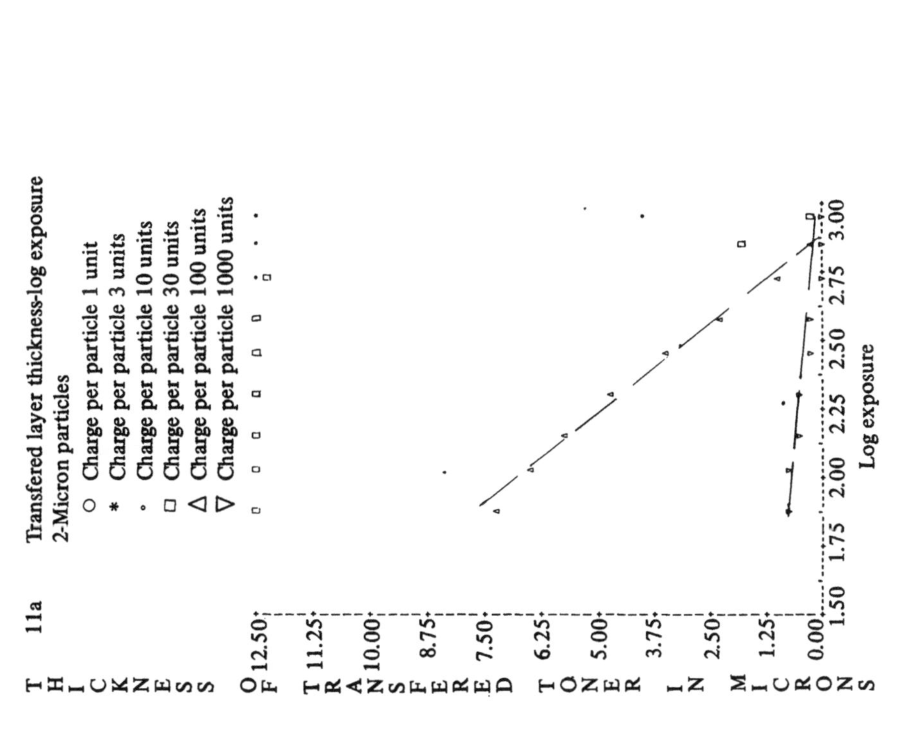

Fig. 11. The toner layer transferred under the Coulomb fall-off law for (a) 2 micron particles, (b) 1 micron particle, (c) 0.5 micron particle and (d) 0.2 micron particle for a toner deposition time of 0.02 time units

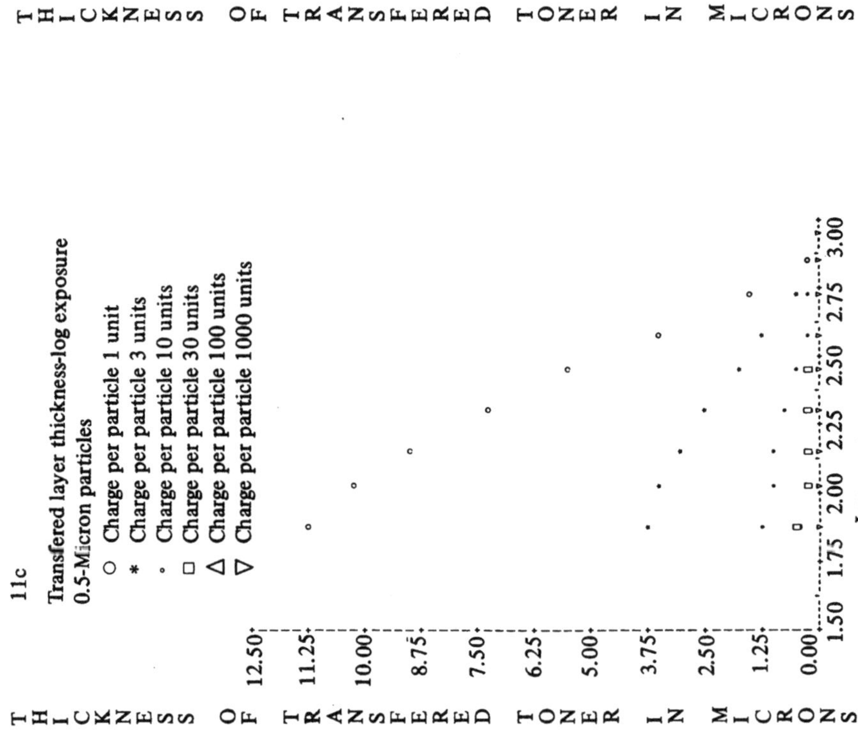
11c
Transfered layer thickness-log exposure
0.5-Micron particles
○ Charge per particle 1 unit
* Charge per particle 3 units
◦ Charge per particle 10 units
□ Charge per particle 30 units
△ Charge per particle 100 units
▽ Charge per particle 1000 units
THICKNESS OF TRANSFERED TONER IN MICRONS
12.50
11.25
10.00
8.75
7.50
6.25
5.00
3.75
2.50
1.25
0.00
1.50 1.75 2.00 2.25 2.50 2.75 3.00
Log exposure

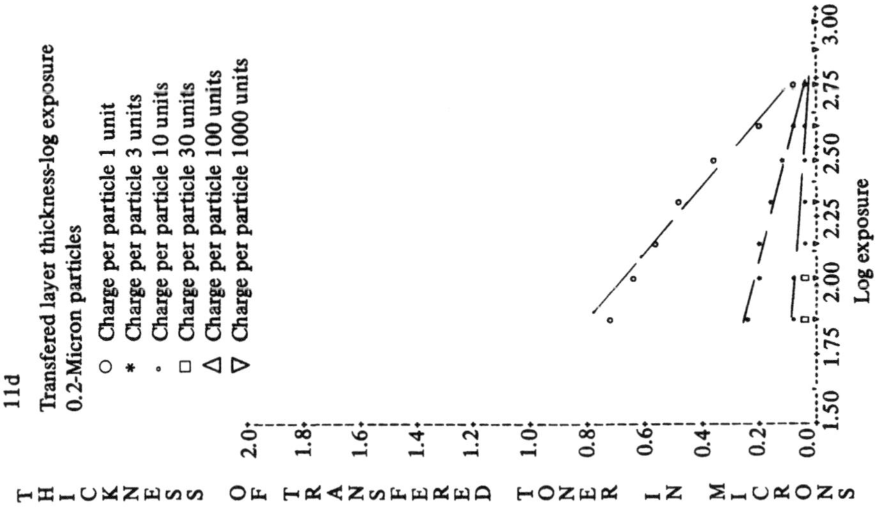
11d
Transfered layer thickness-log exposure
0.2-Micron particles
○ Charge per particle 1 unit
* Charge per particle 3 units
◦ Charge per particle 10 units
□ Charge per particle 30 units
△ Charge per particle 100 units
▽ Charge per particle 1000 units
THICKNESS OF TRANSFERED TONER IN MICRONS
2.0
1.8
1.6
1.4
1.2
1.0
0.8
0.6
0.4
0.2
0.0
1.50 1.75 2.00 2.25 2.50 2.75 3.00
Log exposure

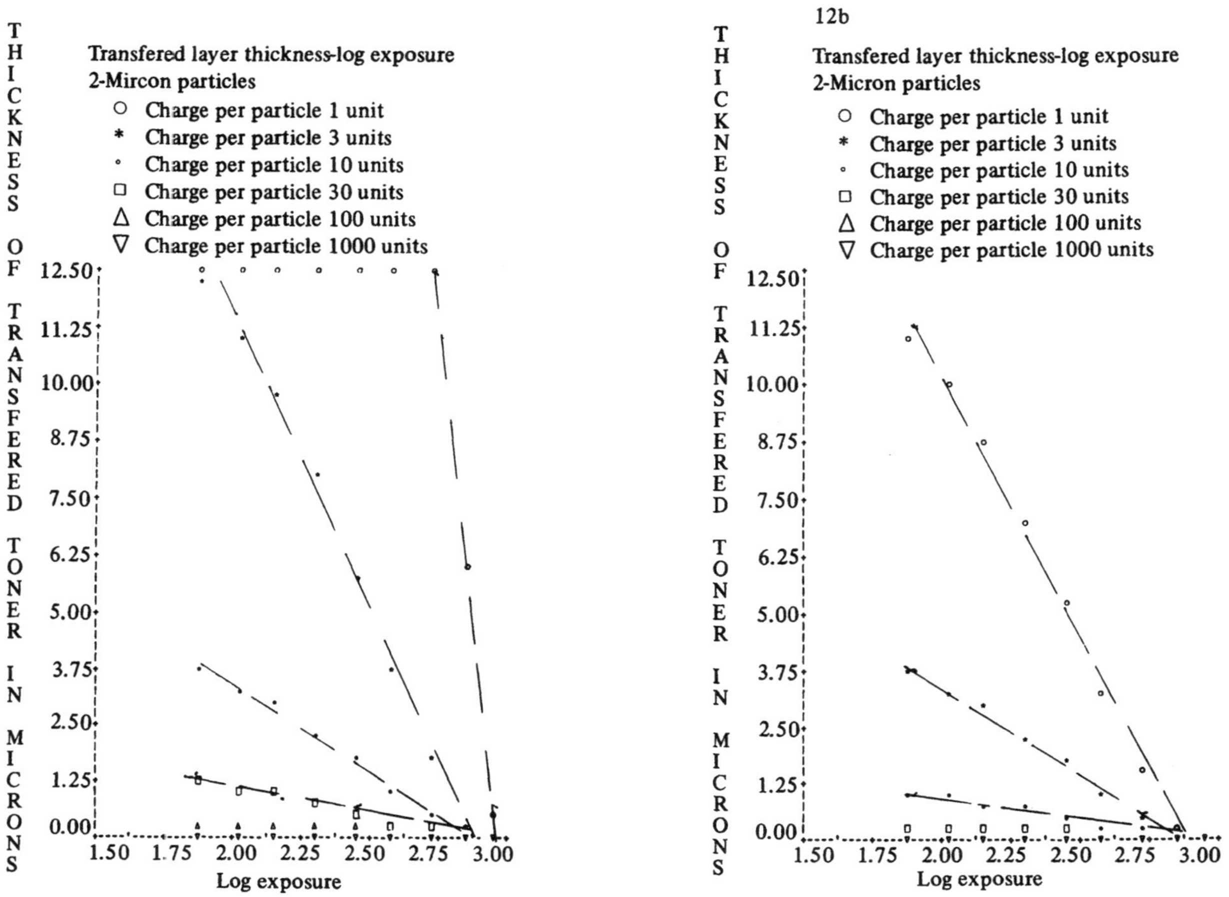

Fig. 12. The transferred toner layer under the coulom law fall-off transfer function for toner deposition times of (a) 0.001 units and (b) 0.0003 units

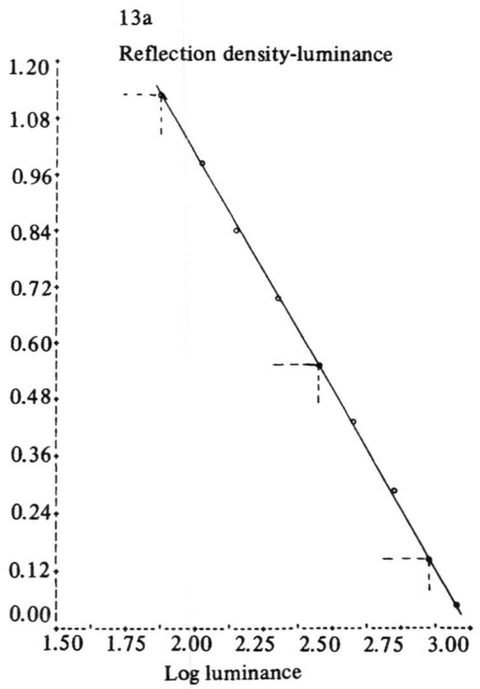

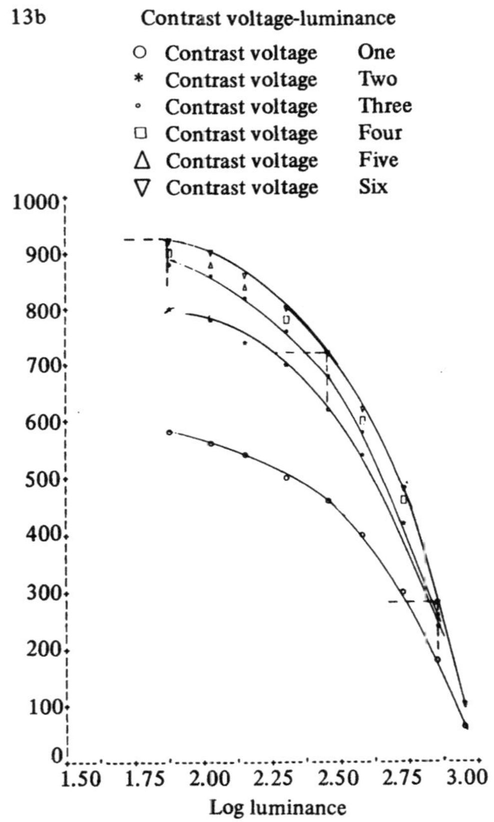

Fig. 13. (a) The input function, (b) the photoconductor transpose of the input function, (c) the developer transpose of the photoconductor output, and (d) the printing transferral transpose of the developer output. Note that (d) is a reasonably good reproduction of (a)

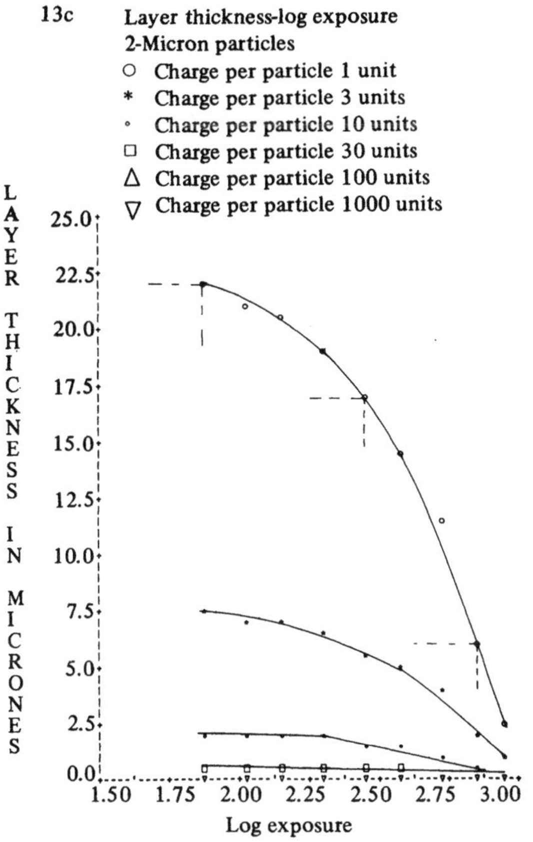

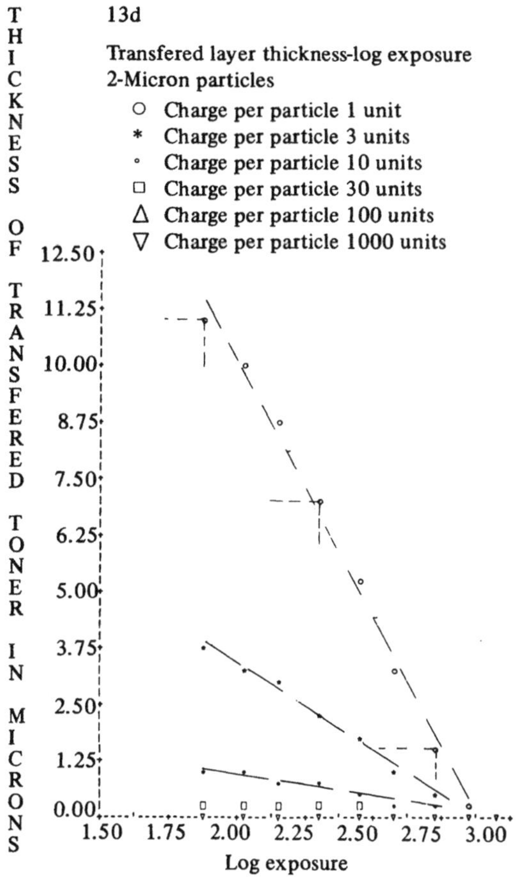

fifteen-thousands the time used in Figure 11. Comparison of Figures 12a, b and 11a show that at a time of 0.001 units, only the 2 micron particles with a charge density of unity have exceeded a transfer thickness greater than 12.5 microns. At a time of 0.0003 units even the 2 micron particles with a charge density of unity have not quite transferred a 12.5 micron thick layer.

Figure 13a is the original input to the system (Fig. 6), Figure 13b is the response of the photoconductor (Fig. 5), Figure 13c is the combined response of the photoconductor and the developer (Fig. 9), Figure 13d is the combined response of the photoconductor, the developer and the transferral step (Fig. 12a). Comparison of Figures 13a and d shows that the model gives reasonably good reproduction.

4. Conclusions

The simple simulation model gives reasonable tonal response. The experiments performed in the computer were experiments that would have been difficult to perform in the laboratory because of the difficulty of obtaining a single uniform size particle and the same charge density on each particle. The data indicate that the developed image on the photoconductor will not give a good tonal response unless the properties of the photoconductor are modified to give a different decay characteristic. The transferred toner layer (if it transfers by the coulombic law transfer funtion) will give fairly good tonal response. Experiments to determine the transfer function are indicated. The background will always be present on the photoconductor; as it is usually the thinnest layer present, it will tend not to transfer to the copy sheet. All the data indicate that good development may be obtained in small fractions of a time constant unit. It appears that the charge density on the toner particles should not be too high. In fact it appears that for particles in the range 2 to 0.2 microns, charge densities below thirty units would be best. (Even thirty may be too high). High charge density and small particle size will lead to inadequate coverage and low observed optical densities on the copy sheet. The images will be gray rather than black.

Experiments with a computer must be carefully defined and programmed to be meaningful. One must always remember that the computer is an idiot. The computer cannot detect errors in the definition and analysis of the problem. It is not a substitute for good rigorous thinking. However, since it frees the investigator from long tedious calculations, it gives him more time for analysis.

5. Future work

The model is not complete;

1. The photoconductor response for selenium and for ZnO-binder should be built into the program.

2. The development electrode should be explicitly included and the effect of variations in spacing between the electrode and the photoconductor investigated as well as the effect of varying the bias voltage.

3. Toner in the dry toner range should be investigated. The simulation model used particle sizes in the range for liquid toners and a unit area of 100 square microns. Dry toners have particle sizes in the range of 20, 15, 10 and 5 microns diameter. Hence the unit area must be increased to 10,000 square microns.

4. The effect on the toner deposition time constant of different charge densities on the particles and of different charge to mass ratios should be explicitly investigated.

5. Distributed toners, a range of particle sizes and of charge density present simultaneously, should be investigated. In the distributed toner, the charge density should vary with the particle size by one of the relationships discussed under (4); i. e., each particle size will have a different charge density. The influence of charge density on the deposition rate must be included in the toner deposition relationship.

References

[1a] Iverson, K. E., "A Programming Language", 1962, John Wiley and Sons, Inc., New York, New York.

[1b] Katzan, H., Jr., "APL Programming and Computer Techniques", 1970, John Wiley and Sons, Inc., New York, New York.

[2a] Gilman, L., and A. J. Rose, "APL\360 An Interactive Approach", 1970, John Wiley and Sons, Inc., New York, New York.

[2b] Berry, Paul, "APL\360 Primer", 1970, IBM Publication #GH 20–0689–1

[2c] Falkoff, A. D., and K. E. Iverson, "APL\360 User's Manual", 1969, IBM Publication #H20–0683.

[2d] "APL/360–OS(5734–XM6) and APL\360–DOS (5736–XM6) General Information Manual", 1970, IBM Publication GH20–0850–1.

[3] Schaffert, R. M., "Electrophotography", 1965, Focal Press, New York, New York.

[4] Dessauer, J. H., and H. E. Clark, "Xerography and Related Processes" p. 202–217, 1965, Focal Press, New York, New York.

[5] Shanin, M. H., "J. Appl. Optics, Suppl 3", 106 (1969).

[6] Gallo, C. F., J. E. Germanos and J. E. Courtney –ibid, 111 (1969).

[7] Hoesterey, D. C., "J. Appl. Phys." **33,** 992 (1962).

[8] Ref. 3, Pages 175–233.

[9] Ref. 4, Pages 65–112.

[10] Li, H. T., and P. J. Regensburger, "J. Appl. Phys." **34,** 1730 (1963).

[11] Tabak, M. D., and P. J. Warter, Jr., "Phys. Rev." **173,** 899 (1968).

[12] Tabak, M. D., "J. Appl. Phys." **Suppl 3,** 4 (1969).

[13] Viscakes, J., V. Gaidelis, A. Matuliones, E. A. Montrimas and A. Satas, **ibid**, 27 (1969).
[14] Viscakas, J., E. A. Montrimas and A. A. Pazeria, **ibid**, 79 (1969).
[15] Warter, P. J., Jr., **ibid**, 65 (1969).
[16a] Batra, I. P., and H. Seki, "J. Appl. Phys." **41**, 3409 (1970).
[16b] Batra, I. P., K. K. Kanazawa and H. Seki, **ibid**, 3416 (1970).
[16c] Batra, I. P., K. K. Kanazawa, B. H. Schechtman and H. Seki, "J. Appl. Phys." **42**, 1124 (1971).
[16d] Batra, I. P., and B. H. Schechtman, "J. Phys. Chem. Solids" **32**, 769 (1971).
[16e] Seki, H., I. P. Batra, W. D. Gill, K. K. Kanazawa and B. H. Schechtman, "I.B.M. J. Res." **15**, 213 (1971).
[17] Ref. 3, Pages 235–262.
[18] Ref. 4, Pages 120–169.
[19] Kiess, H., "J. Appl. Phys." **40**, 4054 (1969).
[20] Williams, R., and A. Willis, **ibid 39**, 3731 (1968).
[21] Amick, J. A., "R.C.A. Review" **20**, 770 (1959).
[22] Williams, R., "J. Appl. Optics" **Suppl 3**, 15 (1969).
[23] Hauffe, K., V. Martinez, H. Pusch, J. Range, R. Schmidt and R. Stechemesser, **ibid**, 34 (1969).
[24] Kikuchi, S., Y. Takahashi and T. Sakata, **ibid**, 42 (1969).
[25] Larach, S., and J. Turkevich, **ibid**, 45 (1969).
[26] Cassiers, P. M., K. Verbille and J. Van Engleland, **ibid**, 83 (1969).
[27] Inoue, E., H. Kokado and T. Yamaguchi, **ibid**, 91 (1969).
[28] Schelfaut, F., and W. Maenhout–Van Der Vorst, **ibid**, 96 (1969).
[29] Kiess, H., **ibid**, 100 (1969).
[30] Ref. 3, Pages 262–285.
[31] Ref. 4, Pages 169–202.
[32] Grossweiner, L., "Organic Photoconductors in Electrophotography", 1970, Most Associates, Marblehead, Mass.
[33] Chiu, T. T., and J. K. Knight, "TAPPI", **54**, 391 (1971).
[34] Fridkin, V. M., and V. Y. Litvinenko, "J. Appl. Optics" **Suppl 3**, 166 (1969).
[35] Ref. 3, Pages 360–380.
[36] Ref. 4, Pages 259–375.
[37] Streifer, W., and H. Stark, "J. Appl. Optics" **Suppl 3**, 149 (1969).
[38] Gygax, H. R. and G. Forgo, **Personal Comm.**
[39] Mohn, E., **Personal Comm.**
[40] Dahlquist, J. A., and J. Brodie, "J. Appl. Phys." **40**, 3020 (1969).
[41] Stark, H. M., and R. S. Menchel, "J. Appl. Phys." **41**, 2905 (1970).
[42] Hutter, E. C., and E. C. Giaimo, "Phot. Sci. Eng." **14**, 197 (1970).
[43] Hutter, E. C., "Phot. Sci. Eng." **15**, 251 (1971).
[44] Ref. 4, Pages 391–406.
[45] Mees and James, "The Theory of the Photographic Process" Chap. 22, 1966, The MacMillan Company.
[46] Ref. 4, Chapter 8.

Discussion

Morrison, S. R.: Is the objective of this study to aid in the design of machines or do you visualize it being used to adjust machine parameters for each individual picture.

Lu Valle, J. E.: Primarily it is to help us understand the electrophotographic system and to aid us in the design of machines.

Boroky, H.: You mentioned that the S-type of discharge curve of commercial ZnO paper is not always the case, the curve sometimes exponential. Would you like to comment on this as to the reason behind the difference?

Lu Valle, J. E.: I don't really know, it has been empirically observed.

Schaffert, R.: I should like to comment about the statement that density is proportional to thickness of the transferred toner layer. This is true only so long as the thickness is less than that required to provide maximum blockness of the used toner.

Lu Valle, J. E.: The computer stops calculation when the layer becomes too thick. This limitation was emposed to prevent the problem you mention.

Boroky, H.: With all commercial machines in liquid toning copies sometimes show a trailing effect., have you had a chance to clear up this phenomena?

Lu Valle, J. E.: Not yet, we hope to understand this when our model is improved.

Struck, H.: Toner looses his properties after a time. Could you please tell about some reasons for this mechanism?

Lu Valle, J. E.: The surfactant oxidizes, the toner agglomerates and settles out because it is not a true collodial suspension.

Electrophotographic Process with Quantum Efficiency More than Unity

A. Matulionis and J. Viščakas

Abstract

An analysis of the discharge mechanisms in different electrophotographic layers usually leads to the conclusion that the photodischarge quantum efficiency defined as the number of elementary charges neutralized by one photon absorbed in a charged layer should be less than unity. In fact, the well-known photocurrent gain process typical for most photoconductors can not be used for electrophotography because of its specific conditions. The carrier multiplication in high electric field, though observed in substances used in electrophotography (ZnO), seemed to be quite improbable because of low mobilities, as usually found in the layers either containing binder or being amorphous. However, the carrier multiplication due to the high field effect is supposed to be the most reasonable explanation for the quantum efficiency that exceeds unity in an electrophotographic process using As_2Se_3 layers.

The mechanism for high quantum efficiency effective at moderate electric fields is of interest as well. The theoretical aspects of one possible mechanism similar to the photocurrent gain process in photoconductors are dealt with. The mechanism, called latensification, is based on the carrier injection currents controlled by an external electrode and by the space charge of a charged layer. The conditions can be chosen so that the injection current in the photodischarged parts of the layer is substantially higher than that in the non-discharged parts. Therefore the latent image is intensified. This process goes on in a self-sustaining manner if it is not limited by carrier capture. This limitation is less severe than that for high-field effect. Moreover the suggested latensification process is efficient for latent images formed by heavily and uniformly absorbed light. The possible scheme of this process occurring in amorphous selenium is discussed.

1. Introduction

Photosensitivity of electrophotography depends on latent image formation and visualization. Recent progress in visualization is not yet able to cover insufficient photosensitivity of electrophotographic layers. Conventional electrophotography operates at moderate electric fields and uses non-injecting substrates. These conditions favor charge storage in the dark, but eliminate any other carriers except photogenerated ones from the photodischarge process. Even if carrier lifetime is long, each photogenerated carrier discharges only one elementary charge in the layer. Consequently the discharge quantum efficiency defined as the number of elementary charges neutralized by one photon absorbed in a charged layer is equal to the quantum efficiency of photogeneration, which in the case of visible light is either equal to or less than unity because of the field-controlled recombination [1]. In order to obtain essentially higher photosensitivity, the additional carriers should play a predominant role. Injection and high-field multiplication are the possible sources for the additional carriers. This paper discusses the progress in this field of research. Main attention is paid for a latensification of electrostatic images. The latensification means the process resulting in the additional discharge at already discharged places of the layer. The proposed latensification is based on injection of carriers and it does not exploit steep non-linear processes.

2. Carrier multiplication in high electric fields

The problem of carrier multiplication relating to electrophotography has been thoroughly discussed by Williams [2]. The competition between the multiplication of photogenerated carriers and the steep increase in dark current is found to become significant at about the same electric fields. The successful measurements using the free surfaces of ZnO crystal charged in air confirm that the carrier multiplication can, in principle, be exploited. In the conventional ZnO binder layers the crystallites utilized are large enough and the electric fields are high, therefore these layers cannot be regarded as unsuitable for carrier multiplication, though there is no evidence that the photosensitivity of either ZnO or any other crystalline-binder layers is contributed by the electron multiplication.

In general, the carrier multiplication is not expected in amorphous layers. Despite of this the photodischarge quantum efficiency for an electrophotographic process on some As_2Se_3 layers exceeds unity [3, 4]. The obtained fast discharge is substantially faster than the dark one. This cannot be ascribed to the illumination-controlled injection. It is reasonable to ascribe it to the carrier multiplication in the high electric field region formed at the surface in the dark. To expect carrier multiplication, the mean free path of several hundred Ångström is required. A mean free path is usually considered to be less. However, the estimation from mobility data may not be valid and the mobility in As_2Se_3 is strongly dependent on the electric field. Therefore longer free paths can be expected. This is confirmed in part by the study of external photoeffect of many amorphous solids. The advantages of multiplication of photogenerated carriers are obvious. However, in the case of a thin region of high electric field, the multiplication of uniformly photogenerated carriers in the whole bulk is small. The simultaneous increase in the dark discharge rate increases the lower limit for the intensities of illumination to be used. A steep non-linearity of multiplication vs. field introduces the problems of control and reproducibility.

3. Latensification

In moderate electric fields, i.e. when the carrier multiplication does not take place, a high quantum efficiency can be achieved by a proper use of injection currents. An example of the illumination induced injection is given in Figure 1 [5]. The data are for the 0.23 mm thick amorphous selenium sandwiched between two aluminum plates kept under the constant voltage. An attenuated Nd laser pulse of 1.06 micron infrared, radiation photogenerated the amount of charge less than on the aluminum electrodes. At the lower voltages (Fig. 1, Curves a, b). the triangular photocurrent pulse, typical for the uniformly photogenerated carriers drifting in a uniform electric field was obtained. The transit time measurements helped to find out that the mobile carriers were holes. A long lasting photocurrent (Fig. 1, Curves c, d) at the higher voltages can be ascribed to the illumination induced injection current. A non-injecting aluminum electrode becomes an injecting one as a result of the drift of the photo-generated holes. It is noteworthy that the obtained photocurrents are substantially higher than the dark current. However, since the constant voltage is prerequisite, the observed phenomena cannot be used in electrophotography unless some modifications are made. The detail of one possible mechanism of latensification of electrostatic images will be given below. In this case a well injecting substrate is prerequisite and long lifetimes of carriers are desirable.

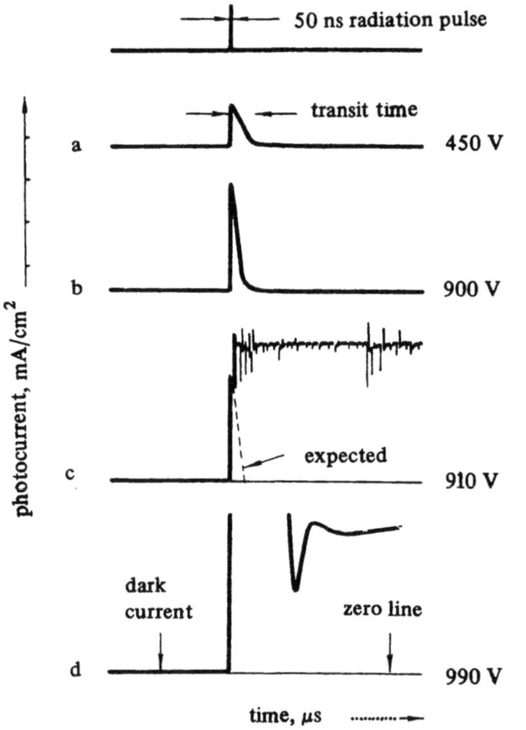

Fig. 1. The photocurrent due to a 50 ns pulse of 1.06 μm radiation in the Al-amorphous Se-Al samples kept under constant voltages [5]: 1–450 V, 2–900 V, 3–920 V, 4–990 V. (layer thickness 0.23 mm; 1 (mA/cm^2)/cm, 10 μs/cm)

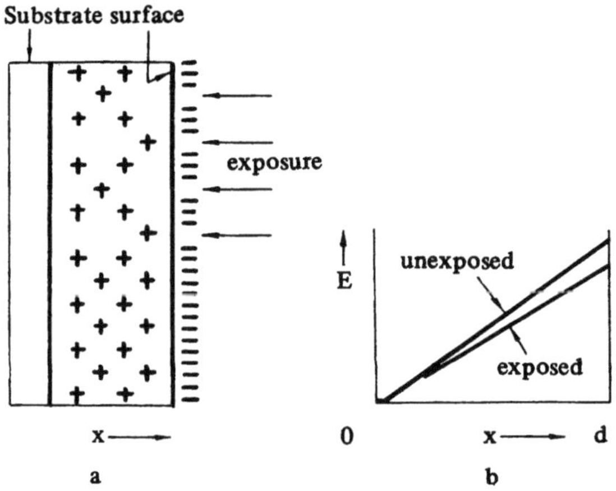

Fig. 2. Charge (a) and electric field (b) distribution for initial latent image.

When a substrate is an injecting one, the surface charge of a charged layer must be compensated by the equal space charge of the opposite sign. Under such conditions the dark discharge by injection currents is prevented by zero field at the substrate [6]. Dark discharge is caused by the free carriers generated in the region of electric field (the surface is included, but the substrate is excluded). The cross-section and the field distribution of the charged layer in the case of uniform space charge is given in Figure 2. The situation remains quite similar when some parts of the layer have been photodischarged in part by a low exposure. The only consequence is slow relaxation of the latent image in the dark.

The latensification starts when the voltage equal to the potential of the unexposed parts of the layer is applied between the substrate and the electrode and the electrode is approaching as close as possible (Fig. 3). No charge accumulates in the substrate and in the electrode at the unexposed parts of the layer, but some does at the photodischarged parts. The density of the accumulated charge is equal to:

$$|\sigma| = \frac{|\Delta\sigma_{s0}|\, C_0}{C_e(1 + \frac{C_0}{C_1})}, \qquad (1)$$

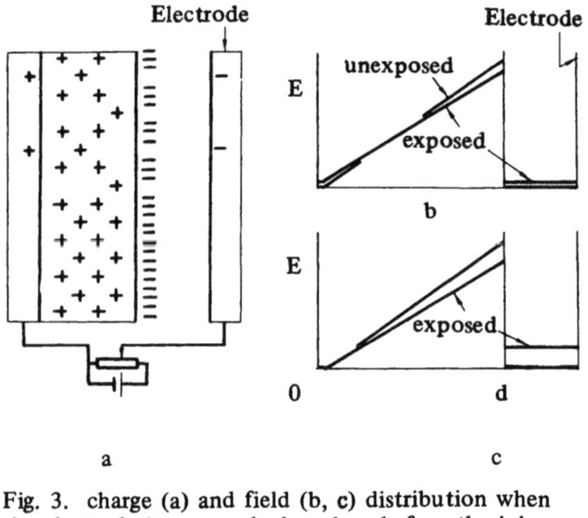

Fig. 3. charge (a) and field (b, c) distribution when the electrode is approached: a, b – before the injection, c – after the injection

where $\Delta\sigma_{s0}$ is the difference in the surface charge density between the unexposed and exposed parts of the layer in absence of the electrode, C_0 and C_1 are the geometric capacitance of the layer and the layer-electrode spacing, $C_l = \frac{\Delta\sigma_s}{\Delta V}$ is the capacitance of the charged layer.

When the substrate is an injecting one and there is a non-zero field at the substrate (Fig. 3b), the holes are injected into the layer. All of them reach the surface if their lifetime is long enough:

$$\tau_h \gg t_{tr} = \int_0^d \frac{dx}{\mu E(x)}\ , \tag{2}$$

where t_{tr} is the hole transit time, d the layer thickness, x – the distance from the substrate, and E(x) – the electric field distribution (Fig. 3b).

When the injection is over, the electric field at the substrate is zero again, but it increased in the spacing (Fig. 3c). If all holes reach the surface, we obtain:

$$|\int_0^d \rho\, dx| = |\sigma_{s0}| - |\Delta\sigma_{s0}| > |\sigma_s| = |\sigma_{s0}| - |\Delta\sigma_{s0}|\ 1 + \frac{C_1}{C_l}\ , \tag{3}$$

where ρ and σ are the densities of space and surface charges, respectively. σ_{s0} is the surface charge density before the exposure. For $C_1 = C_l$ one gets $\Delta\sigma_s = 2\Delta\sigma_{s0}$ which corresponds to the gain of 2. If it is not enough, the electrode must be taken away as far as possible.

Figure 4 illustrates the situation when the electrode is infinitely far away. The density of the charge in the substrate is equal to:

$$|\sigma| = |\Delta\sigma_{s0}|\frac{C_1}{C_l}\ . \tag{4}$$

The electric field has an inversion of its direction somewhere at x_i (Fig. 4a, b):

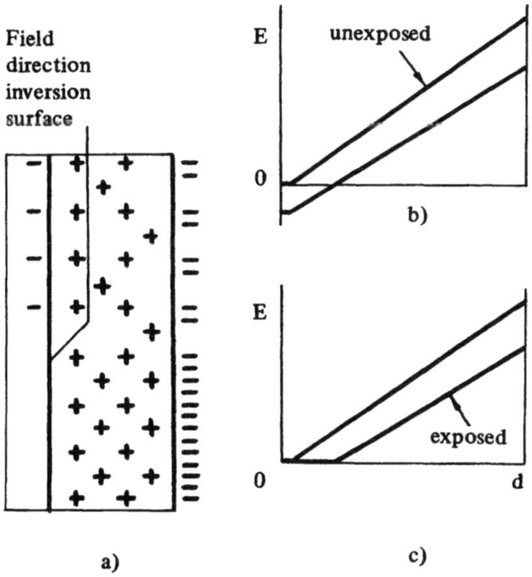

Fig. 4. Charge (a) and field (b, c) distribution when the electrode is away: initial situation (a, b) and the final situation (c).

$$|\int_0^{x_i} \rho\, dx| = |\sigma|\ . \tag{5}$$

The electron injection starts and they neutralize the space charge up till the inversion surface. The necessary conditions for entire neutralization can be found elsewhere [7]. When the electron injection and the neutralization is over, the space charge again becomes equal to the surface charge. The final situation resembles the initial one with an exception

$$\Delta\sigma_s = \Delta\sigma_{s0}\left(1 + \frac{C_1}{C_l}\right) = \Delta\sigma_{s0} k\ . \tag{6}$$

There is no limitation to repeat the operations from the starting point of the latensification as many times as desirable. After n cycles one obtains

$$k = \left(1 + \frac{C_1}{C_l}\right)^n\ . \tag{7}$$

For n = 20, if $C_1/C_l = 0.2$, the gain k = 40. The number of cycles is limited by the dark discharge. The maximum number can be estimated as

$$n_{max} = \tau f , \tag{8}$$

where τ is the time constant of the dark discharge, f–the frequency of the repetition. If $f = 100$ Hz and $\tau = 100$ s, the gain due to latensification is unlimited in the practical sense.

In order to avoid the intensification of the dark discharge the voltage between the electrode and the substrate should be kept equal to the potential of the unexposed parts of the layer. It should not be kept constant in the literary sense.

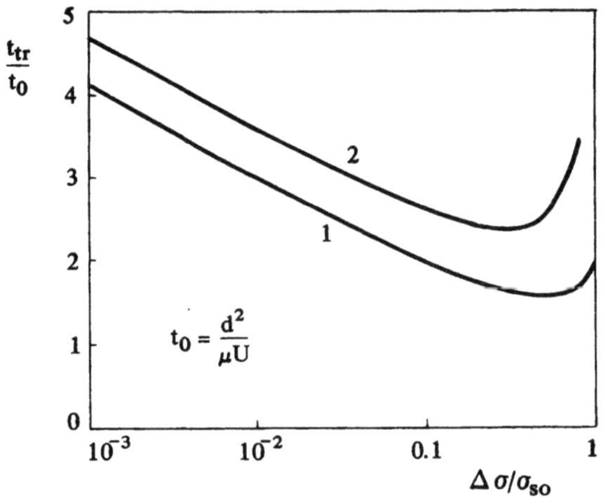

Fig. 5. Transit time vs the discharge for $C_2 = 2C_0$ and for the situation presented in Fig.3b: $1 - C_1 = C_0$, $2 - C_1 = C_0/5$.

Discussion

The condition (2) gives the lower limit for hole lifetime in a very general form. The distribution of the electric field constantly varies because of injection and latensification. Some intermediate situations are given in Figures 2, 3 and 4. If the situation of Figure 3b is chosen, the hole transit time can be calculated for various $\Delta\sigma_s$. The latter may be regarded as the measure of the exposure or the measure of the latensification. In the latter case one can see that if the condition (2) is valid in the very beginning of the latensification, the process goes on in a self-sustaining manner (Fig. 5). Figure 5 also demonstrates that the condition (2) can be easily satisfied when the reduction of the surface charge due to the exposure is greater. If the exposure has been low, the latensification can be succesful only in the substances with long lifetime. For example, if $\frac{\Delta\sigma_{s0}}{\sigma_{s0}} = 0.1\%$, $V = 300$ V, $\alpha = 0.1$ mm, $\mu = 0.2$ cm^2/ V s and $C_0/C_1 = 5$, it is necessary to have $\tau_h > 0.1$ ms. This requirement can be satisfied only in perfect materials, e.g. in a certain type of amorphous selenium, where the hole lifetime exceeds several miliseconds [8].

Since the ratio of the layer thickness to the layer-electrode spacing must be as great as possible, the latensification is favored in thicker layers. Therefore it may become promising in infrared electrophotography, where the radiation is weakly absorbed. On the other hand the high-field multiplication is promising for thin layers and for a highly absorbed light. To exploit the latensification, difficult problems of the vibrating electrode are to be solved. The problem of photographic resolution may also be a serious one.

References

[1] Viščakas, J., V. Gaidelis, A. Matulionis, E. Montrimas, A. Šatas, Proc. IX Int. Conf. Phys. Semiconductors, Moscow, 2, 1290, 1968.
[2] Williams, R., Appl. Optics, Suppl. 3, 15, 1969.
[3] Tauraitienė, S. A. Tauraitis, E. Montrimas, Repts Int. Congress Photographic Sci. Moscow, Section E, 244, 1970.
[4] Montrimas, E., S. Tauraitiene, A. Tauraitis, Report for the present Colloquium.
[5] Matulionis, A., K. Teegarden, Description of invention assigned to Department of Navy U.S.A., 1971.
[6] Montrimas, E., V. Gaidelis, Lievutos Fizikos Rinkinys 7 (1), 175, 1967.

[7] Gaidelis, V., E. Montrimas, A. Pažėra, J. Viščakas, Report for the present Colloquium.
[8] Juška, G. (Private communication); G. Juška, A. Matulionis, A. Sakalas, J. Viščakas, phys. stat. solidi 36, K 121, 1969.

Discussion

Stockman, D. L.: 1. How long was the duration of the triangular current pulse shown in the first slide?

2. What do you consider are the possibilities for controlling the injection over the large areas required for imaging systems?

Matulionis, A.: 1. The current pulse was essentially longer then the light pulse, it was inversely proportional to the voltage applied, it was equal to the transit time of holes in amorphous selenium when the voltage was equal 450 V the transit time was about 5 μs.

2. There is no experimental evidence that the observed current is uniformly distributed.

Pell, E. M.: Have you observed the latensification with long lifetime selenium?

Matulionis, A.: We face some problems of vibration etc.

Berg, W. F.: In analogy to the silver-halide photographic case, one would expect super-additively of exposure with any S-shaped characteristic curve; where there is reciprocity failure, such effects can be increased by choosing suitable times and intensities of the uniform exposure. There should be some practical application of this, if desired, especially on machines where conditions can be rigidly controlled.

Morrison, S. R.: It would appear that there may be resolution related to the large film thicknesses and electrode spacings. Has the resolution to be expected been calculated?

Matulionis, A.: I quite agree with Dr. Morrison that the resolution problem will be a very serious one.

Arneth, R.: 1. Did you use strongly or weakly absorbed light? According to your explanation there must be a difference in the ΔV initial you defined.

2. Even in the case you used a non-absorbing binder, ΔV initial must depend from pigment/binder-ratio.

Matulionis, A.: Weakly absorbed light was used. In case of large amounts of dyes (5–10 mg of eosine per 1 g of ZnO) the same regularities are observed at the edges of sensitization region. A pronounced S-shape is absent at the dye absorption maximum.

Some Remarks to an Analog Model of a ZnO-Binder Layer

H. Camenisch

Abstract

The S-shaped discharge curves of ZnO-binder layers can be represented by considering the layers as being made up of—in this case, four—elementary layers, each characterized by a capacitor and a resistor in parallel, with suitable feedback systems between each pair of elementary layers. Analysis on an analog computer gives the changes of potential at each of the capacitors as a function of time and the S-shaped discharge curve often found on ZnO layers.

1. Introduction

The S-shaped discharge curves of many ZnO-binder layers are, as a rule, explained by atomistic models involving the mobility, trapping, and release of the pairs of charge carriers, e^- and p^+, potential barriers etc. These models involve a good deal of speculation. An attempt to provide a phenomenological model based intirely on gross electrical concepts will be described below, one of its advantages being that it lends itself to evaluation on an analog computer, with all its attendant flexibility.

2. Procedure

In order to simulate a compound ZnO-layer on the analog computer a very simple and rough model was set up by replacing each of the four elementary layers in our paper (p.21) by a capacitor C and a resistance R in parallel, with feedback systems between each adjoining pair of layers.

The time constant t = R x C is determined

(1) by the light intensity as a function of depth,
(2) by the number of generated electron-hole pairs as a function of depth,
(3) by the actual electric field (this has not yet been taken into account, but is planned for evaluation by a hybrid computer) and
(4) by whether the elementary layer is spectrally sensitized or not.

Figure 1 shows the replacement of the four elementary layers by RC-combinations in series. Each layer is connected one to another (a) through a nonlinear feedback to the upper layer, and (b) through another feedback to the lower one. Figure 2 shows the feedback-connections from layer to layer.

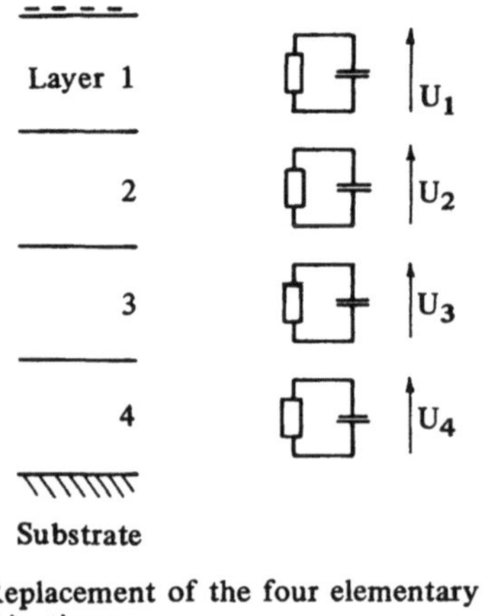

Fig. 1. Replacement of the four elementary layers by RC-combinations

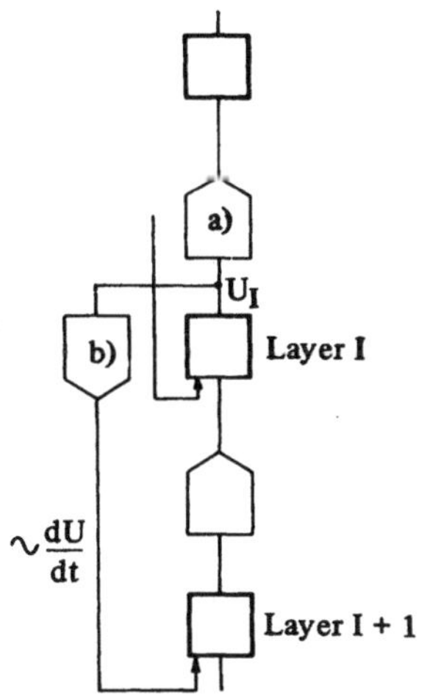

Fig. 2. Feedback-connections between two neighbouring layers

In a first approximation, the feedbacks to the lower layers were chosen to be proportional to the first derivative of the layer potential, for the following reason:

Any change of layer potential is the consequence of generated (e^- und p^+) pairs. That means that only a change of layer potential may produce a feedback to another one. On the assumption that the generated holes recombine straight away with the negative oxygen ions on the top surface, we assume no feedback to the uppermost layer.

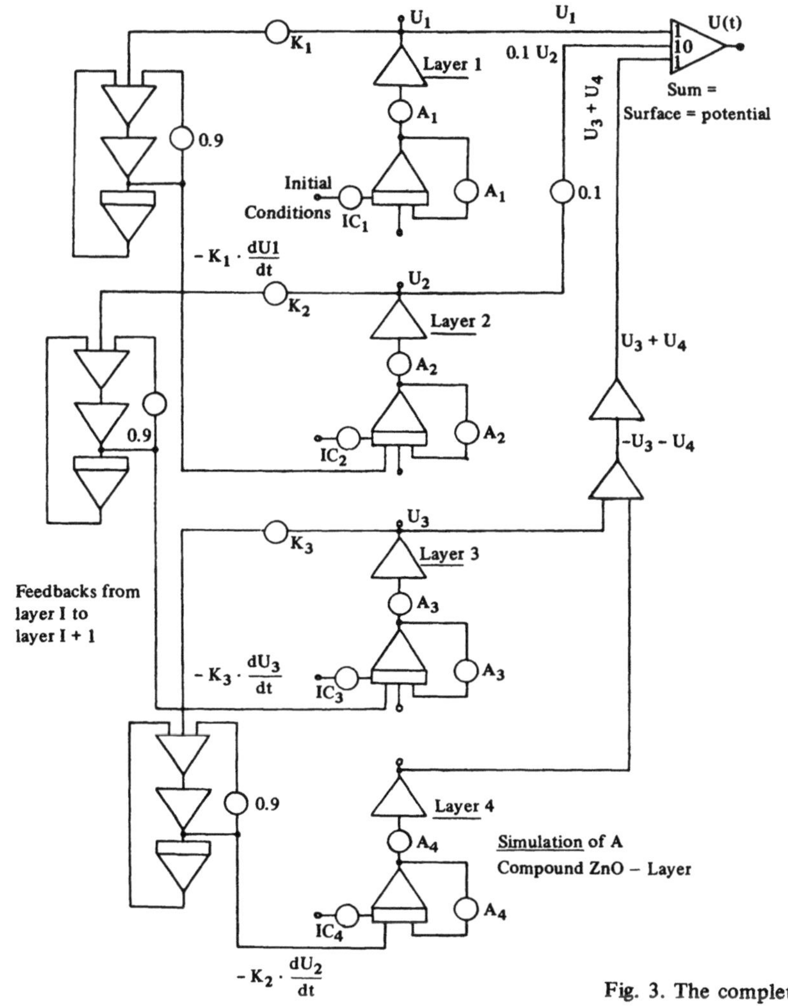

Fig. 3. The complete analog circuit

The free electrons flowing into the deeper layer will charge its capacitor. Depending on the characteristics of that layer, this space charge will be reduced differently, and using the experiments described earlier it was possible to adjust the parameters in the analog model.

Figure 3 shows the complete analog model on the assumption that originally there is no space charge in any of the layers; only the uppermost elementary layer has an "initial condition", in other words, a surface charge.

3. Results

Figure 4 shows the curves of the potential vs. time of exposure on each of the four capacitors representing our four elementary layers.

Following the assumptions and results of the computer model it can be seen that the behaviour of the compound layer

(1) just after exposure is governed approximately by the upper elementary layers;
(2) in the middle of the discharge curve by all the layers; and
(3) for long exposure times by the layers near the substrate.

Figure 5 shows the same results as Figure 4 but drawn in another manner, showing the layer potential over time and depth.

The model fairly represents the actual discharge behaviour of the materials described in the earlier paper (p. 21).

It has not yet been possible to correlate the computer parameters with the real physical parameters of layers. In particular, it has not been possible to confirm experimentally the changes of potential within the layer, but the analog model gives a certain idea of what could happen inside the layer.

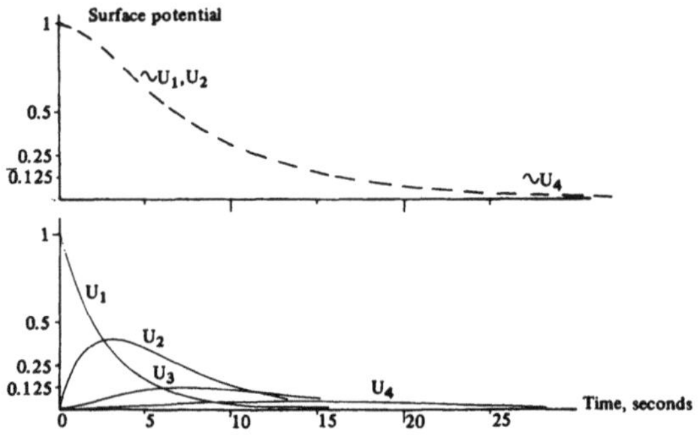

Fig. 4. Layer potentials and surface potential as a function of exposure time

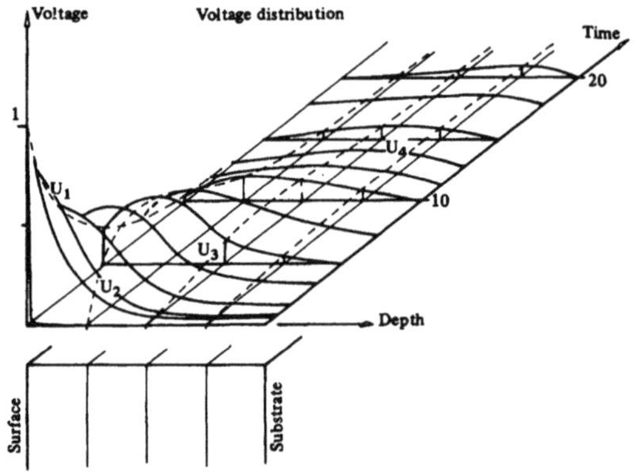

Fig. 5. Potential distribution against layer depth and time

A Comment on the Limited Utility of Surface Potential-Exposure Measurements

David L. Stockman

The use of surface potential-exposure curves in electrophotography is an excellent means of conveying information to the design engineer on the operational characteristics of the photoreceptor. It is of more limited utility, however, in studies of mechanism of generation and transport of charge in photoreceptor layers. This brief communication outlines the problem and shows, for one simple case, the limitations of the method and the need for supplementary measurements for use in mechanistic studies.

In electrophotography, the quantities of interest are; 1) how many carriers are generated?, 2) how far do they move?, 3) how fast did they get there? The elctrophotographic gain, G, of a photoreceptor is normally defined as

$$G = \frac{\text{number of carriers traversing the layer}}{\text{number of incident quanta}} \qquad (1)$$

which for a capacitive layer is

$$G = \frac{\epsilon\epsilon_0 \, (dV/dt)_{t=0}}{e\,Id} \qquad (2)$$

where I is the photon flux, $\epsilon\epsilon_0$ the permittivity, d the thickness, e the electronic charge and $(dV/dt)_{t=0}$ the inital rate of discharge of the voltage. Normally G is a function of field, sample thickness, temperature and exciting wavelength. The definition of G in Eqn. 1 includes steps involving absorption, photophysical relaxation, free carrier generation and the transport and trapping of free charge. Since each of these individual steps can depend upon the field, thickness, etc., it is important to clearly separate the steps for individual study. Surface potential-exposure curves are inadequate for this purpose except in the very simplest cases.

In electrophotography, the initial fields employed are typically 10^5 to 10^6 volts cm^{-1} implying surface charge densities of 10^{11} to 10^{12} charges cm^{-2} for capacitive layers. It is instructive to view the problem of the removal of this charge on the basis of the time scale of the measurement. Measurement times, t_m, must be short compared to dielectric relaxation time, τ_D ($=\rho\epsilon\epsilon_0$) to avoid polarization due to internal generation. For most materials of practical interest, this condition is usually easily satisfied. Once this condition is met, there is a time domain of interest in these transient measurements which is to be compared to the transit time, t_{tr}, of the carriers. In the first instance, $t_m \gg t_{tr}$, which is the usual case in surface potential measurements. Here this measurement can yield information on how many carriers were generated and how far they moved, i. e., the carrier range ($= \mu_D \tau_0$ where μ_D is the drift mobility and τ_0 the lifetime against deep trapping). For times such that $t_m \ll t_{tr}$, μ_D and τ_0 can be separated and thus yield information on how fast the carriers move. The difficulty in most real electrophotographic discharge measurements is to decide in what regime of interest you are for any given material and condition of measurement. This normally requires an independent experiment. The most common one employed in insulator physics is the blocking-electrode current-mode measurement which has been extensively discussed elsewhere [1, 2]. A failure to transport the charge rapidly or completely can appear xerographically to be a failure in free carrier generation. This is most commonly manifested by a rapid decrease in G as the field is lowered.

One of the more recent examples of this type of effect was first published by Regensburger [3] studying the sensitization of PVK*) by selenium. The type of xerographic data reported by Regensburger are shown in Figure 1, plotted in a common fashion as the xerographic gain from the initial discharge rate as a function of the initial field on the photoreceptor. At low fields, the gain falls very rapidly, for all thicknesses of PVK, and seems to approach a threshold value of ~6 x 10^4 volts cm^{-1}. It was originally interpreted as a barrier to injection of the holes from selenium into PVK. Mort [4] recently demonstrated that the fall-off in gain at the low fields is associated with a space charge limited current which occurs because of the very low carrier velocities at low fields.

*) Poly-vinyl carbazole

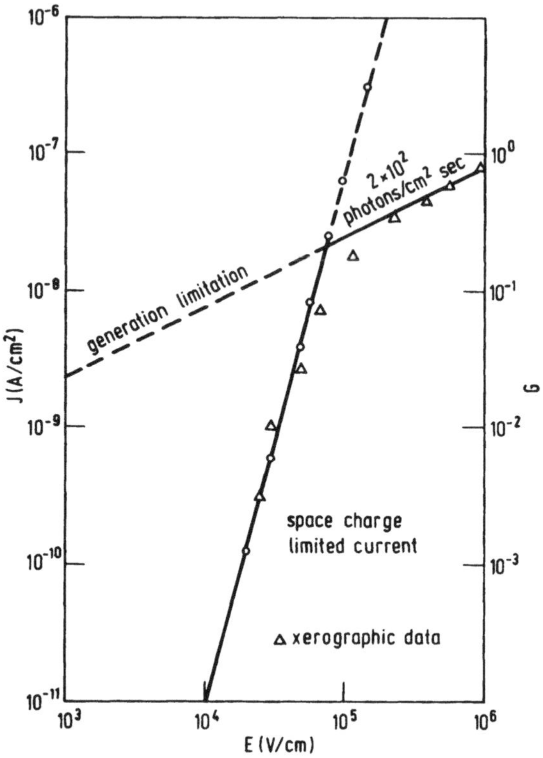

Fig. 1. Xerographic gain as a function of initial field after Regensburger (3).

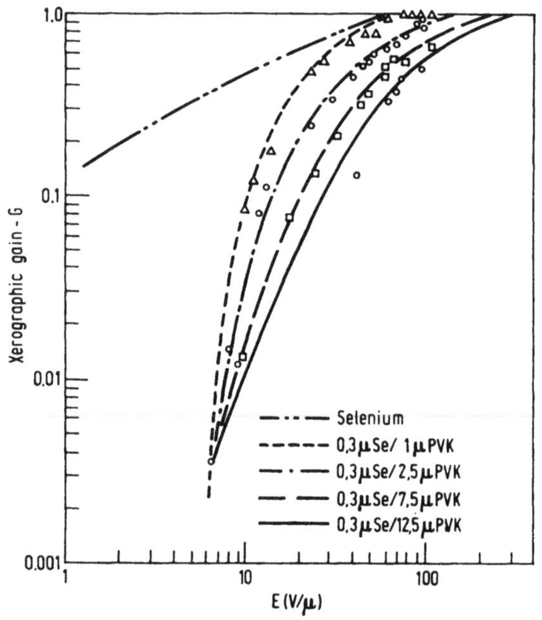

Fig. 2. Comparison of the measured xerographic gain with the composite limitations introduced by transport and generation. After Mort (4).

The data presented by Mort on space charge current flow in such a device is shown in Figure 2 along with typical surface potential data illustrating two regimes of interest. At low fields, the electrophotographic gain is limited by transport due to space charge limited currents while at high fields the gain is determined by the rate of generation of free charge in the selenium.

In complex structures such as sensitized or unsensitized binder plates where the field distributions, the depth of penetration of the light and the free carrier transport properties are not always known accurately, surface potential-exposure curves can be very misleading when used without supplementary measurements. Even in relatively simple organic structures where exciton motion following bulk absorption may result in extrinsic generation at the surface [5], and unambiguous interpretation of surface potential exposure curves is not possible without independent knowledge of the individual steps in the process. Thus the use of surface potential exposure curves by themselves for mechanistic evaluation should be avoided.

References

[1] Helfrich, W., Physics and Chemistry of the Organic Solid State, Vol. 3, p. 2, Ed. by D. Fox, et al., Interscience, John Wiley and Sons, New York, 1967.

[2] Spear, W. E., J. Non-Crystalline Solids, 1, 197 (1969).

[3] Regensburger, P. J., J. Photochem. Photobiol., 8, 429 (1968).

[4] Mort, J., Phys. Rev., to be published.

[5] Stockman, D., Proceedings of this conference, S. 194ff.

Continuous Tone Images with Electrophotographic Layers

Günther Bethe

The problem of producing continuous tone pictures by electrophotographic methods has been discussed already in the early day of electrophotography. In spite of a large effort until today no process is available for making continuous tone reproduction. I return to this problem because I believe to have discovered a new and particularly simple method.

For the better understanding, a brief repetition of well known facts is indispensable. The electrophotographic plate consists of a photoconductor-resin layer which is coated on a conductive support, such as conductive paper or metal. During the charging under a corona the upper layer is charged negatively and the rear side of the layer–at the interface layer/support–is charged positively. In a rough approximation the layer can be considered as a charged capacitor. In large charge areas, however, a significant electric field can be detected only near the border charge/uncharged area. This is schematically represented in Figure 1a. According to the course of the electric field on large charged areas, there is no significant electric field available for toner precipitation in the inner regions of these areas.

One of the first proposals to overcome this difficulty has been the application of an auxiliary electrode at a close distance from the charged layer. This method, however, exhibits large difficulties in the construction of such a device. Another method consists in a construction of an electric screen with small areas of high conductivity and those of highly insulating behavior in the dark and highly photoconductive behavior under illumination. Such an electrophotographic screen can be seen in Figure 1b. The first proposal in this direction stems from Chester Carlson, the inventor of the electrophotography. A technical application was not possible because the preparation of a fine raster was to expensive. Also an admixture of a fine powder of an inert material did not supply satisfactory results. This becomes understandable if one considers that with growing fineness of the electric charge-

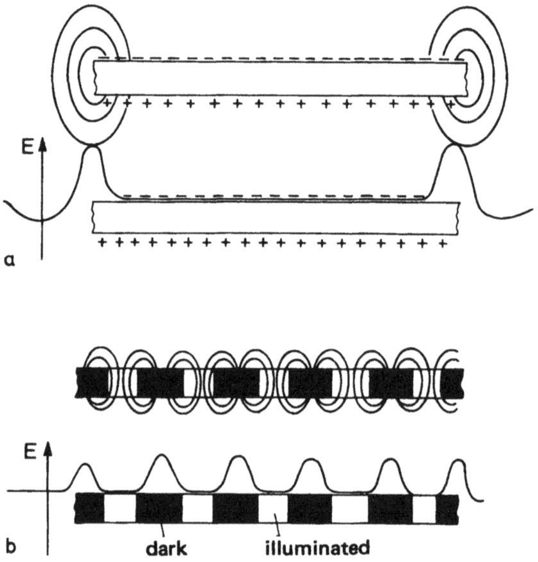

Fig. 1. Schematic representation of the course of the electric field over a large charged area (a) and of an area screened in a large number of insulating (black rectangles) and of conductive (white rectangles) microspots (b).

screen, the field strength at a given distance from the surface of a layer can become too small.

The crucial idea was the application of coarse-grained conductive material, the size of which is comparable with the thickness of the layer. Only with this rather coarse-grained powder as additive to a zinc oxide resin dispersion, the presupposition is fulfilled to prepare small conductive and insulating areas, as represented in Figure 1b. By this technique, the layer is divided into a large number of very small capacitors, the entire surface of which is passed through by electric field because of the very small size of the condenser plates.

The production of these screened layers is rather simple. During the preparation of the zinc oxide-resin slurry, a suitable ratio of the conductive coarse-grained material is added. These layers are particularly useful for the preparation of electrophotographic offset printing plates. No more screening is necessary since it appears itself during the toner process.

Walter de Gruyter
Berlin · New York

Hauffe – Wolkenstein
(Editors)

Symposium on Electronic Phenomena in Chemisorption and Catalysis on Semiconductors

held in Moscow, July 2-4, 1968

With 123 Fig. and 33 Tab. Large-oktavo. XII + 261 pp. 1969.
Paper DM 80,— $ 25.50
ISBN 3 11 000846 7

Contents:

Bei Fragen zur Produktsicherheit wenden Sie sich bitte an:
If you have any questions regarding product safety,
please contact:

Walter de Gruyter GmbH
Genthiner Straße 13
10785 Berlin
productsafety@degruyterbrill.com